Advances and Applications of CO_2 Laser

Advances and Applications of CO_2 Laser

Edited by **Olivia Graham**

CLANRYE INTERNATIONAL

New Jersey

Published by Clanrye International,
55 Van Reypen Street,
Jersey City, NJ 07306, USA
www.clanryeinternational.com

Advances and Applications of CO_2 Laser
Edited by Olivia Graham

International Standard Book Number: 978-1-63240-041-3 (Hardback)

Printed in the United States of America.

Contents

Preface

Over the recent decade, advancements and applications have progressed exponentially. This has led to the increased interest in this field and projects are being conducted to enhance knowledge. The main objective of this book is to present some of the critical challenges and provide insights into possible solutions. This book will answer the varied questions that arise in the field and also provide an increased scope for furthering studies.

The advances as well as applications of carbon dioxide laser are elucidated in this book. It provides comprehensive information regarding the fundamentals of CO_2 laser operation and its upgraded new systems and technologies. Frequency equalization, lasers based on diffusion cooled V-fold geometry, laser producing ultrashort pulses etc. are some of the topics covered in the book. This book also discusses its industrial applications such as welding, cutting, engraving etc. and its clinical applications such as laser treatment, dentistry etc.

I hope that this book, with its visionary approach, will be a valuable addition and will promote interest among readers. Each of the authors has provided their extraordinary competence in their specific fields by providing different perspectives as they come from diverse nations and regions. I thank them for their contributions.

Editor

Part 1

Basic Processes

1

CO_2 Laser Photoacoustic Spectroscopy: I. Principles

Dan C. Dumitras, Ana Maria Bratu and Cristina Popa
Department of Lasers, National Institute for Laser, Plasma, and Radiation Physics, Bucharest Romania

1. Introduction

Laser photoacoustic spectroscopy (LPAS) has emerged over the last decade as a very powerful investigation technique, capable of measuring trace gas concentrations at ppmV (parts per million by volume), or even sub-ppbV (parts per billion by volume) level. Recent achievements in this field have made it possible to fully characterize the method and improve the design of instrument components in view of the task they are expected to fulfill.

1.1 Historical remarks

The photoacoustic (PA) (formerly also known as optoacoustic) effect consisting in sound generation from the interaction of light and matter was discovered by Alexander Graham Bell (Bell, 1880). He noticed that focused intensity-modulated light (chopped sunlight) falling on an optically absorbing solid substance produced an audible sound. In the next year, light absorption was detected through its accompanying acoustic effect not only in solids, but also in liquids and gases by Bell (Bell, 1881), Tyndall (Tyndall, 1881), Röntgen (Röntgen, 1881), and Preece (Preece, 1881). They found the sound was stronger when the substance was placed in a sample cell (then called "photophone" and later "spectrophone"). It was Bell again that first described the resonant amplification of the PA signal (Bell, 1881). The PA effect was also investigated at different light wavelengths. Bell and Preece were among the first to notice a PA signal for an aerosol when they experimented with cigar smoke. The advances of photoacoustic spectroscopy up to the invention of the laser were reviewed by Kaiser in 1959 (Kaiser, 1959).

Over the last five decades, technological developments in the field of lasers and high-sensitivity pressure detection systems (microphones and electronics) have contributed to the substantial progress of photoacoustic spectroscopy. The introduction of laser light sources emitting highly monochromatic and collimated intense beams have opened up new areas of research. Lasers provide the advantage of high spectral power density owing to their intrinsic narrow linewidth. This laser linewidth is usually much smaller than the molecular absorption linewidth (GHz region at atmospheric pressure), and therefore it is not an important issue in most measurements. A true revival of PA spectroscopy was due to Kerr and Atwood (Kerr & Atwood, 1968), who made the earliest experiments with a laser illuminated PA detector in 1968, and Kreuzer (Kreuzer, 1971), who first measured gas concentrations using a PA detector and a laser in 1971. Later experiments by Kreuzer and collaborators (Kreuzer & Patel, 1971; Kreuzer et. al., 1972) effectively demonstrated the extremely high sensitivity attainable by this method. To improve the detection of atmospheric pollutants, Dewey et al. (Dewey et. al., 1973) have used in 1973 an acoustic resonance chamber and have reached amplification factors higher than 100. In 1977, the feasibility of *in situ* aerosol measurements, which were important for atmospheric applications, was first reported by Bruce and Pinnick (Bruce & Pinnick, 1977) and Terhune and Anderson (Terhune & Anderson, 1977). Subsequently, many experimental and theoretical works have been reported in the literature, proving the applicability of the method not only in spectroscopy, but also in various fields of physics, chemistry, biology, medicine, and engineering. The potential of laser photoacoustic spectroscopy has been discussed in several review articles (Patel & Tam, 1981; West, 1983; Hess, 1983; Tam, 1986; Sigrist, 1986, 2003; Meyer & Sigrist, 1990; Harren & Reuss, 1997; Harren et al., 2000; Miklos et al., 2001; Schmid, 2006) and books (Pao, 1977; Rosencwaig, 1980; Zharov & Letokhov, 1986; Hess, 1989; Mandelis, 1992, 1994; Bicanic, 1992; Gusev & Karabutov, 1993; Sigrist, 1994; Mandelis & Hess, 1997).

1.2 Features of a gas sensor

The most important features of a gas sensor include high sensitivity and selectivity, large dynamic range, high accuracy and precision, good temporal resolution, ease of use, versatility, reliability, robustness, and multicomponent capability. Gas chromatographs are neither sensitive nor fast enough. Although there is no ideal instrument that would fulfill all the requirements mentioned above, a spectroscopic method and particularly the simple setup of LPAS provide several unique advantages, notably the multicomponent capability, high sensitivity and selectivity, wide dynamic range, immunity to electromagnetic interferences, convenient real time data analysis, operational simplicity, relative portability, relatively low cost per unit, easy calibration, and generally no need for sample preparation. LPAS is primarily a calorimetric technique and, as such, differs completely from other previous techniques, as the absorbed energy can be determined directly, instead of via measurement of the intensity of the transmitted or backscattered radiation. In conjunction with tunable lasers, *in situ* monitoring of many substances occurring at ppbV or even pptV (parts per trillion by volume) concentrations is a routine task today. PA detection provides not only high sensitivity but also the necessary selectivity for analyzing multicomponent mixtures by the use of line-tunable IR lasers, e.g., CO lasers (Sigrist et al., 1989) or CO_2 lasers (Meyer & Sigrist, 1990).

CO_2 laser photoacoustic spectroscopy offers a sensitive technique for detection and monitoring of trace gases at low concentrations. The CO_2 laser is of special interest, as it ensures high output power in a wavelength region (9-11 μm) where more than 250 molecular gases/vapors of environmental concern for atmospheric, industrial, medical, military, and scientific spheres exhibit strong absorption bands (Hubert, 1983). This laser, however, can be only stepwise tuned when operated in cw (continuous wave). Nevertheless, it is an ideal source to push the sensitivity of PA gas detection into the concentration range of ppbV or even lower. Instruments based on LPAS have nearly attained the theoretical noise equivalent absorption detectivity of 10^{-10} cm^{-1} in controlled laboratory conditions (Harren et al., 1990). This high sensitivity cannot be achieved in real detection conditions due to the coherent photoacoustical background signal and interfering background absorption of normal atmospheric constituents.

At present many research groups are actively involved in the development of LPAS systems for various applications in different disciplines, including nondestructive evaluation of materials, environmental analysis, agricultural, biological, and medical applications, investigation of physical processes (phase transitions, heat and mass transfer, kinetic studies), and many others. Our facility, which was originally designed for ethylene (C_2H_4) analysis at the low ppb level, is adaptable with minor modifications to a broad range of gases and vapors having absorption spectra in the infrared (IR).

2. Basic principles

2.1 Linear laser spectroscopy methods

A gaseous molecule that absorbs electromagnetic radiation is excited to a higher electronic, vibrational or rotational quantum state. The excited state loses its energy by radiation processes, such as spontaneous (fluorescence) or stimulated emission, and/or by collisional relaxation, in which energy is converted into translational energy. Radiative emission and chemical reactions do not play an important role in the case of vibrational excitation, because the radiative lifetimes of the vibrational levels are long compared with the time needed for collisional deactivation at pressures used in photoacoustics (~1 bar), and the photon energy is too small to induce chemical reactions. Thus, in practice the absorbed energy is completely released via either fluorescence (at low pressures) or collisions. The latter give rise to a gas temperature increase due to energy transfer to translation as heat, appearing as translational (kinetic) energy of the gas molecules. The deposited heat power density is proportional to the absorption coefficient and incident light intensity. The nonradiative relaxation process occurs when the relaxation time can compete with the radiative lifetime of the excited energy levels. Radiative decay has a characteristic lifetime of 10^{-7} s at visible wavelengths as compared with 10^{-2} s in IR at 10 μm. For nonradiative decay these values depend on pressure (the decay time τ is inversely proportional to pressure) and may vary strongly at atmospheric pressures (10^{-3}-10^{-8} s) depending on the gas nature and the involved energy level.

There are three techniques of linear laser spectroscopy, based on measurement of different physical quantities:

- the absorption method and the cavity ringdown spectroscopy (intensity);
- the radiative method (fluorescence); and
- the photothermal (calorimetric) method (pressure, temperature).

The most important optical process, as far as spectroscopic trace gas detection is concerned, is based on the extinction of radiation by molecular absorption. The absorption features and strengths specific of each molecule make it possible to identify trace gases and determine their concentrations. Absorption coefficients are typically on the order of 1 cm^{-1} (one wave number). The absorption of trace gas molecules in a gas mixture may be monitored by detecting the attenuation of the laser beam over a fixed absorption path length L. According to the Beer-Lambert law, the transmitted laser power in the absence of saturation is given by:

$$P(L) = P(0)\exp(\alpha_p L) = P(0)\exp(\alpha c L)\,, \tag{1}$$

where $P(0)$ and $P(L)$ are the laser powers before and after the absorption cell, respectively; α_p (cm^{-1}) is the absorption coefficient at a given pressure of the gas at a specific laser wavelength: $\alpha_p = \alpha c$; α (cm^{-1} atm^{-1}) is the gas absorption coefficient (the absorption coefficient normalized to unit concentration), and c (atm) is the trace gas concentration. Also, $\alpha_p = N_{tot}\sigma$, where σ (cm^2) is the absorption cross section per molecule and N_{tot} = 2.5×10^{19} molecules cm^{-3} is the number of absorbing molecules per cubic centimeters at 1013 mbar and 20°C. It results:

$$c = -\frac{1}{\alpha L}\ln\frac{P(L)}{P(0)} = -\frac{1}{\alpha L}\ln\left(1 - \frac{\Delta P}{P(0)}\right) \cong \frac{1}{\alpha L}\frac{\Delta P}{P(0)}, \tag{2}$$

which is valid for $\Delta P/P(0) << 1$ (optically thin sample), where $\Delta P = P(0) - P(L)$. For a given L, the detection limit is given by the smallest relative change $\Delta P_{min}/P(0)$ that can be measured in the transmitted signal. For dilute mixtures and modest absorption path lengths, the desired signal is the small difference between two large values so that high quantitative accuracies in signal intensities are required. The most sensitive method employs frequency modulation and harmonic detection. The sensitivity depends on the linewidth, and for atmospherically broadened lines, Reid et al. (Reid et al., 1978) have reached $\Delta P_{min}/P(0) \cong 10^{-5}$ in a diode laser spectrometer (1050-1150 cm^{-1}). With a path length of 100 m, the result is a sensitivity of 10^{-9} cm^{-1}, which corresponds to concentrations of 3 ppbV of a weakly absorbing molecule such as SO_2 (ν_1 band), or 0.01 ppbV of a strongly absorbing molecule such as CO. Assuming the same detectable attenuation $\Delta P_{min}/P(0) \cong 10^{-5}$, a path length $L = 1$ m, and an absorption coefficient $\alpha = 30.4$ $cm^{-1}atm^{-1}$ (typical of fundamental absorption in the mid-IR), one obtains a minimum detectable absorption coefficient $\alpha_p = 10^{-7}$ cm^{-1}. This number corresponds to a concentration of 3.3 ppbV at atmospheric pressure. Conventional absorption techniques, which require precise measurements of the difference between two nearly equal signals are, however, unable to realize the full potential of the higher power levels now attainable. Improvement may be obtained by: increasing the path length L in a multipass or intracavity arrangement, or using wavelength modulation, i.e., by modulating the wavelength of the incident intensity across a molecular absorption line. In multipass transmission absorption spectroscopy, a multipass transmission cell (White cell) filled with analyte gas with mirrors at each end is used. The beam is folded back and forth through the cell, creating an extended yet defined optical path length within a confined space.

Cavity ringdown spectroscopy confines gas in an optically reflective cavity where laser radiation is introduced. Radiation amplitude decays at a certain rate in the absence of absorption. An absorbing sample gas in the cavity increases the rate of decay, thus indicating the presence of an absorbing species. The advantages of the method are high sensitivity and a small sample volume, while indirect measurement is an important drawback: as the measured parameter is the rate of light intensity decay, decay caused by absorption by the analyte of interest has to be distinguished from the one caused by the mirrors and other cavity-dependent losses.

In linear detection, sensitivity is limited by laser power fluctuations, and a considerable improvement can be obtained by the dark background methods, in which one measures a quantity that is directly proportional to absorption, rather than that part of the laser beam which is absorbed. In the visible, this can be done by monitoring the fluorescence from the upper level of the transition. In the IR, however, the spontaneous emission rate is too low, and most of the excess vibrational energy is converted to heat through inelastic collisions.

The fluorescence method requires that a certain part of the excitation should relax through radiative channels. This condition is fulfilled by detecting atoms and molecules in the UV, visible, and near-IR spectral regions. As a principal advantage of the fluorescence techniques, the observed signal is proportional to the concentration of the measured species and the accuracy, therefore, depends on the magnitude of the signal relative to detector noise. The sensitivity is so high, that it makes it possible to detect single atoms in the laser beam.

The basic principle of all photothermal (PT) techniques is the absorption of light in a sample leading to a change in its thermal state. This may be a change in temperature or another thermodynamic parameter of the sample that is related to temperature. Measurement of either the temperature, pressure or density change that occurs due to optical absorption is ultimately the basis for all PT spectroscopic methods. PT analysis can be considered as an indirect absorption measurement, since the measured quantity is not an optical signal. (It should be noted here that the classical absorption measurement is not a direct measurement either. Though the measured value in this case is an optical one, namely the transmitted light, the absorbed light quantity is derived from the difference of the incident energy and the transmitted one). The sample heating which produces the PT signal is directly correlated to the absorbed electromagnetic energy. Unlike in conventional transmission spectroscopy, neither scattered nor reflected light contributes to the signal. Although a PT effect can be induced by any light source, lasers are nowadays the preferred source of excitation for two reasons: (i) To a first approximation, the PT signal is proportional to the temperature rise in the sample and thus proportional to the absorbed energy. (ii) For many applications, the selectivity of a PT analysis, as with any other absorption method, depends on the tunability of the excitation wavelength.

PA spectroscopy is an indirect technique in that an effect of absorption is measured rather than absorption itself. Hence the name of photoacoustic: light absorption is detected through its accompanying acoustic effect. The advantage of photoacoustics is that the absorption of light is measured on a zero background; this is in contrast with direct absorption techniques, where a decrease of the source light intensity has to be observed. The spectral dependence of absorption makes it possible to determine the nature of the trace components. The PA method is primarily a calorimetric technique, which measures the precise number of absorbent molecules by simply measuring the amplitude of an acoustic signal. In LPAS the nonradiative relaxation which generates heat is of primary importance. In the IR spectral region, nonradiative relaxation is much faster than radiative decay.

PA spectroscopy relies on the PA effect for the detection of absorbing analytes. The sample gas is in a confined (resonant or nonresonant) chamber, where modulated (e.g., chopped) radiation enters via an IR-transparent window and is locally absorbed by IR-active molecular species. The temperature of the gas thereby increases, leading to a periodic expansion and contraction of the gas volume synchronous with the modulation frequency of the radiation. This generates a pressure wave that can be acoustically detected by a suitable sensor, e.g., by a microphone. The advantages of the PA method are high sensitivity and small sample volume; besides, the acoustic measurement makes optical detection unnecessary. The main drawback is caused by the sensitivity to acoustic noise, because the measurements are based on an acoustic signal. A comparison of the linear laser spectroscopy methods is presented in Table 1 (Zharov & Letokhov, 1986).

Method / Characteristics	Absorption	Fluorescence	PA
Spectral range	UV - far IR	UV and visible	UV - far IR
Sensitivity (cm^{-1})	10^{-5} - 10^{-9}	Up to single atoms	10^{-7} - 10^{-10}
Time resolution (s)	1	1 - 10^{-12}	1 - 10^{-3}
Necessary conditions	-	Radiative channels of relaxation	Nonradiative channels of relaxation

Table 1. Comparison of linear laser spectroscopy methods.

The favorable properties of LPAS are essentially determined by the characteristics of the laser. The kind and number of detectable substances are related to the spectral overlapping of the laser emission with the absorption bands of the trace gas molecules. Thus, the accessible wavelength range, tunability, and spectral resolution of the laser are of prime importance. With respect to minimum detectable concentrations (LPAS sensitivity), a laser with high output power P_L is a benefit, because the PA signal is proportional to P_L. The broad dynamic range is an inherent feature of LPAS and therefore is not affected by the choice of the radiation source. In contrast to remote-sensing methods, LPAS is a detection technique applied locally to samples enclosed in a PA cell. In order to still obtain some spatial resolution, either the samples have to be transported to the system, or the system has to be portable. The temporal resolution of LPAS is determined by the time needed for laser tuning and the gas exchange within the cell. Thus, a small volume PA cell and a fast tunable laser are a plus.

The availability of suitable laser sources plays a key role, as they control the sensitivity (laser power), selectivity (tuning range), and practicability (ease of use, size, cost, and reliability) that can be achieved with the photoacoustic technique. The CO_2 laser perfectly fits the bill for a trace gas monitoring system based on LPAS. This IR laser combines simple operation and high output powers. The frequency spacing between two adjacent CO_2-laser transitions range from 1 to 2 cm^{-1}. By contrast, the typical width of a molecular absorption line is approximately 0.05 to 0.1 cm^{-1} for atmospheric conditions. Since this is not a continuously tunable source, coincidences between laser transitions and trace gas absorption lines are mandatory. Fortunately, this does not hamper its applicability to trace gas detection, as numerous gases exhibit characteristic absorption bands within the wavelength range of the CO_2 laser which extends from 9 to 12 μm when different CO_2 isotopes are used. The CO_2 laser spectral output occurs in the wavelength region where a large number of compounds (including many industrial substances whose adverse health effects are a growing concern) possess strong characteristic absorption features and where absorptive interferences from water vapors, carbon dioxide, and other major atmospheric gaseous components may influence the measurements.

2.2 PA effect in gases

The PA effect in gases can be divided into five main steps (Fig. 1):

1. Modulation of the laser radiation (either in amplitude or frequency) at a wavelength that overlaps with a spectral feature of the target species; an electrooptical modulation

device may also be employed, or the laser beam is modulated directly by modulation of its power supply; the extremely narrowband emission of the laser allows the specific excitation of molecular states; the laser power should be modulated with a frequency in the range $\tau_{th} >> 1/f >> \tau_{nr}$, where τ_{th} is the thermal relaxation time, and τ_{nr} the nonradiative lifetime of the excited energy state of the molecule.

2. Excitation of a fraction of the ground-state molecular population of the target molecule by absorption of the incident laser radiation that is stored as vibrational-rotational energy; the amount of energy absorbed from the laser beam depends on the absorption coefficient, which is a function of pressure.
3. Energy exchange processes between vibrational levels (V-V: vibration to vibration transfer) and from vibrational states to rotational and translational degrees of freedom (V-R, T transfer); the energy which is absorbed by a vibrational-rotational transition is almost completely converted to the kinetic energy of the gas molecules by collisional de-excitation of the excited state; the efficiency of this conversion from deposited to translational energy depends on the pressure and internal energy level structure of the molecule; vibrational relaxation is usually so fast that it does not limit the sensitivity; however, notable anomalies occur in the case of diatomic molecules, such as CO, where vibrational relaxation is slow in the absence of a suitable collision partner, and of the dilute mixtures of CO_2 in N_2, where the vibrational energy is trapped in slowly relaxing vibrational states of N_2; the kinetic energy is then converted into periodic local heating at the modulation frequency.
4. Expansion and contraction of the gas in a closed volume that give rise to pressure variation which is an acoustic wave; the input of photon energy with correct timing leads to the formation of a standing acoustic wave in the resonator.
5. Monitoring the resulting acoustic waves with a microphone; the efficiency at which sound is transmitted to the microphone depends on the geometry of the cell and the thermodynamic properties of the buffer gas.

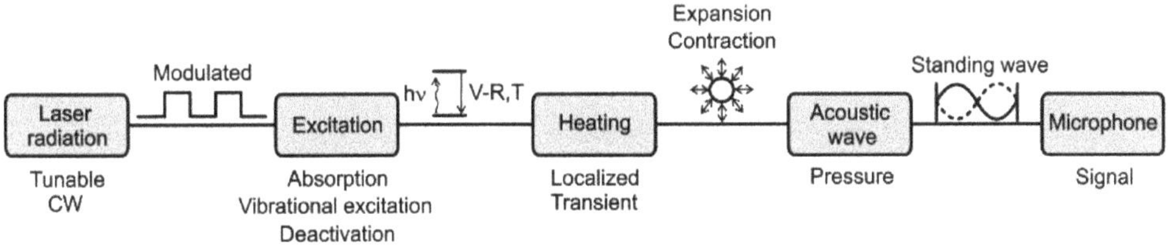

Fig. 1. Schematic of the physical processes occurring during optical excitation of molecules in photoacoustic spectroscopy.

From kinetic gas theory it can be estimated that a molecule performs 10^9-10^{10} collisions per second at 1 bar pressure. This means that at atmospheric pressure the photon energy is transformed into an acoustical signal in about 10^{-5}-10^{-6} s. For most polyatomic molecules signal production is even faster. The time needed by the pressure wave to travel from the laser beam area to the microphone in the acoustic cell is therefore in most cases longer than the vibrational relaxation time. For a distance of a few centimeters this transit time is about 10^{-4} s. The time delay between excitation and detection of the pressure wave, however, is influenced not only by energy transfer processes and the transit time, but also by the response time of the gas-microphone system, being about 10^{-4} s or longer (Hess, 1983).

2.3 Typical laser photoacoustic setup

A typical setup of a resonant LPAS, as used in the authors' laboratory for gas studies, is shown in Fig. 2. The continuous wave laser radiation is amplitude-modulated by a mechanical chopper operating at an acoustic resonance frequency of the PA cell. It is then focused by a lens and directed through the resonant PA cell. The transmitted laser power is monitored with a powermeter (signal P_L in Fig. 2). Inside the cell the radiation produces pressure modulation recorded by microphone as an acoustical signal V, which is processed by a lock-in amplifier locked to the chopper frequency. The normalized absorption can then be deduced as being proportional to V/P_L ratio (Cristescu et al., 1997, Dumitras et al., 2007a).

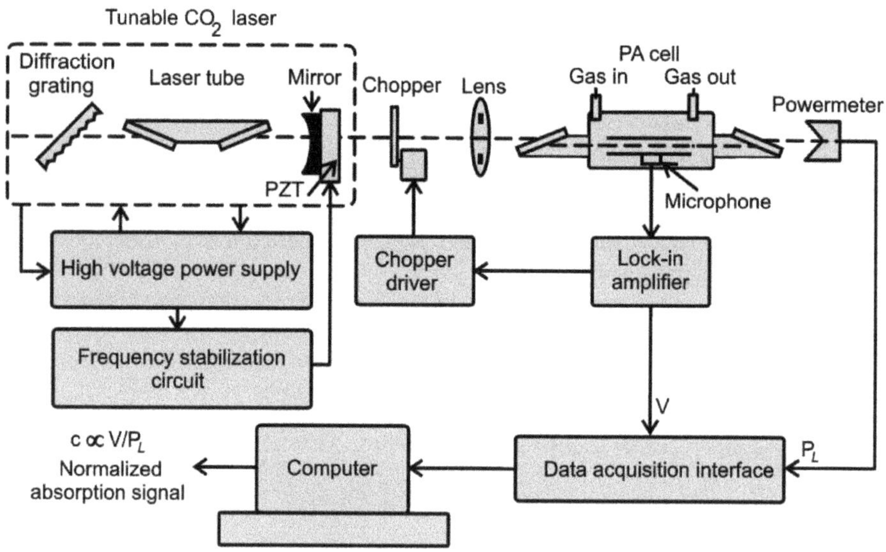

Fig. 2. Typical laser photoacoustic setup for trace gas measurements.

The power reading after beam passage through the PA cell can only be used for "transparent" gas samples. Let us evaluate if this condition is fulfilled. If the absorption is assumed to follow the Beer-Lambert law (Eq. 1), in the case of small absorption, the fractional absorption of the laser beam in the PA cell is given as $\Delta P/P(0) \cong \alpha Lc$ (Eq. 2). The quantity αLc is known as the optical density of the gas in the resonator tube (this quantity is also called absorbance). Therefore, the PA signal proportional to ΔP depends linearly on the absorption coefficient, and its dependence on gas concentration is also linear. At $\alpha Lc = 0.06$, a deviation of ~3% results from the linear behavior (~10% for $\alpha Lc = 0.07$). An optical density of 0.06 (an ethylene concentration of 65 ppmV for $L = 30$ cm, the length of our cell) may thus be regarded as the upper limit of the linear range of a PA detector. Consequently, the PA signal can be modeled as a linear function of concentration in the full range from a few tens of pptV to 65 ppmV ethylene, so that the range spans over 6 orders of magnitude! Taking into account typical values for the absorption coefficients of the species to be measured (e.g., for ethylene at concentrations in the range 1 ppbV-10 ppmV, $\alpha c \cong 3\text{x}10^{-8}$-$3\text{x}10^{-4}$ cm^{-1}) and usual lengths of the PA cells (~30 cm), the fractional absorption is very small (10^{-6}-10^{-2}), which means that in the worst case less than 1% of the incident laser power is absorbed in the sample gas inside the PA cell. It follows that the powermeter measures the real value of the laser power inside the PA cell (we have "transparent" gases).

Another advantage of photoacoustic spectroscopy as a tool for trace gas analysis is that very few photons are absorbed as the laser beam passes through the sample cell. As a result, notwithstanding the losses from absorption in the windows, the transmitted beam typically has sufficient power for analyzing samples in successive cells, via a multiplexing arrangement. A multiplexed photoacoustic sensor can be used to monitor many different samples simultaneously so that one instrument can be deployed to monitor up to 20 different locations within a clean room, industrial plant or other facility (Pushkarsky et al., 2002).

Following the terminology introduced by Miklos et al. (Miklos et al., 2001), the name 'PA resonator' will be used for the cavity in which the resonant amplification of the PA signal takes place. The term PA cell (or PA detector; both terms are used in the literature to describe the device in which the PA signal is generated and monitored) is reserved for the entire acoustic unit, including the resonator, acoustic baffles and filters, windows, gas inlets and outlets, and microphone(s). Finally, PA instrument (PA sensor) stands for a complete setup, including the PA cell, light source, gas handling system, and electronics used for signal processing.

It is interesting to mention that the *reverse* PA effect, called "sonoluminiscence", consists in the generation of optical radiation by acoustic waves, while the *inverse* PA effect is the generation of sound due to optical energy being lost from a sample, instead of being deposited in a sample as in the usual PA effect (Tam, 1986).

3. Photoacoustic signal

3.1 Resonant cells

A PA cell can be operated either in nonresonant mode or at an acoustic resonance frequency specific to the PA resonator. In the so-called nonresonant mode, the modulation frequency is much lower than the first acoustic resonance frequency of the PA resonator. In this case, the wavelength of the generated acoustic wave is larger than the cell dimensions. Thus, the generation of standing acoustic waves is not possible. A nonresonant PA cell lacks any means of energy accumulation in the acoustic wave, i.e., the induced pressure fluctuations are a function of the energy absorbed on that cycle alone and, in fact, any acoustic energy remaining from previous cycles tends only to produce noise on the desired signal. The main drawbacks of the nonresonant scheme are the low modulation frequency, which makes the system susceptible to $1/f$ noise, and the relatively large background signal generated by absorption in the windows of the cell and by radiation scattered to the walls. Nevertheless, the acoustically nonresonant cell has an advantage in low-pressure operation, as the signal, and hence the SNR, remains constant as pressure is decreased, whereas for the resonant cell, it drops almost linearly with decreasing pressure (Fig. 3) (Dumitras et al., 2007b). Also, the background signal, which limits the sensitivity of the nonresonant cell at atmospheric pressure, has been found to depend approximately linearly on pressure and would be less troublesome in low-pressure operation (Gerlach & Amer, 1978).

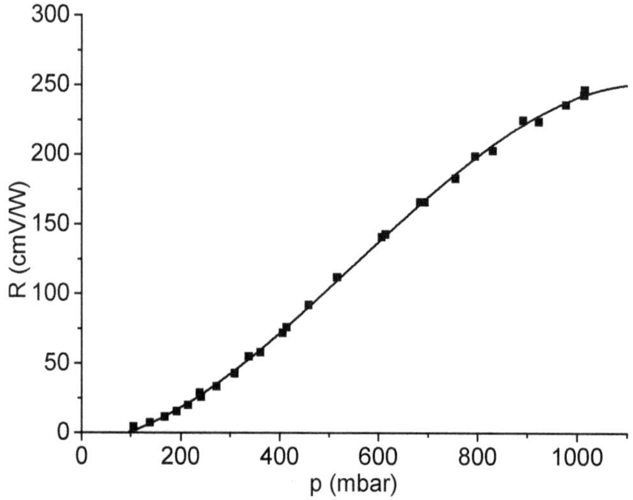

Fig. 3. Dependence of the PA cell responsivity on the total gas pressure in the cell (measured with 1 ppmV ethylene in nitrogen).

Nonresonant operation can compete with enhanced resonant operation only at much lower frequencies and smaller cell volumes; however, a number of practical difficulties have been cited. At low frequencies, gas inlet-outlet ports act as pneumatic short circuits for the induced pressure (Kritchman et al., 1978). Excess acoustic energy in previous cycles of the modulated light can produce noise in the nonresonant signal, while in resonant operation this type of noise is avoided because the energy in each cycle contributes to a standing wave (Kamm, 1976). For the small, nonresonant cell, attachment of the microphone can lead to difficulties in extracting the optimum pressure response signal (Dewey, 1977).

With increasing modulation frequency, the acoustic wavelength equals the cell dimensions at a certain point, and the resonant modes of the cell can be excited, leading to an amplification of the signal. The signal can be boosted manifold by: a) designing the sample cell as an acoustic resonance chamber, so that the pressure fluctuations produced by spatially and temporally nonuniform excitation contribute to standing acoustic waves within the chamber, and b) minimizing dissipation of the acoustic energy and modulating the laser beam spatially and temporally at a frequency which coincides with one of the natural resonant acoustic frequencies of the chamber. The system becomes an acoustic amplifier in the sense that the energy existing in the standing wave is many times higher than the energy input per cycle, and the signal is amplified by a quality factor Q. The final signal amplification obtainable depends on the resonator losses. After an initial transient state, during which energy is accumulated in the standing acoustic wave, a steady state is reached in which the energy lost per cycle by various dissipation processes is equal to the energy gained per cycle by absorption of IR laser photons. Resonance properties mainly depend on the geometry and size of the cavity. For an acoustically resonant PA cell, important parameters will include gas characteristics such as heat capacity, thermal conductivity, viscosity, energies and relaxation times of the molecular vibrations and the sound velocity which determines the resonant frequencies of the cavity. Other parameters govern the loss mechanisms that determine the quality factors of the resonances and also cause small shifts in the resonant frequencies.

A comparison of the microphone signals for nonresonant operation at 100 Hz and resonant operation at 564 Hz is depicted in Fig. 4 (a) and (b), respectively, together with the chopper waveforms. For nonresonant operation, the laser beam was amplitude-modulated with a duty cycle (pulse duration divided by the pulse period) of 25%, and the PA signal exhibits ringing at the resonant frequency on top of the 100-Hz square wave. For resonant operation, the laser beam was amplitude-modulated with a duty cycle of 50% and the microphone output was simply a coherent sine wave. In Fig. 4 (b), the data were recorded with a concentration of 1 ppmV of ethylene in the PA cell.

The resonant cells can be adequatelly characterized by a model based on an acoustic transmission line (Cristescu et al., 2000).

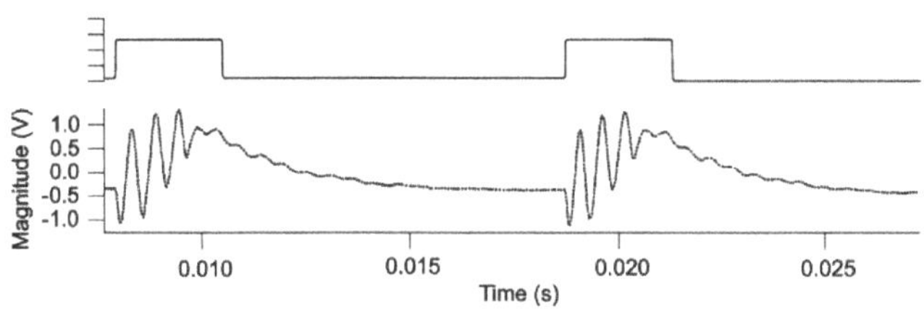

(a)

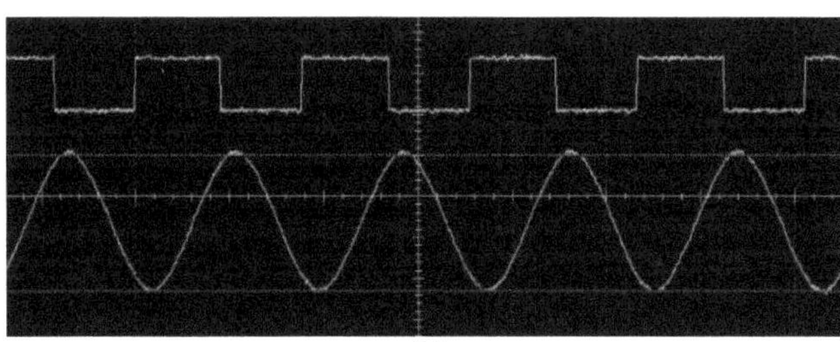

(b)

Fig. 4. Microphone signals for: (a) nonresonant operation of the PA cell (Pushkarsky et al., 2002); (b) resonant operation of the PA cell (our cell), recorded with a Tektronix DPO 7104 Digital Oscilloscope, horizontal scale 1 ms/div, vertical scales 1 V/div. (rectangle wave) and 0.5 V/div. (sine wave amplified x 100).

3.2 Resonance frequencies

Several distinct resonances can be generated if the dimensions of a cavity are comparable with the acoustic wavelength. The standing wave patterns and resonance frequencies depend on the shape and size of the PA resonator. The most frequently used resonator is the cylinder, the symmetry of which coincides well with that of a laser beam propagating along the cylinder axis. The natural acoustic resonance frequencies of a lossless cylindrical resonator (fully reflecting walls) are determined as a solution of the wave equation in cylindrical coordinates (Hess, 1983):

$$f_{kmn} = \frac{v_s}{2}\left[\left(\frac{k}{L}\right)^2 + \left(\frac{\alpha_{mn}}{\pi r}\right)^2\right]^{1/2}, \tag{3}$$

where v_s is the sound velocity, L and r are the length and radius of the cylinder, the k, m, n indices (non-negative integers) refer to the values of the longitudinal, azimuthal, and radial modes, respectively, and α_{mn} is the n-th root of the derivative of the m-th Bessel function:

$$\frac{dJ_m(z)}{dz} = 0 \tag{4}$$

($\alpha_{00} = 0$, $\alpha_{01} = 3.8317$, $\alpha_{02} = 7.0153$, $\alpha_{10} = 1.8412$, $\alpha_{11} = 5.3311$, $\alpha_{12} = 8.5360$, etc.). For the first longitudinal mode, $k = 1$, $m = 0$, $n = 0$ and $f_{100} = f_0 = v_s/2L$.

In deducing Eq. (3), it was assumed that there was no phase shift on reflection of the pressure wave from the cavity walls caused either by wall compliance or boundary layer effects. If we depart from the assumption of complete wall rigidity, the boundary layer can be seen to cause significant frequency deviations from the above formula. To evaluate the frequency from Eq. (3), we must know the sound velocity, which may vary with frequency and pressure due to molecular relaxation effects and the nonideal behavior of the gas.

In reality, frequencies at the resonances are somewhat smaller. The corresponding resonance frequencies for PA resonators with open-open ends can be obtained from the following expression (Morse & Ingard, 1986):

$$f_0 = \frac{v_s}{2(L + \Delta L)}, \tag{5}$$

where the quantity ΔL is the so-called end correction, which should be added to the length of the pipe for each open end. Physically, the end correction can be understood as an effect of the mismatch between the one-dimensional acoustic field inside the pipe and the three-dimensional field outside that is radiated by the open end. The end correction can be approximated by the following expression: $\Delta L \cong 0.6r$, where r is the radius of the pipe (Miklos et al., 2001). More precisely, the end correction slightly decreases with frequency; therefore the resonance frequencies of an open pipe are not harmonically related but slightly stretched. In our experimental setup, the resonance frequency for 0.96 ppmV of ethylene in pure nitrogen is 564 Hz at L = 30 cm. By taking v_s = 343 m/s in nitrogen at 22°C (the sound velocity in nitrogen of 330 m/s at 0°C was corrected for the room temperature), we have $\Delta L \cong 0.2$ cm for the two open ends of our PA resonator and $\Delta L \cong 0.6r$ (r = 0.35 cm).

In an ideal gas, the sound velocity is given by:

$$v_s = (\gamma RT / M)^{1/2}, \tag{6}$$

where $\gamma = C_p/C_v$ is the ratio of specific heats at constant pressure and volume, R is the ideal-gas constant, T is the absolute temperature, and M is the molecular weight. The sound velocity in an ideal gas only depends on temperature and remains unchanged at pressure modifications if γ is constant. In the case of ideal gases, $\gamma = 1.4$ for diatomic gases and $\gamma = 1.33$ for triatomic gases. Experimentally, the following values have been measured: 1.404 for N_2, 1.401 for O_2, 1.404 for CO, 1.32 for H_2O, 1.31 for NH_3, 1.31 for CH_4, and 1.25 for C_2H_4.

For nonideal gases, the sound velocity can be approximately calculated by the following formula:

$$v_s = \left[\gamma(RT + 2Bp) / M\right]^{1/2}, \tag{7}$$

where B is the second virial coefficient and p is the pressure.

Little attention has been given to the role of the buffer gas (defined as the optically nonabsorbing gaseous component in photoacoustic detectors). In principle, the molecular weight and the thermodynamic and transport properties of the buffer gas should have a significant impact on the photoacoustic signal. One would also expect the energy transfer between the absorbing species and the buffer gas to play an important role in PA detection (Thomas III et al., 1978; Gondal, 1997). In a mixture of ideal gases, the sound velocity $\overline{v}_s$ and consequently the resonant frequencies of a PA resonator depend on the effective specific heat ratio and the average mass of the mixture:

$$\overline{v}_s = \left(\overline{\gamma} RT / \overline{M}\right)^{1/2}, \tag{8}$$

where the specific heat ratio $\overline{\gamma}$ and the average molecular weight $\bar{M}$ are:

$$\overline{\gamma} = \frac{xC_p^b + (1-x)C_p^a}{xC_v^b + (1-x)C_v^a}, \tag{9}$$

$$\bar{M} = xM^b + (1-x)M^a. \tag{10}$$

Here C_p^b, C_v^b, C_p^a and C_v^a are the heat capacities of the buffer and absorbing gases, respectively; M^b and M^a are their molecular weights; and x is the fractional concentration of the buffer gas. When the molecular weight of the buffer gas is increased, the resonance frequency of the PA resonator shifts to lower values. In conclusion, the resonance frequency is a sensitive function of temperature and gas composition, both of which influence the speed of sound.

At a fixed temperature, v_s also depends on the water content in the air (Rooth et al., 1990):

$$v_s = v_{s0}'\left[1 - \frac{p_w}{p_{air}}\left(\frac{\gamma_w}{\gamma_{air}} - \frac{5}{8}\right)\right]. \tag{11}$$

Here γ_w and γ_{air} are the ratios of the specific heats of water vapor and air. The partial pressures of water and air are denoted as p_w and p_{air}. The sound velocity in dry air is written as v_{s0}'. The increase of the resonance frequency of a 30-cm long longitudinal resonator at ambient temperature is 0.90 Hz for 1% of water vapors added to the gas. For all practical purposes, the variation of the resonance frequency with the CO_2 concentration is negligible: -0.15 Hz per 1000 ppmV. For a given water vapor concentration, the resonance frequency provides information about the gas temperature inside the resonator. In most cases, the PA cell resonance frequency has to be determined experimentally.

Since the resonance frequency is proportional to the sound velocity, the temperature dependence of the sound velocity is directly mirrored by the resonance frequency. The sound velocity in air has a temperature coefficient of about 0.18%/°C, thus a frequency shift $\delta f \cong 0.0018 f_0 \Delta T$ is expected for a temperature change ΔT (°C) (Miklos et al., 2001). The true resonance frequency may therefore deviate from the fixed modulation frequency by δf. Then the PA signal will not be excited at the peak of the resonance, but slightly to one side. Since a detuning from the resonance peak by 0.46 (Δf(10%)/Δf(FWHM) = 16/35) results in a 10% drop of the PA signal (see Fig. 5), the detuning should be smaller than $\pm 0.23 f_0/Q$ for 10% signal stability. This stability can be ensured, if the condition $Q\Delta T \leq 128$ ($\delta f \leq 0.23\Delta f$ or $0{,}0018 f_0 \Delta T \leq 0.23\Delta f$ or $Q\Delta T \leq 0.23/0.0018$) is fulfilled ($\Delta T \leq 7.9$°C in our case for Q =16.1). The corresponding condition for PA signal stability of 2% can be written as $Q\Delta T \leq 64$ ($\Delta T \leq 4$°C in our case). These examples clearly show that low-Q photoacoustic resonators are not sensitive to temperature variations and consequently do not need temperature stabilization or active tracking of the resonance to adjust the modulation frequency.

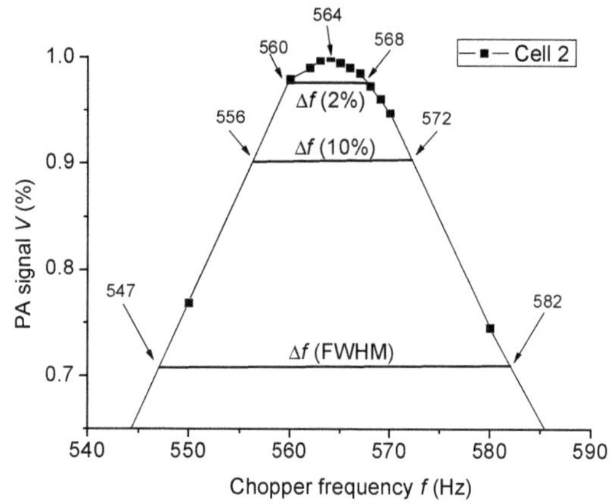

Fig. 5. Resonance curve of our PA cell showing the full width at half maximum (FWHM) and the full widths for a signal drop of 2% and 10% from its maximum.

3.3 Dissipation processes

The various dissipation processes occurring in an acoustic cavity were first discussed at length by Kamm (Kamm, 1976). The energy accumulation attainable in a standing wave of a resonant cavity is many times larger than the energy loss occurring during a single period of an acoustic oscillation. This acoustical amplification effect is limited, however, by various dissipation processes. The losses can be divided into surface effects and volumetric ones (Johnson et al., 1982). The surface losses are due to the interaction of the standing wave with the internal resonator surface and may be subdivided into the following dissipation processes:

1. compliance of the chamber walls;
2. dissipation at the microphone diaphragm;
3. losses due to wave scattering at surface obstructions such as gas inlet/outlet, microphones, and windows;
4. viscous and thermal dissipation in the boundary layer at the smooth internal surfaces.

In a carefully designed high quality resonator, the contribution of the first three effects can be minimized. The dominant contribution is caused by the viscous and thermal boundary layer losses. Throughout the major portion of the resonator volume the expansion and contraction of the gas occur adiabatically. We neglect heat conduction and viscous losses in the volume of the gas because the acoustic power loss from these effects is very small. However, the wall consists of a material with a thermal conduction coefficient much greater than that of the gas. Thermal dissipation occurs because expansion and contraction of the gas do not proceed adiabatically near the walls, where the process will change to isothermal. The temperature variation changes exponentially from the adiabatic propagation regime in the gas to a zero value at the wall. This leads to heat conduction within a transition region (thermal boundary layer), which is responsible for the thermal dissipation process. Outside a thin boundary layer with thickness d_h, near the wall, the thermal losses can be neglected:

$$d_h = \left(\frac{2K}{\rho \omega C_p} \right)^{1/2} . \quad (12)$$

Here K is the thermal conductivity of the gas, ρ is the density of mass, C_p is the molar heat capacity, and $\omega = 2\pi f$ is the angular frequency.

The viscous dissipation can be explained by the boundary conditions imposed by the walls. At the surface, the tangential component of the acoustic velocity is zero, whereas inside the cavity, it is proportional to the acoustic pressure gradient. Thus, viscoelastic dissipation occurs in a transition region with a thickness d_v, which is called the viscous boundary layer:

$$d_v = \left(\frac{2\mu}{\rho \omega} \right)^{1/2} , \quad (13)$$

where μ is the viscosity coefficient.

Equations (12) and (13) are only valid if d_h and d_v are much smaller than the radius of the PA resonator (d_h, d_v << r), which yields a lower frequency limit. The upper frequency limit is reached when the wavelength of sound is comparable to the cross-sectional dimensions of the tube ($\lambda_s = v_s/f_0 \cong r$). The magnitude of d_h and d_v can be calculated by using the properties of nitrogen at standard pressure (p = 1 atm) and room temperature: $K = 2.552\text{x}10^{-2}$ W/(m K), $\mu = 1.76\text{x}10^{-5}$ Pa s, $C_p = 1.04\text{x}10^3$ J/(kg K), $\rho = 1.142$ kg/m^3 and $\gamma = C_p/C_v = 1.4$. As a result, the values for the thermal and viscous boundary layer thicknesses are, respectively: $d_h = (2K/\rho\omega_0 C_p)^{1/2} = 2.6(f_0)^{-1/2}$ (mm Hz$^{1/2}$) $\cong$ 0.110 mm (at f_0 = 564 Hz) and $d_v = (2\mu/\rho\omega_0)^{1/2} = 2.2(f_0)^{-1/2}$ (mm Hz$^{1/2}$) $\cong$ 0.093 mm (at f_0 = 564 Hz). Therefore, at atmospheric pressure and audio frequencies, both boundary layers are only a fraction of a millimeter thick.

The volumetric or bulk losses are caused by processes that tend to establish equilibrium in the propagating wave. These damping processes are:

1. free space viscous and thermal dissipation;
2. diffusion effects;
3. radiation effects; and
4. relaxational damping (dissipative relaxation processes within polyatomic gases).

Friction due to compressional motion and the transformation of organized energy into heat due to temperature gradients are responsible for the free space viscous and thermal losses.

These two processes are often called Stokes-Kirchhoff losses and are small compared with surface damping. Diffusion and radiation effects are normally very small. Nevertheless, radiation losses through openings, e.g., pipes connecting the resonator to buffer volumes, cannot be neglected. The radiation losses can be reduced by increasing the acoustic input impedance of the openings. This is achieved by terminating the cavity resonator at the openings with acoustic band-stop filters, which prevent sound escape from the resonator. Relaxational effects can add a significant contribution in diatomic and polyatomic molecules. The reason for the relaxational losses is the phase difference between gas pressure and density in the dispersion region, leading to an irreversible conversion of sound energy into thermal energy.

3.4 Quality factor

The amount of signal enhancement that occurs when the laser is modulated at a resonance frequency is determined by the quality factor. At resonance, the amplitude of the PA signal is Q times larger than the amplitude far from the resonance frequency, i.e., the amplification is equal to the value of the Q factor. The quality factor of the system, Q, is the ratio between the energy stored in a specific mode (the acoustic wave) and the energy losses per cycle of this acoustic wave:

$$Q = \frac{2\pi \text{ accumulated energy}}{\text{energy lost over one period}}. \tag{14}$$

For high Q values the quality factor can be deduced dividing the resonance frequency by its bandwidth at the 0.707 amplitude point:

$$Q = \frac{f_0}{\Delta f} = \frac{\omega_0}{\Delta\omega}, \tag{15}$$

where f_0 and Δf are the resonance frequency and the full-width value of the resonance profile ($\omega_0 = 2\pi f_0$ and $\Delta\omega = 2\pi\Delta f$). The full width is measured between the points where the amplitude of the resonance profile is at $1/\sqrt{2}$ the peak value amplitude (half-maximum values of the intensity). Therefore, Δf is also called the full width at half maximum (FWHM). Q is typically between 10 and 50 for longitudinal resonators, but can be as high as 1000 for spherical cavities.

Also, the quality factor can be calculated as (Kamm, 1976; Bernegger & Sigrist, 1987):

$$Q = \frac{2S}{2\pi r\left[d_v + (\gamma - 1) d_h\right]}, \tag{16}$$

where S stands for the cross section of the resonator tube and r for the radius of the tube. By introducing the radius (r = 3.5 mm) of the PA resonator we used and the values for the thermal and viscous boundary layer thicknesses determined in the previous section, Eq. (16) yields Q = 14.2, in agreement with the experimentally determined value (Q = 16.1).

The overall Q factor for a resonance may be found by summing all the losses, expressed as $1/Q_i$:

$$\frac{1}{Q_{tot}} = \sum_i \frac{1}{Q_i}, \tag{17}$$

In practice, we only include three contributions: viscous and thermal dissipation in the boundary layer at the smooth internal surfaces (surface loss), free space viscous and thermal dissipation, and relaxation losses (volumetric losses).

3.5 Pressure amplitude

The transformation of the absorbed laser energy into heat is usually modeled by a simple relaxation process, while the well-known acoustic-wave equation is applied to calculate the sound-pressure field. The laws of fluid mechanics and thermodynamics can be used to model the acoustic and thermal wave generation in gases. The governing physical equations are the laws of conservation of energy, momentum, mass, and the thermodynamic equation of state. The physical quantities characterizing the acoustic and thermal processes are the temperature T, pressure p, density ρ, and the three components of the particle velocity vector $\mathbf{v}$. By eliminating the variables T, ρ, and $\mathbf{v}$ (and by neglecting the influence of the thermal and viscous interactions of the gas), a linear (inhomogeneous) wave equation can be derived for the acoustic pressure changes, p (Miklos et al., 2001):

$$\frac{\partial^2 p(\mathbf{r},t)}{\partial t^2} - v_s^2 \nabla^2 p(\mathbf{r},t) = (\gamma - 1)\frac{\partial H(\mathbf{r},t)}{\partial t}, \tag{18}$$

where $H(\mathbf{r},t)$ is the heat density deposited in the gas by light absorption. The term on the right-hand side of the equation describes the heat input changes over time. When the heat input is constant, this term is zero and no pressure wave is generated. Thus, the heat input must be modulated, which requires that the laser radiation be also amplitude or frequency modulated. A modulated laser beam generates periodic sound due to the periodic localized heating of the gas. From an acoustic point of view, the PA cell is a linear acoustic system, which responds as a whole to the disturbance generated by light absorption. The differential equation (18) is not valid for capillary tubes with a small diameter ($2r \approx d_v, d_h$) nor for gases with exceptionally high viscosity or heat conductivity.

When the absorbing gas can be modeled by a two-level system consisting of the vibrational ground and the excited state, Meyer and Sigrist (Meyer & Sigrist, 1990) found that the amplitude of heat production rate, H, is given by:

$$H(\mathbf{r},t) = \alpha_p I(\mathbf{r},t), \tag{19}$$

where $I(\mathbf{r},t)$ is the intensity of the laser beam. This equation is valid only when the laser beam is slowly chopped in the kHz range or below, and in the absence of optical saturation.

If the cross-sectional dimensions of a resonator are much smaller than the acoustic wavelength, the excited sound field develops a spatial variation only along the length of the resonator, i.e., a one-dimensional acoustic field is generated. A narrow pipe (or tube) can be regarded as a one-dimensional acoustic resonator. A pressure wave propagating in the pipe will be reflected by an open end with the opposite phase. Through multiple reflections a standing wave pattern with pressure nodes will be formed. Therefore, open pipes should have resonances when the pipe length is equal to an integer multiple of the half wavelength.

Bernegger and Sigrist (Bernegger & Sigrist, 1987) proved that the plane acoustic wave propagation can be modeled by the one-dimensional analogue of the electrical current flow in a transmission line. According to this theory, a cell constant C (Pa cm/W) only dependent on the geometry of the cell (it includes the losses of the PA resonator), which relates the pressure amplitude p with the absorbed laser power P_L, can be defined at resonance frequency:

$$p = C(\omega = \omega_0)\alpha_p P_L , \tag{20}$$

where p (N/m^2 = Pa) is the pressure response of the cell, α_p (cm^{-1}) is the absorption coefficient at a given pressure of the gas at the laser wavelength, and P_L (W) is the laser power. The units of C are given in Pa cm/W based on the usual dimensions of p, α_p, and P_L. Here, the angular frequency is $\omega_0 = 2\pi f_0$, where f_0 is the resonance frequency; for a longitudinal resonant cell, the first resonance frequency is $f_0 = v_s/2L$ (Eq. 3), so that $\omega_0 = \pi v_s/L$. C is usually determined by calibration measurements, where one single absorbing substance with known absorption spectrum is investigated.

Equation (20) implies that for a reasonably small laser power (no saturation), slow modulation frequency ω_0 ($\omega_0\tau << 1$, where τ is the thermal relaxation time characteristic for the cooling of the gas to equilibrium), and small absorption ($\alpha_p L << 1$), the sound pressure amplitude depends linearly on the absorption coefficient and the laser power.

3.6 Cell constant

For a given PA cell geometry ("high-Q" case), Kreuzer (Kreuzer, 1977) deduced that:

$$C(\omega_0) = \frac{(\gamma - 1)LQG}{\omega_0 V}, \tag{21}$$

where V is the volume of the PA resonator and G a geometrical factor (depending on the transverse beam profile but not on the cell length) on the order of 1 Pa m^3/W s (Bijnen et al. (Bijnen et al., 1996) found a value G = 1.2 Pa m^3/W s for their specific experimental conditions). Since the quantities in Eq. (21) are independent of the laser power and absorption coefficient, these factors can be regarded as characteristic setup quantities for PA resonators. The quantity C describes the sensitivity of the PA resonator at a given resonance frequency. It is widely known as the 'cell constant'. It depends on the size of the resonator, the frequency, and the Q factor of the resonance selected for PA detection. It also depends on the spatial overlap of the laser beam and the standing acoustic wave pattern. Its 'cell constant' name is therefore misleading, as it characterizes the complete measurement arrangement (including the acoustic resonator with a selected resonance, microphone position, and laser beam profile with spatial location) rather than the mere PA cell. Moreover, it depends on frequency, and its value differs for different resonance modes. Therefore, it would more appropriately be called a 'PA setup constant' (Miklos et al., 2001) rather than a 'cell constant'. However, since the name 'cell constant' is already established in the literature, we will continue to use it hereinafter.

As the cell constant is inversely proportional to an effective cross section defined by $S_{eff} = V/L$ and $\omega_0/Q = \Delta\omega$ (Eq. 15), it follows that:

$$C(\omega_0) = \frac{(\gamma - 1)G}{\Delta\omega S_{eff}}. \tag{22}$$

Based on this formula, we can estimate the magnitude of the cell constant. By introducing in Eq. (21) or Eq. (22) the values for our medium-Q resonator (r = 3.5 mm, L = 30 cm, Q = 16.1 and f_0 = 564 Hz; $S_{eff} \cong 0.4$ cm^2, Δf = 35 Hz), it follows C = 4720 Pa cm/W, which is almost twice as much as the experimentally measured value (2500 Pa cm/W). If an open resonator is built into a closed PA cell, then the pressure generated by the PA effect will be distributed over the entire closed volume. Therefore, the total volume of the PA cell must be taken into account instead of the volume of the resonator. A PA resonator optimized for high-Q performance (S_{eff} ~80 cm^2, Q ~1000 at f_0 = 1 kHz) has a cell constant of about 800 Pa cm/W. The cell constant of a low-Q resonator is a complicated function of several parameters, and therefore cannot be determined with sufficient accuracy by calculation. It has to be determined experimentally by calibration measurements using certified gas mixtures.

The possibilities for improving the cell constant of acoustic resonators are limited (Miklos et al., 2001). The only parameter that can really be changed over a broader range is the effective cross section of the cell. A reduction of the cell diameter will increase the cell constant. A lower limit is set by the diameter and divergence of the laser beam employed. The cell constant for modulated measurements is inversely proportional to the FWHM value of the resonance profile. Unfortunately, the half width cannot be reduced indefinitely, because it scales approximately with the surface-to-volume ratio of the resonator. As the cross section of the cell is reduced, the surface-to-volume ratio increases. It is therefore impossible to achieve small cross sections and a small bandwidth (high Q) simultaneously. The smallest diameter used in practical systems is several millimeters, the largest about 10 cm.

Combining Eqs. (20) and (21), we have:

$$p = \frac{(\gamma - 1)\alpha_p L P_L Q G}{\omega_0 V}. \tag{23}$$

It should be noted that the amplitude of the PA pressure signal is a function of (1) the heat capacity of the mixture (γ), (2) laser power (P_L), (3) modulation frequency (ω_0), (4) vibrational relaxation times of the absorbing gas, and (5) damping effects of the buffer gas (Q). The first four factors contribute to the power going into the sound wave, and the last mechanism determines the Q of the resonances. From Eq. (23) it follows that the amplitude of the pressure wave (the PA pressure signal) is proportional to the absorption coefficient and laser power, but inversely proportional to the modulation frequency and effective cross section V/L of the PA resonator. Thus, the signal increases with decreasing resonator dimensions and modulation frequency. As the noise increases with a decrease of these parameters, there is a maximum in the SNR for a certain combination of cell size and modulation frequency.

For resonant operation, the modulation frequency is tuned to one of the resonance modes of the PA resonator, i.e., $\omega = \omega_m$. Not only the m-th mode, but all the other modes of the acoustic resonator are excited as a result. The resonance amplitude is proportional to Q, while the amplitudes of the other resonances are inversely proportional to the quantity $\omega^2 - \omega_m^2$. Therefore, distant resonances will not be excited efficiently. Certain resonances can be suppressed for special symmetry conditions, e.g., azimuthal modes cannot be excited if the cylindrical laser beam propagates exactly along the cylinder axis.

The measured PA signal also depends on the exact position of the microphone in the resonator. The signal detected by the microphone is proportional to the integral average of the pressure over the microphone membrane. Since mostly miniature microphones are applied in photoacoustics, the integral can be approximated by the value of the pressure amplitude at the microphone location.

Angeli et al. (Angeli et al., 1992) reported a dependency of the cell constant on the kind of calibration gas. They concluded that the cell constant could not be determined unambiguously by a calibration measurement using a single absorbing species, indicating the "nonabsolute" character of photoacoustic spectroscopy. This result would have severe implications and would render analyses of multicomponent gas mixtures very difficult or impossible. Fortunately, Thöny and Sigrist (Thöny & Sigrist, 1995) proved that detailed investigations including a number of different gases and measurements on numerous laser transitions contradict those observations and revealed the expected independence of the cell constant within the measurement errors.

3.7 Optimization of the PA cell geometry

Since the PA signal is inversely proportional to the cell volume and modulation frequency, high PA signal levels can be obtained by taking a small cell volume (< 10 cm^3) and low modulation frequencies (< 100 Hz). However, noise sources (intrinsic noise of the microphone, amplifier noise, external acoustic noise) show a characteristic $1/f$ frequency dependence, and therefore the SNR of such a gas-microphone cell is usually quite small. The SNR of a PA cell can be increased by applying higher modulation frequencies (in the kHz region) and acoustic amplification of the PA signal. For this reason, resonant PA cells operating on longitudinal, azimuthal, radial, or Helmholtz resonances have been developed. Furthermore, resonant cells can be designed for multipass or intracavity operation.

A qualitative behavior for Q and ω_0 can be derived from simple geometrical considerations. So, for Q, the energy stored in a specific mode is proportional to the cell volume ($\propto r^2L$), while the energy losses per cycle of the acoustic wave are proportional to the cell surface ($2\pi rL$) and to the thicknesses $d_h \approx d_v = d \propto \omega_0^{-1/2} \propto L^{1/2}$. Therefore:

$$\omega_0 \propto L^{-1}, \tag{24}$$

$$Q(\omega_0) \propto (r^2L/rLL^{1/2}) \propto rL^{-1/2}, \tag{25}$$

and

$$C(\omega_0) \propto (L)(rL^{-1/2})/(L^{-1})(r^2L) \propto r^{-1}L^{1/2}. \tag{26}$$

which is represented graphically in Fig. 6. These equations show that the product $Q(\omega_0)C(\omega_0)$ is nearly independent of the cell dimensions for any kind of resonant PA cell. The operation of the cell in a longitudinal mode is more advantageous because it makes it possible to optimize the resonance frequency and the Q-factor independently, which cannot be achieved in the case of radial resonance.

Cell geometries with large diameter-to-length ratios designed to excite the resonance in the radial or azimuthal acoustic modes possess high Q values and high resonance frequencies, but have low cell constants. PA cells with high Q values are sensitive to long-term drifts (e.g., due to thermal expansion if the temperature is not carefully controlled), so that they require an active locking of the modulation frequency on the resonance frequency of the cell.

In a longitudinally excited resonator, a smaller acoustic gain, as a consequence of a relatively low Q value, is compensated for by the signal gain due to the smaller diameter.

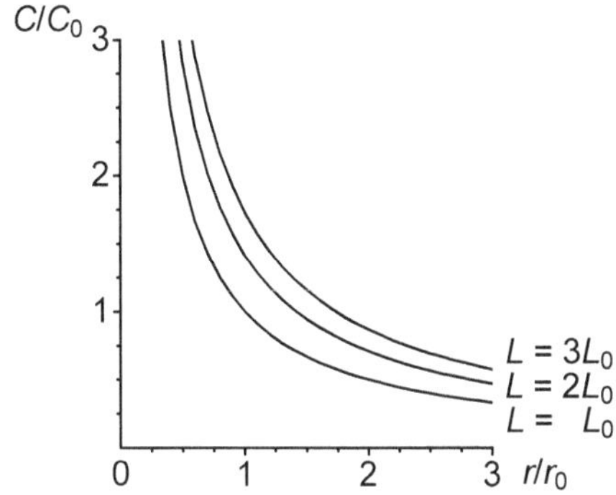

Fig. 6. Graphical representation of Eq. (26): dependence of the normalized cell constant on tube radius and resonator length.

According to Eqs. (24-26), to obtain a higher acoustic signal in a longitudinally excited resonator with a low Q-factor (a higher C), it is necessary to have a resonator with a large length and a small diameter. Yet, narrowing the tube diameter and increasing the tube length are restricted by the divergence of the laser beam over the length of the cell. The maximum length is limited by the minimum frequency at which the cell is to be operated or by the maximum absorption coefficient α_{max} that is to be detected ($L << 1/\alpha_{max}$). The minimum possible diameter is set by the beam diameter or the volume-to-surface ratio that is needed to minimize adsorption and desorption at the cell walls. A too small diameter of the PA cell gives rise to high PA background signals due to absorption of the wings of the gaussian laser beam profile. On the other hand, a high quality factor is required in order to decrease the background signal caused by window heating. In conclusion, the optimization of the PA cell geometry depends on the specific experimental conditions and the application for which it is designed.

3.8 Voltage signal

When the resonance contributions are included, the photoacoustic voltage signal can be obtained at a given operating frequency simply by multiplying the pressure response (Eq. 20) by the microphone responsivity ($V = pS_M$ and $\alpha_p = \alpha c$):

$$V = \alpha C S_M P_L c \,, \tag{27}$$

where: V (V) is the photoacoustic signal (peak-to-peak value); α (cm^{-1} atm^{-1}), the gas absorption coefficient at a given wavelength; C (Pa cm W^{-1}), the cell constant; S_M (V Pa^{-1}), the microphone responsivity; P_L (W), the cw laser power (unchopped value; 2x measured average value); and c (atm), the trace gas concentration (usually given in units of per cent, ppmV, ppbV or pptV). This equation reveals that the photoacoustic signal is linearly dependent on laser power. Thus, sensitive measurements benefit from using as much laser power as is reasonably available. Moreover, the signal is directly dependent on the number of molecules in the optical path (trace gas concentration), which means that this technique is truly a "zero-baseline" approach, since no signal will be generated if the target molecules are not present.

Equation (27) is valid as long as absorption is small ($\alpha_p L << 1$), and the modulation frequency is higher than the inverse of the molecular diffusion time but lower than the inverse of the molecular relaxation time. The PA signal is linearly dependent on the absorption coefficient, cell constant, microphone responsivity, incident laser power, and absorbent trace gas concentration. Thus, by doubling Q (and consequently C), or the microphone responsivity, or the laser power, or the number of absorbing molecules in the optical path, the voltage will also double. The peak-to-peak value of the signal is obtained by multiplying by $2\sqrt{2}$ the rms voltage amplitude measured by the lock-in amplifier. As a rule, another parameter is used to characterize the PA cell, namely:

$$R = CS_M, \tag{28}$$

where R (V cm/W) is the (voltage) responsivity of the PA cell or the calibration constant. The cell constant C is multiplied by the responsivity of the microphone given in V/Pa units. A comparison of different PA cells can be made independently of the application in terms of this figure of merit. However, the cell characterization can be used only if a calibrated microphone is available. In this way, Eq. (27) becomes:

$$V = \alpha R P_L c\,. \tag{29}$$

To increase the detection sensitivity, we have to ensure: a) a cell constant as large as possible (optimization of the PA resonator); b) a large microphone responsivity; c) a laser power as high as possible, provided that saturation does not become a limiting factor; d) a narrow bandwidth of the lock-in amplifier, and e) a high absorption coefficient of the trace gas to be measured at the laser wavelength. It is also useful to increase the number n of microphones (connected in series), but this number is limited by the dimensions of the PA cell. The summation of the signals from the single microphones results in an n-times higher effective PA signal, because the total responsivity $S_{M\,tot}$ is increased n-fold, i.e.:

$$S_{M\,tot} = nS_M\,. \tag{30}$$

On the other hand, the incoherent noise only increases by $\sqrt{n}$. One thus obtains:

$$\mathrm{SNR}_{\mathrm{tot}} = \sqrt{n}\,\mathrm{SNR}\,. \tag{31}$$

The minimum measurable voltage signal $V = V_{min}$ is obtained at SNR = 1, where the minimum detectable concentration $c = c_{min}$ can be recorded:

$$c_{\min} = \frac{V_{\min}}{\alpha P_L R}\,. \tag{32}$$

The sensitivity of PA instruments increases with the laser power, as $V \propto \alpha P_L$. However, the voltage signal does not depend on the length of the absorption path. Furthermore, in contrast to other techniques based on absorption spectroscopy, the response of the acoustic detector is independent of the electromagnetic radiation wavelength as long as the absorption coefficient is fixed. According to theoretical considerations, extremely low detection limits on the order of $\alpha_{min} = \alpha c_{min} \cong 10^{-10}$ cm^{-1} for 1 W incident laser power have been predicted (Sigrist, 1986) and experimentally proved (Harren et al., 1990). Such sensitivity makes it possible to detect many trace constituents in the sub-ppbV range. Theoretical calculations (see Section 2) predict the linearity of the signal response over a concentration range as broad as 7 orders of magnitude. This wide dynamic range, characteristic of LPAS, is important for air pollution monitoring, as it helps conduct measurements in polluted areas at the source (emission) as well as in rural areas (immission) (Sigrist et al., 1989).

3.9 Saturation effects

A PA signal may become saturated due to either a large concentration of the measured analyte or high laser power levels. We showed in Section 2 that, in the case of ethylene, the signal starts to saturate at a concentration of 65 ppmV. As a matter of fact, Thöny and Sigrist (Thöny & Sigrist, 1995) observed weak saturation effects on 10P(14) CO_2 laser transition for a concentration of 100 ppmV of ethylene. The degree of saturation is gas dependent. We found (Section 2.3) that a deviation of ~3% from linear behavior resulted in an optical density $\alpha Lc = 0.06$.

By increasing laser intensity, the excitation pumping rate of the molecules grows higher, and a molecule is more likely to absorb a nearby photon before it relaxes to the ground state. So, as the molecules in the excited state increase in numbers, the number of molecules which can absorb laser radiation is reduced. The gas actually becomes as though more transparent to laser radiation, and the effective absorption coefficient per unit laser power is lowered; this is called laser power saturation. Saturation due to nonlinear absorption of the laser power only occurs in focused high-power laser beams or when the PA cell is placed intracavity in a laser, so that the laser power can be on the order of tens of watts or even higher than 100 W. The pumping rate to a higher vibrational-rotational level is proportional to the laser light intensity; in the case of saturation it exceeds the collisional de-excitation rates.

Harren et al. (Harren et al., 1990) studied the saturation effects by placing the PA cell intracavity of a waveguide CO_2 laser. Extracavity, the ratio between 10P(14) and 10P(16) lines is 5.96 ± 0.2. Intracavity, this ratio becomes 2.8 ± 0.3 (47% from its extracavity value) at an intracavity laser power of 130 W (for a laser beam waist of 0.282 mm, that is at a laser intensity higher than 200 kW/cm^2). By lowering the intracavity laser power, this ratio increases to its extracavity value. This effect is caused by saturation of the transitions in C_2H_4 at the 10P(14) CO_2 laser line. Depletion from the vibrational excited level (ν_7) via other vibrational levels (e.g., ν_{10}) through collisions becomes slow in comparison with the pump rate due to the high intracavity power. When the laser beam waist is increased to 1.02 mm (laser intensity is decreased to 15.9 kW/cm^2), the ratio of the absorption coefficients of C_2H_4 on the 10P(14) and 10P(16) CO_2 laser lines increased to 4.7 ± 0.5 (78% of its extracavity value). To compensate for the saturation effect, these authors used an absorption coefficient of 23.7 cm^{-1}atm^{-1} (78% of 30.4 cm^{-1}atm^{-1} at an intracavity power of 100 W) for C_2H_4 at the 10P(14) CO_2 laser line.

By using an intracavity arrangement where the CO_2 laser power was varied between 10 and 70 W, Groot (Groot, 2002) measured the saturation parameter of ethylene for several laser lines. The relation of the effective absorption coefficient α_e to the intrinsic absorption coefficient α is given by $\alpha_e = \alpha/(1 + P/P_s)$, where P_s (W) is the laser power saturation parameter and represents a measure for the relaxation rate. At $P = P_s$, the absorption coefficient decreases to half its initial value. The following values were obtained for P_S: 178 W at 10P(8) line; 102 W at 10P(10); 112 W at 10P(12); 51.8 W at 10P(14); 101 W at 10P(16); 128 W at 10P(18), and 112 W at 10P(20). The strongest saturation effect was observed at 10P(14) line, where the absorption coefficient is the largest. The saturation for this line at a laser power of 130 W corresponds to the equivalent absorption coefficient $\alpha_e = 0.285\alpha$. The stronger saturation in this case compared with the results of Harren et al. (Harren et al., 1990) could be accounted for by a tighter focusing of the laser beam (smaller beam waist). As a matter of fact, saturation is determined by the laser beam intensity (irradiance) rather than the laser power. Power saturation does not depend on the gas concentration in the PA cell (if the absorbing gas concentration is not too high).

4. Noises and limiting factors

4.1 Noises

In order to obtain an optimum SNR, noise control and interfering signals have to be taken into account (Dutu et al., 1994a; Dutu et al., 1994b). These limiting factors are discussed in the following two sections.

Noise plays an important role in all photoacoustic measurements and is of particular importance in the detection of ultralow gas concentrations, because the noise level limits the ultimate sensitivity. In the photoacoustic literature, the detection level is usually defined by the SNR, where the noise is given by the microphone signal measured with the laser light blocked. However, when light hits the PA cell, an additional background signal is generated which exists even when the absorbing species are not present in the detector. The background signal is often larger than the noise signal, and therefore the detection limit or sensitivity has to be defined by the signal-to-background ratio (SBR) in most experiments. Unfortunately, it is common practice to consider only the SNR. This procedure yields an extrapolated detection limit that may be far too small. The background signal is usually determined with a nonabsorbing gas, such as nitrogen, in the PA detector. It is influenced by many system properties, such as the pointing stability, the beam divergence, and the diameter of the laser beam.

For photoacoustic spectroscopy, "noise" often has a structure that is coherent with the signal from the target species, and therefore should more appropriately be treated as a background signal, not as noise. The background signal can be determined by measuring the acoustic signal in the absence of absorbers (i.e., with pure nitrogen), but with the same flow and in the same pressure conditions as those used for the sample gases.

The sensitivity-limiting factors which are encountered in LPAS can be classified into three categories:

a. Electrical noise, by which we mean any random fluctuation, whether electronic or acoustic, which does not have a fixed phase relation with the modulation of the laser intensity. It determines the ultimate detector sensitivity.
b. Coherent acoustic background noise, meaning a signal caused by the modulation process, but not attributable to the presence of the light beam in the PA cell. This signal is at the same frequency as, and locked in phase with respect to, the laser intensity modulation.

c. Coherent photoacoustic background signal. This signal, which is always present in the PA detector, is caused by the laser beam, yet not by light absorption in the bulk of the gas. Rather it is due to laser beam heating of the windows and of the absorbates at their surfaces, and heating of the PA resonator walls by the reflected or scattered light owing to imperfections of the focusing lens, windows and inner walls of the PA resonator. This signal is in phase with, and at the same frequency as, the laser intensity modulation. Therefore, it is not filtered out by the lock-in amplifier connected to the microphone. Thus, a background signal proportional to the laser power becomes the main factor that limits sensitivity.

The background signal in the PA cells may arise from several sources, some of which are listed below (Gerlach & Amer, 1980):

1. Window surface absorption: the molecules absorbed on the window surface and/or the window surface itself absorb the modulated laser radiation, and the resulting gas heating in the cell generates a pressure pulse.
2. Window bulk absorption: even the highest quality ZnSe window substrates exhibit a residual window absorption of $\sim 10^{-3}$ cm^{-1}.
3. Off-axis radiation within the cell: light scattered from the windows and at the edge of the chopper blade may strike the inside walls of the PA resonator, where it may be absorbed and produce a signal.
4. Light scattering or absorption due to microaerosols.
5. Small amounts of contamination that may outgas from the cell materials, seals, and so forth.

The detection limit of the PA cell is determined by the combined effect of the intrinsic stochastic noise of the microphone, acoustic background noise, and photoacoustic background signal. Background signals are deterministic, and to the extent that they can be quantified and minimized, do not reduce the performance of the cell significantly. The detection limit is defined either at a signal-to-noise ratio of unity (SNR = 1) or at a signal-to-background ratio of unity (SBR = 1).

The amplifier input noise and microphone noise are gaussian in nature, that is, the amount of noise is proportional to the square root of the bandwidth in which the noise is measured. All of these noise sources are incoherent. The input noise of the SR830 lock-in amplifier used in our experiments is about 6 nV (rms)/ $\sqrt{Hz}$. Microphone noise, which is manifested as a noise voltage present at the microphone output terminals, can be expressed as a product between the normalized noise pressure value owing to both thermal agitation of the diaphragm and cartridge responsivity at the corresponding frequency and the square root of the measurement bandwidth. The electrical noise of Knowles EK models electret microphones is 40 nV (rms)/ $\sqrt{Hz}$. The overall random noise of multiple sources is determined by taking the square root of the sum of the squares of all the individual incoherent noise figures. For gaussian noise, the peak-to-peak value is about 5 times the rms noise value, while for the two other types of noises, the rms value must be multiplied by a factor of $2\sqrt{2} \cong 2.8$ to obtain the peak-to-peak amplitude. Electrical noise usually has a broadband frequency spectrum and can be reduced efficiently by narrowband filtering of the signal, as is done in the phase sensitive detection. A detection bandwidth of 0.25 Hz was set (a time constant of 1 second) in all of our measurements. Electrical noise can be reduced by using state-of-the-art (and therefore very expensive) lock-in amplifiers and/or by using longer time averaging (the noise decreases with the square root of the averaging time) at the cost of longer measurement times.

The two types of coherent background, however, are extremely narrowband signals at the same frequency as the modulation and hence cannot be filtered out. In addition, since the signal and the coherent photoacoustic background signal are both proportional to laser power, no improvement will be achieved as the laser power is increased.

Table 2 shows the magnitudes of these limiting factors in the case Brewster windows are used. We expressed each factor in several different sets of units (Dutu et al., 1994b): voltage, pressure amplitude, equivalent absorption coefficient that would give the same pressure amplitude, and the concentration of ethylene that would be required to give that much absorption.

Noise type	Root-mean-square (rms) value	Equivalent pressure[a]	Equivalent absorption[b]	Equivalent C_2H_4 concentration[c]
Electrical noise, V_N^e	0.15 $\mu V/\sqrt{Hz}$	5.3×10^{-6} $Pa/\sqrt{Hz}$	1.5×10^{-9} $W\ cm^{-1}/\sqrt{Hz}$	4.9×10^{-11} $W/\sqrt{Hz}$
Coherent acoustic background noise[d], V_N^{ac}	2.6 μV	9.2×10^{-5} Pa	2.6×10^{-8} $W\ cm^{-1}$ [f]	8.6×10^{-10} W [g]
Coherent photoacoustic background signal[e], V_N^b	2.7 μV/W	9.5×10^{-5} Pa/W	2.7×10^{-8} cm^{-1} [h]	8.9×10^{-10} [i]

[a] The equivalent peak-to-peak pressure was obtained by dividing the peak-to-peak noise level to microphone sensitivity: $2\sqrt{2}\ V_N^i / S_M$, where V_N^i is either V_N^e, V_N^{ac} or V_N^b (in our case, $S_M = 8 \times 10^{-2}$ V/Pa)
[b] The equivalent absorption was obtained by dividing the peak-to-peak noise level to cell responsivity: $2\sqrt{2}\ V_N^i / R$, where V_N^i is either V_N^e, V_N^{ac} or V_N^b (in our case, $R = 280$ V cm/W)
[c] The equivalent C_2H_4 concentration was obtained by dividing the equivalent absorption to the C_2H_4 absorption coefficient at one atmosphere pressure of the gas at the 10P(14) laser wavelength ($\alpha_p = 30.4\ cm^{-1}$)
[d] We expressed the coherent acoustical background noise in V independent of the bandwidth, as did Gerlach and Amer (Gerlach & Amer, 1980) and Beck (Beck, 1985) (and not in $V/\sqrt{Hz}$ as used by Harren et al. (Harren et al., 1990)); our measurements show that the acoustical background noise was independent of the lock-in bandwidth when the equivalent noise bandwidth (ENBW, the effective bandwidth for gaussian noise) of the low pass filter was varied between 0.08 Hz and 8 Hz (the lock-in time constant T was changed between 0.3 and 30 s, where ENBW = $1/(4T)$ for a slope of 6 dB/oct)
[e] The coherent photoacoustical background signal was measured in pure nitrogen at atmospheric pressure (1011 mbar) and at a temperature of 22°C: 12 μV at a laser power of 4.4 W for 10P(14) line of the CO_2 laser; this signal was the same both in a static gas or at a flow rate of 50 sccm (standard cubic centimeters per minute)
[f] The same as the limiting sensitivity of the cell, S_{cell}
[g] With a laser power $P_L = 4.4$ W, the minimum measurable concentration of ethylene was 0.2 ppbV, the same as the limiting measurable concentration of ethylene, c_{lim} (Table 2, Part II)
[h] The same as the minimum measurable absorption coefficient, α_{min} (Table 2, Part II)
[i] The same as the minimum detectable concentration, c_{min} (Table 2, Part II)

Table 2. Noises measured in our PA system.

The limiting electrical noise measured at resonance frequency was $V_N^e = 0.15\ \mu V/\sqrt{Hz}$. At atmospheric pressure, the acoustic background noise was $V_N^{ac} = 2.6$ μV (at resonance frequency) under normal working conditions. A photoacoustic background signal of V_N^b (nitrogen) = 2.7 μV/W was observed, in phase and at resonance frequency, as the cell was filled with pure N_2 at atmospheric pressure.

To determine the actual levels of noises that would be observed in practice, the random electrical noise level of $0.15\ \mu V/\sqrt{Hz}$, for example, must be multiplied by $B^{1/2}$. Also, to get the noise-equivalent absorption one must multiply 1.5x10^{-9} W cm^{-1}/ $\sqrt{Hz}$ by $B^{1/2}P^{-1}$, where B is the bandwidth and P the laser power. In order to get an idea of the sensitivity that can be achieved for a representative trace gas, the equivalent ethylene concentration that would give the same signal level is also tabulated. To get the noise-equivalent ethylene concentration, multiply 4.9x10^{-11} W / $\sqrt{Hz}$ by $B^{1/2}P^{-1}$.

Our coherent acoustical background was 2.6 μV or 9.2x10^{-5} Pa, equivalent to an absorption of 2.6x10^{-8} W cm^{-1}. To get the equivalent absorption coefficient divide the latter number by P_L (5.9x10^{-9} cm^{-1}). This background signal is dependent on the location of the PA cell in relation to the sound sources associated with the modulation process.

The coherent photoacoustic background was 2.7 μV/W, or 9.6x10^{-5} Pa/W, assuming the beam was optimally aligned. This is equivalent to an absorption coefficient of 2.7x10^{-8} cm^{-1}, or an ethylene concentration of about 0.89 ppbV, independent of the laser power. Since the noise and coherent acoustical background can be made negligible by using high laser power, as is done in intracavity operation, the coherent photoacoustic background will be the ultimate limit of sensitivity.

In order to obtain a maximized SNR, a resonance frequency under 1 kHz is necessary. Under 1 kHz, the noise level is determined by the $1/f$ amplifier noise, showing a frequency behavior of $1/\omega$. Together with $C \propto \omega_0^{-1/2}$ (because $C \propto Q/\omega_0$, $Q \propto (d_v, d_h)^{-1}$ and $(d_v, d_h) \propto \omega_0^{-1/2}$, see Sections 3.6, 3.4 and 3.3), we get a SNR proportional to $\omega_0^{1/2}$. Above 1 kHz, where the $1/f$ amplifier noise is negligible, the SNR is proportional to $\omega_0^{-1/2}$.

Below 1 kHz, the $1/f$ amplifier noise is the main source. Above 1 kHz, the frequency independent Brownian noise takes over. Since the pressure amplitude is inversely proportional to the square root of the resonance frequency ($p \propto C$, Eq. 20), a convenient resonance can be found between 500 and 1500 Hz. This limits the choice to a cell length of 100-300 mm. If optimal signal enhancement were the only argument, one would rather choose a large (300 mm) resonator ($C \propto L^{1/2}$, Eq. 26). Shorter resonator lengths are necessary in the case of an intracavity setup due to the limited space inside the cavity. Also, a fast time response of the cell requires a short cell length.

When the sample gas is flown continuously through the detector, acoustical noise can be produced, if the gas flow is turbulent, if acoustical noise from the surroundings is coupled directly into the detector sample space or into the tubes connected to the detector and then propagated into the detector, or if acoustic disturbances from the pump running the sample gas through the detector are propagated through the tubes. Thick detector and tube walls, small flow rates, mounting of the cell and chopper in separate sound insulating boxes, etc. must be chosen to suppress these noise contributions.

The background signal can be minimized by placing the windows at nodes of the mode being excited and by introducing buffer volumes at both ends of the cell. The ratio of buffer to resonator diameters must be large enough, and the buffer length has to be equal to one-fourth of resonator length.

4.2 Gas interference

Interference of other absorbing substances may impair the theoretical detection limit in a multicomponent analysis of the real atmosphere. Such interference may be caused by other molecular systems present in the environment or substances that are entrained by the carrier flux. If an interfering species is present in the environment, its effect can be minimized by either the introduction of scrubbers and cryogenic traps or the use of dual beam techniques using two PA cells. Sample-entrained interfering species present a more serious problem, since they will be present only near the source and therefore cannot be eliminated by dual beam spectroscopy.

In ambient air, one finds CO_2 concentrations of 330-365 ppmV (0.033%-0.0365%) (Sigrist, 1986; Harren et al., 1990; Rooth et al., 1990). This level may rise to about 1% in the practical conditions of an agricultural application. This poses a serious practical problem. The CO_2 molecule possesses absorption vibrational band transitions $v_1 \rightarrow v_3$ ($10^00 \rightarrow 00^01$) and $2v_2 \rightarrow v_3$ ($02^00 \rightarrow 00^01$) which are weak, while the lower levels are barely populated at room temperature (~1%). However, due to the exact coincidence of these vibrational-rotational transitions with the CO_2 laser lines and the relatively high concentration of CO_2 in comparison with trace gases like C_2H_4, carbon dioxide is inevitably excited by CO_2 laser radiation, and the related photoacoustic signal may exceed the trace signal by many orders of magnitude. The absorption coefficient increases strongly with temperature, but is independent of the CO_2 concentration over a wide range. A 1.5% concentration of CO_2 has an absorption strength comparable to 1 ppmV of C_2H_4 (for CO_2 at the 10P(14) laser line, $\alpha(CO_2) = 2.1\text{x}10^{-3}$ atm^{-1}cm^{-1} and $c(C_2H_4) = c(CO_2)\alpha(CO_2)/\alpha(C_2H_4) = 10^{-6}$ atm = 1 ppmV). At the 10P(14) line of CO_2, 360 ppmV of CO_2 has an absorption coefficient equal to that of 24.8 ppbV of C_2H_4. Similarly, at the 9R(30) line of CO_2 at 21°C, the same concentration of CO_2 has an absorption coefficient equal to that of 13.5 ppbV of NH_3. Water vapor exhibits a broad continuum with occasional weak lines in the frequency range of the CO_2 laser (for H_2O at the 10P(14) laser line, $\alpha(H_2O) = 2.85\text{x}10^{-5}$ atm^{-1}cm^{-1}). The two dominant peaks are the absorption lines on 10R(20) and the most favorable for ambient air measurement, the 10P(40) laser transition. The absorption of 1% of water vapor in air (50% relative humidity at 18°C) is about the same as that of 9.4 ppbV of C_2H_4 at the 10P(14) line or 5 ppbV of NH_3 at the 9R(30) line of CO_2 (Rooth et al., 1990). However, at a constant temperature the absorption coefficient $\alpha(H_2O)$ depends on the water vapor concentration x and appears to obey the relation: $\alpha(H_2O) = \alpha_0 x$, where α_0 is a constant. The natural unpolluted atmosphere contains H_2O at a concentration level of ~1.5%.

Ammonia (a colorless, poisonous gas with a characteristic smell and well solvable in water) is vibrationally excited to the ν_2 state, usually by means of the $saR(5, K)$ transitions at λ = 9.22 μm. These levels can be excited by the 9R(30) line of the CO_2 laser, where the absorption coefficient $\alpha(NH_3)$ has a value of 56 cm^{-1} atm^{-1} (Rooth et al., 1990). Ammonia is present in the atmosphere in concentrations ranging from below 0.1 ppbV over open water up to several tens of ppbV in areas with intensive livestock breeding.

Due to the additive character of the photoacoustic signal under normal atmospheric conditions, the presence of a large amount of water vapor and carbon dioxide impedes C_2H_4 detection in the low-concentration range (ppbV). Consequently, some means of selective spectral discrimination is required if ethylene is to be detected interference free in the matrix of absorbing gases. There are several ways to overcome this problem. The first is to remove CO_2 from the flowing sample by absorption on a KOH (potassium hydroxide)-based scrubber inserted between the sampling cell and the PA cell (a specific chemical reaction results: KOH $\rightarrow$ K_2CO_3 and water). In this way, concentrations below 1 ppmV CO_2 (equivalent to a concentration of 0.07 ppbV of C_2H_4) can be achieved without influencing the C_2H_4 concentration.

Another method is a multicomponent analysis approach. The PA spectrum of an arbitrary gas mixture is represented by a linear combination of the absorption spectra of all constituents. Hence, the absorption spectra of all expected constituents that contribute to the total absorption have to be determined prior to the analysis of a multicomponent gas mixture. Let us assume a nitrogen atmosphere including a mixture of n absorbing gases at unknown concentration levels $c_1, c_2, \ldots, c_n$, low enough to assure linearity. The PA signal $V(\lambda)$ of the n absorbing compounds j ($j = 1, 2, \ldots, n$) with their concentration c_j and their wavelength-dependent absorption coefficients $\alpha_j(\lambda)$ is the sum of the individual signals from each compound. In some cases, the calculation becomes more complicated due to different phases of photoacoustic signals generated by the individual constituents of the mixture. For some components, e.g. CO_2, a temporal delay in the production of the PA signal may occur. This effect is known as kinetic cooling and results in a phase shift of the PA signal. So, the PA signal has to be considered as the vector sum of the individual signals from each compound:

$$V(\lambda) = \sum_{j=1}^{n} V_j(\lambda) = RP_L(\lambda) \sum_{j=1}^{n} c_j \alpha_j(\lambda), \tag{33}$$

where R (V cm/W) is the cell responsivity. By sequentially tuning the laser to m different wavelengths (discrete CO_2-laser transitions) λ_i, $i = 1, 2, \ldots, m$, we obtain m measured photoacoustic signals V_i from which we derive a set of m linear equations for the unknown concentration levels c_j:

$$V_i(\lambda_i) = RP_i \sum_{j=1}^{n} \alpha_{ij}(\lambda_i) c_j = \sum_{j=1}^{n} a_{ij} c_j, \tag{34}$$

where $P_i = P_L(\lambda_i)$, $a_{ij} = RP_i\alpha_{ij}$ (a constant for a given gas, a given wavelength and a given laser power), and $m \geq n$ (it should be noted that the system of linear equations is only well defined if the number m of laser transitions is higher than the number of gas components j); α_{ij} is the absorption coefficient of the j-th trace absorbant gas at wavelength λ_i, while P_i is the laser power at that wavelength. The measurements result in PA signals V_i from all components j in the gas mixture which absorb on the wavelength of the laser transition, λ_i. The minimum number of measurements at different laser wavelengths must be equal to at least the number of unknown trace gases, $m = n$. In this case we can define:

$$B = \begin{pmatrix} b_{11} \ldots b_{1n} \\ \vdots \quad\quad \vdots \\ b_{n1} \ldots b_{nn} \end{pmatrix} = \begin{pmatrix} a_{11} \ldots a_{1n} \\ \vdots \quad\quad \vdots \\ a_{n1} \ldots a_{nn} \end{pmatrix}^{-1}. \tag{35}$$

Therefore,

$$c_j = \sum_{i=1}^{n} b_{ji} V_i . \tag{36}$$

for $j = 1, 2, \ldots, n$. The coefficients b_{ji} have units of atm V^{-1}. If the number m of CO_2 laser lines used to carry out the analysis of a gas mixture is equal to or higher than the number n of absorbing components in the sample (whose CO_2 laser absorption coefficients are known), the unknown concentrations of each n component can be determined with the proper selection of laser lines. The solution of sets of simultaneous equations is generally required to estimate the concentrations of each species in a multicomponent mixture and select the optimal wavelengths for a fixed number of laser lines. The selection of CO_2 laser wavelengths for the optimum detection of a single species in the presence of interferences can usually be carried out by comparing the corresponding laser absorption profiles.

This method was used by Perlmutter et al. (Perlmutter et al., 1979), who observed a minus sign of the calculated CO_2 concentration level. The minus sign stems from the fact that the absorption coefficients of CO_2 were taken to be positive in the numerical analysis. Actually the absorption coefficients of CO_2 present in nitrogen at low concentration levels (up to ~0.5%), at CO_2 laser transitions, are of negative sign. The absolute values are unchanged. This minus sign is associated with the kinetic cooling effect. They found experimentally that in a longitudinal resonant PA cell (chopping frequency = 1 kHz) the CO_2 gives a $180^o \pm 10^o$ out-of-phase PA signal relative to operation with normal gases like ethylene. This is true when CO_2 is present at concentration levels up to ~0.5% in nitrogen. At concentration levels higher than ~0.5%, the kinetic cooling phase deviation does not exceed ~180° and highly depends on concentration, thus leading to an increasing PA signal level.

A crucial feature of photoacoustics on gas mixtures is the molecular dynamics involved in the conversion of internal molecular energy to heat (Olaffson et al., 1989; Henningsen et al., 1990). This is particularly important when dealing with mixtures involving CO_2 and N_2. The near degeneracy between the fundamental asymmetric stretch of CO_2 (2349 cm^{-1}) and the N_2 v = 1 vibration (2331 cm^{-1}) leads to a large cross section for resonant energy transfer. In a CO_2 laser, this mechanism is used to advantage by adding N_2 to the gas mixture in order to increase the pump rate by energy transfer from the vibrationally excited N_2 to CO_2 in the ground state. In our case, the situation is reversed. Thus, following absorption of CO_2 laser radiation, an excited CO_2 molecule transfers its excitation energy to N_2, where it resides for a long time owing to the metastable character of the excited N_2 levels (the lifetime of the vibrational level $v = 1$ is $\cong$ 1 ms at 1 atm; 1 atm = 101.325 kPa). Since the CO_2 molecule was initially taken out of an excited state, and the transition was a hot band transition, there now is a non-equilibrium situation among the CO_2 vibrational levels, and equilibrium is eventually restored at the expense of translational energy. Thus, following radiation absorption, a transient cooling of the CO_2 gas takes place, and the effect is therefore referred to as kinetic cooling. In trace gas detection, this means that the photoacoustic phase of the CO_2 signal will be significantly different from the phase of the trace gas signal, where no kinetic cooling is involved. The situation is further complicated if water is present in the gas mixture, since water molecules are effective in deexciting the metastable N_2 levels and hence in reducing the phase contrast. The presence of 1% water vapors speeds up the relaxation of vibrationally excited N_2, and this effect reduces the phase contrast to about 135° down from 180°. This phase contrast is a very important aid in the analysis of mixtures where one of the components is strongly dominant, since a quantitative analysis of the phase contrast may provide information about the H_2O concentration.

The presence of H_2O and CO_2 will always influence the measurement of C_2H_4 and NH_3 concentrations. These background gases absorb CO_2 laser radiation and produce simultaneously occurring photoacoustic signals. A comparison of the predicted amplitude and phase of the photoacoustic signal with experimental data is given in Fig. 7 (Rooth et al., 1990).

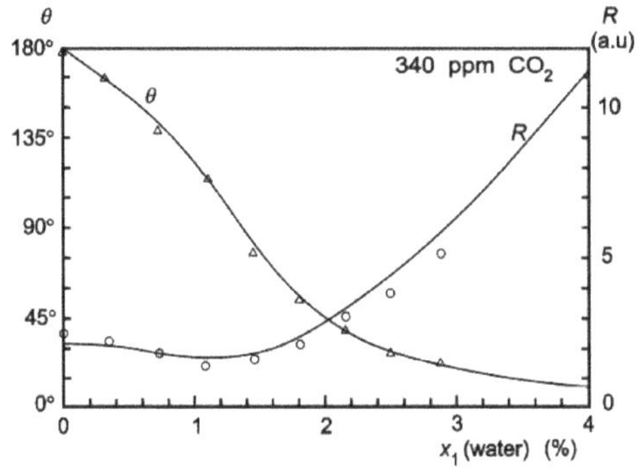

Fig. 7. Predicted amplitude R and phase θ for air with 340 ppmV CO_2 as a function of water vapor concentration. The corresponding experimental data are plotted for a frequency of 560 Hz and an excitation at 9R(28) CO_2 line (9.23 μm) (Rooth et al., 1990).

In Fig. 8, the phase of the calculated heat production rate for a CO_2 - N_2 - O_2 - H_2O mixture is plotted as a function of the concentrations c_{CO_2} and c_{H_2O} [19]. The data used for this plot were those for 10R(20) CO_2 laser transition, i.e., I_0 = 20 W/cm^2, σ_{H_2O} = 3.5x10^{-23} cm^2 and σ_{H_2O} = 1.0x10^{-22} cm^2, and a chopper frequency f = 2650 Hz. As demonstrated in Fig. 8, the phase reversal only occurs within rather narrow concentration ranges. Thus, a heat-rate phase different from 0° or 180° is rarely expected for low H_2O and CO_2 concentrations.

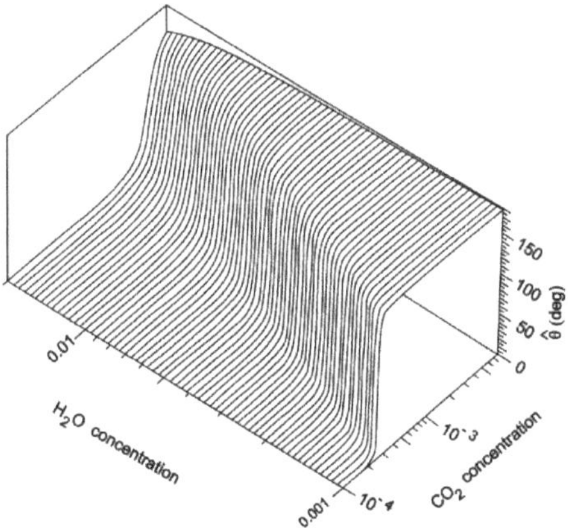

Fig. 8. Calculated phase of heat production rate for a CO_2 - N_2 - O_2 - H_2O mixture as function of the concentrations c_{CO_2} and c_{H_2O} and for c_{N_2} = 0.8 and c_{O_2} = 0.2 (Meyer & Sigrist, 1990).

The multicomponent analysis can utilize the phase information of the photoacoustic response to suppress the CO_2 signal. A high concentration of CO_2 yields a phase shift of the signal with respect to the acoustic signal of ethylene. A combined signal for a CO_2-C_2H_4 mixture is less than the sum of both individual amplitudes (vectorially added). The zero phase of the two-phase vector lock-in amplifier is adjusted to pure C_2H_4 absorption, and thus a mixture of CO_2 and C_2H_4 in air is measured on two CO_2 laser transitions. One obtains four pieces of information, i.e. the CO_2-C_2H_4 mixture phase shift and absorption coefficients for both lines. From this, with known absorption coefficients for both lines, the CO_2 and C_2H_4 concentrations can be extracted (Rooth et al., 1990). A good estimate is obtained from the difference between the two measured signals V_a and V_b. Putting $V_a = R_a\exp(i\theta_a)$ and $V_b = R_b\exp(i\theta_b)$ the magnitude of the difference is found with the cosine rule:

$$|V_a - V_b| = \left[R_a^2 + R_b^2 - 2R_aR_b\cos(\theta_a - \theta_b)\right]^{1/2}. \quad (37)$$

Here it is only the difference between the two phase angles that is required, so absolute calibrations can be avoided. This approach has the advantage that a high laser power can be used, and no partial failure of the scrubber can falsify the C_2H_4 concentration.

In a multicomponent mixture, this effect can be taken into account by measuring the amplitudes V_i of the PA signal at the laser transitions i as well as its phases θ_i, where the number i = 1... m stands for the discrete CO_2 laser transitions with the powers $P(\lambda_i) = P_i$. Thus, similar to Eq. (34) we have the following equation for the PA signal amplitude:

$$V_i\cos\theta_i = RP_i\sum_{j=1}^{n} c_j\alpha_{ij}(\lambda_i)\cos\theta_{ij}, \quad (38)$$

with i =1... m, j = 1... n, $n \le m$, where c_j is the concentration of the gas component j and α_{ij} is the absorption coefficient of the gas compound j at the laser transition i. The phase θ_{ij} is a mathematical aid for easy calculation. It is nearly independent of the laser transition i for a certain gas component and can thus be written as θ_j. In our wavelength range it is only CO_2 which shows a phase $\theta_j = \pi$, whereas all the other gases studied so far show a phase $\theta_j = 0$. In real measurements small deviations of the phases from the predicted ones occur due to measurement errors. Nevertheless, the approximation $\theta_{ij} = \theta_j = 0$ for all the other air compounds is well justified. It should be noted that the system of linear equations is only well defined if the number of gas components j is smaller than the number m of laser transitions, i.e., for $n \le m$. Based on measurements of the signals V_i, phases θ_j, and laser powers P_i, and knowing the absorption coefficients α_{ij} from literature data or calibration measurements, the unknown concentrations c_j can be derived by solving the above equation system. The algorithm of the data analysis has been described by Meyer and Sigrist (Meyer & Sigrist, 1990). For multicomponent mixtures an algorithm, e.g., a nonlinear Levenberg-Marquardt fit (Moeckli et al., 1998) is employed to fit the measured spectrum on the basis of calibration spectra of the individual compounds.

The concentrations of C_2H_4, CO_2, and H_2O in nitrogen at atmospheric pressure can be determined by measuring the PA signals using three CO_2 laser transitions, e.g., 10P(14), 10P(20) and 10R(20) (Sigrist et al., 1989). The 10P(14) and 10R(20) transitions coincide with sharp peaks of the IR spectra of C_2H_4 (α = 30.4 cm^{-1}atm^{-1}) and H_2O (α = 8.36x10^{-4} cm^{-1}atm^{-1}), respectively. The 10P(20) line that is used to measure CO_2 concentration (α = 2.2x10^{-3} cm^{-1}atm^{-1}) could be replaced by many other transitions without much change in sensitivity, because CO_2 is relatively spectrally flat.

Rooth et al. (Rooth et al., 1990) determined the H_2O, CO_2, and NH_3 contents in ambient air by using four laser transitions (10R(20) is used to compute water vapor concentration using the absorption coefficient at the actual gas temperature; the influence of the CO_2 absorption on the measurement of H_2O is also taken into account by using the 9R(18) and 9R(28) lines;

the difference in the two signals yields the CO_2 concentration with the help of Eq. (37); the fourth line, 9R(30) together with 9R(28) provides the NH_3 concentration). Nägele and Sigrist (Nägele & Sigrist, 2000) recorded the PA signal on two transitions for each compound, carefully selected for maximum absorption, minimum absorption interference, and good laser performance. In addition, they measured the PA signal on two laser transitions (10P(12), 10P(40)), for which all of the investigated gases exhibit negligible absorption, to verify the constant background signal. Therefore, the spectra to monitor ethylene (10P(14), 10P(16)), ethanol (9P(8), 9P(32)), methanol (9P(34), 9P(36)), and CO_2 (10P(20), 9R(20)) comprise ten different transitions. Thus cross references are possible and the background signal, which is the same for these lines, can be subtracted. This extension to several laser lines yields better detection limits and selectivity, although the time for one full measurement increases with the number of lines.

5. Conclusions

Based on previous discussion, PA spectroscopy, performed with tunable CO_2 lasers as radiation sources, offers the following main characteristics relevant to *in situ* trace gas monitoring:

i. Its high sensitivity makes it possible to measure absorption coefficients on the order of 10^{-8} cm^{-1}, corresponding to densities of $\mu g/m^3$ or concentrations of ppbV (10^{-9} atm) for most substances;
ii. The PA cell responsivity is independent of radiation wavelength;
iii. A large number of gases and vapors are measurable with the same instrument;
iv. It has high selectivity, meaning that it can clearly distinguish among various compounds;
v. The experimental setup is rather simple, immune to interference, and, for example, does not require cryogenic cooling of the IR detectors, etc.;
vi. Relative portability for *in situ* measurements (carried on mobile trailers in the troposphere or on balloon-borne systems to the stratosphere);
vii. Operational simplicity and real time data analysis make it capable of performing quasi continuous measurements;
viii. The calibration with certified gases and gas mixtures is straightforward and reliable;
ix. Detection linearity and a wide dynamic range of at least 6 orders of magnitude are offered (from several fractions of ppbV to tens of ppmV), i.e., the same apparatus can be used for low (immission) and high (emission) concentrations.
x. Specially designed PA cells can perform continuous measurements on flowing gas mixtures, i.e., a much better temporal resolution can be achieved than the one provided by, *inter alia*, gas chromatography.

The outstanding features of the PA cell, most importantly its small size, simplicity, and robustness, cannot be fully utilized unless it is combined with a suitable laser source. The recent commercial availability of sealed-off, medium-power (50 W), grating-tunable CO_2 lasers has paved the way to the development of instrumentation with excellent sensitivity and compact footprints that can be readily deployed in industrial or medical settings. Further improvements are possible by using resonant PA cells with high Q factors (limited, though, by fluctuations of the modulation frequency), multipass, or intracavity arrangements.

Other laser sources were successfully applied in photoacoustic spectroscopy. Recent developments in compact near-infrared (NIR) and IR all-solid-state tunable lasers, such as the tunable semiconductor lasers, quantum-cascade lasers, and devices utilizing non-linear optical mixing in non-linear crystals (OPOs, difference frequency generators (DFGs), optical parametric amplifiers (OPAs)), have significantly advanced the application of photoacoustic techniques in sensitive trace gas analysis.

NIR diode lasers are becoming more and more popular due to recent development of cheap high-quality, compact sources having a spectral emission which falls in the absorption range of many molecules of great practical interest. The range of the available NIR diode lasers spans from about 0.8 to 2.1 μm. Many gases like methane, acetylene, CO, and CO_2 exhibit overtone absorptions in the 1.55-1.65 μm wavelength region which can be covered by a conventional external cavity diode laser (ECDL). Detection limits are lower compared with measurements in the fundamental absorption region but still sufficient for many applications. A combination of diode lasers with PA detection was used by various authors to detect ammonia (Fehér et al., 1994; Schmohl et al., 2002), methane (Schäfer et al., 1998), elemental carbon (Petzold et al., 1995), toluene (Beenen et al., 1998), and water vapor (Bozóki et al., 1999). Different experimental arrangements such as external cavity diode lasers (Sneider et al., 1997) or intracavity PA cells (Bozóki et al., 1996) were tested. At present, 1.6 μm diode lasers, coinciding with the first vibrational overtones and combination bands of molecules containing a CH bond, are those that - in the NIR range - provide the best trade-off between cost and molecular detection efficiency (Boschetti et al., 2002).

Recent progress with quantum-cascade lasers makes them attractive sources in the important 3- to 5-μm spectral range. This area is important not only because the characteristic absorption bands of, among others, CO, N_2O, HCl and CH_2O, lie herein, but also because there is an atmospheric transparent window in this range. Soon after the first appearance of these lasers, gas monitoring applications using various detection schemes were reported (Sharpe et al., 1998; Kosterev et al., 2008). Quantum-cascade lasers were used to detect ammonia and water vapor at 8.5 μm (Paldus et al., 1999), NO at 5.2 μm (Menzel et al., 2001), $^{12}CH_4$, $^{13}CH_4$ and N_2O isotopomers at 8.1 μm (Gagliardi et al., 2002), trace gases (CH_4, N_2O, H_2O) in laboratory air at 7.9 μm (Kosterev & Tittel, 2002), carbon dioxide, methanol and ammonia at 10.1/10.3 μm (Hofstetter et al., 2001), CH_4 and NO at 7.9 μm and 5.3 μm (Grossel et al, 2006; Grossel et al., 2007) and simultaneously CO and SO_2 at 4.56 μm and 7.38 μm (Liu et al., 2011). In contrast to semiconductor (diode) lasers, quantum-cascade lasers are unipolar light sources based on only one type of carrier, usually electrons, making intraband transitions between confined energy levels within the conduction band. The term "cascade" comes from the fact that the confined energy levels are arranged the way of a waterfall, so that electrons undergoing lasing transitions travel from one stage to the next, just like water does in a multiple-step water cascade. Therefore, one electron can emit sequentially up to n photons when n steps are present. The emission wavelength of a quantum-cascade laser is determined not by the semiconductor bandgap but by the quantum confinement in the quantum wells created by the quantum-well material and the barrier material. Therefore, quantum-cascade lasers can span a wide wavelength range using the same material system. Quantum-cascade lasers with wavelengths from 3.5 to 13 μm have been fabricated by use of the same material system (InGaAs wells and InAlAs barriers).

Widely tunable, narrowband optical parametric oscillator (OPO)-based laser sources were used for trace gas spectroscopy in the fundamental C-H stretch vibration region (3-5 μm) (Bohren et al., 1997), to detect ethane at 3.34 μm (Kühnemann et al., 1998; van Herpen et al., 2002), N_2O at 2.86 μm (Costopoulos et al., 2002), or methane at 3.39 μm (Miklós et al., 2002). Most of today's commercial OPOs are based on BaB_2O_4 crystals (BBO) pumped by the third harmonic of a Nd:YAG laser and their IR tuning range is limited to approximately 2 μm. Difference frequency generation (DFG) is certainly the most promising technique for the extension of the tuning range of an existing tunable laser to the mid IR (2.5-4.5 μm) (Fischer & Sigrist, 2002). Spectrometers using DFG were applied to monitoring, for example, formaldehyde in ambient air at 3.53 μm (Rehle et al., 2001) and volcanic gases (CH_4, CO_2, HCl, SO_2, H_2O vapor) at 3.3-4.4 μm (Richter et al., 2002).

6. References

Angeli, G.Z.; Solyom, A.M.; Miklos, A. & Bicanic, D.D. (1992). Calibration of a Windowless Photoacoustic Cell for Detection of Trace Gases. *Anal. Chem.*, Vol.64, No.2, (January 1992), pp. 155-158, ISSN 0003-2700

Beck, S.M. (1985). Cell Coatings to Minimize Sample (NH_3 and N_2H_4) Adsorption for Low-Level Photoacoustic Detection. *Appl. Opt.*, Vol.24, No.12, (June 1985), pp. 1761-1763, ISSN 0003-6935

Beenen, A. & Niessner, R. (1998). Development of Photoacoustic Gas Sensor for In-Situ and On-Line Measurement of Gaseous Water and Toluene. *Analyst*, Vol.123, No.4, pp. 543-545, ISSN 0003-2654

Bell, A.G. (1880). On the Production and Reproduction of Sound by Light. *Am. J. Sci.*, Vol.XX, pp. 305-324

Bell, A.G. (1881). Upon the Production of Sound by Radiant Energy. *Phil. Mag. J. Sci.*, Vol.XI, pp. 510-528

Bernegger, S. & Sigrist, M.W. (1987). Longitudinal Resonant Spectrophone for CO-laser Photoacoustic Spectroscopy. *Appl. Phys. B*, Vol.44, No.2, (October 1987), pp. 125-132, ISSN 0946-2171

Bicanic, D. (Ed.) (1992). *Photoacoustic and Photothermal Phenomena III*, Springer Series in Optical Sciences, Vol.69, Springer, ISBN 978-0-387-55669-9, Berlin, Germany

Bijnen, F.G.; Reuss, J. & Harren, F.J.M. (1996). Geometrical Optimization of a Longitudinal Resonant Photoacoustic Cell for Sensitive and Fast Trace Gas Detection. *Rev. Sci. Instrum.*, Vol.67, No.8, (August 1996), pp. 2914-2923, ISSN 0034-6748

Bohren, A. & Sigrist, M.W. (1997). Optical Parametric Oscillator Based Difference Frequency Laser Source for Photoacoustic Trace Gas Spectroscopy in the 3 μm Mid-IR Range. *Infrared Phys. Technol.*, Vol.38, No.7, (December 1997), pp. 423-435, ISSN 1350-4495

Boschetti, A.; Bassi, D.; Iacob, E.; Iannotta, S.; Ricci, L. & Scotoni, M. (2002). Resonant Photoacoustic Simultaneous Detection of Methane and Ethylene by Means of a 1.63-μm Diode Laser. *Appl. Phys. B*, Vol.74, No.3, (February 2002), pp. 273-278, ISSN 0946-2171

Bozóki, Z.; Sneider, J.; Gingl, Z.; Mohácsi, A.; Szakáll, M. & Szabó, G. (1999). A High-Sensitivity, Near- Infrared Tunable-Diode-Laser-Based Photoacoustic Water-Vapor-Detection System for Automated Operation. *Meas. Sci. Technol.*, Vol.10, No.11, (November 1999), pp. 999-1003, ISSN 0957-0233

Bozóki, Z.; Sneider, J.; Szabó, G.; Miklós, A.; Serényi, M.; Nagy, G. & Fehér, M. (1996). Intracavity Photoacoustic Gas Detection with an External Cavity Diode Laser. *Appl. Phys. B*, Vol.63, No.4, (October 1996), pp. 399-401, ISSN 0946-2171

Bruce, C.W. & Pinnick, R.G. (1977). In Situ Measurements of Aerosol Absorption with a Resonant cw Laser Spectrophone. *Appl. Opt.*, Vol.16, No.7, (July 1977), pp. 1762-1765, ISSN 0003-6935

Costopoulos, D.; Miklós, A. & Hess, P. (2002). Detection of N_2O by Photoacoustic Spectroscopy with a Compact, Pulsed Optical Parametric Oscillator. *Appl. Phys. B,* Vol.75, No.2-3, (September 2002), pp. 385-389, ISSN 0946-2171

Cristescu, S.; Dumitras, D.C. & Dutu, D.C.A. (2000). Characterization of a Resonant Cell Using the Acoustic Transmission Line Model. *Proc. SPIE SIOEL '99: Sixth Symposium on Optoelectronics,* Vol.4068, T. Necsoiu, M. Robu, D.C. Dumitras (Eds.), pp. 263-272, SPIE, ISBN 978-0-819-43705-1, Bellingham, WA, USA

Cristescu, S.; Dumitras, D.C.; Dutu, D.C.A. & Mujat, C. (1997). Real-Time Detection System for Laser Photoacoustic Applications. *Rom. Rep. Phys.,* Vol.49, No.8-9-10, pp. 757-768, ISSN 1221-1451

Dewey Jr. C.F.; Kamm, R.D. & Hackett, C.E. (1973). Acoustic Amplifier for Detection of Atmospheric Pollutants. *Appl. Phys. Lett.,* Vol.23, No.11, (December 1973), pp. 633-635, ISSN 0003-6951

Dewey, C.F. (1977). Design of Optoacoustic Systems. In *Optoacoustic Spectroscopy and Detection,* Ch. 3, Y.-H. Pao (Ed.), 47-77, Academic, ISBN 978-0-125-44159-9, New York, NY, USA

Dumitras, D.C.; Dutu, D.C.; Matei, C.; Magureanu, A.M.; Petrus, M. & Popa, C. (2007a). Improvement of a Laser Photoacoustic Instrument for Trace Gas Detection. *U. P. B. Sci. Bull., Series A,* Vol.69, No.3, pp. 45-56, ISSN 1223-7027

Dumitras, D.C.; Dutu, D.C.; Matei, C.; Magureanu, A.M.; Petrus, M. & Popa, C. (2007b). Laser Photoacoustic Spectroscopy: Principles, Instrumentation, and Characterization. *J. Optoelectr. Adv. Mater.,* Vol.9, No.12, (December 2007), pp. 3655-3701, ISSN 1454-4164

Dutu, D.C.A.; Cristescu, S. & Dumitras, D.C. (1994b). Measurement of Photoacoustic Signal and Noises in a Sensitive Spectrophone with a Frequency Stabilized CO_2 Laser. In *Proc. SPIE ROMOPTO'94-Fourth Conference in Optics,* Vol.2461, V.I. Vlad (Ed.), pp. 308-316, SPIE, ISBN 978-0-819-41813-5, Bellingham, WA, USA

Dutu, D.C.A.; Dumitras, D.C.; Cristescu, S. & Sarkozy, L. (1994a). Evaluation of Photoacoustic Signal and Noises in a Sensitive Spectrophone Irradiated by a CO_2 Laser Beam. *Rom. Rep. Phys.,* Vol.46, No.7-8, pp. 639-644, ISSN 1221-1451

Fehér, M.; Jiang, Y.; Maier, J.P. & Miklós, A. (1994). Optoacoustic Trace-Gas Monitoring with Near-Infrared Diode Lasers. *Appl. Opt.,* Vol.33, No.9, (March 1994), pp. 1655-1658, ISSN 0003-6935

Fischer, C. & Sigrist, M.W. (2002). Trace-Gas Sensing in the 3.3-μm Region Using a Diode-Based Difference-Frequency Laser Photoacoustic System. *Appl. Phys. B,* Vol.75, No.2-3, (September 2002), pp. 305-310, ISSN 0946-2171

Gagliardi, G.; Tamassia, F.; De Natale, P.; Gmachl, C.; Capasso, F.; Sivco, D.L.; Baillargeon, J.N.; Hutchinson, A.L. & Cho, A.Y. (2002). Sensitive Detection of Methane and Nitrous Oxide Isotomers Using a Continuous Wave Quantum Cascade Laser. *Eur. Phys. J. D,* Vol.19, No.3, (June 2002), pp. 327-331, ISSN 1434-6060

Gerlach, R. & Amer, N.M. (1978). Sensitive Optoacoustic Detection of Carbon Monoxide by Resonance Absorption. *Appl. Phys. Lett.,* Vol.32, No.4, (February 1978), pp. 228-231, ISSN 0003-6951

Gerlach, R. & Amer, N.M. (1980). Brewster Window and Windowless Resonant Spectrophones for Intracavity Operation. *Appl. Phys.A,* Vol.23, No.3, (November 1980), pp. 319-326, ISSN 0947-8396

Gondal, M.A. (1997). Laser Photoacoustic Spectrometer for Remote Monitoring of Atmospheric Pollutants. *Appl. Opt.*, Vol.36, No.15, (May 1997), pp. 3195-3201, ISSN 1559-128x

Groot, T. (2002). *Trace Gas Exchange by Rice, Soil and Pears. A Study Based on Laser Photoacoustic Spectroscopy*, PhD Thesis, University of Nijmegen, ISBN 978-9-090-15380-2, Nijmegen, The Netherlands

Grossel, A.; Zeninari, V.; Joly, L.; Parvitte, B.; Courtois, D. & Durry, G. (2006). New Improvements in Methane Detection Using a Helmholtz Resonant Photoacoustic Laser Sensor: A Comparison Between Near-IR Diode Lasers and Mid-IR Quantum Cascade Lasers. *Spectrochim. Acta A*, Vol.63, No.5, (April 2006), pp. 1021-1028, ISSN 1386-1425

Grossel, A.; Zeninari, V.; Joly, L.; Parvitte, B.; Durry, G. & Courtois, D. (2007). Photoacoustic Detection of Nitric Oxide with a Helmholtz Resonant Quantum Cascade Laser Sensor. *Infrared Phys. Technol.*, Vol.51, No.2, (October 2007), pp. 95-101, ISSN 1350-4495

Gusev, V.E. & Karabutov, A.A. (1993). *Laser Optoacoustics*, American Institute of Physics, ISBN 978-1-563-96036-9, Melville, NY, USA

Harren, F.J.M. & Reuss, J. (1997). Spectroscopy, Photoacoustic, In *Encyclopedia of Applied Physics*, Vol.19, G.L. Trigg (Ed.), 413-435, VCH Publishers, ISBN 978-3-527-40478-0, New York, USA

Harren, F.J.M.; Bijnen, F.G.C.; Reuss, J.; Voesenek, L.A.C.J. & Blom, C.W.P.M. (1990). Sensitive Intracavity Photoacoustic Measurements with a CO_2 Waveguide Laser. *Appl. Phys. B*, Vol.50, No. 2, (February 1990), pp. 137-144, ISSN 0946-2171

Harren, F.J.M.; Cotti, G.; Oomens, J. & te Lintel Hekkert, S. (2000). Photoacoustic Spectroscopy in Trace Gas Monitoring, In *Encyclopedia of Analytical Chemistry*, Vol.3, R.A. Meyers (Ed.), 2203-2226, Wiley, ISBN 978-0-471-97670-7 Chichester, UK

Henningsen, J.; Olaffson, A. & Hammerich, M. (1990). Trace Gas Detection with Infrared Gas Lasers. In *Applied Laser Spectroscopy*, W. Demtröder, M. Inguscio (Eds.), 403-416, Plenum Press, ISBN 978-0-306-43717-1, New York, NY, USA

Hess, P. (1983). Resonant Photoacoustic Spectroscopy, In *Topics in Current Chemistry*, Vol.111, F.L. Boschke (Ed.), 1-32, Springer, ISBN 3-540-16403-0, Berlin, Germany

Hess, P. (Ed.). (1989). *Photoacoustic, Photothermal and Photochemical Processes in Gases*, Topics in Current Physics, Vol.46, Springer, ISBN 978-0-387-51392-2, Berlin, Germany

Hofstetter, D.; Beck, M.; Faist, J.; Nägele, M. & Sigrist, M.W. (2001). Photoacoustic Spectroscopy with Quantum Cascade Distributed-Feedback Lasers. *Opt. Lett.*, Vol.26, No.12, (June 2001), pp. 887-889, ISSN 0146-9592

Hubert, M.H. (1983). *Laser Optoacustic Detector Measurement of Signatures of a Selection of Environmental Contaminants.* Rep. No. 83-715-1, Ultra Lasertech Inc., Mississauga, Canada

Johnson, R.H.; Gerlach, R.; Thomas III, L.J. & Amer, N.M. (1982). Loss Mechanisms in Resonant Spectrophones. *Appl. Opt.*, Vol.21, No.1, (January 1982), pp. 81-89, ISSN 0003-6935

Kaiser, R. (1959). On the Theory of the Spectrophone. *Can. J. Phys.*, Vol.37, No.12, (December 1959), pp. 1499-1513, ISSN 0008-4204

Kamm, R.D. (1976). Detection of Weakly Absorbing Gases Using a Resonant Optoacoustic Method. *J. Appl. Phys.*, Vol.47, No.8, (August 1976), pp. 3550-3558, ISSN 0021-8979

Kerr, E.L. & Atwood, J. (1968). The Laser Illuminated Absorptivity Spectrophone: A Method for Measurement of Weak Absorptivity in Gases at Laser Wavelengths. *Appl. Opt.*, Vol.7, No.5, (May 1968), pp. 915-921, ISSN 0003-6935

Kosterev, A.A. & Tittel, F.K. (2002). Chemical Sensors Based on Quantum Cascade Lasers. *IEEE J. Quantum Electron.*, Vol.38, No.6, (June 2002), pp. 582-591, ISSN 0018-9197

Kosterev, A.A., Wysocki, G.; Bakhirkin, Y.A.; So, S.; Lewicki, R.; Fraser, M.; Tittel, F. & Curl, R.F. (2008). Application of Quantum Cascade Lasers to Trace Gas Analysis. *Appl. Phys. B*, Vol.90, No.2, (February 2008), pp. 165-176, ISSN 0946-2171

Kreuzer, L.B. & Patel C.K.N. (1971). Nitric Oxide Air Pollution: Detection by Optoacoustic Spectroscopy. *Science*, Vol.173, No.3991, (July 1971), pp. 45-47, ISSN 0036-8075

Kreuzer, L.B. (1971). Ultralow Gas Concentration Infrared Absorption Spectroscopy. *J. Appl. Phys.*, Vol.42, No.7, (June 1971), pp. 2934-2943, ISSN 0021-8979

Kreuzer, L.B. (1977). The Physics of Signal Generation and Detection. In *Optoacoustic Spectroscopy and Detection*, Ch.1, Y.-H. Pao (Ed.), 1-25, Academic, ISBN 978-0-125-44150-9, New York, NY, USA

Kreuzer, L.B.; Kenyon, N.D. & Patel, C.K.N. (1972). Air Pollution: Sensitive Detection of Ten Pollutant Gases by Carbon Monoxide and Carbon Dioxide Lasers. *Science*, Vol.177, No.4046, (July 1972), pp. 347-349, ISSN 0036-8075

Kritchman, E.; Shtrikman, S. & Slatkine, M. (1978). Resonant Optoacoustic Cells for Trace Gas Analysis. *J. Opt. Soc. Am.*, Vol.68, No.9, (September 1978), pp. 1257-1271, ISSN 1084-7529

Kühnemann, F.; Schneider, K.; Hecker, A.; Martis, A.A.E.; Urban, W.; Schiller, S. & Mlynek, J. (1998). Photoacoustic Trace-Gas Detection Using a cw Single-Frequency Parametric Oscillator. *Appl. Phys. B*, Vol.66, No.6, (June 1998), pp. 741-745, ISSN 0946-2171

Liu, W.; Wang, L.; Li, L.; Liu, J.; Liu, F.-Q. & Wang, Z. (2011). Fast Simultaneous Measurement of Multi-Gases Using Quantum Cascade Laser Photoacoustic Spectroscopy. *Appl. Phys. B*, Vol.103, No.3, (June 2011), pp. 743-747, ISSN 0946-2171

Mandelis, A. & Hess, P. (Eds.) (1997). *Progress in Photothermal and Photoacoustic Science and Technology*, Vol.3, SPIE Press Book, ISBN 978-0-819-42450-1, Bellingham, WA, USA

Mandelis, A. (Ed.) (1992). *Progress in Photothermal and Photoacoustic Science and Technology*, Vol.1, Elsevier, ISBN 978-0-819-42450-1, New York, USA

Mandelis, A. (Ed.) (1994). *Progress in Photothermal and Photoacoustic Science and Technology*, Vol.2, Prentice Hall, ISBN 978-0-131-47430-8, Englewood Cliffs, NJ, USA

Menzel, L.; Kosterev, A.A.; Curl, R.F.; Titel, F.K.; Gmachl, C.; Capasso, F.; Sivco, D.L.; Baillargeon, J.N.; Hutchinson, A.L.; Cho, A.Y. & Urban, W. (2001). Spectroscopic Detection of Biological NO with a Quantum Cascade Laser. *Appl. Phys. B*, Vol.72, No.7, (May 2001), pp. 859-863, ISSN 0946-2171

Meyer, P.L. & Sigrist, M.W. (1990). Atmospheric Pollution Monitoring Using CO_2-Laser Photoacoustic Spectroscopy and Other Techniques. *Rev. Sci. Instrum.*, Vol.61, No.7, (July 1990), pp. 1779-1807, ISSN 0034-6748

Miklós, A.; Hess, P. & Bozoki, Z. (2001). Application of Acoustic Resonators in Photoacoustic Trace Gas Analysis and Metrology. *Rev. Sci. Instrum.* Vol.72, No.4, (April 2001), pp. 1937-1955, ISSN 0034-6748

Miklós, A.; Lim, C.-H.; Hsiang, W.-W.; Liang, G.-C.; Kung, A.H.; Schmohl, A. & Hess, P. (2002). Photoacoustic Measurement of Methane Concentrations with a Compact Pulsed Optical Parametric Oscillator. *Appl. Opt.*, Vol.41, No.15, (May 2002), pp. 2985-2993, ISSN 0003-6935

Moeckli, M.A.; Hilbes, C. & Sigrist, M.W. (1998). Photoacoustic Multicomponent Gas Analysis Using a Levenberg-Marquardt Fitting Algorithm. *Appl. Phys. B*, Vol.67, No.4, (October 1998), pp. 449-458, ISSN 0946-2171

Morse, P.M. & Ingard, K.U. (1986). *Theoretical Acoustics*, Princeton University Press, ISBN 978-0-691-02401-4, Princeton, NJ, USA

Olaffson, A.; Hammerich, M.; Bülow, J. & Henningsen, J. (1989). Photoacoustic Detection of NH_3 in Power Plant Emission with a CO_2 Laser. *Appl. Phys. B*, Vol.49, No.2, (August 1989), pp. 91-97, ISSN 0946-2171

Paldus, B.A.; Spence, T.G.; Zare, R.N.; Oomens, J.; Harren, F.J.M.; Parker, D.H.; Gmachl, C.; Cappasso, F.; Sivco, D.L.; Baillargeon, J.N.; Hutchinson, A.L. & Cho, A.Y. (1999). Photoacoustic Spectroscopy Using Quantum-Cascade Lasers. *Opt. Lett.*, Vol.24, No.3, (February 1999), pp. 178-180, ISSN 0146-9592

Pao, Y.-H. (Ed.) (1977). *Optoacoustic Spectroscopy and Detection*, Academic, ISBN 978-0-125-44150-6, New York, USA

Patel, C.K.N. & Tam, A.C. (1981). Pulsed Optoacoustic Spectroscopy of Condensed Matter. *Rev. Mod. Phys.*, Vol.53, No.3, (July-September 1981), pp. 517-554, ISSN 0034-6861

Perlmutter, P.; Shtrikman, S. & Slatkine, M. (1979). Optoacoustic Detection of Ethylene in the Presence of Interfering Gases. *Appl. Opt.*, Vol.18, No.13, (July 1979), pp. 2267-2274, ISSN 0003-6935

Petzold, A. & Niessner, R. (1995). Novel Design of a Resonant Photoacoustic Spectrophone, for Elemental Carbon Mass Monitoring. *Appl. Phys. Lett.*, Vol.66, No.10, (March 1995), pp. 1285-1287, ISSN 0003-6951

Preece, W.H. (1881). On the Conversion of Radiant Energy into Sonorous Vibrations. *Proc. R. Soc.*, Vol.31, pp. 506-520

Pushkarsky, M.B.; Weber, M.E.; Baghdassarian, O.; Narasimhan, L.R. & Patel, C.K.N. (2002). Laser-Based Photoacoustic Ammonia Sensors for Industrial Applications. *Appl. Phys. B*, Vol.75, No.4-5, (April 2002), pp. 391-396, ISSN 0946-2171

Rehle, D.; Leleux D.; Erdelyi, M.; Tittel, F.; Fraser, M. & Friedfeld, S. (2001). Ambient Formaldehyde Detection with a Laser Spectrometer Based on Difference-Frequency Generation in PPLN. *Appl. Phys. B*, Vol.72, No.8, (June 2001), pp. 947-952, ISSN 0946-2171

Reid, J.; Shewchun, J.; Garside, B.K. & Balik, E.A. (1978). High Sensitivity Pollution Detection Employing Tunable Diode Lasers. *Appl. Opt.*, Vol.17, No.2, (January 1978), pp. 300-307, ISSN 0003-6935

Richter, D.; Erdelyi, M.; Curl, R.F.; Tittel, F.K.; Oppenheimer, C.; Duffell, H.J. & Burton, M. (2002). Field Measurements of Volcanic Gases Using Tunable Diode Laser Based Mid-Infrared and Fourier Transform Infrared Spectrometers. *Opt. Lasers Eng.*, Vol.37, No.2-3, (March 2002), pp. 171-186, ISSN 0143-8166

Röntgen, W.C. (1881). Über Töne, welche durch intermittierende Bestralung eines Gases entstehen. *Ann. der Phys. und Chem.*, Vol.1, pp. 155-159

Rooth, R.A.; Verhage, A.J.L. & Wouters, L.W. (1990). Photoacoustic Measurement of Ammonia in the Atmosphere: Influence of Water Vapor and Carbon Dioxide. *Appl. Opt.*, Vol.29, No. 25, (September 1990), pp. 3643-3653, ISSN 0003-6935

Rosencwaig, A. (1980). *Photoacoustics and Photoacoustic Spectroscopy*, Chemical Analysis Vol.57, Wiley, ISBN 978-0-471-04495-4, New York, USA

Schäfer, S.; Mashni, M.; Sneider, J.; Miklós, A.; Hess, P.; Pitz, H.; Pleban, K.-U. & Ebert, V. (1998). Sensitive Detection of Methane with a 1.65 μm Diode Laser by

Photoacoustic and Absorption Spectroscopy. *Appl. Phys. B,* Vol.66, No.4, (April 1998), pp. 511-516, ISSN 0946-2171
Schmid, T. (2006). Photoacoustic Spectroscopy for Process Analysis. *Anal. Bioanal. Chem.,* Vol.384, No.5, (March 2006), pp. 1071-1086, ISSN 1618-2642
Schmohl, A.; Miklós, A. & Hess, P. (2002). Detection of Ammonia by Photoacoustic Spectroscopy with Semiconductor Lasers. *Appl. Opt.,* Vol.41, No.9, (March 2002), pp. 1815-1823, ISSN 0003-6935
Sharpe, S.W.; Kelly, J.F.; Hartman, J.S.; Gmachl, C.; Capasso, F.; Sivco, D.L.; Baillargeon, J.N. & Cho, A.Y. (1998). High-Resolution (Doppler-Limited) Spectroscopy Using Quantum-Cascade Distributed-Feedback Lasers. *Opt. Lett.,* Vol.23, No.7, (September 1998), pp. 1396-1398, ISSN 0146-9592
Sigrist, M.W. (1986). Laser Generation of Acoustic Waves in Liquids and Gases. *J. Appl. Phys.,* Vol.60, No.7, (October 1986), pp. R83-R121, ISSN 0021-8979
Sigrist, M.W. (2003). Trace Gas Monitoring by Laser Photoacoustic Spectroscopy and Related Techniques. *Rev. Sci. Instrum.,* Vol.74, No.1, (January 2003), pp. 486-490, ISSN 0034-6748
Sigrist, M.W. (Ed.) (1994). *Air Monitoring by Spectroscopic Techniques,* Vol.127, Wiley Chemical Analysis Series, Wiley, ISBN 978-0-471-55875-3, New York, USA
Sigrist, M.W.; Bernegger, S. & Meyer, P.L. (1989). Atmospheric and Exhaust Air Monitoring by Laser Photoacoustic Spectroscopy, In *Topics in Current Physics "Photoacoustic, Photothermal and Photochemical Processes in Gases"*, Ch.7, Vol.46, P. Hess (Ed.), 173-211, Springer, ISBN 978-3-540-51392-2, Berlin, Germany
Sneider, J.; Bozóki, Z.; Szabó, G. & Bor, Z. (1997). Photoacoustic Gas Detection Based on External Cavity Diode Laser Light Sources. *Opt. Eng.,* Vol.36, No.2, (February 1997), pp. 482-486, ISSN 0091-3286
Tam, A.C. (1986). Applications of Photoacoustic Sensing Techniques. *Rev. Mod. Phys.,* Vol.58, No.2, (April-June 1986), pp. 381-431, ISSN 0034-6861
Terhune, R.W. & Anderson, J.E. (1977). Spectrophone Measurements of the Absorption in Visible Light by Aerosols in the Atmosphere. *Opt. Lett.,* Vol.1, No.2, (August 1977), pp. 70-72, ISSN 0146-9592
Thomas III, L.J.; Kelly, M.J. & Amer, N.M. (1978). The Role of Buffer Gases in Optoacoustic Spectroscopy. *Appl. Phys. Lett.,* Vol.32, No.11, (June 1978), pp. 736-738, ISSN 0003-6951
Thöny, A. & Sigrist, M.W. (1995). New Developments in CO_2-Laser Photoacoustic Monitoring of Trace Gases. *Infrared Phys. Technol.,* Vol.36, No.2, (February 1995), pp. 585-615, ISSN 1350-4495
Tonelli, M.; Minguzzi, P. & Di Lieto, A. (1983). Intermodulated Optoacoustic Spectroscopy. *J. Physique (Colloque C6),* Vol.44, No.10, (October 1983), pp. 553-557, ISSN 0449-1947
Tyndall, J. (1881). Action of an Intermittent Beam of Radiant Heat upon Gaseous Matter. *Proc. R. Soc.,* Vol.31, pp. 307-317
van Herpen, M.; te Lintel Hekkert, S.; Bisson, S.E.; Harren, F.J.M. (2002). Wide Single-Mode Tuning of a 3.0-3.8-μm, 700-mW, Continuous Wave Nd:YAG-Pumped Optical Parametric Oscillator Based on Periodically Poled Lithium Niobate. *Opt. Lett.,* Vol.27, No.8, (April 2002), pp. 640-642, ISSN 0146-9592
West, G.A. (1983). Photoacoustic Spectroscopy. *Rev. Sci. Instrum.,* Vol.54, No.7, (July 1983), pp. 797-817, ISSN 0034-6748
Zharov, V.P. & Letokhov, V.S. (1986). *Laser Optoacoustic Spectroscopy,* Vol.37, Springer, ISBN 978-3-540-11795-4, Berlin, Germany

2

CO_2 Lasing on Non-Traditional Bands

Vladimir Petukhov and Vadim Gorobets
Institute of Physics of National Academy of Sciences of Belarus
Belarus

1. Introduction

Construction of powerful and efficient laser sources, lasing in various IR ranges, is of importance for further development of a number of trends, e.g., spectroscopy, laser chemistry, sounding of the atmosphere, and metrology. The most natural way to solve this problem is to use unconventional (nontraditional) transitions to produce lasing in commonly used CO_2 lasers. The spectral range of CO_2 lasers is greatly increased in lasing on transitions of the so-called "hot" band $01^11\text{-}11^10$, whose P-branch is in the range of 10.9-11.3 μm. Thorough investigations of gain, vibrational temperatures (T_1, T_2, T_3), and output parameters on lines of the hot band made it possible to achieve efficient lasing both for pulse TEA and for cw longitudinal-discharge CO_2 lasers.

In studying the lasing spectrum of hot transitions in TEA CO_2 lasers some lines not belonging to the $01^11\text{-}11^10$ band. We suggested, that these lasing lines belong to higher level transitions, e.g., $10^01\text{-}20^00$ (04^00), which were called "doubly hot," i.e., transitions in which compared to hot transitions two deformation quanta or one symmetric quantum rather than one deformation quantum is added both to the upper and to the lower energy level.

In the present work lasing in both a TEA laser and a low-pressure laser with longitudinal discharge on some transitions of the CO_2 molecule in the range of 11.0-11.6 μm is reported. The rather high resolution of the spectral equipment used and calculation of transition frequencies on the basis of recent spectroscopic constants made it possible to identify definitively the lasing lines obtained as belonging to the doubly hot bands $02^21\text{-}12^20$ and $10^0l\text{-}20^00$ and the sequence hot band $01^12\text{-}11^11$. To find optimum conditions for lasing on the aforementioned bands experimental studies of vibrational temperatures in active media of a TEA CO_2 laser and a low-pressure laser with longitudinal discharge were carried out.

Earlier the lasing on the $02^00(10^00)\text{-}01^10$ band of the CO_2 molecule has been obtained in the specific systems at cryogenic temperatures under the lowest efficiency. The optimization of the active medium and its electrical discharge pumping conditions based on the original technique of the temperature model allowed to obtain in the simple TE CO_2 laser with UV preionization the powerful lasing on the $02^00\text{-}01^10$ band at room temperature. The dependencies of the output and spectral performances of the 16 (14) micrometers lasing vs. a content of the active medium, pumping parameters and cavity characteristics have been carried out.

To increase the power performances of the 16 (14) microns CO_2 laser the possibility of lasing on the $02^01(10^01)\text{-}01^11$ band have been experimentally and theoretically investigated under the combined (electrical + optical) excitation of the active medium. The conditions for obtaining effective lasing at the rotational-vibrational transitions of the 02^01-01^11 (λ = 16.4 μm) and 10^01-01^11 (λ = 14.1 μm) bands of the CO_2 molecule are examined. To obtain population inversion in the indicated channels one should initially populate the 00^02 vibrational level, considerable population of which can be accomplished comparatively simply, for example, in an electric discharge. Then a powerful two-frequency radiation resonant with the 00^02-02^01 (10^01) and 01^11-11^10(03^10) transitions acts on the medium excited in such a way. We will discuss by what means such a scheme of lasing in one active medium can be accomplished.

The lidar complex of equipment based on CO_2 laser specially designed for atmospheric sensing, with tuning on generation lines in the spectral ranges 9-11.3 and 4.5-5.6 μm will be described. Considerable extension of the spectral range to the short-wave region is attained due to effective CO_2 laser second harmonic generation in nonlinear crystals. Taking into the real potentialities of the lidar complex in hand, using a package of spectroscopic data HITRAN, computer simulation of atmospheric transmission has been made. On this basis, by the method of differential absorption a method has been elaborated for measuring of small concentrations of a number of gases.

2. Effective oscillation of a cw CO_2 laser in the range of 11 μm (01^11-11^10 band)

The CO_2 laser oscillation spectrum expansion to the long-wave region is of interest for various scientific and practical applications, for example, for spectroscopy, atmosphere monitoring, etc. From this point of view, the use of the P-branch of the hot 01^11- 11^10 band (10.9−11.4 μm) (see Fig. 1) has considerable promise. Weak hot band lasing was registered in the middle of the 60s in specific long tube (2−4 m) laser systems. The problem of obtaining the hot band oscillation in commercially available cw CO_2 lasers is associated with the low gain realized under conventional conditions. Therefore, the effective hot band cw CO_2 laser oscillation demands, first of all, as in the pulsed TEA CO_2 system, comprehensive study of excitation and active medium composition effects on the hot band gain.

In this work we present experimental results of searching for optimal conditions of the hot band line lasing in a cw CO_2 laser with a commercial 1.2 m sealed-off tube. The hot band gain and output optimization was carried out depending on the active medium composition as well as on the discharge current.

The gain in the active medium was measured by small signal probing using the compensation method. We used as a probing laser a specially developed cw stabilized CO_2 laser tunable over many of the hot band, sequence 00^02-10^01(02^01) band and regular 00^01-10^00(02^00) band lines.

Analysis of the results obtained was carried out on the basis of the universally accepted CO_2-molecule vibrational temperature model (Petukhov et. al.., 1985). The vibrational temperatures of the asymmetric (ν_3) and bound symmetric-bend ($2\nu_2 \approx \nu_1$) mode, T_3 and T_2, respectively, have been determined from the following expressions (Petukhov et. al.., 1985):

$$T_3 = -\frac{3380}{\ln\frac{K_s}{2.1\cdot K_t} - \frac{36}{T}},\ T_2 = -\frac{960}{\ln\frac{K_h}{K_t} - \frac{18}{T}}, \qquad (1)$$

where K_r, K_s, K_h are measured small signal gains of the corresponding regular, sequence and hot band lines; T is the translational temperature determined from the gain distribution over the regular band lines (Petukhov et. al.., 1985). The lock-in amplifier and box-car integrator used in the recording system allowed us to achieve a better than 2% measurement gain accuracy for all bands.

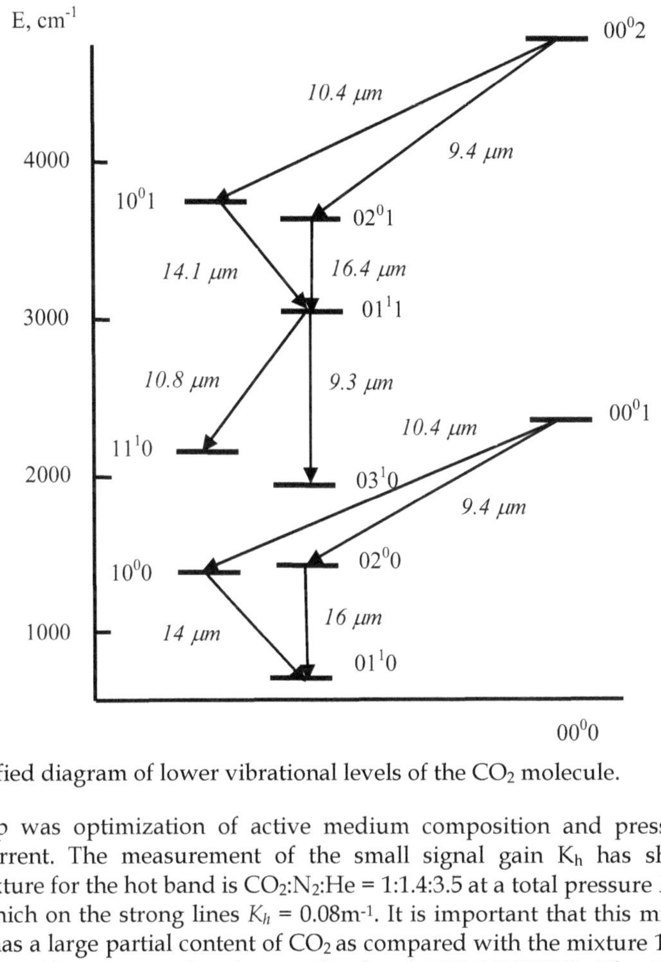

Fig. 1. Simplified diagram of lower vibrational levels of the CO_2 molecule.

The first step was optimization of active medium composition and pressure P and of discharge current. The measurement of the small signal gain K_h has shown that the optimum mixture for the hot band is CO_2:N_2:He = 1:1.4:3.5 at a total pressure P = 11 Torr (I = 15 mA) in which on the strong lines K_h = 0.08m^{-1}. It is important that this mixture contains less He and has a large partial content of CO_2 as compared with the mixture 1:1.6:6.5 (P = 15 Torr, I = 10 mA) optimum for the regular band 00^{0}1-10^{0}0(02^{0}0). The theoretical and experimental investigations of vibrational temperatures of a CO_2 molecule has shown that to obtain considerable hot band gain K_h of a CO_2 molecule it is necessary to heat up the $v_2(v_1)$ mode characterized by vibrational temperature T_2 along with the excitation of the v_3 mode (temperature T_3). For the conventional values of $T_3 \approx 1600 - 2200$ K realized in an electric discharge the K_h gain is shown to achieve its maximum if $T_2 \sim 1/3T_3$ (Bertel et. al.., 1983). Such a relationship between T_2 and T_3 can be reached in gas mixtures with greater CO_2 and lesser He contents, as compared to those optimal for the regular band oscillation. In addition, an increased specific energy input is also required.

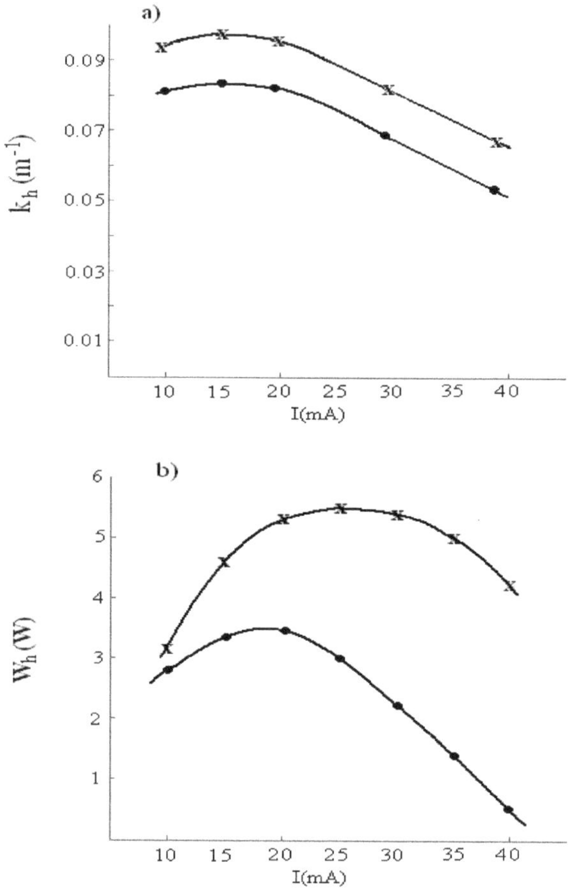

Fig. 2. A typical pattern of the hot band gain K_h (a) and output power W_h (b) as functions of the disharge current I for the CO_2:N_2:He - 1:1.4:3.5 (P - 11 Torr) mixture without Xe (•) and with optimal Xe content (0.3 Torr) (×).

It is known for the CO_2 laser regular band that addition of Xe to the active medium sometimes results in an output power increase (Gorobets et al., 1990). We have also investigated the influence of Xe on the characteristics of the active medium and the lasing parameters (Fig. 2). It has been found that small additions of Xe to the mixture (~ 30% of the CO_2 content) increase K_h by 25% and the lasing power in the hot band by a factor of 1.5. The analysis of vibrational temperatures shows that this is due to the increase in excitation efficiency of vibrations of N_2 and the υ_3 asymmetric mode of CO_2 in electric discharge [Fig. 3(a)]. Besides, using the reconstruction method of K_h, K_s and K_r gains we have found from the experimental values of T_3, T_2 and T that addition of Xe reduces the CO_2 molecule dissociation in the discharge [Fig. 3(b)]. This also results in an output increase. It is noteworthy that addition of Xe considerably improves the output parameters only for the low gain transitions. The study of the Xe effect on the laser output for the regular band where gain is fairly high (K_r≈0.6 m^{-1}) has shown that in this case an output power increase is not large (~15%).

After optimization of the gas content, pressure and discharge current we optimized the laser resonator. In the optimal case, the laser resonator was formed by a flat 100 lines/mm^{-1} grating and a totally reflecting concave mirror (R = 3m). The resonator length was 1.5 m. About 6% of the radiation was extracted through the grating zeroth order.

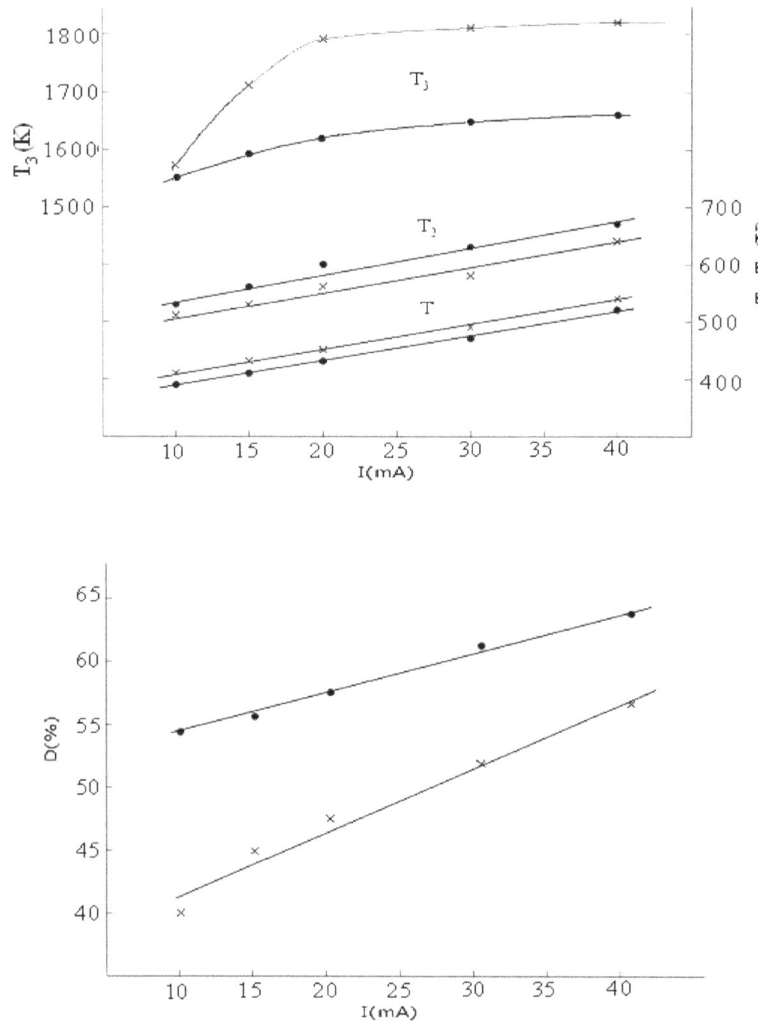

Fig. 3. Vibrational T_3, T_2 and transitional T temperatures (a) and CO_2 molecules dissociation degree D (b) vs the discharge current I for the CO_2:N_2:He - 1:1.4:3.5 (P - 11 Torr) mixture without Xe (•) and with optimal Xe content (0.3 Torr) (×).

Thus, after the above improvements the commercially available sealed-off laser (LG-22) (FSUE RPC "Istok", Fryasino, Russia) oscillates on more than 30 lines of the P-branch of the 01^11- 11^10 band in the 10.9—11.3 μm range with output power no less than 0.5 W. On strong lines (P(16)—P(26)) output power was ~ 6W at efficiency ~3% which makes up ~40% of analogous laser parameters in the case of oscillation on the lines of regular bands 00^01-10^00 (02^00) under optimum conditions.

3. New laser transitions of the CO_2 molecule in the wavelength range of 11.0-11.6 μm

Construction of powerful and efficient laser sources, generating in various IR ranges, is of importance for further development of a number of trends, e.g., spectroscopy, laser chemistry, isotope separation, sounding of the atmosphere, and metrology. The easiest and most natural way to solve this problem is to use unconventional transitions to produce lasing in commonly used CO_2 lasers (Churakov et al., 1987). The spectral range of CO_2 lasers is greatly increased in lasing on transitions of the so-called "hot" band 01^11-11^10, whose P-branch is in the range of 10.9-11.3 μm. Thorough investigations of gain, vibrational temperatures, and output parameters on lines of the hot band made it possible to achieve efficient lasing both for pulse TEA and for cw longitudinal-discharge CO_2 lasers.

In studying the lasing spectrum of hot transitions in TEA CO_2 lasers (Bertel at al., 1983). some lines not belonging to the 01^11-11^10 band occurred in the spectral range of 875-882 cm^1 that were not identified due to the poor resolution of the monochromator used and the lack of reliable spectroscopic data in the literature at that time. It was suggested, that these lasing lines belong to higher level transitions, e.g., $10^\circ1$-$20^\circ0$(04^00), which were called 'doubly hot," i.e., transitions in which compared to hot transitions two deformation quanta or one symmetric quantum rather than one deformation quantum is added both to the upper and to the lower energy level.

In the present work lasing in both a TEA laser and a low-pressure laser with longitudinal discharge on some transitions of the CO_2 molecule in the range of 11.0-11.6 μm is reported for the first time. The rather high resolution of the spectral equipment used and calculation of transition frequencies on the basis of recent spectroscopic constants made it possible to identify definitively the lasing lines obtained as belonging to the doubly hot bands 02^21-12^20 and $10^\circ1$-20^00 and the sequence hot band 01^12-11^11 (Fig. 1).

Within the scope of a commonly used model of vibrational temperatures (Gordiets at al. 1980, Smith and Thompson, 1981) let us analyze what gain coefficients for a weak signal can be realized for the aforementioned bands in electrical-discharge lasers. Estimates showed that to achieve a suitable gain at the new transitions it is necessary, together with the heating up of the asymmetric type of oscillations, characterized by the vibrational temperature T_3, to strongly excite the connected deformation and symmetric modes (T_2). Moreover for the sequence hot band 01^12-11^11 there is one additional condition - the excitation of the asymmetric mode must be at the same high level as for the sequence bands 00^02-10^01 ($02^\circ1$) (Petukhov et. al.., 1985).

To find optimum conditions for lasing on the aforementioned bands experimental studies of vibrational temperatures in active media of a TEA CO_2 laser and a low-pressure laser with longitudinal discharge were carried out using to the techniques described in (Bertel at al.,

1983, Gorobets et al., 1990). According to the measurements, mixtures of the composition CO_2:N_2:Ne = 1:1:1, in which with an increased specific energy contribution the gain coefficients for a weak signal are – 0.2 m^{-1} for doubly hot transitions and –0.3 m^{-1} for sequence hot ones, are optimum for TEA CO_2 lasers. The vibrational temperatures T_3 and T_2 must have values of –2000 K and –650 K, respectively. For low-pressure lasers with longitudinal discharge mixtures of the compositions CO_2:N_2:He:Xe = 1:1.2:2.5:0.4 (doubly hot bands) and 1:1.5:2.5:0.4 (sequence hot) are optimum. The gain coefficient in such mixtures for the aforementioned transitions can reach –0.04 m^{-1} (T_3–1800 K, T_2–600 K). It should be noted that at these transitions the gain is considerably lower than that at the ordinary (~20 times) and hot (~4 times) bands, and consequently a high-Q cavity, lack of harmful losses, and careful selection of the active medium and the conditions of its excitation are necessary to obtain lasing on transitions with such a low gain.

The lasing mode on the new transitions was studied first on a TEA CO_2 laser with UV preionization. The distance between electrodes that were 4 cm wide was 2 cm. The length of the discharge gap was 70 cm. The main charge and the UV preionization were energized from a battery of low-inductance capacitors with a total capacitance of 0.25 μF, charged to a voltage of 30 kV. The design and of the laser and its performance are described in detail in (Gorobets et al., 1995). A two-transmission three-mirror resonator was used to increase the length of the active medium to 140 cm. A planar grating with 150 lines/mm working in the first order according to an autocollimation scheme with a reflection coefficient not less than –90%, was one of the end mirrors of the resonator. Radiation from the resonator (~5%) was extracted through the zero order of the grating. The other two mirrors were spherical (R = 10 m) with a highly reflective coating. The active medium of the laser was a mixture of gases of the composition CO_2:N_2:He = 0.8:1.0:1.2 with a total pressure of 200 Torr, which is close to the optimum found from experimental investigations of vibrational temperatures. For this mixture lasing was achieved at more than 10 lines in new bands of the 11.3-11.6 μm range. At the strongest lines the energy in the pulse exceeded 150 mJ. The peak power with a pulse length at half-height of ~0.5 μs attained ~0.3 MW. More thorough investigations of the lasing spectrum of the new transitions were done in the present work for a low-pressure CO_2 laser.

Experiments were performed with a GL-501 production-type gas-discharge tube of an LG-22 commercial laser (FSUE RPC "Istok", Fryasino, Russia). The inner diameter of the tube–15 mm, length of the discharge gap –1.2 m. The tube, which worked in the sealed-off mode, was filled with a gas mixture of the composition CO_2:N_2:He:Xe = 1:1.2:2.5:0.4 under the total pressure of 13.5 Torr. The total reflection spherical mirror (R = 3 m) of the commercial laser was not replaced, and a diffraction grating, which worked according to an autocollimation scheme in the first order, was used instead of the output mirror. The emission was extracted through the zero order. The cavity base was 1.5 m. Most of the new lines were obtained with a grating with 100 lines/mm (reflection coefficient - 95%, extraction of emission - 3%). A number of lines in the range of 11.0-11.4 μm, where comparatively strong hot transition are located, were successfully obtained with a more selective grating with 150 lines/mm (93 and 3%, respectively). In addition, to increase the Q-factor of the cavity the germanium etalon was placed before the grating (perpendicular to the output radiation), which not only increased the Q-factor of the grating, returning 75% of the radiation back to the cavity, but also increased its selectivity substantially. As a whole, this device, consisting of a grating and an etalon, was a highly selective output mirror with a reflection coefficient of 97.5% for a grating with 100 lines/mm and 95.5% for a grating with 150 lines/mm. It should be noted, that this original technique made it possible to separate weak lines of the new transitions from closely positioned ones in some regions of the spectrum of stronger hot lines.

Line	λ_{meas}, μm	λ_{cal}, μm	Intensity, W	Line	λ_{meas}, μm	λ_{cal}, μm	Intensity, W
$02^{2}1$-$12^{2}0$				$10^{0}1$-$20^{0}0$			
P(12)	11.2513	11.251536	10	*P(14)*	11.0358	11.036728	10
*P(14)**	11.2729	11.273239	25	*P(16)*	11.0590	11.058304	10
P(15)	11.2834	11.284243	15	*P(18)*	11.0794	11.080312	15
P(16)	11.2944	11.295321	16	*P(20)*	11.1027	11.102758	20
P(17)	11.3064	11.306522	24	*P(22)*	11.1260	11.125648	20
*P(18)**	11.3173	11.317785	35	*P(24)*	11.1488	11.148987	16
P(19)	11.3300	11.329186	30	*P(28)*	11.1961	11.197036	20
P(20)	11.3410	11.340633	28	*P(30)*	11.2223	11.221757	17
P(21)	11.3517	11.352241	25	*P(32)*	11.2467	11.246953	20
P(22)	11.3634	11.363870	20	*P(34)*	11.2729	11.272628	25
P(23)	11.3754	11.375690	20	*P(36)*	11.2978	11.298791	20
P(24)	11.3872	11.387499	30	*P(38)*	11.3251	11.325449	10
P(25)	11.3991	11.399538	25	$01^{1}2$-$11^{1}1$			
P(26)	11.4117	11.411523	18				
P(27)	11.4241	11.423790	16	*P(23)*	11.0408	11.041722	25
P(28)	11.4357	11.435948	15	*P(24)*	11.0512	11.050460	18
P(29)	11.4494	11.448451	15	*P(25)*	11.0650	11.064261	25
P(30)	11.4596	11.460775	15	*P(26)*	11.0733	11.072789	15
P(31)	11.4730	11.473525	9	*P(28)*	11.0953	11.095487	30
P(32)	11.4855	11.486009	10	*P(29)*	11.1096	11.110543	30
P(33)	11.4981	11.499017	8	*P(30)*	11.1176	11.118556	25
P(34)	11.5112	11.511655	12	*P(32)*	11.1425	11.142003	28
P(35)	*11.5245*	11.524934	4	*P(33)*	11.1591	11.158465	20
P(36)	11.5375	11.537715	4	*P(34)*	11.1667	11.165831	30
				P(36)	11.1908	11.190045	30
				*P(46)**	11.3173	11.317088	35
$01^{1}1$-$11^{1}0$							
P(47)	11.3075	11.307393	40				
P(49)	11.3352	11.334278	35				
P(50)	11.3601	11.359850	40				
P(51)	11.3610	11.361589	18				
P(53)	11.3907	11.389331	20				

*The lines are not identified ambiguously (they may belong to the both lines).

Table 1. Measured and calculated values of wavelengths and experimental values of intensities for new transitions

The lasing spectrum in the range of 11.0-11.6 μm was studied in detail with the gas-discharge tube being energized from a pulsed source. It was found experimentally that the following pumping parameters are optimum for lasing at the new transitions: a pulse rate of

375 Hz, a length of the excitation pulse of ~5O μs, an average current of 8.5 mA. Under these conditions with careful adjustment of the diffraction grating and the etalon we managed to obtain more than *50* new lasing lines (see Table 1). Lasing wavelengths were measured with an SPM-2 monochromator (Carl Zeiss Jena, Germany) with a highly selective diffraction grating, whose resolution was not worse than 0.0005 μm. Absolute calibration of the monochromator was done using the technique described in (Gorobets et al., 1992), which is based on a search for a line with an anomalously high gain, e.g., the line *P(23)* of the hot band. In addition, correction calibration against known wavelengths of hot transitions was done on virtually the entire investigated spectrum. New lines were identified by comparing measured and calculated values of transition wavelengths. Calculations were done using standard methods. Values of the constants G, B, D, H were well known (Witteman, 1987).

The peak power (intensity) on the strongest lines of the new bands with a lasing pulse length at half-height of –*50* ns was – 30 W. The average output power reached –0.2 W. Lasing was achieved at a number of new transitions and in the continuous mode with the discharge tube being energized from a dc power supply. More than 25 new lasing lines with λ = 11.1-11.4 μm, belonging to all the aforementioned bands, were observed in this mode in the spectral range studied. The output power on strong lines attained 0.25 W.

The characteristics of the output radiation given in the present work are not the best attainable. Optimization of the active medium composition, the conditions of its excitation, and the cavity parameters will make it possible to increase the efficiency of lasing at the new transitions. However at present the large number of new lasing lines obtained substantially broadens the potentialities of simple laser systems on CO_2 for various applications.

4. Optimisation of a cw CO_2 laser output

4.1 Optimisation technique

To the present time the number of optimization methods of CO_2 laser power parameters is developed. However, the known methods are either complex, since they are based on the calculations calling for a knowledge of a great number of parameters or by virtue of sufficiently rough approximations, not always provide the necessary accuracy, as in the development of laser systems generating on the nonregular transitions - 00^02-10^01, 02^01 (sequence bands); 01^11- 11^10 (hot band); 02^21-12^20, 02^01-12^00 (double hot bands). The gain on these transitions is much weaker than on the regular transitions 00^01-10^00, 02^00 and hence a careful optimization of the active medium composition and of the resonator and pumping parameters is required to provide the lasing on them. Therefore until now remains to actual search new and perfecting of known methods of optimization CO_2 lasers.

We have developed and experimentally tested the method of optimizations of the cw CO_2 lasers energy parameters. To realize it, it is necessary to known the vibrational temperatures of the symmetrical (T_1), bending (T_2) and asymmetrical (T_3) modes of the CO_2 molecule vibrations. At the present time, the generally recognized fact is that the knowledge of these temperatures as well of the gas temperature (T) of the gas mixture makes it possible to determine all the most important characteristics of the active medium (population of the energy levels, the energy accumulated in different modes of CO_2, the efficiency of excitation, and so on). Next, the main energy characteristics of the laser system can be calculated based on the information about the temperatures of the medium.

In our works we used the method of determination of vibrational temperatures, which is based on the measurements of the gain on separate vibrational lines of regular and nonregular CO_2 bands (Petukhov et. al.., 1985). The advantages of this method are the possibility of determination of all vibrational and translational temperatures at once, a relative simplicity and sufficiently high accuracy as compared with other known methods. Besides, a knowledge of the absolute values of the gain factors and of the active medium composition is not needed here, which is sometimes very important.

It would appear reasonable that within the limits of the model of vibrational temperatures the output power (P) for every above-indicated bands is dependent only on temperatures. The experimental investigations performed by us show that for a typical low-pressure CO_2 laser with a longitudinal continuous discharge the active medium in the lasing regime differs significantly from that in the absence of lasing in only the value of the asymmetric vibration temperatures T_3, while for the other temperatures T_2, T_1 and T the difference is insignificant (less than 10 %). Such a temperature approximation is predominantly due to fact that the energy capacitance for vibrations of symmetric and bending modes of CO_2 is much greater than that for vibrations of the asymmetric mode as well as due to the constant effective heat abstraction from the low laser levels. This approximation may be thought of as by true for laser system with on efficiency of transformation of the energy contributed to the discharge to the lasing energy of -10 % or less percent, which is characteristic of all real continuous CO_2 lasers. In this case, using the ratio between the temperature of the asymmetric mode and average number of vibrational quanta accumulated in this mode we can write the following simple expression for the output power in every above-indicated band:

$$\varepsilon_3 = \exp\left(-\frac{h\upsilon_3}{kT_3}\right) \Big/ \left[1-\exp\left(\frac{h\upsilon_3}{kT_3}\right)\right], \tag{2}$$

$$P = A \cdot \frac{K_{loss}^{us}}{K_{loss}^{h} - K_{loss}^{us}} \cdot \left(\frac{\exp\left(-\frac{h\upsilon_3}{kT_3}\right)}{\left[1-\exp\left(-\frac{h\upsilon_3}{kT_3}\right)\right]} - \frac{\exp\left(-\frac{h\upsilon_3}{kT_3^*}\right)}{\left[1-\exp\left(-\frac{h\upsilon_3}{kT_3^*}\right)\right]} \right), \tag{3}$$

where A is the proportionality factor dependent on the CO_2 content and independent on the lasing band; $K_{loss}{}^{us}$ is the useful loss factor; $K_{loss}{}^{h}$ is the harmful loss factor; T_3 and $T_3{}^*$ are the vibrational temperatures of the asymmetric mode of the CO_2 molecule in the regime of amplification and lasing, respectively.

The temperature T_3 as well as T_2, T_1 and T can be found if the gain factors of the weak signal in different bands are known (Petukhov et. al.., 1985). To determine the temperature $T_3{}^*$ we will draw on the fact that in the regime of lasing (continuous) the gain factor is equal to the total loss factor:

$$K_g^*\left(T_3^*, T_2, T_1, T\right) = K_{loss}^{h} - K_{loss}^{us}, \tag{4}$$

Then, using the dependence of the gain factor on the difference in the population of the upper and low laser levels, expressed through vibrational temperatures. We can easily obtain expressions for determining T_3 for the above-indicated bands. For example, for the 00^02-10^01 band it has the form

$$\left[1-\exp\left(-\frac{hv_3}{kT_3^*}\right)\right]\cdot\left\{\exp\left(-\frac{hv_3}{kT_3^*}\right)-\exp\left(-\frac{hv1}{kT_1}\right)\exp\left(-\frac{hv_3}{kT_3^*}\right)\right\}=$$
$$=\frac{K_{loss}^{h}-K_{loss}^{us}}{K_g^s}\cdot\left[1-\exp\left(-\frac{hv_3}{kT_3^*}\right)\right]\cdot\left\{\exp\left(-\frac{hv_3}{kT_3^*}\right)-\exp\left(-\frac{hv1}{kT_1}\right)\exp\left(-\frac{hv_3}{kT_3^*}\right)\right\}, \quad (5)$$

Similar expressions are also true for other bands. Thus, if the all temperatures in the regime of amplification and the loss factors are known, it is an easy matter to calculate the output power for different bands. The temperatures T_3, T_2, T_1 and T can be determined from the measurement of the gain factor of a weak signal on the lines of different bands by the method described in (Petukhov et. al.., 1985).

4.2 Experimental setup

Figure 4 shows a block diagram of the experimental setup for CO_2 laser optimization. A sealed off a cw CO_2 laser was the source of probe radiation. It could be tuned over the vibrational-rotational lines of the regular (00^01-10^00, 02^00) bands, the sequence (00^02-10^01, 02^01) bands, the hot (01^11- 11^10) band or the new (02^21-12^20, 02^01-12^00....) bands.

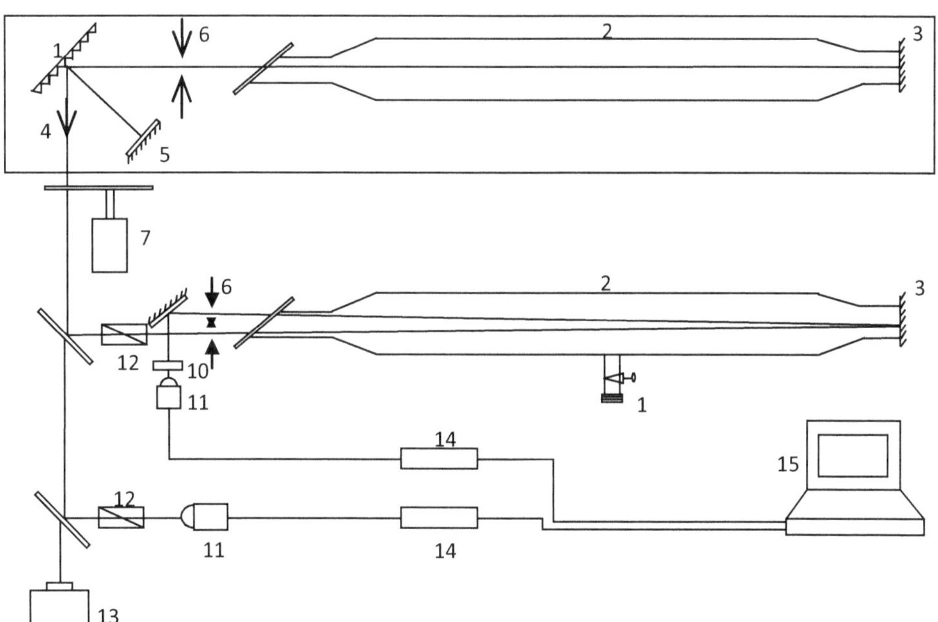

1 - probing laser; 2 - discharge tube; 3 - 100% reflection mirror; 4 - grating; 5 - additional mirror; 6 - iris diaphragm; 7 - chopper; 8 - ZnSe plane-parallel plate; 9 - mirror; 10 - interference filter; 11 - photo detector; 12 - polarizer; 13 - spectrum analyzer; 14 - ADC; 15 - computer; 16 gas valve.

Fig. 4. Experimental setup for CO_2 laser optimization

A production-type water-cooled sealed-off gas-discharge tube of GL-501 type (FSUE RPC "Istok", Fryasino, Russia) was used as an active element of the probe laser. It has discharge-gap length of ~1 .2 m, and the inside diameter is of 15 mm. The tube was filled with a CO : N_2 : He :Xe mixture in a proportion of 1.0:1.6:4.0:0.6 at a total pressure of *13.5* Torr. The laser cavity was formed by ~100% reflecting mirror with a curvature radius of 3 m built in the tube, a plane diffraction grating and an additional mirror with a large curvature radius.

We used a nonconventional scheme of the laser cavity. The diffraction grating operated in the first diffraction order in the nonLittrow scheme. Laser radiation was extracted from the cavity through the zero order. Our studies have shown that the diffraction grating with 150 lines/mm and a reflectance of >95%, combined with the additional mirror with a curvature radius of 10 m, are optimal for obtaining the necessary high spectral resolution with a sufficiently high output power. A more detailed description of the construction of the probe laser is given in the next part.

A signal from the probe laser passed a two times through the active medium under study in the discharge tube and was recorded by a liquid nitrogen cooled photo detector. This discharge tube was similar to one used as the active element of the probe laser. In addition to the measuring signal, we used a reference signal that does not pass along the investigated active medium and appears as a result of reflection of a portion of radiation from the ZnSe plane-parallel plate (see Fig. 4). This portion of radiation was directed to the another liquid nitrogen cooled Ge:Au photodetector of the reference channel, which makes automatically possible to account for the possible instability of the output laser radiation by way of normalization of the measuring signal to reference one.

The electric signals from two photodetectors were fed into an two-channel digital registration system on the base of PC. Lasing wavelengths were measured with SPM-2 spectrum analyzer (Carl Zeiss Jena, Germany) with a highly selective diffraction grating, whose resolution was not worse than 0.0005 μm.

4.3 Results and discussions

To test the method proposed we have performed experimental investigations and calculations of the output power (*P*) dependence on the discharge current (*I*) for a cw CO_2 laser, operated on four different bands (00^01-10^00, 00^02-10^01, 1^11- 11^10 and 02^21-12^20). The cw CO_2 laser was similar to one used as the probe laser. The only distinction is the using of the appropriate diffraction grating with optimum Q-factor for each band.

The temperatures T_3, T_2 and T (see Fig. 5a), used in the calculations, were determined from the measurement of the gain factor of a weak signal by the method described in (Petukhov et. al.., 1985). For our experiments T_1 is approximately equal T_2. According to our calculations the loss factors for different bands have the following values:

$$for\ P(18)\ \ 00^01-10^00-K^h_{loss}=4.8\times10^{-4}\,cm^{-1};K^{us}_{loss}=5.3\times10^{-4}\,cm^{-1},$$

$$for\ P(19)\ \ 00^02-10^01-K^h_{loss}=4.8\times10^{-4}\,cm^{-1};K^{us}_{loss}=5.3\times10^{-4}\,cm^{-1},$$

$$for\ P(19)\ \ 01^11-11^10-K^h_{loss}=2.3\times10^{-4}\,cm^{-1};K^{us}_{loss}=1.9\times10^{-4}\,cm^{-1},$$

$$for\ P(19)\ \ 02^21-12^20-K^h_{loss}=2.1\times10^{-4}\,cm^{-1};K^{us}_{loss}=1.1\times10^{-4}\,cm^{-1},$$

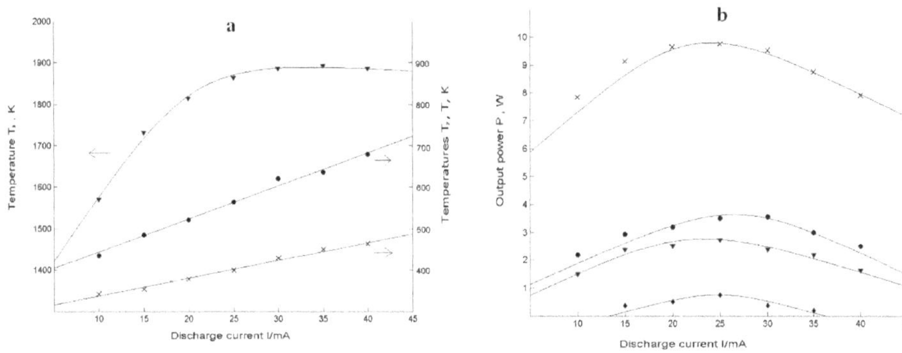

Fig. 5. Dependencies of temperatures T_3 (∇), T_2 (O), and T (×) (a) and output power P (b) from a discharge current (× - for P(18) $00^{0}1$-$10^{0}0$, O - for P(19) $00^{0}2$-$10^{0}1$, ∇ - for P(19) $01^{1}1$-$11^{1}0$, ◊ - for P(19) $02^{2}1$-$12^{2}0$)

Figure 5b shows the calculated curves and experimental values of the output power for the considered bands. At first we calculated the value of P/A for every band in accordance with discharge current. Then, we determined the proportionality coefficient A from the experimental data for the $00^{0}1$-$10^{0}1$ band at I=25 mA. As this takes place, the coefficient A has a common value for all lasing bands. Next, the dependence P on I was constructed.

The method of output optimization of cw CO_2 lasers has been developed. The method is based on vibrational and translational temperatures determination by gain measurements on the ro-vibrational lines of regular ($00^{0}1$-$10^{0}0$, $02^{0}0$) and nonregular ($00^{0}2$-$10^{0}1$,$02^{0}1$; $01^{1}1$-$11^{1}0$; ($02^{2}1$-$12^{2}0$, $02^{0}1$-$12^{0}0$...) bands of CO_2 molecule. To test the validity of the method, the experiment realization has been done for a low pressure CO_2 laser with the cw longitudinal discharge, that can oscillate on the lines of regular and nonregular lines. The good agreement between calculation and experiment data has been observed. Thus, a good agreement between the calculated and experimental data, which is observed, as a whole, for all the investigated bands, is demonstration that this method can be applied to the optimization of the output power of cw CO_2 lasers. This method can be also successfully used for the optimization of the output parameters depending on the pumping and Q-factor of the resonator of the lasers generating only on the regular transitions $00^{0}1$-$10^{0}0$ and $00^{0}1$-$02^{0}0$.

5. 16(14) microns TE CO_2 laser working on the $02^{0}0(10^{0}0)$-$01^{1}0$ and $02^{0}1(10^{0}1)$-$01^{1}1$ bands

Earlier the lasing on the $02^{0}0(10^{0}0)$-$01^{1}0$ band of the CO_2 molecule (see Fig. 1) has been obtained in the specific systems at cryogenic temperatures under the lowest efficiency (Wexler, 1987). The optimization of the active medium and its electrical discharge pumping conditions based on the original technique of the temperature model (gain measurements on the several bands: $00^{0}1$-$10^{0}0$, $00^{0}2$-$10^{0}1$, $01^{1}1$- $11^{1}0$ of CO_2 molecule) allowed to obtain in the simple TE CO_2 laser with UV preionization (the active media length is 65 cm the width of electrodes is 2.5 cm, the interval between electrodes is 1.8 cm, (the voltage on the 0.2 micro Farad capacitor is 6.5 kV) the powerful lasing on the $02^{0}0(10^{0}0)$-$01^{1}0$ bands at the room temperature. The output pulse energy of 57 mJ and the peak power of some tens kWatt have been reached. The dependencies of the output and spectral performances of the 16 (14) micrometers lasing vs. a content of the active medium, pumping parameters and cavity characteristics have been carried out.

To increase the power performances of the 16 (14) microns CO_2 laser the possibility of lasing on the $02^01(10^01)$-01^11 band have been experimentally and theoretically investigated under the combined (electrical + optical) excitation of the active medium. The conditions for obtaining effective lasing at the rotational-vibrational transitions of the 02^01-01^11 (λ = 16.4 μm) and 10^01-01^11 (λ= 14.1 μm) bands of the CO_2 molecule are examined. To obtain population inversion in the indicated channels one should initially populate the 00^02 vibrational level, considerable population of which can he accomplished comparatively simply, for example. in an electric discharge (Petukhov et. al.., 1985). Then a powerful two-frequency radiation resonant with the 00^02-$02^01(10^01)$ and 01^11-11^10 (03^10) transitions, saturating an individual rotational-vibrational transition, acts on the medium excited in such a way. As a result of this the first electromagnetic field, resonant with the 00^02-$02^01(10^01)$ transition, populates the upper laser level $02^01(10^01)$, while simultaneously the second field, resonant with the 01^11-11^10 (03^10) transitions, depopulates the lower level 01^11 which also leads to inversion of the populations in the 02^01 (10^01)-01^11 16(14) μm channel. In (Churakov et al., 1987). we have discussed by what means such a scheme of lasing in one active medium can be accomplished.

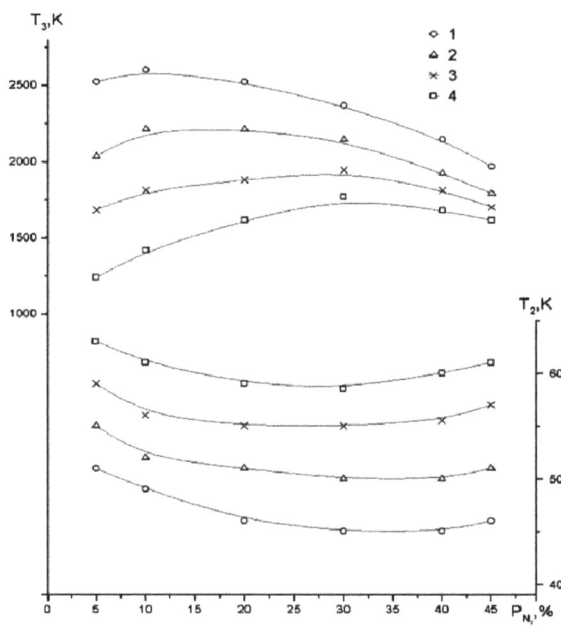

Fig. 6. Vibrational temperatures T_2 and T_3 vs N_2 content at fixed values of P_{CO2} 5(1), 10 (2), 20 (3), and 30% (4) in the mixture CO_2:N_2:He (P = 200 Torr, U= 30 kV, C = 0.25 μF)

Let us examine formation of inversion on the $02^01(10^01)$-01^11 transition using the example of an ordinary pulsed TEA CO_2 laser. For this we employ a temperature model which at the present time is widely used to describe processes occurring in the active media of CO_2 lasers and amplifiers. According to this model, population of the vibrational levels is unambiguously connected with the vibrational temperature of the symmetric (T_1), bending (T_2). and asymmetric (T_3) modes of the CO_2 molecule. We performed experimental investigations of the vibrational temperatures in the active medium of the TEA CO_2 laser, directed toward a search for the optimum conditions for lasing in the 16(14) μm channel. The vibrational temperatures T_3 and T_2 (T_1=T_2 for conditions examined) were determined from the ratios of the measured amplification coefficients of a weak signal at the individual rotational-vibrational lines of the 00^01-10^00, 00^02-10^01 and 1^11-11^10 bands by the procedure described early.

Let us examine what kind of the small gain and the output energy can be attained in the TEA CO_2 laser on the $02^01(10^01)$-01^11 transitions. On the basis of the experimentally determined vibrational temperatures T_3 and T_2 (see Fig. 6) using the well-known expression (Gordiets et al., 1980) we calculated the small gain. The calculations shown that the small gain in the $02^01(10^01)$- 01^11 band can attain a significant value (>$1m^{-1}$). The necessary conditions for the effective lasing have been determined. It is shown that in optimum conditions the output energy can reach *1.3* J/l at the peak power 5 MW and at the full efficiency of 2 %.

6. A stabilized cw CO_2 laser automatically switched between generations lines

This part describes a cw CO_2 (CO) laser with stabilized output parameters that can be automatically switched from line to line. The laser generates 115 vibration-rotation CO_2 lines between 9.15 and 11.3 μm and 100 CO lines between 5.3 and 6.4 μm. The laser is switched from CO_2 operation to CO operation by replacing a sealed laser tube. Then computerized control of the laser spectrum is described.

Although there are many publications on tunable lasers (Gorobets et al., 1992) it is premature to think that all design and operation problems of tunable CO_2, and especially of CO lasers, have been resolved. Computer control over the tuning of the generation wavelength is required (Gorobets et al., 1992). Fully computerized CO_2, and CO lasers could be extensively used to monitor active media to improve lidar systems, in the spectroscopy and analysis of gases.

We have described the design of a laser head with a sealed tube and separate units (a high-voltage power supply, unit for tuning the lasing wavelength, an AFT unit, and a modulator) of an actively stabilized a cw CO_2 (CO laser) that can he automatically switched between generation lines. The laser is switched from CO_2 to CO operation by replacing the discharge tube.

Further improvements include an electro-mechanical drive for the diffraction grating, an electronic control unit compatible with various computers, interfaces, and a control algorithm to link the laser to the computer.

6.1 Laser structure

The laser structure is shown in Fig. 7. The GL 501 (CO_2) or GL-509 (CO) (FSUE RPC "Istok", Fryasino, Russia) commercial discharge tubes 1 are used because they have similar structures and the same discharge distance - 1.2 m. The laser cavity is formed by a 100% end mirror *2* with a curvature radius of ~3 m that is integrated in the tube and a flat diffraction grating *3* with 100 lines/mm set on a rotating arm 4 and a PZT drive *5*. The grating reflects in the first diffraction order when operating with CO_2 and in the second order with CO (the reflectivity is about 90% in both cases). There is an additional mirror 6 on the rotating arm which forms a corner reflector (Gorobets et al., 1992) to keep the direction of the output beam unchanged when the laser is switched from line to line. This optic layout is particularly suited to the laser tubes we were employing. Besides, when the output beam has zero-order reflection at the grating, the optic losses are much lower, and more laser lines can he generated, which is essential for CO tubes whose gains are relatively small.

The diffraction-grating rotating arm and the tube braces 7 are fixed on three invar rods *8*. The end plate *9* is rigidly fixed to the laser frame, and unit *4* is attached to the lower rod via two bearings. As a result, thermal expansion in the rods does not lead to any misalignment of the grating. The use of invar rods and the high rigidity of the structure leads to a good passive stability of the cavity length.

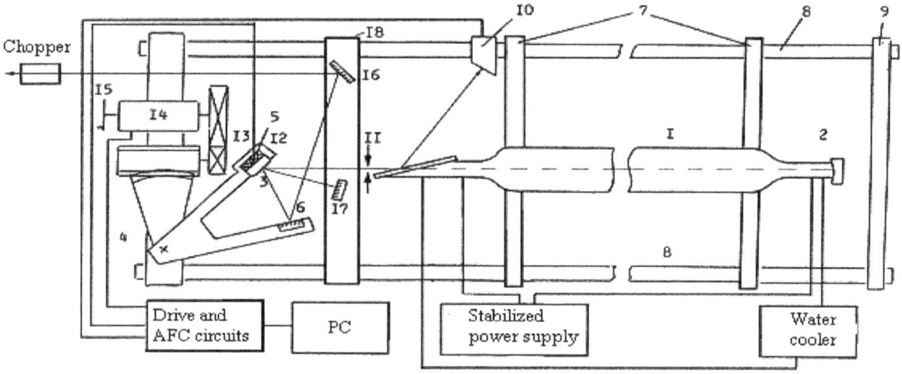

1 - laser tube; 2 - end mirror; 3 - diffraction grating; 4 - rotating arm stage; 5 - piezo-electric drive; 6 - turning mirror; 7 - laser tube braces; 8 - invar rods; 9 - end plate; 10 - pyro-electric detector; 11 - iris diaphragm; 12 - holder of the diffraction grating; 13 - gear-box with a worm wheel; 14 - stepper motor; 15 - graduated wheel; 16 - turning mirror

Fig. 7. Diagram of the laser structure

The radiation frequency is stabilized by coupling the output power to the wavelength and an appropriate curve is used in the stabilization system. The laser is in fact stabilized by automatically tuning the cavity length with the PZT drive *5* to which the diffraction grating *3* is fixed. The AFC circuit is similar to that of the Edinburgh instruments lasers. The signal for feedback loop is taken from a pyro-electric detector 10 which is exposed to the radiation reflected from the GaAs Brewster window of the discharge tube (see Fig. 7). The laser only generates the fundamental transverse mode because of the iris diaphragm 11 and the AFC system.

The key element in the laser is the system for tuning the lasing wavelength. Structurally, it is the rotating arm *4* of the diffraction grating and driven by the electronics driving the AFC system. The laser is switched between the lines by turning the grating with respect to the cavity axis.

The rotating arm *4* comprises the grating holder *12* which aligns the grating in vertical and horizontal planes, a two-stage reduction gear-box *13* with a worm-wheel (the transfer ratio is 1/1620), and a small stepper motor *14*. The precise reduction box, which has split gears, can rotate the grating and the corner mirror 6 through 40° and can set the grating angle to within 10". The laser spectrum can be also tuned manually with a calibrated wheel 15. When active, the control unit feeds pulses to the stepper motor, and the counter displays the number of motor steps on a front panel. The turning rate can be set to between 50–500 step/sec (up to 5 lines/sec). It can be also made to go in single steps. The motion of the arm is limited in both directions by limit switches.

An initial-state indicator with a low-voltage spark discharger acts as one limit switch and was designed to set the rotation arm in the starting position. At a constant voltage of several volts, the spark discharge in air occurs at a very small gap width (~1.5 μm) and, hence, the discharger generates a signal. The uncertainty the gap width at which the discharger generates a signal is less than 1 μm, with corresponds to a grating turn of ±4". Thus the reproducibility and accuracy of the initial position is good and corresponds to one step of the system. Note that an error of one step is not significant when switching the laser to a particular line since the distance between neighboring lines in CO_2 and CO lasers in terms of the grating turning angle is ~ 240", which is equivalent to 60 steps of the motor.

The laser tuning system can be linked to computers through appropriate interfaces and software. The turning angles of the diffraction grating with respect to the cavity axis corresponding to each laser line are leaded into the computer. The turning angles in terms of motor steps are derived from the grating pitch. When the laser is operating, the calculated angles may differ from the real values by some constant. This difference may be due to composition variations of the gas in the tube as a result of the electric discharge. Even so the intervals between neighboring spectral lines, and hence the distance between the grating positions in terms of the turning angle remain unchanged. The correction to the calculated positions should been found experimentally for all lines in terms of motor steps. This experimental correction also takes into account the uncertainty of the position of the zero-angle discharger.

The correction can be determined in two ways using the laser tuning software. The first method is semiautomatic and requires an external spectral device (a monochromator or a gas cell with a known absorption spectrum, e.g., NH_3). The position of the grating for the selected reference line is determined, and the difference between this value and the calculated posit ion is fed into computer as the correction. In the second method, the correction is determined automatically by finding a reference line without an external spectral reference (Gorobets et al., 1992). The correction is determined using bright lines which can be easily identified in the output spectrum. In the case of the CO_2 laser a good line is P(56) in the 00^01-10^00 band, which coincides with the P(23) line of the 01^11-11^10 band. The algorithm for finding this line was described in (Gorobets et al., 1992) and it reliably determines the correction. A similar method is possible for the CO laser.

Once the experimental correction is found, the laser is tuned to the selected line. The correction is added to the angle corresponding to the selected line, and the computer moves the grating to the correct angle with respect to the previous position, then it activates the AFC system.

The temperature of the liquid cooling the laser tube should be kept constant. This is particularly important for the CO laser since the number of lasing lines, especially in the short-wave band, depends on the gas temperature in the tube (Aleinikov and Masychev, 1990). We used a standard water cooler with a closed cycle to remove the heat from the laser tube. It cools the tube with distilled water at a temperature between 2 and 10 °C and keeps it constant to within 1 °C.

The stabilized power unit is standard for CO_2 lasers and an additional current stabilizer built around a vacuum tube. The current stabilizer suppresses current oscillations by several orders of magnitude, especially those at the mains frequency of 50 Hz. The current through the tube can be tuned between 10 and 40 mA.

To modulate the laser power, we used a electromechanical chopper. It was built around a electric motor. A thin precisely made disk with sixteen slits made from titanium foil 0.1 mm thick was mounted on the motor axis. The signal for the feedback of the active frequency control was taken from an optic couple. The electronics drive the modulator at 125, 250, 500, and 1000 Hz, and it can be detuned by ±5.9% from these frequencies. With a crystal oscillator and automatic control of the modulation frequency its very stable (the frequency usually differs from the preset value by less than 0.01%).

6.2 Output laser parameters

The laser characteristics have been measured on an optic bench using traditional techniques (Gorobets et al., 1995). We first consider the spectral and energy parameters.

Since the diffraction grating has a reflectivity of 90% the laser with a CO_2 tube, generates about 90 lines between 9.15 and 10. 95 μm (00^01-10^00, 02°0 bands) The output power in the fundamental mode reaches 10 W for strong lines and is over 1 W side lines. Using the same grating, the laser generates 25 lines in the P-branch of he hot 01^11–11^10 band of the CO_2 molecule. In this case the spectrum is shifted to the red to 10.94 –11.25 μm. The output power at strong lines was 3-4 W and 0.5 W at the band edge. However the conditions needed for the hot baud, where the gain is considerably smaller, are not optimum. When the laser tube is filled with CO_2, N_2, He and Xe, output was considerably higher and, the number of lines was larger (see 1-3 parts).

When the CL-509 tube is inserted, the laser efficiently generates about 100 vibration-rotation lines of the CO molecule between 5.28 and 6.43 μm. Output powers at the strongest lines were ~1 W in the fundamental mode at the optimum discharge current. The lines were identified using the data in (Aleinikov and Masychev, 1990).

Note that the parameters of the grating are better when generating CO and hot of CO_2 lines, where the gain is smaller than in the more conventional 00^01-10^00,02^00 bands. The lines in the conventional bands will clearly be stronger in a cavity with a lower Q factor.

6.3 Instability of the output laser power

The long-term instability of the output power was checked using a laser calorimeter whose signal was fed to a chart-recorder. Figure 8 shows a typical plot of the laser output power over one hour. The instability in the laser output power on the P(16) line of the 00°1-02°0 band over one hour was ±1.1%. Similar measurements with other lines of CO and CO_2 molecules demonstrated that the long-term instability of the laser power is less than ±1.25%. However the time constant of the calorimeter is long and the short-term instability could not be monitored. The instability over the measurement time needs to be known in many applications. For example, when determining contaminants by the differential method (at the absorption line and off the line), the measurement may last from several seconds to a minute. The output power instability in this time interval was measured as follows.

The laser beam was modulated by the chapter and fed to a light detector cooled by liquid nitrogen (Gorobets et al., 1995). The detector's electric signal was processed by a lock-in amplifier, digitized and sent to a computer. One measurement, including the signal processing in the ADC took about 0.7 s. Measurements lasting over 70 s demonstrated that the output power of the laser generating at the P(24) line of the 00^01-10^00 band of the CO_2 molecule varied by ±1.6% around the mean with the AFC system on and by 12.5% with the AFC system off. The larger instability measured by the second method may be due to longer time constant of the calorimeter. The short-term power instability was also measured using an digital oscilloscope (band-width of about 1MHz) connected to the Ge:Au detector. The instability over times of the order of microseconds was estimated to be several times smaller than the long-term instability quoted above. Several lasers have been used to monitor the atmosphere and to obtain spectroscopic measurements over a long time. They have proven to the reliable devices with a long service life.

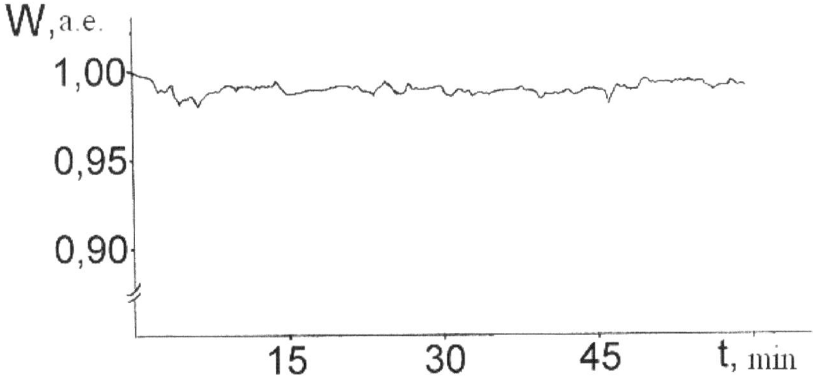

Fig. 8. Trace of the output power over 1 h at the P(16) line of the CO_2 molecule's 00^01-02^00 band

7. Detection of small N_2O concentrations using a frequency doubling $^{12}C^{18}O_2$ laser

The destruction of the protective ozone layer of the Earth (so called "ozone holes") can result in a global environmental and climatic catastrophe showing for many years a continuous unflagging. It is well known that the products of human activity such as freons and nitrous oxide (N_2O) are responsible for "ozone holes". Freons appear as a result of manufacturing some kinds of plastics and using refrigerators. Nitrous oxides penetrate into the atmosphere primarily due to microbiological changes in soil caused by agricultural human activity. Moreover, (Crutzen, 1996) determined that there is a direct coupling between the life of microorganisms in soil and the ozone layer thickness.

The conservation of the ozone envelope depends on many factors. However it is beyond doubt that modern reliable techniques monitoring the atmosphere for the presence of freons and nitrous oxide would assist greatly in a solution of this serious global problem. Laser gas-analysis methods are well suited to this task. They are capable of working with high speed, i.e. practically in real time mode. The ability to determine extremely low gas concentrations (for laser photoacoustics on the level of 0.1 ppb) and to cover extensive areas of the earth from a single point of observation (for lidars - about 10 km) give them unquestionable advantages as compared to other known diagnostic methods. There is reliable and effective laser procedure based on CO_2 laser for the detection of prevailing freons, the strong absorption bands of which overlap with emission lines of the laser.

Spectral analysis of N_2O has shown that the characteristic feature of this molecule consists of the absorption now low in the ranges where known effective lasers can operate. There are only the complex and (or) inefficient multitasked systems with nonlinear frequency conversation (generation of harmonics with the subsequent frequency summation), parametric oscillators, and tunable diode lasers. Therefore, the development of reliable and efficient laser methods for N_2O sensing remains a topical problem. An additional difficulty arising with the development of such methods applies to the necessity to detect low concentrations of nitrous oxide (background content of this gas in the atmosphere is 0.2—0.4 ppm).

The main goal of the present investigation is the development of a reliable and high-efficiency laser method for detecting low concentrations of N_2O. The other goal of the work is the test of this method as a remote gas analyzer. The procedure is based on the idea of using a nonlinear frequency-doubled CO_2 laser operating on the isotopic carbon dioxide modification $^{12}C^{18}O_2$. Such a powerful laser system would emit neighboring lines both coinciding well and adjacent to N_2O absorption lines. This fact allows one to apply the highly accurate technique which uses corresponding on/off line pairs for the differential absorption. The high efficiency of the system and strong absorption of N_2O molecules (we use the strongest band in the range of $\lambda \sim 4.5$ μm) would give a possibility to measure low gas concentrations both in short and long (~10 km) measurement paths.

7.1 Active medium optimization of the $^{12}C^{18}O_2$ laser

It is known that the use isotopically substituted carbon dioxide molecules make it possible to increase substantially the number of lines and to extend the spectral range of CO_2 lasers. That is important for different applications, in particular for atmospheric gas detection. The use of $^{12}C^{18}O_2$ as molecules of the active medium of lasers is of special interest, since for this molecule the maximum gain lies at wavelength ~9.4 μm and not ~10.6 μm, as for $^{12}C^{16}O_2$, and, consequently. there is a possibility for efficient lasing in a shorter wavelength range down to 8.9 μm. The doubled emission frequencies of $^{12}C^{18}O_2$ laser well coincide with absorption lines of many molecules including nitrous oxide.

However, for a number of reasons, and particularly because of the much higher price, CO_2-lasers based on isotopically substituted carbon dioxide molecules are not in wide use. For $^{12}C^{18}O_2$, molecules there is also the problem of the isotoporeplacement of $^{18}O_2$ with $^{16}O_2$ as these molecules are active in discharge plasma. The electrode surface, discharge chamber and tubes walls accumulate with time ordinary oxygen $^{16}O_2$. Then, under the discharge conditions, $^{16}O_2$ replaces (isotopically) $^{18}O_2$, in the active medium. This results in rapid degradation of the $^{12}C^{18}O_2$ active medium. This fact is especially important for TEA CO_2-

lasers the discharge chambers of which are much more difficult to pump out as compared to low-pressure sealed-off lasers. Another problem is the necessity of proper choice of materials which accumulate less ordinary oxygen. Therefore, we have performed detailed investigations aimed at active medium optimization for high-energy parameters with a simultaneous decrease in the price of the active medium based on isotopically substituted molecules $^{12}C^{18}O_2$ at the expense of its dilution with inexpensive carbon dioxide $^{12}C^{16}O_2$.

This section gives the results of our spectral investigations of the gain and the lasing for the TEA CO_2 laser operating both on $^{12}C^{18}O_2$ and, just for comparison, on ordinary $^{12}C^{16}O_2$. The analysis of the conditions required for efficient lasing in the range of 9 μm is given too.

Experiments were performed with a UV-preionized TEA module specialty developed for lidar systems (Gorobets et al., 1995). The module had a working volume of 70 × 2.5 × 2 cm. The distance between electrodes was 2 cm. Both main and auxiliary discharges were fed from low-inductance capacitors having a total capacity of 0.2 μF charged up to the 25 kV voltage. The discharge duration was ~500 ns.

The isotopically substituted form of carbon dioxide $^{12}C^{18}O_2$ with an ^{18}O enrichment factor of 80% obtained as a gas mixture containing 4% $^{12}C^{16}O_2$, 32% $^{12}C^{16}O^{18}O$, and 64% $^{12}C^{18}0_2$ was used in the experiments. It is much more expensive to prepare a mixture with a higher factor of enrichment for $^{12}C^{18}O_2$. The first measurements and calculations concerned the gain. For example, Fig. 9 shows respective gain for some lines of the P-branch of 00°1-10°0 band of the $^{12}C^{18}O_2$ molecule (λ~9.4 μm) and of the $^{12}C^{16}0_2$ molecule (λ~10.6 μm) for the mixture $^{12}C^{16}O_2$: $^{12}C^{16}O^{18}O$: $^{12}C^{18}O_2$: N_2 : He = 10:3:7:20:60 (the manufactured mixture was diluted with $^{12}C^{16}O_2$) with a total pressure of 500 Torr. Measurements carried out at t = 4 μs after the start of the discharge when the highest gains were realized. Gain measurements were performed by probing the active medium with a cw laser, lasing on the corresponding lines.

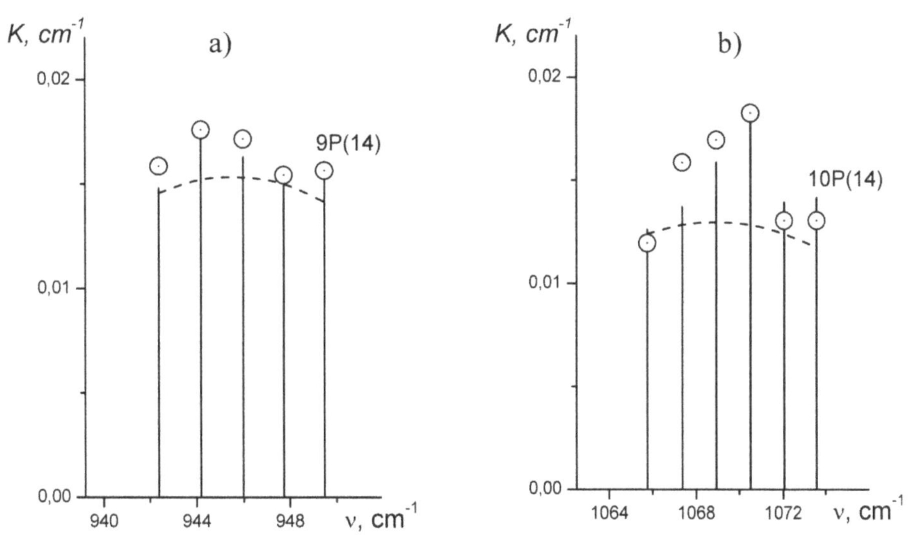

Fig. 9. Measured (O) and calculated (|) values of the gain at the lines of the 10P($^{12}C^{16}O_2$) (a) and 9P($^{12}C^{18}O_2$) (b). Dashed lines - the gain calculated without lines overlapping.

After passing the medium in question, the probing emission came to a photodetector producing a signal to a digital oscilloscope, connected to a personal computer where the measurements were stored and averaged. The calculations were performed based on a vibrational temperature model (Petukhov et. al.., 1985) that is in wide use for simulations of the active medium of CO_2, lasers. The measurements have shown that under the above conditions: $T= 350 \pm 10$ K, $T_1 = T_2 = 420\pm20$ K, and T_3=1475, 1510 and 1550, ±15 for molecules $^{12}C^{16}O_2$, $^{12}C^{16}O^{18}O$, and $^{12}C^{18}O_2$, accordingly. T designates the gas temperature. T_1 and T_2 are the temperatures of symmetric and deformation vibrations that are virtually the same for the whole scope of carbon dioxide variations in question. T_3 is the temperature of asymmetric vibrations. The technique used and apparatus to measure gains and to determine vibrational temperatures were described early. The gain spectrum was calculated using the scheme described in (Gordiets et al., 1980). Additionally the overlap of individual vibration– rotation lines in bands 00^01-10^00, 00^01-02^00, 00^02-10^01, 00^02-02^01, 01^11-11^10, 01^11-03^10 for molecules $^{12}C^{16}O_2$ and $^{12}C^{18}O_2$ and bands 00^01-10^00 , 00^01-02^00 for molecule $^{12}C^{16}O^{18}O$ and spectroscopic data given in (Witteman,.1987) were allowed for.

As is evident in Fig. 9, the gain maximum in the P-branch of the band 00^01-10^00 of the $^{12}C^{16}O_2$ molecule in the mixture of isotopic forms of carbon dioxide $^{12}C^{16}O_2$, $^{12}C^{16}O^{18}O$ and $^{12}C^{18}O_2$, occurred at lines P(18) and P(20). These lines are especially affected by the overlap with lines from other bands. The P(18) line of the 00°1-10°0 band for the $^{12}C^{18}O_2$, molecule is shifted from lines P(3) 00°1-02°0 $^{12}C^{16}O^{18}O$, R(8) 00°1-02°0 $^{12}C^{16}O_2$ and P(26) 01^11-11^10 $^{12}C^{18}O_2$, at 0.046, 0.045 and 0.021 cm^{-1}, and line P(20) 00^01-10^00 $^{12}C^{18}O_2$-from lines P(5) 00°1-02°0 $^{12}C^{16}O^{18}O$, R(6) 00°1-02°0 $^{12}C^{16}O_2$, and P(28) 01^11- 11^10 $^{12}C^{18}O_2$ at 0.007, 0.072, 0.085 cm^{-1}, respectively. The homogeneous width of the line at half-height under our conditions is ~0.05 cm^{-1}. We note that the calculation and experimental data are in a good agreement. For instance, the gain calculated for line P(18) 00^01-10^00 $^{12}C^{18}O_2$ is 1.83×10^{-2} cm^{-1} while the measured value is $(1.81 \pm 0.06) \times 10^{-2}$ cm^{-1}

It's also worth noting that higher gains are realized for the $^{12}C^{18}O_2$ molecule than for the $^{12}C^{16}O_2$ molecule. As seen from Fig. 9, the gains on the strongest lines of the P-branches for both isotopes are approximately equal, though this mixture contains 1.8 times less $^{12}C^{18}O_2$ than $^{12}C^{16}O_2$. There are three reasons for this. First, the limiting gain for the lines of the P-branch of the band 00^01-10^00 of $^{12}C^{18}O_2$ is approximately 1.5 times higher than that for the P-branch of the band 00^01-10^00 of $^{12}C^{16}O_2$ (Witteman,.1987); second, for the lines of the P-branch of this band of the $^{12}C^{18}O_2$ molecule in the $^{12}C^{16}O_2$+$^{12}C^{16}O^{18}O$ mixture the effect of the overlap with other bands is essential; and, finally, under the conditions of dynamic equilibrium between the v_3 modes of both isotopic forms the temperature T_3 for $^{12}C^{18}O_2$, is higher than for $^{12}C^{16}O_2$.

We also measured the effect of the dilution of molecules $^{12}C^{18}O_2$ by the usual form of carbon dioxide $^{12}C^{16}O_2$, on the output energy. The lasing pulse energies measured in the ranges of 9.4 μm ($^{12}C^{18}O_2$) (1) and 10.4 μm ($^{12}C^{16}O_2$) (2) are given in Fig. 10. These data were acquired for the TEA-module described above using a non-selective cavity composed of a non-transmitting and germanium output (R=~50%) mirrors. The energies emitted in the ranges 10.4 and 9.4 μm are approximately equal to each other at the 30 Torr content of $^{12}C^{18}O_2$, approaching to the content of $^{12}C^{18}O_2$ at which the corresponding gains get equal. Also note that the total emission energy ($E_{9.4} + E_{10.6}$) is independent in this case of isotoporeplacement of carbon dioxide and reaches the value of 5 J.

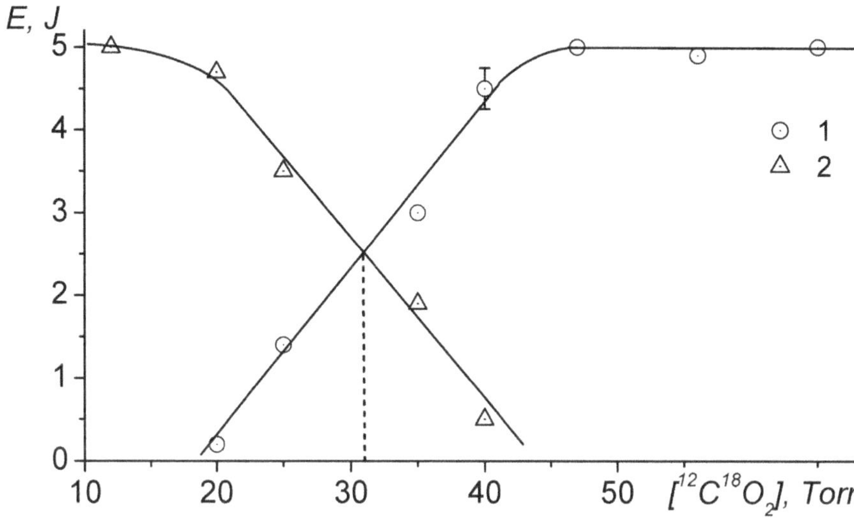

Fig. 10. Lasing energy measured in the ranges of 9.4 μm for $^{12}C^{18}O_2$ (O) and 10.4 μm for $^{12}C^{16}O_2$ (Δ) versus $^{12}C^{18}O_2$ content

Thus, the results of the investigations demonstrate that efficient lasing of the CO_2 laser even with non-selective cavity is possible in the range of 9.4 μm when the $^{12}C^{18}O_2$, content is 30% of total carbon dioxide amount. When the laser operates with selective cavity such as that based on a diffraction grating, the percentage of $^{12}C^{18}O_2$, can go down to ~20%. This fact is key for reduction of the price of the active medium.

An experimental series performed as well by us was oriented toward providing long-duration autonomous laser operation without active mixture replacement and noticeable degradation of its composition. To this end, the discharge chamber was well evacuated (no more than 0.2 Torr/day inleakage) and the material it was made of was properly selected. The best results were achieved with a glass-epoxy cylinder when the operation was virtually quasi-sealed-off, i.e. without replacement of the active medium on $^{12}C^{18}O_2$ during 1–2 months (more than 10^5 pulses) without noticeable decrease of the laser energy. In addition, after long operation and before working gases replacement. the old mixture was pumped through liquid nitrogen traps to recover carbon dioxide for repeated use.

7.2 Using a nonlinear crystal etalon for second harmonic generation from CO_2 lasers

The poor efficiency of the frequency conversion attributable on the whole to mid IR lasers can be compensated to a large degree by application of non-traditional optical schemes. Therefore, the problem remaining topical in the mid IR range along with creation of higher-quality crystals applies to development of novel nonlinear conversion schemes and search for the laser operation modes which are optimal for frequency conversion. To this end and with the aim of reaching high conversion efficiency, we have performed some investigations on the basis of which it was possible to realize original optical frequency conversion schemes including intracavity versions.

In this work we used an original high-efficiency optical nonlinear conversion scheme without focusing optics developed by ourselves. An $AgGaSe_2$ nonlinear crystal was acting as an Fabry-Perot etalon. In this case it could be placed in a laser cavity without reflection-reducing coating. However sonic problems connected with the spacing of its Fabry-Perot transmittance bandwidths and laser cavity modes can arise from it. To clarify these problems some calculations have been made (see Fig. 11).

The line shape function in Figs. 11a and 11b were calculated using the Foight and Lorentz expression, correspondingly. The intervals between the longitudinal modes for the lasers were calculated according with the condition *c/2L*, where *L* is the cavity length and c is the velocity of light. The curves in Fig. 11c are the Fabry-Perot etalon transmittance bandwidths calculated for different temperatures according with Airy's Formula for the etalon made from the $AgGaSe_2$, crystal with the length of 17 mm.

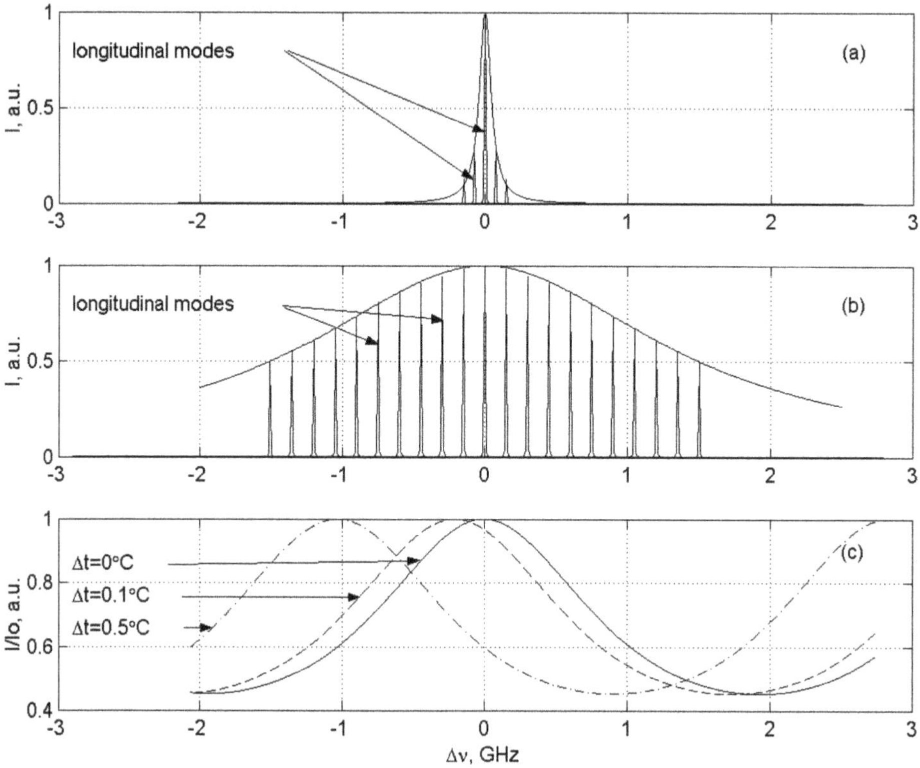

Fig. 11. Line shape function for a low-pressure longitudinal-discharge CO_2 laser (a) and for a TEA CO_2 laser (b) and a transmittance bandwidth of the $AgGaSe_2$ crystal acting as an Fabri-Perot etalon for different temperature variations (c).

In spite of the fact that the pressure broadened line width strongly differs for the TEA CO_2-laser and the low-pressure longitudinal-discharge CO_2 laser. For the first the full width at half maximum of the line shape function is about 3 GHz, and for the second ~0.1 GHz. At the fixed temperature the etalon practically does not influence on oscillations of both lasers, as its bandwidth is much wider than longitudinal modes and comparable to the pressure broadened line width for the TEA laser. Therefore, it is easy to achieve a concurrence between maxima of the line shape function (for TEA and low-pressure longitudinal-discharge lasers) and transmittance bandwidth of the Fabry-Perot etalon by a small angular tuning of the crystal.

However in real conditions a nonlinear crystal has a temperature drift due to the $d(nl)/dt$ thermal expansion. For the used $AgGaSe_2$ crystal according with the dates of Clevelend Crystals Inc. the factor of linear expansion (α) is 15x10^6 °C^{-1} and the thermo-optical factors (dn_0/dt and dn_e/dt) are ~50x10^{-6} °C^{-1}. Our calculations (see Fig.11c) demonstrate the rather strong temperature drift of the crystal. It is especially important in the case of crystal operating for a low-pressure longitudinal-discharge CO_2 laser. In our experiments the crystal was supported by a massive metal holder heated by a thermoelement and it was stabilized with accuracy <0.1°C. Besides it should be noted, that in our experiments the crystals having, the very small absorption (~0.01 cm^{-1}) were used. This fact essentially simplifies process of temperature stabilization. Possible changes of the cavity losses because of the crystal temperature drift for a low-pressure longitudinal-discharge laser did not influence strongly on its output from the fact that we used a pulse-periodic regime of the lasing.

7.3 Second harmonic generation from a TEA $^{12}C^{18}O_2$ laser

The optical crystal was used as a nonlinear output mirror of the TEA $^{12}C^{18}O_2$, laser. The cavity of the laser was formed by a 150 line/mm grating and an $AgGaSe_2$ nonlinear crystal. A plate made of LiF was used to select second harmonic emission generated in the nonlinear crystal. A high-quality monocrystal sample made of $AgGaSe_2$, with a 12 x 10-mm section (L = 19 mm) was used as a nonlinear output mirror of the laser. The working faces of the crystal were mechanically polished and were not coated. The highly parallel faces (better than 10") caused the sample to operate as a Fabri-Perot etalon. The angle Θ (phase matching angle) was adjusted near 46°, and φ = 45°. The angle Θ is such that the highest efficiency of the second harmonic oscillation is observed at a normal incidence at line 9P(32) (λ = 9.06 μm) and at the neighboring line 9P(34) (λ = 9.05 μm) of the isotopic modifications $^{12}C^{18}O_2$ of carbon dioxide.

The cavity length was 1.1 m. Before the output mirror ($AgGaSe_2$ crystal) there was in the cavity an iris diaphragm (diameter ~8 mm). In the case when the TEA module was filled with the mixture $^{12}C^{16}O_2$: $^{12}C^{16}O^{18}O$: $^{12}C^{18}O_2$: N_2 : He = 104 : 32 : 64 : 200: 600 at a total pressure 500 Torr and a nonlinear crystal was used as an output mirror (~60% reflection), the output energy at lines 9P(32) and 9P(34) of carbon dioxide isotope $^{12}C^{18}O_2$, was ~0.8 J while the peak power was ~4MW. The diameter of the output beam was ~7 mm. The lasing spot had a good spatial distribution. The energy density of the output emission was ~2 J/cm^2 (~10 MW/cm^2).

Under above conditions the second harmonic generation energy ($E_{2\omega}$) was 52 mJ and the peak power ($P_{2\omega}$) was ~2 MW. The conversion efficiency reached almost 15%, and at the peak power it was ~50%. The efficiency was calculated by the standard method ($\eta = E_{2\omega} / E_\omega$ – *for* the per pulse energy and $\eta = P_{2\omega} / P_\omega$ – for peak power). E_ω,, P_ω and $E_{2\omega}$, $P_{2\omega}$ – are energy parameters of the laser emission in the ranges 9 and 4.5 μm, respectively.

7.4 Low-pressure longitudinal-discharge $^{12}C^{18}O_2$ laser with frequency doubling in $AgGaSe_2$ crystal

There are two types of electric-discharge CO_2, lasers which are promising to detect N_2O content along a measurement path. First of them – low-pressure longitudinal-discharge excitation – is more efficient for small and average paths (L = 0.1 - 2.0 km). As a rule, it must operate with the laser beam reflection by a so-called corner reflector . The second–TEA–is suitable for long paths (L > 2 km) when the lidar operates either using the backscattering signal or pulses reflected by a topographic target]. This section considers the CO_2 laser intended for gas analysis in small and average paths.

The laser the optical system of which is shown in Fig. 12 is automatic tunable and output stabilized as described early. The active element was a sealed-off gas-discharge tube like the industrial GL-501 (see Fig. 7) with the discharge gap of 1.2 m. Our experiments were performed with the $^{12}C^{18}O_2$ isotopic forms of carbon dioxide with the low enrichment factor with respect to $^{18}O_2$ described earlier. It is known that when a gas-discharge tube is fed by a pulsed power supplier, the peak power in optimal regime may go up more than by one order as compared to cw electric pumping. This is of especial importance for lasers used in lidar systems. First, the length of the probing path increases up; secondly, pulse-periodic lasing at an optimal repetition rate (~1 kHz) are suitable for receiving and processing of optical and electrical signals and do not require additional devices for modulation.

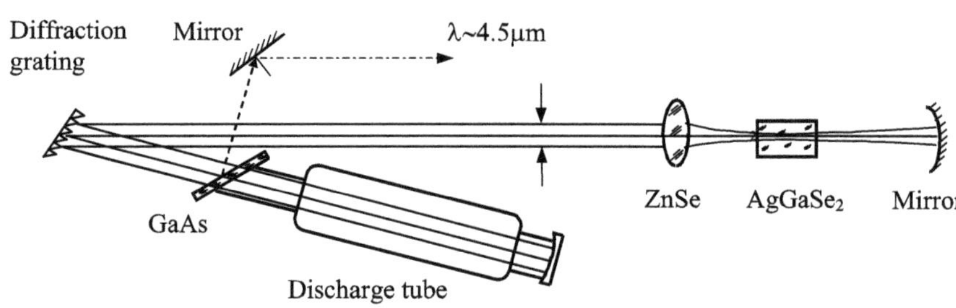

Fig. 12. Optical scheme of the $^{12}C^{18}O_2$ laser with intracavity frequency doubling by non-linear crystal.

Application of the pulse-periodic regime is of especial importance for second harmonic generation in nonlinear crystals. In this case the benefit in the conversion efficiency considerable at the peak power. Our experiments showed the output peak power of the laser to go up to ~100 W (almost by one order as compared with cw lasing) at lines 9P(32) and 9P(34) of the $^{12}C^{18}O_2$ molecule (at each individual line in single-mode operation) when a pulsed power supply is applied.

It is very difficult for longitudinally excited CO_2 lasers to obtain efficient frequency conversion in nonlinear crystals, as their output power is several orders lower than the peak value attainable in pulsed TEA CO_2-systems. Thus, for the $AgGaSe_2$ crystal, for instance, with a mean length (l~ 2 cm) the second harmonic conversion efficiency attained by us with cw discharge was a little more than a tenth of a percent (in case of pulse-periodic discharge it was ~1%), which was a record-breaking value for such laser sources. Therefore, it is even more attractive for such lasers than for powerful TEA systems to place a nonlinear crystal into the cavity.

In our experiments a monocrystal sample made of $AgGaSe_2$ and having a high optical quality (absorption factor ~0.0l cm^{-1}) was used. The crystal was uncoated and had the rectangular 3.5 x 8.5-mm^2 section. The length of the crystal *(l)* was 17mm. Highly parallel (~.4") working faces provided for a possibility to use the crystal as a Fabri-Perot etalon. The phase-matching angle was ~46°. When the incident pumping emission was normal to the crystal, the highest efficiency of second harmonic generation occurred at lines 9P(32) and 9P(34) which were emitted by the laser.

With the optimal radius of the spherical mirror and the focus of the coated lens it was possible to decrease the diameter of the laser beam passing through the crystal by more than one order and, therefore, to increase considerably the pumping density. Along with this we have provided for the pumping beam to be quasi-parallel in the nonlinear crystal. To couple out second harmonic (~75%), we used a Brewster window (GaAs). As the second harmonic polarization was orthogonal to pumping emission, its output was much more than that attainable in a typical conversion system.

A characteristic feature of the proposed system consists of application of the absolutely reflecting (with no out coupling) cavity for pumping emission, which even more increases its intensity. Our experiments showed the highest second harmonic peak power output of the cavity had been equal to 2 W. This is more than one order higher than the analogous parameters reached with the same laser operating with typical optical systems. It is important here to use high quality lens coating with a high damage threshold. This intracavity optical system is simply adjusted and provides with high stable output.

7.5 Laser detection of N_2O

The lidar apparatus complex is shown in Fig. 13. The CO_2 laser was either TEA or low pressure as described previously. All optical elements CO_2 laser, receiving telescope with the objective, photo-detectors, beam-splitting plates, etc. were fixed on a massive metal base to provide good repeatability of the experimental results. The load carrying base of the lidar complex was placed on a construction equipped with mechanisms rotating the system in horizontal (360°) and vertical (45°) directions for immediate and reliable targeting. The target is made visible using an optical sight (12 x 50). Comparatively low dimensions (1.4 x 0.7 x 1.2 m) and mass (~200 kg) give a possibility for development of a mobile version.

A schematic drawing of the receiver/transmitter is given in Fig. 14. The CO_2 laser pulses are output through the beam splitter and are directed into the atmosphere using a transmitting telescope. The emission passed through the atmosphere and trapped by the telescope of the Cassegrain type and with a 250 mm aperture. It consists of two mirrors: front and rear made of sittal with a reflecting aluminum coating protected by corrosion-preventive film. The system uses additionally a ZnSe coated multi-lens objective. The optical configuration of the objective telescope provides for focusing in the focal plane of the detected emission with a 250-mm cross-section down to a diameter ~0.5 mm. After that the detected emission gets on the photodetector which has the sensitive area with a diameter ~1 mm. Optical signals are measured in two channels. Besides the measurement signal, there is a reference one not transmitted through the atmosphere.

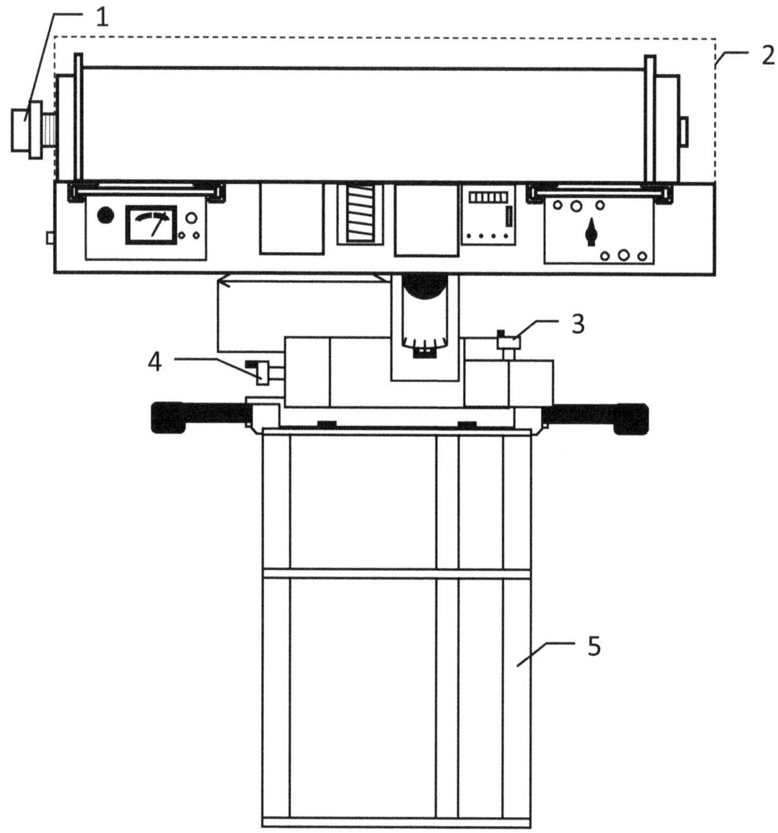

1 – output mirror of CO_2-laser; 2 – CO_2-laser; 3 – horizontal rotation unit; 4 – vertical rotation unit; 5 – metallic base; 6 – receiver/transmitter

Fig. 13. Lidar complex.

The reference signal is produced using a 1-mm-thickness beam-splitter made of BaF_2. Such a thin plate located at an angle ~50° introduces minimum loss for polarized radiation of the TEA CO_2 laser. The loss in the reflection by both faces was no more than 2%. At the same time, only small portion of the emission would be reflected to provide the reference channel. Passing the focusing lens, the reflected portion comes to the sensitive area of the photodetector of the reference channel so organized that this allows automatic account of possible instability of the laser output by normalizing the measurement signal on the reference one, which essentially increases the measurement accuracy and reliability.

As photodetectors nitrogen-cooled photoresistors based on CdHgTe, InSb or germanium doped with gold (Ge:Au) were used. To increase the sensitivity in the measurement channel we used an amplifier. The photodetectors were placed on alignment units. This provided accurate adjustment of the sensitive area of the detector with respect to incident emission. Photodetector signals either arrive to a two-channel ADC or digital oscilloscope and then, via an interface unit, into a computer.

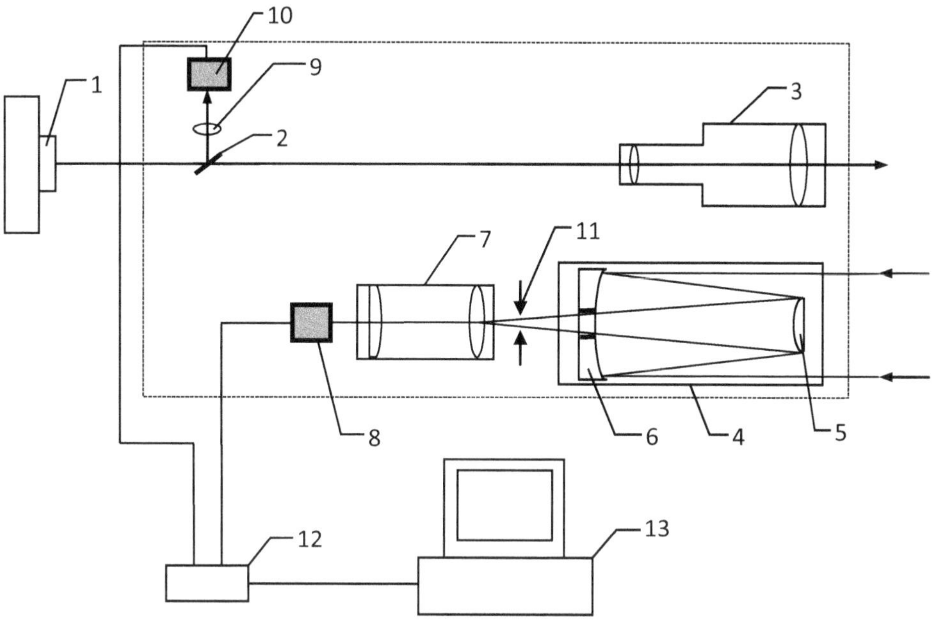

1 – output mirror of TEA CO_2 laser; 2 – BaF_2 beam-splitter; 3 – transmitting telescope; 4 – receiving telescope; 5, 6 – mirrors; 7 – receiving objective (ZnSe); 8 – photodetector (CdHgTe) of measurement channel; 9 – lens (ZnSe) of reference channel; 10 – photodetector (Ge:Au) of reference channel; 11 – iris; 12 – ADC; 13 – personal computer

Fig. 14. Schematic drawing of the receiver/transmitter of the lidar complex.

For monochromatic radiation propagating in homogeneous medium containing several absorbing gases, the transmission T_λ is described by the Buger law :

$$T_\lambda = I_\lambda / I_0 \times exp(-\tau_\lambda)\,, \tag{6}$$

where I_0 and I_λ, are the intensities of the emission with the wavelength λ before and after its passage through a gas layer with length L, $\tau_\lambda = L\sum_i k_{\lambda i} c_i$ is the optical thickness, $k_{\lambda i}$ and c_i are accordingly the absorption coefficient at the wavelength λ_i and the concentration of the ith absorbing gas.

The analysis of the optical characteristics of the detected gases was performed using the differential absorption technique that is in wide use now for laser atmospheric probing. The probing is made at an on/off pair of laser emission lines. "On" line has the maximally possible resonance absorption, and "off"—minimal. The two-frequency differential absorption technique takes useful information only in the resonance absorption by a gas in question. The effects of such factors as water vapour continuum, non-resonance molecular and aerosol absorption, dust, smog, etc. scattering, atmospheric turbulence will be virtually absent due to the comparatively weak monotonic spectral dependence when "on" and "oil" lines are located near each other.

Prior to measuring N_2O, the lidar complex was tested and calibrated for CO and H_2O measurements using CO_2 lasers both TEA and low pressure operating on the ordinary

$^{12}C^{16}O_2$ with frequency doubling by a nonlinear $AgGaSe_2$ crystal. Measurements of the CO and H_2O concentrations also allowed us to account these gases as background for the N_2O measurements.

The theoretical analysis of the absorption lines of CO and background gases (particularly, H_2O) for a path with 2L = 200 m has shown that it is reasonable to select "on" line among the doubled frequencies of the $^{12}C^{16}O_2$, laser such as 9R(30) (at λ = 4.6099 μm the absorption is 50%/ppm). 9R(18) (λ= 4.6412 μm—45%/ppm) and 9P(24) (λ= 4.7931 μm—37%/ppm). Accordingly. the most suitable "off" line belongs to the same laser and are 9R(28) (λ= 4.6148 μm). 9R(20) (λ = 4.6357 μm) and 9P(26) (λ = 4.8018 μm) at which the absorption by carbon dioxide and background gases is virtually absent.

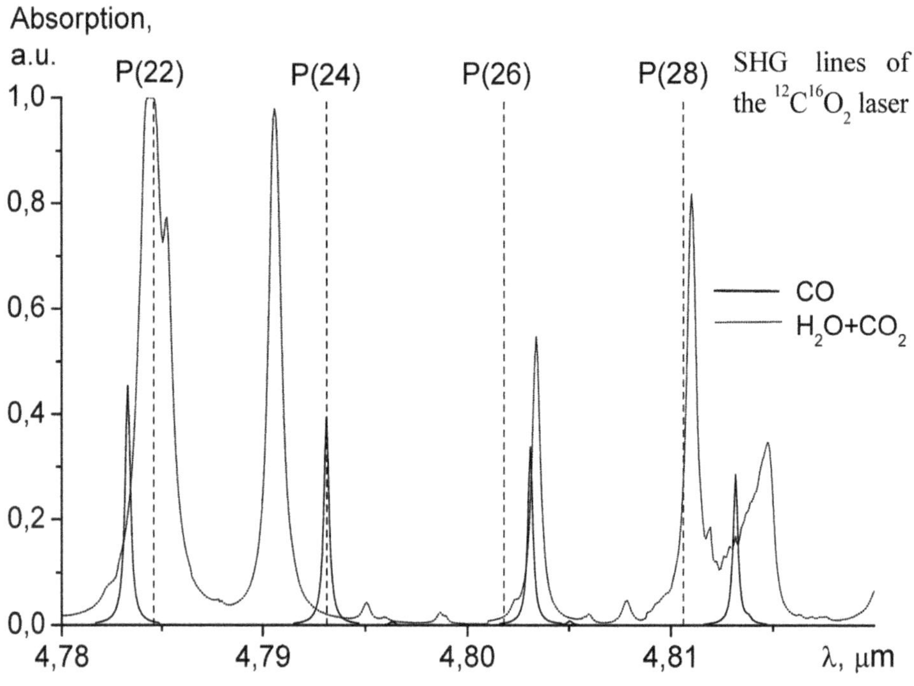

Fig. 15. Absorption spectrum of CO and background gases (H_2O and CO_2) in the range of 4.78 - 4.82 μm. The conditions: path length (2L)=200 m, P=1 atm, T=287 K, gas contents: CO - 1 ppm, H_2O - 10000 ppm, CO_2 - 330 ppm.

Fig. 15 shows a calculated absorption spectrum of the atmospheric gases in the range of our investigations. We select this spectral range due to the following advantages. One of the four selected laser lines (9P(22)— 9P(28)), namely 9P(24) ("on line"), coincides well with the absorption peak of CO. Two of them (9P(22) and 9P(28)) coincides sufficiently well with the absorption lines of H_2O, while 9P(26) demonstrating no absorption of both CO and H_2O is quite suitable as the "off" line. In addition, there is no noticeable absorption by other atmospheric gases (CO_2, for instance) at these lines. Then, carrying out consecutive measurements at these lines, it will be possible to measure concentrations of CO and H_2O.

Then, from the known concentration of H_2O determined independently (for instance, using psychometric devices), it will be possible to calibrate the technique, i.e. obtain the evidence that the results of the laser atmospheric probing are reliable. We used the line 9P(22) that was almost fully absorbed by H_2O as a reference one to check laser tuning at the selected lines. Based on such an original technique we have measured carbon dioxide and water vapour near a highway at a height about 10 m over the cart!, surface. The laser emission was reflected by a plywood sheet painted with a metallic color. The length of probing was 2L = 0.2 km.

The carbon monoxide concentration measured in autumn (5^{00}-6^{00} p.m.) has varied from 0.8 to 1.2 ppm. The measured mean concentration of CO was ~1 ppm. The measurement accuracy determined from H_2O calibrations was ~5%.

N_2O measurements were performed with the same path *(2L* = 0.2 km) using tile low-pressure $^{12}C^{18}O_2$ laser with frequency doubling by a nonlinear crystal. As in the previous case, the emission was reflected by a metallized plywood sheet. Fig. 16 shows a calculated spectrum of the atmospheric gases absorption in the range of 4.5—4.55 μm. We select this spectral range as there are some doubled frequencies of efficient tines of the $^{12}C^{18}O_2$ laser. It is reasonable to select frequency doubled R(32) or R(40) as "on' line, and doubled frequencies of the neighboring R(34) or R(38)—as "off" line. It is important that the indicated lines do not coincide with the absorption lines of background gases H_2O and CO which are always present in the atmosphere. In this way we carried out a number of measurements of N_2O concentration along a researched path at various seasons and times of day. The analysis of the received data has shown that N_2O content in the atmosphere varied considerably, and it is mainly caused by intensity of the transport movement. For example, our experiments performed in autumn in different times during a few days have shown that the N_2O concentration in the path was from 0.35 to 0.5 ppm. The measurement accuracy is estimated to be better than 15%.

We also have measured N_2O for a longer path *(2L* = 1.4 km) using the frequency-doubling TEA $^{12}C^{18}O_2$ laser described earlier. In this case, the laser beam was reflected by a building wall. The averaged content of N_2O was in a good agreement with the value obtained for the shorter path.

The experimental investigations and the calculations carried out have proved conclusively the promising character of the technique developed for the determination of low nitrous oxide concentrations. The technique is based on the use of $^{12}C^{18}O_2$, lasers with effective frequency doubling in nonlinear crystals.

The research carried out has given a reliable technique for laser atmospheric probing of nitrous oxide and effective laser systems to implement this procedure. It is of importance that the path probing is made with a powerful molecular gas laser. Such lasers have narrow emission lines and high stability of spectral and energy output. These characteristics are achieved, as distinct from semiconductor and solid state lasers, naturally without any additional devices. Thus the laser system is simplified and the measurement accuracy increases. The $^{12}C^{18}O_2$ laser system with effective nonlinear frequency-doubling is much promising for global network of lidar stations for atmosphere monitoring.

A reliable procedure or remote high-accuracy laser detection of N_2O as one of the principal destroyers of the protective ozone layer of the Earth has been developed. The procedure is based on using a CO_2 laser system emitting efficiently in the ~4.5 μm range. In this case lasing from isotopic modification $^{12}C^{18}O_2$ of carbon dioxide with its subsequent frequency doubling by a nonlinear crystal is used. With the object of reducing the price the composition of the active medium (both for TEA laser and low-pressure longitudinal-discharge excitation laser) has been optimized. New high-efficiency intracavity frequency doubling schemes based on nonlinear $AgGaSe_2$ crystals have been developed for CO_2 lasers of both types. Low concentrations of N_2O and concentrations of the principal background gases CO and N_2O have been measured under real atmosphere conditions with the aid of the lidar complex built around these lasers.

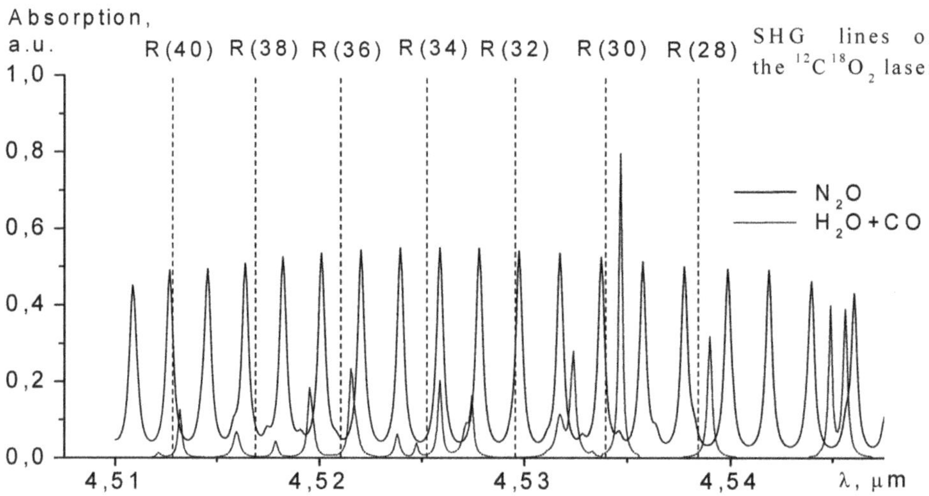

Fig. 16. Absorption spectrum of N_2O and background gases (CO and H_2O). The conditions: path length (2L)=0.2 km, P=1 atm, T=287 K, gas contens: N_2O – 0.4 ppm, H_2O – 10 000 ppm, CO – 1 ppm

8. Conclusion

Optimization of the gas content, pressure, discharge current and the cavity of a low-pressure laser with longitudinal discharge were carried out. Thus, after the above improvements the commercially available sealed-off laser oscillates on more than 30 lines of the P-branch of the 01^11- 11^10 band in the 10.9–11.3 μm range with output power no less than 0.5 W. On strong lines [P(16)–P(26)] output power was ~6W at efficiency ~3% which makes up ~ 40% of analogous laser parameters in the case of oscillation on the lines of regular bands 00^01-10^00 (02^00) under optimum conditions.

The peak power on the strongest lines of the new bands (10°1-20°0(04^00) with a lasing pulse was 30 W. The average output power reached 0.2 W. Lasing was achieved at a number of new transitions. More than 25 new lasing lines with λ = 11.1-11.4 μm, belonging to all the aforementioned bands, were observed in the spectral range studied.

The method of output optimization of cw CO_2 lasers has been developed. The method is based on vibrational and translational temperatures determination by gain measurements on the ro-vibrational lines of regular (00^01-10^00, 02^00) and nonregular (00^02-10^01, 02^01; 01^11-

11^10; (02^21-12^20, 02^01-12^00...) bands of CO_2 molecule. To test the validity of the method, the experiment realization has been done for a low pressure CO_2 laser with cw longitudinal discharge, that can oscillate on the lines of regular and nonregular lines. The good agreement between calculation and experiment data has been observed

We examined what kind of the small gain and the output energy can be attained in the TEA CO_2 laser on the 16(14) μm $02^01(10^01)$-01^11 transitions. On tile basis of the experimentally determined vibrational temperatures T_3 and T_2 we calculated the small gain. The calculations shown that the small gain in the $02^01(10^01)$-01^11 band can attain a significant value (>1 m^{-1}). The necessary conditions for the effective lasing have been determined. It is shown that in optimum conditions the output energy can reach *1.3* J/l at the peak power 5 MW and at the full efficiency of 2 %.

The experimental investigations and the calculations carried out have proved conclusively the promising character of the technique developed for the determination of low nitrous oxide concentrations. The technique is based on the use of $^{12}C^{18}O_2$, lasers with effective frequency doubling in nonlinear crystals. The research carried out has given a reliable technique for laser atmospheric probing of nitrous oxide and effective laser systems to implement this procedure. It is of importance that the path probing is made with a powerful molecular gas laser. Such lasers have narrow emission lines and high stability of spectral and energy output. They were much promising for global network of lidar stations for atmosphere monitoring.

9. References

Aleinikov V.S. and Masychev V.I. (1990). *Carbon monoxide Lasers,* Radio Svyaz, Moscow, (in Russian)

Bertel I.M., Petukhov V.O., Stepanov B.I., Trushin S.A., Churakov V.V. (1982). Investigation of the vibrational temperature kinetics in a TEA CO_2 laser. *Sov. J Quantum Electron.,* v. 12, No 8. pp. 1044-1049

Bertel, I.M.; Petukhov V.V. et al. (1983) Diagnostics of active mediums of CO_2 lasers with the use of nontraditional transitions of molecule CO_2, *Nonequilibrium Processes in Gas Dynamics,* (in Russian), Minsk, Belarus

Bertel, I.M.; Petukhov, V.O.; Trushin, S.A. and Churakov, V.V. (1983). Study of the Gain and the Conditions of Efficient Lasing on Lines of the Hot Band in a TEA CO_2 laser, *Preprint of the IF AN BSSR* No. 289, (in Russian), Minsk, Belarus

Churakov, V.V.; Petukhov, V.O. and Tochitsky S.Ya. (1987). Two-color TEA CO_2 laser oscillation on the lines of regular and hot hands. *Appl. Phis. B,* v .42, pp. 245-249.

Crutzen, P.J. (1996). My life with O_3, N_2O, and other compounds, *Nobel Lecture,* Angew. Chem. Int. Ed. Engl. 35 1758–1777.

Gordiets B.F., Osipov A.I., and Shelepin L.A., (1980) Kinetic *Processes in Gases and Molecular Gas Lasers,* (in Russian), Moscow, Russia

Gorobets, V.A,; Petukhov, V.O.; Tochitsky S.Ya. and V. V. Churakov. (1992) Method of Tuning a CO_2 laser on a Chosen Lasing Line, *Author's Certificate 1771367 SSSR*; MKI H 01 S 3/22

Gorobets, V.A.; Petikhov, V.O.; Tochitsky S.Ya. and Churakov V.V. (1995). Transversely excited CO_2 lidar laser tunable over lines of regular and nontraditional bands. *Quantum Electronics,* v. 25, No *5,* pp. 489-493.

Gorobets, V.A.; Petukhov, V.O. and Churakov V.V. (1990) Optimization of the Output Power of a Continuous CO_2 Lasing on Unconventional Transitions, *Preprint of the IF AN BSSR No. 608,* (in Russian), Minsk, Belarus

Petukhov, V.O. et al. (1985). Fizicheskaya Gazodinamika: Eksperimental 'noe Modelirovanie, *Diagnostika (Physical Gas Dynamics: Experimental Modeling and Diagnostic).* IHMT AS BSSR, (in Russian), Minsk, Belarus

Smith, K. and Thompson R. (1981). *Numerical Modeling of Gas Lasers,* (in Russian), Moscow, Russia

Wexler, B.E.; Manuccia T.J. and Waynant R. (1987). CW and improved pulsed operation of the 14 μm and 16 μm CO_2 lasers, *Appl. Phys. Lett,* v.31, No 11, pp. 730-732.

Witteman, W. (1987). The CO_2 Laser, *Springer-Verlag,* Berlin, Heidelberg, New York, London, Paris, Tokyo

3

CO_2 Laser Photoacoustic Spectroscopy: II. Instrumentation and Applications

Dan C. Dumitras, Ana Maria Bratu and Cristina Popa
Department of Lasers, National Institute for Laser, Plasma, and Radiation Physics, Bucharest Romania

1. Introduction

In this chapter, the main components of an instrument based on laser photoacoustic spectroscopy (LPAS) principles are described in detail. Special emphasis is laid on the home-built, frequency-stabilized, line-tunable CO_2-laser source and the resonant photoacoustic cell. All of the parameters that are characteristic to the photoacoustic cell, including the limiting sensitivity of the system, are measured and compared with the best results reported by other authors. Approaches to improve current sensor performance are also discussed. Other aspects of a functional photoacoustic instrument, such as the gas handling system and data acquisition and processing, are outlined.

Two experimental set-ups were designed and characterized with the photoacoustic (PA) cell in an external configuration: the first one with a low power CO_2 laser where the saturation effects are negligible, and a second one with a high power CO_2 laser where the saturation effects are important. In the first case, the minimum detectable concentration was 0.9 ppbV (parts per billion by volume), while in the second case this parameter was improved to 0.21 ppbV. Comparing with the best results published previously in the literature, our minimum detectable concentration is better by a factor of 4.2 in the first case and by a factor of 18 in the second case.

The next section is dealing with several applications developed in our laboratory. We present a precise measurement of the absorption coefficients of ethylene and ammonia at CO_2 laser wavelengths. For ethylene, the values obtained at 10-μm band excellently agree with other measurements reported in the literature, while important differences were found for the absorption coefficients at 9-μm band. Other applications in plant physiology and medicine (lipid peroxidation and measurement of human biomarkers) are briefly reviewed.

Exhaled breath air analysis represents an attractive and promising novel approach for noninvasive detection of human biomarkers associated with different diseases. Due to extremely low level of the substances of interest in exhaled breath air and the interference of many components at a given laser wavelength, we investigated several measures to increase the accuracy for a single trace gas measurement: a) We studied the efficiency of absorptive trapping and cryogenic trapping to remove carbon dioxide and water vapors from exhaled breath samples. As a result, we found the minimum volume for the KOH trap and the optimum flow rate for transferring gas samples from collecting bags to the photoacoustic (PA) cell. b) We refined breath sample collection procedures from patients under medical treatment: alveolar collection vs. mixed expiratory collection; collecting bags; preparation of patients (antiseptic mouthwash, avoiding food for at least 12 hours); clean transfer of the gas samples from disposable bags to the PA cell in less than 6 hours. c) We measured the photoacoustic signal at different CO_2 laser wavelengths to distinguish the influence of various absorbent gas components in the total content.

One study involved assessment of breath ethylene and ammonia levels in patients with renal failure receiving hemodialysis (HD) treatment. Our measurements demonstrated that HD determines simultaneously a large increase of ethylene concentration in the exhaled breath (due to the oxidative stress) and a reduction of the ammonia concentration, correlated to the blood urea nitrogen level. Analysis of ethylene and ammonia traces from breath may provide insight into severity of oxidative stress and metabolic disturbances and give information for determining efficacy and endpoint of HD.

2. Experimental arrangement

2.1 General schematic

The block diagram of the laser photoacoustic spectrometer was presented in the previous chapter (Fig. 2, Part I). The cw, tunable CO_2-laser beam is chopped, focused by a ZnSe lens, and introduced in the PA cell. After passing through the PA cell, the power of the laser beam is measured by a laser radiometer Rk-5700 from Laser Probe Inc. with a measuring head RkT-30. Its digital output is introduced in the data acquisition interface module together with the output from the lock-in amplifier. All experimental data are processed and stored by a computer (Dumitras et al., 2007). The frequency stabilized, line tunable CO_2 lasers (low power and high power, respectively) will be described in the next section. Both CO_2 lasers are used in two parallel measuring lines, where two independent experiments can be conducted simultaneously. A view of the two parallel measurement lines with laser photoacoustic sensors is shown in Fig. 1.

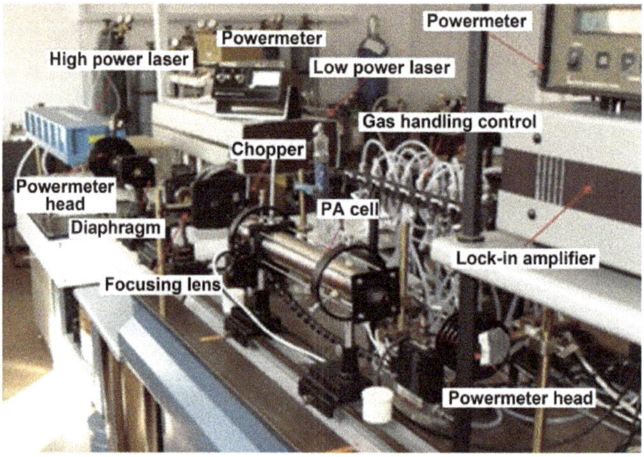

Fig. 1. General view of the PA sensors (two parallel measurement lines).

We decided to use an extracavity arrangement because it has several advantages. In spite of a lower laser power available to excite the absorbing gas in the PA cell, a smaller coherent photoacoustic background signal makes it possible to increase the overall sensitivity of the instrument. This was proved by comparing our results with those obtained with intracavity arrangements (see Table 3). Also, the dynamic range of the PA method is considerably reduced by intracavity operation. Optical saturation may occur for molecules with high absorption cross section while uncontrollable signal changes may be obtained at higher overall absorption in the PA cell, because the loss of light intensity influences the gain of the laser. This effect may cause erroneous results when the sample concentration changes are large. Therefore, high-sensitivity single- and multipass extracavity PA detectors offer a simpler alternative to intracavity devices.

Various modulation methods are applied in PA spectroscopy. It is necessary to distinguish between the modulation of incident radiation and that of sample absorption (Sigrist et al., 1989). The first schemes include the widely used amplitude modulation of the incident radiation by mechanical choppers, electro-optic and acousto-optic modulators as well as the modulation of the laser emission itself by pulsed excitation, Q-switching, and mode-locking. On the other hand, frequency or wavelength modulation of the incident radiation provides the advantage of eliminating the continuum background PA signal caused by wavelength-independent absorption, e.g., by the cell window. The absorption characteristics of the sample can be modulated based on the Zeeman or Stark effect, i.e., by applying modulated magnetic or electric fields to the sample. Consequently, the absorption wavelength of the sample is varied, which corresponds to a wavelength modulation method. The continuum background is suppressed as a result. For example, a reduction of the background by a factor of 500 was achieved by Stark modulation compared with the one obtained in the same PA cell with conventional amplitude modulation by a chopper (Kavaya et al., 1979). However, it should be noted that the application of the Stark modulation scheme in trace gas detection is restricted to molecules with a permanent electric dipole moment like ammonia (NH_3), nitric oxide (NO), etc. Nevertheless, a considerable increase in sensitivity and, even more important, in selectivity in multicomponent mixtures can be achieved.

The light beam was modulated with a high quality, low vibration noise and variable speed (4-4000 Hz) mechanical chopper model DigiRad C-980 or C-995 (30 slot aperture) operated at the appropriate resonant frequency of the cell (564 Hz). The laser beam diameter is typically 5 mm at the point of insertion of the chopper blade and is nearly equal to the width of the chopper aperture. An approximately square waveform was produced with a modulation depth of 100% and a duty cycle of 50% so that the average power measured by the powermeter at the exit of the PA cell is half the cw value. By enclosing the chopper wheel in a housing with a small hole (10 mm) allowing the laser beam to pass, chopper-induced sound vibrations in air that can be transmitted to the microphone detector as noise interference are reduced. A phase reference signal is provided for use with a lock-in amplifier.

The generated acoustic waves are detected by microphones mounted in the cell wall, whose signal is fed to a lock-in amplifier locked to the modulation frequency. The lock-in amplifier is a highly flexible signal recovery and analysis instrument, as it is able to measure accurately a single-frequency signal obscured by noise sources many thousands of times larger than itself. It rejects random noise, transients, incoherent discrete frequency interference and harmonics of measurement frequency. A lock-in measures an ac signal and produces a dc output proportional to the ac signal. Because the dc output level is usually greater than the ac input, a lock-in is termed an amplifier. The lock-in can also gauge the phase relationship of two signals at the same frequency. A demodulator, or phase-sensitive detector (PSD), is the basis for a lock-in amplifier. This circuit rectifies the signals coming in at the desired frequency. The PSD output is also a function of the phase angle between the input signal and the amplifier's internal reference signal generated by a phased-locked loop locked to an external reference (chopper). We used a dual-phase, digital lock-in amplifier Stanford Research Systems model SR 830 with the following characteristics: full scale sensitivity, 2 nV - 1 V; input noise, 6 nV (rms)/ $\sqrt{\text{Hz}}$ at 1 kHz; dynamic reserve, greater than 100 dB; frequency range, 1 mHz - 102 kHz; time constants, 10 μs - 30 s (reference > 200 Hz), or up to 30000 s (reference < 200 Hz).

The diverging IR laser beam is converged by a ZnSe focusing lens (f = 400 mm). In this way, a slightly focused laser beam is passed through the photoacoustic cell without wall interactions. The laser beam diameter $D = 2w$ (or its radius, w), which is a very important issue in LPAS, was calculated at different locations on the beam's propagation path (Fig. 2). A too large beam compared to the inner diameter of the resonant tube could increase the coherent photoacoustical background signal to impracticable values. The calculation was made in three steps: a) inside the laser cavity; b) between the laser coupling mirror and the focusing lens, and c) after the focusing lens (including at the center of the PA cell and at the Brewster windows of the cell).

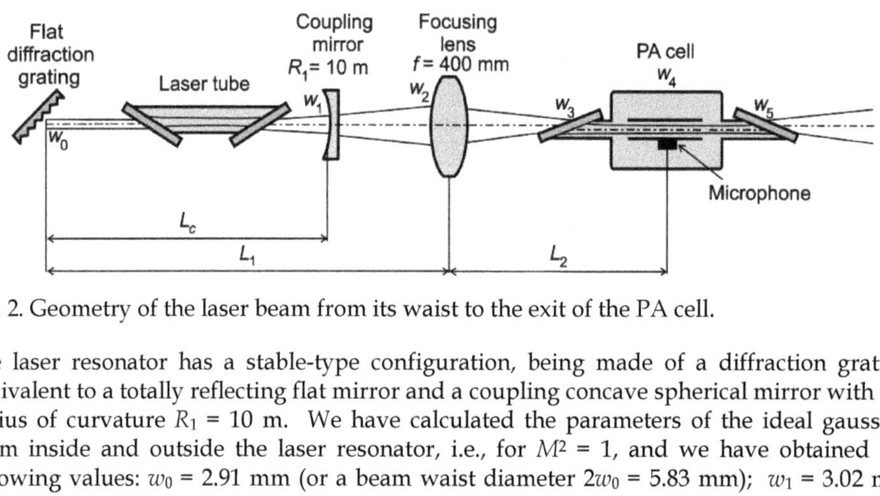

Fig. 2. Geometry of the laser beam from its waist to the exit of the PA cell.

The laser resonator has a stable-type configuration, being made of a diffraction grating equivalent to a totally reflecting flat mirror and a coupling concave spherical mirror with the radius of curvature R_1 = 10 m. We have calculated the parameters of the ideal gaussian beam inside and outside the laser resonator, i.e., for $M^2 = 1$, and we have obtained the following values: w_0 = 2.91 mm (or a beam waist diameter $2w_0$ = 5.83 mm); w_1 = 3.02 mm ($2w_1$ = 6.04 mm); w_2 = 3.16 ($2w_2$ = 6.32 mm); w_3 = 1.75 mm ($2w_3$ = 3.51 mm); w_4 = 0.51 mm ($2w_4$ = 1.03 mm) and w_5 = 2.26 mm ($2w_5$ = 4.53 mm) (for L_c = 690 mm, L_1 = 1060 mm and L_2 = 450 mm). It follows that as the laser beam travels along the PA cell, its diameter is small enough compared to the diameter of the resonant tube (7 mm) to avoid wall absorptions, which ensures that the chosen geometry minimizes the coherent photoacoustical background signal, as intended.

The pertinent questions we set out to answer in assessing the performance of our PA cell were whether the chosen geometry minimized the coherent background signal, and how sensitive the background and calibration were to slight beam deviations from the intended path. Obviously, if the calibration would vary significantly with small movements of the beam, the accuracy of measurements made with the PA cell would be adversely affected unless the cell alignment was very carefully adjusted and rigidly fixed.

The variation of the background and calibration with transversal or longitudinal translation of the PA cell against the propagation beam direction is very small as long as the laser beam does not strike the walls of the cell. The optimum position of the PA cell against the focusing lens (the center of the resonant tube, i.e. the position of the microphones) is 450 mm (for a focusing length of 400 mm). On extending this optimum distance by 30 mm, the signal is decreased by 0.14%, while when shortening the distance by 60 mm, the signal decreases by 0.61%. The dependence of both signal and background signal on the transversal position of the resonant tube relative to the laser beam is similarly low. We found that the cell calibration and background were virtually invariable for reasonably small longitudinal or transversal movements of the cell.

As the angle of incidence deviates from the Brewster angle θ_B (θ_B = arctan$n$ = 67.38^o for ZnSe, where n is the refraction index of the material), the window reflectivity becomes nonzero, and the reflected beam can heat the walls of the cell, making a further contribution to the background signal. Nevertheless, for a ± 10% variation of the angle of incidence relative to the Brewster angle, the reflectance increases from zero to only 2.7% and 1.7%, respectively, meaning that a small deviation from the Brewster angle will not change dramatically the reflectance or the angle of refraction.

Another variable we investigated was the polarization angle of the beam. The cell response in terms of background signal displays a rather broad flat minimum, provided the incidence angle on the window ensures a minimum reflection loss. The data indicate that a deviation of several degrees from vertical polarization can be tolerated. In conclusion, we found that the calibration and background signal were not extremely sensitive to slight misalignments of the beam.

Several authors employed an iris diaphragm in close proximity of the PA cell entrance window to provide spatial filtering in order to reduce the background noise signal caused by off-axis radiation impinging on the internal wall of the chamber. With such a diaphragm, we found that the background signal increased significantly owing to laser beam diffraction at the edges of the aperture. Using high quality optical components (diffraction grating, coupling mirror, lens and windows) together with a well controlled laser beam makes the insertion of an iris diaphragm in the PA instrument unnecessary.

2.2 CO_2 lasers

We have designed, constructed and optimized a rugged sealed-off CO_2 laser (named LIR-25 SF), step-tunable on more than 60 vibrational-rotational lines and frequency stabilized by the use of plasma tube impedance variations detected as voltage fluctuations (the optovoltaic method) (Dumitras et al., 1981; Dumitras et al., 1985; Dutu et al., 1985). The glass tube has an inner diameter of 7 mm and a discharge length of 53 cm. At both ends of the tube we attached ZnSe windows at Brewster angle. The laser is water cooled around the discharge tube. The dc discharge is driven by a high-voltage power supply. The end reflectors of the laser cavity are a piezoelectrically driven, partially (85%) reflecting ZnSe mirror at one end and a line-selecting grating (135 lines/mm, blazed at 10.6 μm) at the other. Piezoelectric ceramics such as lead zirconate titanate (PZT) can be used.

A free-running or unstabilized laser is subject to many perturbations of its frequency. First of all, changes in cavity length affect the frequency of an oscillating mode ($\Delta\nu/\nu = \Delta L/L$), so keeping a constant length is the prime objective in frequency stabilization schemes. Possible perturbations of the cavity length can be divided into two groups: external effects (thermal variations of the spacer material, changes in atmospheric conditions, mechanical vibrations, variations in the position of optical components and in magnetic fields) and internal effects, which are generally related to the discharge noise.

Active power stabilization based on a piezo-driven out-coupling mirror is used in our laser. The principle of the stabilization schemes is based on a comparison between the frequency of a single frequency laser (single-mode, single-line) and some stable point of reference. If the laser frequency is different from that of the benchmark, an error-sensing discriminant is used to derive a signal proportional to the deviation. This error signal is used to control the laser oscillating frequency and retune it to the reference one. Such a servo-loop (closed loop feedback) locks the laser frequency to that of the reference. For moderate stability, the CO_2 laser line profile can be used as the discriminating curve (Dumitras et al., 1981). This method is more appropriate for the CO_2 laser than for other lasers because the CO_2 vibrational-rotational line profile is narrow and has much steeper slopes than, for example, that of the neon line in a He-Ne laser. The error signal is produced by allowing the laser resonance cavity to "ride" around the steep part of the line profile slope, and its amplitude is dependent on the change in cavity mirror separation. This scheme requires internal frequency modulation (jittering) of the laser in order to sense the sign of the derived error signal. Stabilization is then obtained by re-establishing the required separation with a servo-system.

The CO_2 laser is frequency stabilized to the center of the curve representing its output power *versus* frequency (the molecular resonance) upon the variation of plasma tube impedance, when the optical power extracted from the medium is modulated (Dutu et al., 1985). In this closed-loop active stabilization, the cavity length is controlled by a piezoelectrically driven mirror along the cavity axis, which responds to the sum of a dc control voltage, plus a small jitter signal at some convenient frequency (~500 Hz). As can be seen in Fig. 3, where a curve of laser line gain *versus* frequency is drawn, the small cavity jitter induces a sinusoidal variation in laser output as the cavity mode scans across the transition gain profile, which is compared in phase to the jitter voltage.

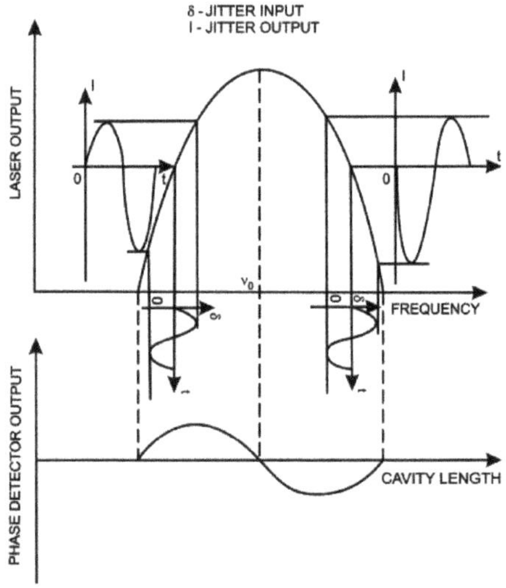

Fig. 3. Generation of the error-frequency signal.

The error signal generated by the phase detector can serve to drive the cavity resonance to the center of the laser gain curve. In this way, the electronic feedback loop seeks the center of the lasing gain profile, the lock-in point being the zero crossing of the phase detector response. If the mean mode frequency is lower than the line center frequency, the phase of the observed laser intensity variation is opposite to the one we have where the mode frequency is higher than the line center frequency. The amplitude of the jitter output increases with the frequency offset from the line center.

In lieu of using an IR detector to sense the laser intensity variation, the cavity length is adjusted using the optovoltaic effect. As the internal laser radiation field intensity is altered by changing the resonant cavity alignment, the discharge impedance, which is proportional to the slope of the curve of laser output power *versus* frequency, is also modulated. The impedance variation is determined by exciting the plasma tube with a high-speed current-regulated power supply and measuring the resulting variation in the voltage drop across the plasma tube (the optovoltaic effect). An intensity variation of 1% is sufficient to change the discharge impedance significantly (~ 0.1%). By using a current regulated power supply, the voltage impedance fluctuation is detected as an ac component of the voltage drop across the plasma tube.

Before any attempt is made to stabilize the frequency of a laser, a single frequency output must be ensured. For this purpose, a laser operating in the lowest transversal mode (TEM_{00}) must be designed (Dumitras et al., 1976). The single line operation of the CO_2 laser is achieved with a dispersive element (diffraction grating). The cavity length of L_c = 690 mm corresponds to a separation between two longitudinal modes of $\Delta\nu = c/2L_c$ = 217 MHz. Calculating the collisional broadening in a mixture of CO_2, N_2, He, Xe, and H_2 at a total pressure of 34 mbar gives a collisional full linewidth at half maximum (FWHM) $\Delta\nu_c$ = 119 MHz. We therefore conclude that a single frequency operation is obtained when a longitudinal mode is tuned on the top of the gain curve ($\Delta\nu > \Delta\nu_c/2$).

To increase the number of oscillating lines, especially those with a smaller gain, and obtain reliable long term operation at a single specific wavelength, some form of wavelength selection introduced in the optical cavity is generally required. As optical dispersion is incorporated by using a diffraction grating or Brewster-angle prisms within the laser cavity, the laser can be made to oscillate on only one vibrational-rotational line, otherwise the particular transition on which the CO_2 laser operates depends on the length of the resonator. That is why the total reflecting mirror must be replaced by a diffraction grating, which is tilted about its groove axis to the blaze angle and acts as a frequency selective reflector. Light diffracted into the first order maximum is returned along the optical axis and taken as laser output, while light in other orders as well as any other wavelength is returned off-axis and gets lost. Another advantage of a laser resonator with a grating is that the laser can be tuned over the entire oscillating linewidth from the line center.

We used a flat diffraction grating with 135 grooves/mm, blazed at 10.6 μm and having a peak efficiency of 96%, mounted in a Littrow configuration. With such a grating, the vibrational-rotational lines emitted by the CO_2 laser in the range P(50) - 10.9329 μm and R(44) - 9.1549 μm can be selected by controlling the grating angle in the range 47°33′32″ to 38°10′02″, which can be set to the desired laser transition with a micrometric screw. This grating presents a good dispersion, as the P(18) and P(20) lines (10.4 μm band) are separated by a 6′38″ angular difference (as compared with 2′49″ for a diffraction grating with 75 grooves/mm).

To meet the frequency stability requirements, the laser cavity must be so constructed as to reduce the effects of ambient vibration and thermal variations on the output frequency of the laser. This places less stringent demands on the performance of the servo-system controlling the laser frequency. To minimize the thermal length changes ($\Delta L/L = \alpha\Delta T$) in the mirror support structure, a material with a low expansion coefficient has to be used for the spacers between the endplates of the cavity which carry the mirrors. Such a material is Invar, which has an expansion coefficient $\alpha = 1.26\text{x}10^{-6}/^{o}\text{C}$. To obtain a passive instability of $3\text{x}10^{-8}$ for the laser frequency, the temperature variation must not exceed 0.024^{o}C. Such constant temperature is hard to maintain, especially in longer lasers, where high power inputs and high heat dissipation cause large temperature instabilities.

Stiffness is a most desirable attribute for minimizing fractional changes in the cavity length. Special measures were taken against mechanical vibrations (by eliminating high frequency vibrations), variations in the position of optical components (by supporting rigidly any intracavity element) as well as to prevent magnetic fields and acoustically borne vibrations, which can be reduced by shielding the laser with some form of enclosure. The design of the remaining structure was chosen so as to avoid the lowering of the basic first resonance. The joints between the elements of the structure, especially the joints perpendicular to the laser axis, were so designed that they did not have any low frequency resonances. To have a spring constant of the joint high enough and to avoid joints using only the spring force of a few screws to connect significant masses, we used large contact areas under compressional stress. We chose a cylindrical shape for the mechanical structure, including the housing, because of its high resistance to bending deformation. A section through the laser cavity assembly is given in Fig. 4 and a photo of the mechanical structure, laser tube, and control panel is presented in Fig. 5.

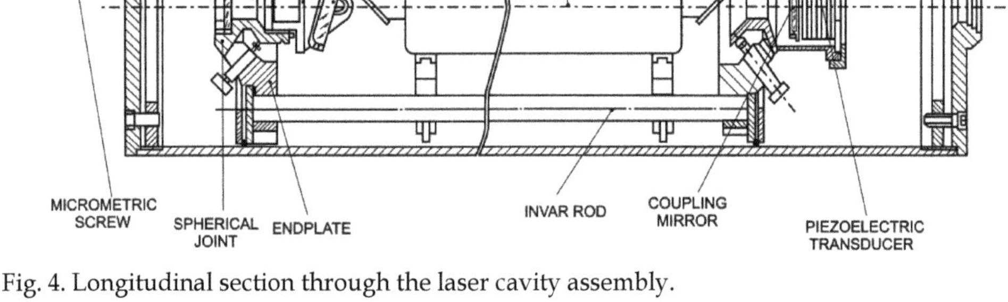

Fig. 4. Longitudinal section through the laser cavity assembly.

To keep the weight unchanged and still maintain the required thermal characteristics, the endplates of the cavity which carry the mirror and the diffraction grating were primarily constructed of aluminum. The three invar rods are potted into the aluminum frame so as to ensure intimate contact between the invar rods and the rigid aluminum structure. To remove the problems associated with weak spring-type controls, the mirror holders were designed ruggedly. The mirrors can be adjusted in angle by sliding two stainless steel spherical joints with respect to one another. As the mirrors are adjusted into their final position, the adjustment screws clamp the mirror holder tightly so that no inadvertent movement is possible.

Fig. 5. Homebuilt frequency stabilized CO_2 laser model LIR-25 SF.

The diffraction grating is mounted in a similar holder and is rotated by a micrometric screw (0 - 25 mm), as shown in Fig. 6. The laser tube was rigidly mounted in the cavity with two concentric rings, the inner one being adjusted by three screws. This system allows the laser tube axis to fall into line with the mirror centers and provides good transverse stability.

Fig. 6. Diffraction grating-drive mechanism.

To achieve active stabilization, an automatic frequency control circuit (lock-in stabilizer) is used to maintain an axial mode on the top of the gain curve. The block diagram of the frequency stabilization system based on plasma tube impedance variations is shown in Fig. 7. A sinusoidal signal with a rough frequency of 500 Hz derived from the pilot oscillator is applied to the piezoelectric transducer, resulting in a simultaneous frequency and amplitude modulation of the laser output, dependent on the amplitude of the sinusoidal signal.

The ac voltage drop across the tube is passed through a tuned amplifier and then synchronously detected. The demodulation of the ac signal is performed in the phase-sensitive detector, with phase (continuously adjustable) determined by the phase shift circuit. The demodulator output is processed by the operational integrator and a high voltage dc amplifier. The dc bias together with this error correction signal is applied to the piezoelectric transducer, thus closing the feedback loop. The error signal generated by the phase detector serves to drive the cavity resonance to the center of the laser gain curve and compensates for the effects of slow drift.

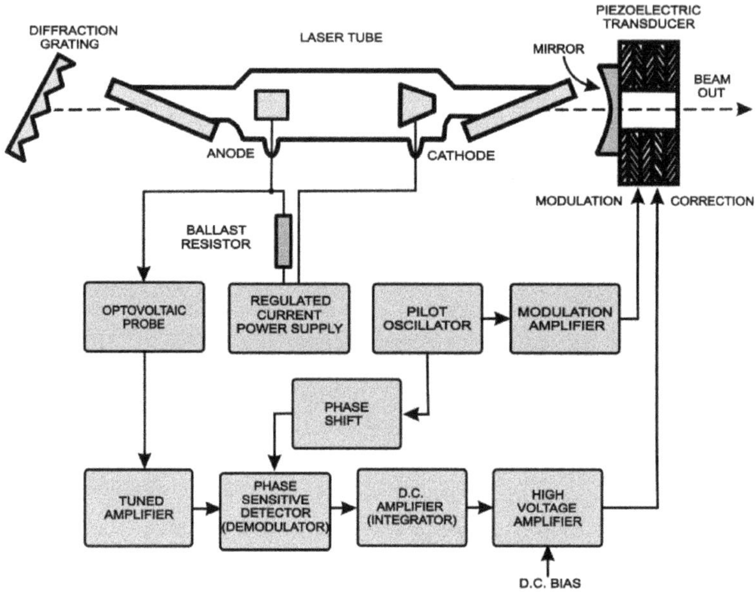

Fig. 7. Block-diagram of the automatic frequency control electronics.

There are fifteen functional modules grouped into three main blocks in accordance with the specific function they play in the operation of the frequency stabilized CO_2 laser:

1. a high speed current regulated power supply, which excites the cw CO_2 laser tube;
2. a servo control system, which controls the length of the resonant cavity;
3. a logic electronic system which allows an efficient operative control and monitoring of the frequency stabilized CO_2 operation.

The high-speed current regulated power supply includes three modules namely: the power supply converter - PSC, the power supply feedback - PSF, and the power supply rectifier PSR. The high frequency of the converter makes it possible to minimize the electronic components that are used in making the high voltage transformer and the low pass filter of the high voltage rectifier.

The modulation signal that is needed to scan the laser line profile is generated by a pilot oscillator and applied to the piezoelectric transducer through a modulation amplifier. Both circuits are placed on the same functional module OMA (Oscillator & Modulation Amplifier). For cw CO_2 lasers excited by current regulated power supplies, the modulation signal which appears in the emitted radiation can be measured according to the optovoltaic effect using a simple band pass filter like that noted with OVP (Optovoltaic Probe). The detected optovoltaic signal is amplified through a two-stage tuned ac amplifier - ACA. The amplified optovoltaic signal is introduced in the phase sensitive detector (PSD module) together with a phase adjustable component of the reference signal from the modulation oscillator.

The error signal is obtained in the output of the electronic dc amplifier module DCA, by processing the output signal of the phase sensitive detector. The DCA module has two modes of operation, namely: (a) gain of ten amplifier with integrating time constant of 1 second; (b) high gain integrator with output slewing rate of about 6 V/s per volt of input. The first mode allows the observation of smoothed output of the demodulator (stabilization discriminator), while the second one is used in closed loop stabilization to observe the error signal. A high voltage amplifier (HVA) that works on the principle of a switching high voltage power supply drives the piezoelectric transducer under the control of the error signal.

The logic electronic system includes four specialized modules, namely: a low voltage stabilized power supply (SPS), a logic drive circuit (LDC), a power meter circuit for laser radiation (PMC), and a beam controller (BCS). Every parameter and state that are monitored or selected in the operation of the frequency stabilized CO_2 laser LIR-25 SF are displayed on the rear panel of the laser (Fig. 5) by means of nine light emitting diodes (LED) driven by the panel display circuit (PDC). The output beam of the laser can be sent to either the powermeter head or the exit by a two-position reflecting mirror, controlled by a beam shutter circuit and a beam selector switch placed on the rear panel.

The long-term frequency instability was $\Delta\nu/\nu = 3\text{x}10^{-8}$ or $\Delta\nu = 1$ MHz ($3\text{x}10^{-5}$ cm^{-1}).

A supplementary power stabilization is possible by using an additional feedback loop (Dumitras et al., 2006). The principle used for power stabilization took into account the dependence of the output power on discharge current. The new feedback loop (Fig. 8) modifies the discharge current so to maintain the output power at a constant value imposed by the reference.

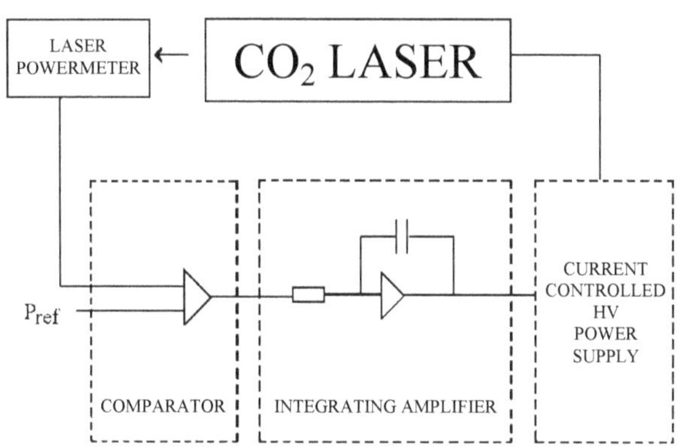

Fig. 8. Supplementary feedback loop used for laser power stabilization.

The performance of the two feedback loops used for frequency and power stabilization of the CO_2 laser were evaluated by using computer data acquisition. It was monitored simultaneously the output power, the discharge current and the temperature inside laser cavity, while the power instability was calculated for a period of 2 minutes. The importance of the two feedback loops to reduce the output power instability can be remarked from these measurements. Thus, when both the frequency stabilization loop and the power stabilization loop are opened and forced hot air perturbation is introduced, the power instability is 10.2%, a value too large for many applications; when the frequency stabilization loop is closed and the power stabilization loop is opened (no perturbation), the power instability is reduced to 1.06%; when the frequency stabilization loop is closed and the power stabilization loop is opened and forced hot air perturbation is introduced into laser cavity, the power instability increases only a little, to 1.57%; and when both the frequency stabilization loop and the power stabilization loop are closed and forced hot air perturbation is introduced, the power instability is reduced drastically, to 0.28%. If there is no perturbation and the two feedback loops are closed, the power instability is only 0.23%. In conclusion, the power instability was reduced by four times with this supplementary feedback loop, from 1% to as low as 0.23% for a period of 2 minutes.

The tunability of our CO_2 laser is presented in Fig. 9. We observed the oscillation of 62 different vibrational-rotational lines in both the 10.4 μm and 9.4 μm bands. In this way, the laser was line tunable between 9.2 μm and 10.8 μm with powers varying between 1 and 6.5 W depending on the emitted laser transition. More than 20 lines had output powers in excess of 5 W.

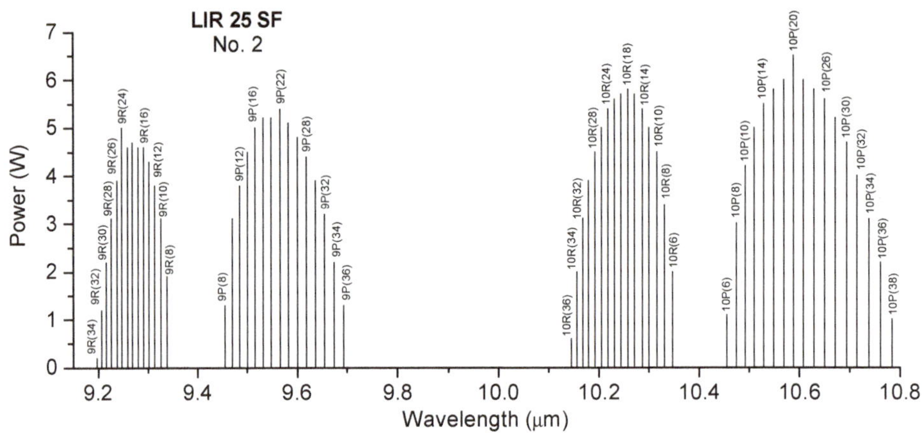

Fig. 9. Tunability of the low power CO_2 laser with diffraction grating: P(max) = 6.5 W; Tunability: 9R(8) - 9R(34); 9P(8) - 9P(36); 10R(6) - 10R(36); 10P(6) - 10P(40); No. of lines: 62.

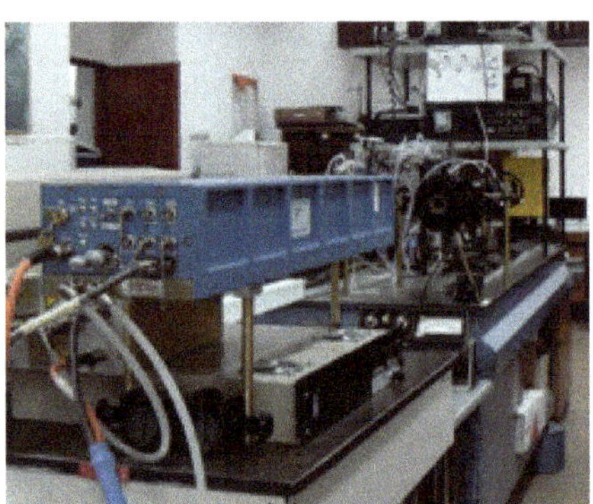

Fig. 10. Coherent GEM SELECT 50™ CO_2 laser in the experimental setup.

To investigate the possibility of using a high power laser in an extracavity configuration, we introduced in the experimental set-up a commercial CO_2 laser (Coherent GEM SELECT 50™ laser) (Fig. 10) with output power till 50 W and tunable on 73 different lines (Fig. 11). When this laser is tuned on 10P(14) line, the maximum power delivered after chopper and focusing lens is 14.5 W.

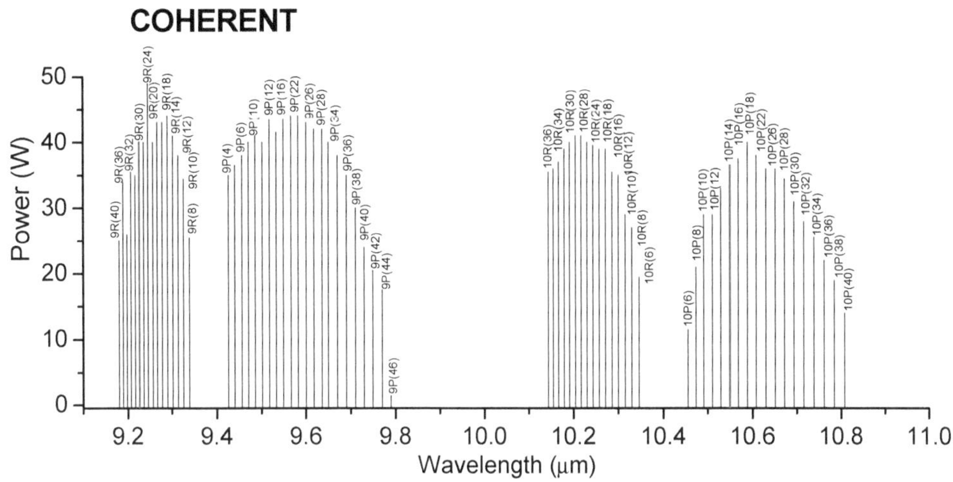

Fig. 11. Tunability of the high power CO_2 laser with diffraction grating: P(max) = 50 W; Tunability: 9R(8) - 9R(40); 9P(4) - 9P(46); 10R(6) - 10R(36); 10P(6) - 10P(40); No. of lines: 73 (10P(14): $\lambda = 10.53$ μm; 9R(30): $\lambda = 9.22$ μm).

2.3 Photoacoustic cell

In the literature the PA cells are often characterized as "nonresonant" or "resonant". This terminology is misleading, because any PA cell can be operated at an acoustic resonance or far from its resonance. Thus, it is preferable to label the system in terms of its nonresonant or resonant mode of operation.

To design an optimum acoustically resonant PA cell to be used in CO_2 laser photoacoustic spectroscopy, the following requirements have to be met:

i. the fraction of laser energy absorbed by the gas must be maximized by increasing either the incident laser power (but maintaining a large SNR) or the optical density of the gas (Eq. 2, Part I);
ii. cell responsivity needs to be as high as possible, because the voltage response is proportional to it (Eq. 29, Part I);
iii. the microphone responsivity has to be as high as possible, and the use of many microphones is advisable (Eqs. 30 and 31, Part I);
iv. the design must make it possible to operate the cell at an acoustic resonance, and the resonance frequency must lie between 400 and 1000 Hz, where the microphone noise is minimal;
v. the quality factor Q of the acoustic resonance must not exceed 50 in order to decrease the influence of small deviations from the resonance frequency;

vi. the electrical noise and the coherent acoustic background noise must be as low as possible; this can be done by using low noise microphones, good acoustic and vibration isolation, low noise electronics, and good electronic isolation (no ground loops, proper shielding);
vii. the coherent photoacoustic background signal due to the heating of the walls and windows must be minimized by using optical components of very high quality and introducing acoustic baffles;
viii. the cell must enable continuous gas flow operation, and consequently not only the cell windows, but also the gas inlets and outlets have to be positioned at pressure nodes of the resonance;
ix. the cell must have low gas consumption and fast response, and the cell volume has to be sufficiently small to prevent prohibitive dilution when the produced trace gas is flowed through the cell volume by a continuous gas stream;
x. the adsorption and desorption rates on the surfaces in direct contact with the sample gas that can influence particularly measurements on sealed-off samples must be minimized by using special cell materials and reducing the surface-to-volume ratio;
xi. the effect of the loss mechanisms which we can control must be minimized by an appropriate system design.

Various ways to design (cylindrical geometry, H geometry, T geometry, or using a Helmholtz resonator) and operate (longitudinal, azimuthal, radial, or Helmholtz resonances) resonant PA cells have been studied (Zharov & Letokhov, 1986). Furthermore, PA cells for multipass (Koch & Lahmann, 1978; Nägele & Sigrist, 2000) or intracavity operation (Fung & Lin, 1986; Harren et al., 1990a) were designed. The effect of window heating in the amplitude modulation schemes has been minimized by introducing acoustic baffles (Dewey, 1977), developing windowless cells (Gerlach & Amer, 1980; Miklos & Lörincz, 1989; Angeli et al., 1992), or using tunable air columns (Bijnen et al., 1996). In many cases the window-heating signal can be markedly reduced by positioning the entrance and exit of the light beam at nodes of the mode being excited.

A cylindrical cell operated at a radial resonance and having Brewster windows mounted at the pressure nodes of the first radial mode, as presented by Gerlach and Amer (Gerlach & Amer, 1980), does not fulfill all these requirements. Therefore, an open resonant cell excited in its first longitudinal acoustic mode was developed to fulfill most of these requirements.

The H-type longitudinally resonant cell was chosen to form the core of our measuring instrument. Dividing the PA cell into a central chamber and two buffer chambers adjacent to the Brewster windows, a design which lowered significantly the coherent photoacoustic background noise, was first proposed by Tonelli et al. (Tonelli et al., 1983). The characteristics of this type of PA cell have been discussed by Nodov (Nodov, 1978), Kritchman et al. (Kritchman et al., 1978), and Harren et al. (Harren et al., 1990a). Its main advantages are: (a) stable operation at a relatively low frequency; a quality factor of about 20, i.e., much lower than that of a radial resonator, which makes it less sensitive to environmental changes; the efficient conversion of radial to longitudinal modes and the relatively long wavelength guarantee a sufficiently high photoacoustic amplitude; (b) a longitudinal resonator is not noticeably influenced by the gas flow at the desired flow rate of several L/h; noise by gas flow phenomena is negligible for properly positioned inlet and outlet ports; (c) window noise is minimal if the windows are located at a quarter wavelength from the ends of the resonator tube; (d) the construction is rugged and simple and can be achieved with low adhesion materials.

Following these guidelines, a PA cell was designed, constructed, and tested. An H-type cylindrical cell designed for resonant photoacoustic spectroscopy in gases is shown in Fig. 12. The longitudinal resonant cell is a cylinder with microphones located at the loop position of the first longitudinal mode (the maximum pressure amplitude). Some general considerations imply that the coherent photoacoustic background signal caused by window heating is decreased if the beam enters the cell at the pressure nodes of the resonance. The advantage of mounting the windows at the pressure nodes is well demonstrated, and the window heating signal is decreased by the Q factor. The laser beam enters and exits the cell at the Brewster angle. It is more advantageous to have the beam pass through the windows at the Brewster angle (θ_B), as θ_B is nearly constant over a wide range of wavelengths, and variations of θ_B with wavelength can be tolerated since reflectivity increases very slowly for small deviations from θ_B.

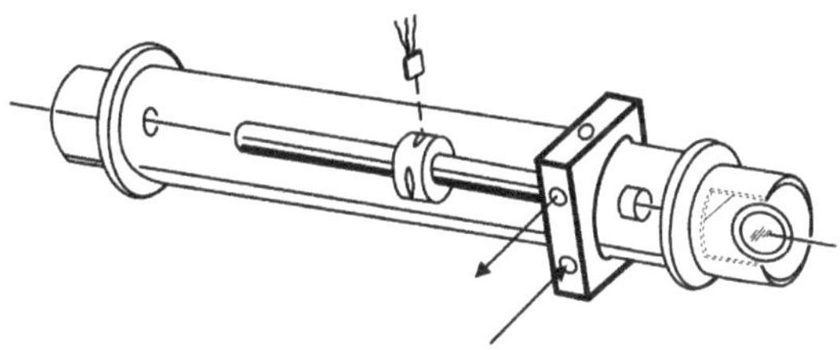

Fig. 12. Schematic of the PA cell designed for the first longitudinal resonance mode.

The influence of scattered light onto the PA background signal can be minimized by using a highly reflecting polished material, with a good thermally conducting substrate. Bijnen et al. (Bijnen et al., 1996) investigated different materials for the resonant tube and found that the background signal decreased for polished stainless steel, polished brass, and polished, gold-coated copper in a ratio of 6:2:1, respectively. In the case of the CO_2 laser, the best performance was obtained by employing a copper tube with a polished gold coating as resonator material. Because of the excellent heat-conducting properties, the absorbed heat can be quickly dispersed in the copper tube. The gold coating was used not only to optimize laser radiation reflection, but also to obtain a noncorrosive surface to withstand aggressive gases.

Many polar compounds (e.g. ammonia) are highly adsorptive and produce an error in real time concentration measurements by adhering to the detector surfaces. These molecules interact strongly with most metals and many insulating materials. Ammonia is a good model compound for these molecules as it shows the characteristic adsorptive behavior that is not a health hazard at low concentrations. The rate of ammonia adsorption on the gas handling surfaces depends on the surface material and temperature, and on the mixture concentration, flow rate, and pressure. Comparing the ammonia results with those for ethylene, which interacts weakly with most surfaces, provides a measure of the cell-sample interaction. Beck (Beck, 1985) evaluated the suitability of several surface materials for minimizing sample adsorption loss. Four materials-304 stainless steel, gold, paraffin wax, and Teflon-were tested using ammonia as a sample. The results show that both metals interact strongly with the sample. Teflon coating (thickness <25 μm) was found to provide accurate real time response for ammonia sample flows. Also, no signal decay is observed following flow termination. Additionally, the coatings must not degrade the acoustic response of the cell. The Teflon coating actually increases the cell Q by a small amount (1 percent). This is attributed to the smooth slick surface obtained by Teflon coating which would decrease any surface frictional or scattering loss of acoustic energy. Rooth et al. (Rooth et al., 1990) tested the following wall materials - stainless steel 304, gold (on Ni-coated stainless steel), Teflon PTFE, and Teflon PFA - in contact with the gas. Stainless steel proved to be an almost unsaturable reservoir for ammonia at pptV levels. The number of stored molecules exceeded by a factor of 10 or more the number of potential locations on the total geometric surface. Despite its inferior properties in terms of adsorption, Olafsson et al. (Olafsson et al., 1989) used a stainless steel cell for detecting NH_3 and found that an operating temperature of 100°C combined with water vapors led to a very significant reduction of NH_3 adsorption. Apparently, the water molecules stick to the walls even more efficiently than NH_3, and the cell walls are effectively coated with water. Later on, the sample cell was constructed with Teflon as wall material (Olafsson et al., 1992).

Since an open pipe efficiently picks up and amplifies noise from the environment, it should be surrounded by an enclosure. In order to ensure high acoustic reflections at the pipe ends, a sudden change of the cross section is necessary. Therefore, the resonator pipe should open up into a larger volume or to buffers with a much larger cross section. The buffers can be optimized to minimize flow noise and/or window signals. The length of the two buffers accounting for half the resonator length is chosen such as to minimize the acoustic background signal originating from absorption by the ZnSe windows. Open pipes were introduced for PA detection as early as 1977 (Zharov & Letokhov, 1986), and the most sensitive PA detectors currently used are based on open resonant pipes. In resonant cells, window signals can be diminished by using $\lambda/4$ buffers next to the windows. These buffers, placed perpendicular to the resonator axis near the windows, are tuned to the resonator frequency and act as interference filters for the window signals (the coupling of the window signals into the resonator is reduced by large buffer volumes that act as interference dampers).

It was found both theoretically and experimentally that the signal amplitude decreases drastically when buffer length $L_{buf} < \lambda/8$ (Bijnen et al., 1996); the resonance frequency and the quality factor, for both the window and gas signals, are not much affected by changing the buffer length. The length of the buffer is optimal for window signal suppression when $L = 2L_{buf}$ ($L >> r$) (L and r stand for resonator length and radius). The dependence of the gas absorption p_g and window absorption pressure p_w on the ratio of the buffer and resonator radii is:

$$p_g \propto \frac{\sqrt{L}}{r}\left(1-\frac{r^2}{r_{buf}^2}\right), \tag{1}$$

$$p_w \propto \frac{r}{r_{buf}\sqrt{L}}. \tag{2}$$

The ratio between the gas absorption signal and window signal then becomes:

$$\frac{p_g}{p_w} \propto \left(\frac{r_{buf}}{r}\right)^2 L. \tag{3}$$

The optimal buffer length, resulting in an optimal suppression of the photoacoustic background signal, is $\lambda/4$. Choosing a buffer length of $\lambda/8$ has the advantage of a shorter cell, and a good, though not optimal, suppression of the window signal is still possible in this case. If the volume and overall size of the buffers pose no problem, their radii have to be large for optimal operation. A practical radius can be deduced to be $r_{buf} \approx 3r$. The gas absorption amplitude is barely influenced above this value. A small radius for the resonator of the resonant cell is advantageous as it enhances gas absorption. The limit for the resonator radius is mainly determined by the wings of the gaussian laser beam profile hitting the wall; r should roughly be at least three times the radius of the laser beam to reduce wall absorption to an acceptable level.

Our PA cell design, based on the above considerations, is shown in Fig. 13 in cross sectional longitudinal view. The PA cell is made of stainless steel and Teflon to reduce the outgassing problems and consists of an acoustic resonator (pipe), windows, gas inlets and outlets, and microphones. It also contains an acoustic filter to suppress the flow and window noise. ZnSe windows at the Brewster angle are glued with epoxy (Torr-Seal) to their respective mounts. The resonant conditions are obtained as longitudinal standing waves in an open tube (resonator) are placed coaxially inside a larger chamber. We use an open end tube type of resonator, excited in its first longitudinal mode. To achieve a larger signal (eqs. 26 and 27, Part I), we chose a long absorption path length (L = 300 mm) and an inner diameter of the pipe of $2r$ = 7 mm ($r \cong 1.5w_5$, see Fig. 2). The fundamental longitudinal wave, therefore, has a nominal wavelength $\lambda_s = 2L$ = 600 mm (and a resonance frequency f_0 = 564 Hz). The effective wavelength is ~1% longer than the nominal value, in accordance with predictions of Eq. (5, Part I). The two buffer volumes placed near the Brewster windows have a length L_{buf} = 75 mm ($\lambda/8$) and a diameter $2r_{buf}$ = 57 mm ($r_{buf} \cong 8r$). The inner wall of the stainless steel resonator tube is polished. It is centered inside the outer stainless steel tube with Teflon spacers. A massive spacer is positioned at one end to prevent bypassing of gas in the flow system; the other is partially open to avoid the formation of closed volumes. Gas is admitted and exhausted through two ports located near the ends of the resonator tube. The perturbation of the acoustic resonator amplitude by the gas flow noise is thus minimized.

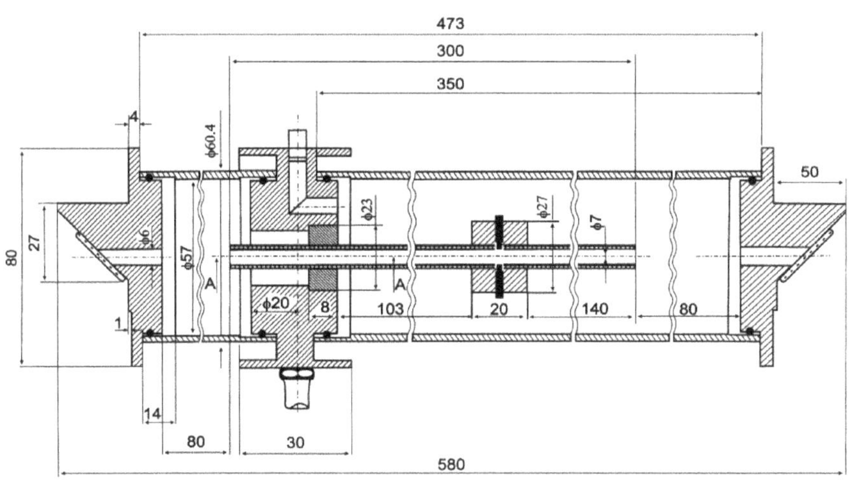

Fig. 13. A resonant photoacoustic cell with buffer volumes.

The total cell volume is approximately 1.0 dm³ (total length 450 mm, and inner diameter 57 mm, minus inner mechanical parts). For flowing conditions, however, it is advantageous to reduce the active volume of the cell. Especially if the flow rate is smaller than 1 L/h (16.6

sccm - standard cubic centimeters per minute), the replenish time for the 1.0 dm^3 cell becomes impractical. The buffer volume at the entrance port of the cell affects the renewal time τ considerably. The buffer volume is approximately 200 cm^3, yielding a time constant of 12 min ($\tau = V/R_{flow}$ = (200 cm^3)/(16.6 cm^3/min) = 12 min). By reducing the buffer volume to 24 cm^3, with $r_{buf} \cong 3r$ (diameter 20 mm, length 75 mm), a τ of 1.5 min is obtained. However, an increased acoustical noise level was observed, due to the gas flow.

In photoacoustic measurements in the gas phase, microphones are usually employed as sensing elements of the acoustic waves generated by the heat deposition of the absorbing molecules. Although high-quality condenser microphones offer the best noise performance, they are rarely used in photoacoustic gas detection because of their large size, lower robustness, and relatively high price. The most common microphones employed are miniature electret devices originally developed as hearing aids. The choice of a miniature microphone is particularly advantageous since it can be readily incorporated in the resonant cavity without significantly degrading the Q of the resonance. The frequency response of electret microphones extends beyond 10 kHz, and the response to incident pressure waves is linear over many orders of magnitude.

In our PA cells there are four Knowles electret EK-3033 or EK-23024 miniature microphones in series (sensitivity 20 mV/Pa each at 564 Hz) mounted flush with the wall. They are situated at the loops of the standing wave pattern, at an angle of 90° to one another. The microphones are coupled to the resonator by holes (1 mm diameter) positioned on the central perimeter of the resonator. The battery-powered microphones are mounted in a Teflon ring pulled over the resonator tube (Fig. 14). It is of significant importance to prevent gas leakage from inside the resonator tube along the Teflon microphone holder, since minute spacing between the holder and resonator tube produces a dramatic decrease of the microphone signal and the Q value. The electrical output from these microphones is summed and the signal is selectively amplified by a two-phase lock-in amplifier tuned to the chopper frequency.

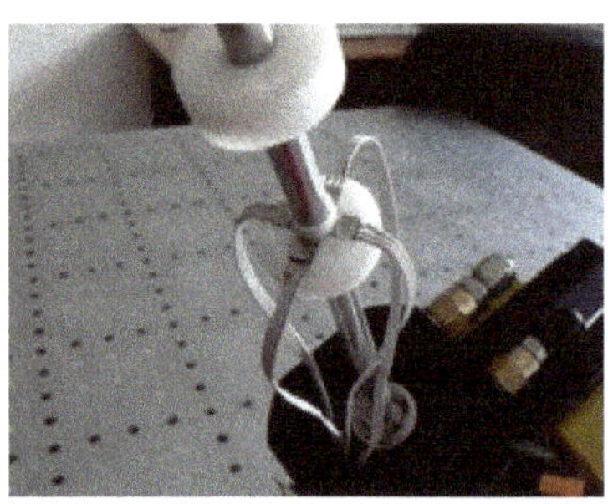

Fig. 14. Teflon rings used to mount the microphones flush with the tube wall.

The resonance curve of our PA cell (cell response in rms volts) was recorded as a function of laser beam chopping frequency and the results are plotted in Fig. 15. An accurate method is to construct the resonance curve point by point. In this case, the acoustic signal is measured at different fixed frequencies thus avoiding potential problems arising from the slow formation of a steady state standing wave in the resonator and the finite time resolution of the lock-in amplifier. It is evident from these data that the cell resonance curve is fairly broad, implying that the absorption measurements would not be considerably affected if the frequency of the laser modulation or the cell resonance itself were to shift by a few hertz. PA cells exhibiting narrow resonances (as in the case of excitation of the radial modes) require tight control of both temperature and laser modulation frequency to avoid responsivity losses during the experiment.

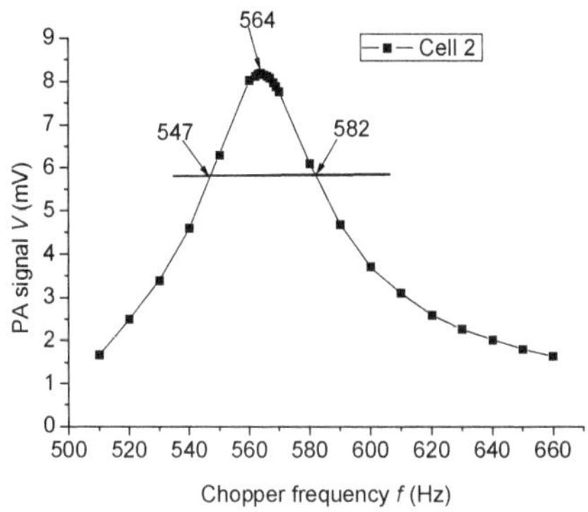

Fig. 15. Resonance curve for the first longitudinal mode of the PA cell.

The acoustic resonator is characterized by the quality factor *Q*, which is defined as the ratio of the resonance frequency to the frequency bandwidth between half power points. The amplitude of the microphone signal is $1/\sqrt{2}$ of the maximum amplitude at these points, because the energy of the standing wave is proportional to the square of the induced pressure. The quality factor was measured by filling the PA cell with 1 ppmV of ethylene buffered in nitrogen at a total pressure of 1 atm and by tuning the modulation frequency in 10 Hz increments (2 Hz increments near the top of the curve) across the resonance profile to estimate the half width, as described above. For this PA cell, the profile width at half intensity was 35 Hz, yielding a quality factor $Q = 16.1$ (Eq. 15, Part I) at a resonance frequency $f_0 = 564$ Hz. The experimentally determined resonance is not completely symmetric, as the curve rises steeply on one side and becomes less steep on the other side of the maximum. This asymmetry is caused by a coherent superposition of the standing acoustic waves in the detection region of the microphones (Karbach & Hess, 1985).

The calibration of the PA system is usually performed with a reference gas. We calibrated our PA cell with the widely used reference gas ethylene, whose absorption coefficients are accurately known at CO_2-laser wavelengths. Ethylene is well suited for this purpose, since it interacts only weakly with common cell surface materials. Ethylene is chemically inert, has the same molecular weight as nitrogen and possesses no permanent dipole moment which means negligible adsorption on the cell walls. Furthermore, its spectrum within the CO_2-laser wavelength range is highly structured. In particular it exhibits a characteristic absorption peak at the 10P(14) laser transition at 949.49 cm^{-1} which is caused by the proximity of the Q branch of the ν_7 vibration of C_2H_4 centered at 948.7715 cm^{-1}. We used a commercially prepared, certified mixture containing 0.96 ppmV C_2H_4 in pure nitrogen throughout our investigations. For calibration we examined this reference mixture at a total pressure p of approximately 1013 mbar and a temperature $T \cong 23^{o}C$ and using the commonly accepted value of the absorption coefficient of 30.4 $cm^{-1}atm^{-1}$ at the 10P(14) line of the $^{12}C^{16}O_2$ laser. For shorter time intervals (by changing the calibration gas mixture after a careful vacuum cleaning of the PA cell), the variation of the cell constant was smaller than

2%. The calibration also depends to some extent on the modulation waveform, since only the fundamental Fourier component of that waveform is resonant with, and hence significantly excites, the first longitudinal mode.

Using this PA cell and an optimized experimental arrangement we measured a cell responsivity R = 280 V cm/W. With the total responsivity of the four microphones $S_{M\,tot}$ = 80 mV/Pa (20 mV/Pa each) (Eq. 30, Part I), a cell constant C = 3500 Pa cm/W can be calculated (Eq. 28, Part I). A comparison of these PA cell parameters with other results reported in the literature is presented in Table 1. Different photoacoustic resonator designs such as longitudinal organ pipe resonators excited in the first longitudinal mode, closed longitudinal resonators excited in the second longitudinal mode, and cylindrical resonators excited in the first radial mode (Fig. 16) were used by various authors. As can be noticed from the table, the cell responsivity we obtained is one of the best values that have been reported up to now.

Authors	**Excitation mode**	**PA cell characteristics**	***R* (V cm/W)**	***C* (Pa cm/W)**
(Dumitras et al., 2007) - Our results	First longitudinal mode	L = 30 cm, f_0 = 564 Hz, 4 microphones, S_M = 80 mV/Pa, Q = 16.1	280	3500
(Crane, 1978)	Nonresonant operation	L = 20 cm, f = 33.3 Hz, 1 microphone	16.3	
(Gerlach & Amer, 1980)	First radial mode	6.54x15.56 cm^2, f_0 = 2.7 kHz, 1 microphone, S_M = 11mV/Pa + preamplif.x10, Q = 560	26.5	241
(Hubert, 1983)	Nonresonant operation	L = 20 cm, f = 695 Hz, 1 microphone	121[a]	
(Ryan et al., 1983)	First longitudinal mode	L = 30 cm, f_0 = 695 Hz, 1 microphone, Q = 17.4	56	
(Gandurin et al., 1986)	Nonresonant operation	L = 15 cm, $2r$ = 1.5 cm, 1 microphone, S_M = 50 mV/Pa	1-10	(20-200)[b]
(Bernegger & Sigrist, 1987)	Second longitudinal mode	L = 60 cm, f_0 = 555 Hz, 1 microphone, Q = 52		1990
(Sauren et al., 1989)	First longitudinal mode	L = 10 cm, f_0 = 1608 Hz, 1 microphone, S_M = 10 mV/Pa	(39)	3900
(Rooth et al., 1990)	First longitudinal mode	L = 30 cm, f_0 = 556 Hz, 4 microphones	114	
(Harren et al., 1990b)	First longitudinal mode	L = 30 cm, f_0 = 560 Hz, 4 microphones, S_M = 40 mV/Pa, Q = 16.4	(200)	5000
(Harren et al., 1990a)	First longitudinal mode	L = 30 cm, f_0 = 560 Hz, 4 microphones, S_M = 40 mV/Pa, Q = 20	(160)	4000
(Harren et al., 1990a)	First longitudinal mode (intracavity operation)	L = 10 cm, f_0 = 1653 Hz, 1 microphone, S_M = 10 mV/Pa, Q = 31.8	(37)	3700

Authors	Excitation mode	PA cell characteristics	*R* (V cm/W)	*C* (Pa cm/W)
(Meyer & Sigrist, 1990)	First radial mode	4.18x16 cm^2, f_0 = 2.65 kHz, 1 microphone, S_M = 12.5 mV/Pa, Q = 340	3.5	(280)
(Thöny & Sigrist, 1995)	Resonance in vicinity of the first radial mode	4.18x16 cm^2, f_0 = 2.86 kHz, 2 microphones, S_M = 25 mV/Pa, Q = 168	1.72	(69)
(Bijnen *et al.*, 1996)	First longitudinal mode	L = 15 cm, f_0 = 1030 Hz, 1 microphone, S_M = 1000 mV/Pa, Q = 40	(2000)	2000
(Fink et al., 1996)	First longitudinal mode (intracavity operation)	f_0 = 1986 Hz, 1 microphone, S_M = 26 mV/Pa	(52)	2000
(Harren et al., 1997)	Second longitudinal mode	L = 45 cm, f_0 = 555 Hz, 1 microphone, S_M = 1000 mV/Pa, Q = 43	(1.64)	1640
(Harren et al., 1997)	First longitudinal mode	L = 10 cm, f_0 = 1600 Hz, 3 microphones, S_M = 60 mV/Pa, Q = 32	(270)	4500
(Henningsen & Melander, 1997)	Nonresonant operation, pulsed excitation	Cell volume = 1 cm^3, f = 1600 Hz	1.45	
(Nägele & Sigrist, 2000)	First longitudinal mode in a multipass resonant PA cell	5x12 cm^2, f_0 = 1250 Hz, 16 microphones, S_M = 160 mV/Pa, Q = 70	260	(1625)
(Pushkarsky et al., 2002)	First longitudinal mode	f_0 = 1915 Hz, 1 microphone, S_M = 22 mV/Pa, Q = 49.7	(71.5)	3250[c]

[a] The value of 50 V cm/W reported for a rms signal of 162 mV was corrected for a peak-to-peak signal of 402 mV.
[b] In brackets are the values we deduced using the authors' specifications for microphone responsivity.
[c] These authors calculated the cell constant by using a PA signal $\sqrt{2}$ rms instead of a $2\sqrt{2}$ x rms signal; as a result, their cell constant (and responsivity) value is half the one we obtained by our methodology of calculus.

Table 1. Comparison of our results with different PA cells.

Based on the measured noises, background signals, and cell responsivity, all parameters characterizing the PA instrument can be evaluated (see Table 2). Some of them depend on the CO_2 laser and the PA cell, while others are determined by either the coherent acoustic background noise or the coherent photoacoustic background signal.

Several factors influence the lowest levels at which selected compounds can be detected by CO_2 laser spectroscopic techniques, within prescribed confidence limits, in a multicomponent mixture. These factors, some of which are interdependent, include (1) the sensitivity or minimum detectable absorptivity α_{min} (cm^{-1}) of the particular CO_2 laser detection technique employed; (2) the absorption coefficients and spectral uniqueness of both the compounds of interest and interferences; (3) the total number of compounds that absorb within the wavelength regions of the CO_2 laser output; (4) the wavelengths and number of CO_2 laser lines used for monitoring; and (5) the output power of the laser at each of these lines when the photoacoustic technique is employed. The minimum concentrations of various vapors that can be detected under interference-free conditions by the CO_2 laser photoacoustic technique are given by the relationship $c_{min} = \alpha_{min}/\alpha(\lambda)$. Here α_{min} is the minimum detectable absorptivity value for the photoacoustic detection system in units of cm^{-1}, and $\alpha(\lambda)$ is the absorption coefficient in units of $cm^{-1}atm^{-1}$ of the vapor of interest at the CO_2 laser monitoring wavelength. In order to determine the concentrations of the various gas mixture components, it is necessary to know the absolute absorption coefficients for every gas component at the laser wavelengths. Our CO_2 laser photoacoustic system with a α_{min} value of 2.7×10^{-8} cm^{-1} should provide interference-free minimum detectable concentrations between 0.9 ppbV and 270 ppbV for vapors with usual absorption coefficients of 30-0.1 $cm^{-1}atm^{-1}$.

Parameter/units	Value
Resonance frequency, f_0 (Hz)	564
Quality factor[a], Q	16.1
Cell responsivity[b], R (V cm/W)	280
Microphone responsivity[c], S_M (V/Pa)	$4 \times 20 \times 10^{-3} = 8 \times 10^{-2}$
Cell constant[d], C (Pa cm/W)	3.5×10^{3}
Pressure amplitude response[e], p/P_L (Pa/W)	10^{-1}
Limiting sensitivity of the cell[f], S_{cell} (W cm^{-1})	2.6×10^{-8}
Limiting sensitivity of the system[g], S_{sys} (cm^{-1}) (at 4.4 W laser power)	5.9×10^{-9}
Limiting measurable concentration of ethylene[h], c_{lim} (ppbV)	0.2
Minimum measurable signal in nitrogen[i], V_{min} (μV) (root mean square)	12
Minimum detectable pressure amplitude[j], p_{min} (Pa)	4.2×10^{-4}
Minimum detectable concentration[k], c_{min} (ppbV)	0.89
Minimum detectable absorptivity[l], α_{min} (cm^{-1})	2.7×10^{-8}
Minimum detectable absorption cross-section per molecule[m], σ_{min} (cm^2)	1.1×10^{-27}
Cell sensitivity for 1 ppbV of C_2H_4 at 1 W of unchopped laser power[n], V_{ppb} (μV at 1 ppbV)	3.0

[a] This quality factor value corresponds to a full bandwidth at the 0.707 amplitude points of $\Delta f = f_0/Q \cong 35$ Hz

[b] The cell responsivity is the signal per unit power per unit absorption coefficient; in our case, the signal per unit power is 11.6 mV/4.0 W = 2.9×10^{-3} V/W (rms value) or 8.2×10^{-3} V/W (peak-to-peak value) for 0.96 ppmV of C_2H_4 (the absorption coefficient $\alpha' = 30.4$ cm^{-1} atm^{-1} x 0.96×10^{-6} atm = 2.92×10^{-5} cm^{-1}, where $\alpha = 30.4$ cm^{-1} atm^{-1} is the absorption coefficient of C_2H_4 at 10P(14) line of the CO_2 laser), so that R = 8.2×10^{-3} V/W/2.92×10^{-5} $cm^{-1} \cong 280$ V cm/W; the same responsivity was obtained with the etalon mixture of 10 ppmV of C_2H_4 in N_2: $R = 8.4 \times 10^{-2}$ V/W/3×10^{-4} $cm^{-1} \cong 280$ V cm/W

[c] The microphone responsivity is determined from the Knowles data-sheet for the microphone type 3033: 54 dB (≅500) attenuation at 564 Hz from 1 V/0.1 Pa, leading to $S_M \cong 20$ mV/Pa

[d] The cell constant is the pressure amplitude per unit absorption coefficient per unit power: $C = R/S_M$

[e] The pressure amplitude response per unit incident power for 1 ppmV of C_2H_4 is $p/P_L = C\alpha'$

[f] The limiting sensitivity of the cell is $S_{cell} = 2\sqrt{2}\ V_N^{ac} / R$ (several authors, e.g., (Harren et al., 1990a) used the rms value of the voltage instead of its peak-to-peak value, resulting in a limiting sensitivity of the cell and of the system and the limiting measurable concentration of ethylene lower by a factor of 2.84; other authors, e.g., (Kosterev et al., 2005), used a parameter named "sensitivity to absorption",

defined as $D = \alpha_{\min}^{el} P_L / \sqrt{\Delta f} = S_{cell} / \sqrt{\Delta f}$, where $\alpha_{\min}^{el}$ is the absorption coefficient corresponding to the coherent acoustical background noise-equivalent signal V_N^{ac} , P_L is the laser excitation power, and Δf is the detection bandwidth).

[g] The limiting sensitivity of the system or the microphone noise-limited minimum detectable absorption coefficient (the equivalent bulk absorption coefficient in the gas) is $S_{sys} = S_{cell}/ P_L$ (the minimum detectable absorption strength is defined as the strength that gives a SNR at the transducer output equal to one).

[h] The limiting sensitivity of the system gives a limiting measurable concentration of ethylene of $c_{lim} = S_{sys}/ \alpha$.

[i] The minimum measurable signal in nitrogen, as determined by the coherent photoacoustical background signal, is $V_{min} = V_N^b$ (nitrogen) x P_L.

[j] The minimum detectable peak-to-peak pressure amplitude can be determined by dividing the minimum measurable signal in nitrogen by the responsivity of the microphone: $p_{min} = 2\sqrt{2}\, V_{min}/ S_M$.

[k] The minimum measurable signal in nitrogen, limited by the synchronous background signal, gives a minimum detectable concentration $c_{min} = 2\sqrt{2}\, V_{min}/ \alpha 2P_{meas}R$, where P_{meas} is the measured laser power after chopper: $P_L = 2P_{meas}$.

[l] The minimum detectable absorption coefficient or the equivalent absorption coefficient of C_2H_4 for a minimum detectable concentration (the absorption coefficient corresponding to the synchronous background signal) is $\alpha_{min} = c_{min}\alpha = 2\sqrt{2}\, V_N^b / R$ (or the ratio between the peak-to-peak value of the synchronous background signal and the cell responsivity).

[m] The minimum detectable absorption cross-section per molecule of ethylene determined by the synchronous background signal is the ratio between the equivalent absorption coefficient of C_2H_4 for minimum detectable concentration and the number of absorbing molecules per unit volume: $\sigma_{min} = \alpha_{min}/ N_{tot}$, where N_{tot} is the number of absorbing molecules per cubic centimeter ($N_{tot} = 2.5 \times 10^{19}$ cm^{-3} at 1013 mbar and 20°C).

[n] Knowing the cell responsivity, we can determine the sensitivity of the photoacoustic cell (the rms voltage amplitude measured by the lock-in amplifier) to measure 1 ppb of a given gas at a given laser frequency with 1 W of unchopped laser power: $2\sqrt{2}\, V_{ppb} = \alpha P_L R c$ ($\alpha = 30.4$ cm^{-1} atm^{-1}, $P_L = 1$ W, $R = 280$ V cm/W, and $c = 10^{-9}$ atm).

Table 2. PA cell parameters.

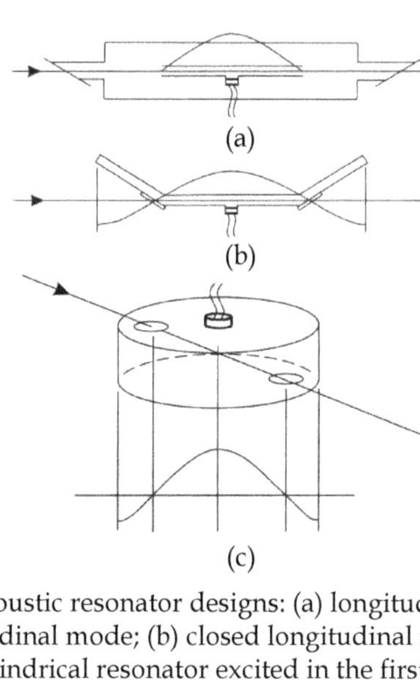

Fig. 16. Different photoacoustic resonator designs: (a) longitudinal organ pipe resonator, excited in the first longitudinal mode; (b) closed longitudinal resonator excited in the second longitudinal mode; (c) cylindrical resonator excited in the first radial mode.

A comparison of the results we obtained by using an extracavity PA cell and the experimental parameters measured using intracavity PA cells is given in Table 3. If we take into consideration only the parameters determined by the coherent acoustical background noise (SNR = 1), then the minimum detectable absorptivity S_{sys} (cm^{-1}) is much lower (1-2 orders of magnitude) in the intracavity arrangements due to the increased laser power. Unfortunately, only SNR is often considered in the literature, which yields an extrapolated detection limit that may be considerably too small. In reality, other background signals such as window absorption limit the ultimate sensitivity. These background signals must always be taken into consideration, as they can only be reduced, but not eliminated. In real PA instruments, the minimum measurable signal V_{min} is higher in extracavity PA cells (3 times in our case) and much higher in intracavity PA cells (hundreds or even thousands of times) than the coherent acoustic background noise V_N^{ac}. From this table it clearly follows that the best sensitivity is obtained with our extracavity PA instrument, with α_{min} (cm^{-1}) being better by one or two orders of magnitude than in intracavity arrangements.

Parameter	Our results (Dumitras et al., 2007)	(Harren et al., 1990a)	(Fink et al., 1996)
PA cell and CO_2 laser			
R (V cm/W)	280	37	52
$S_{M\,tot}$ (mV/Pa)	80	10	26
C (Pa cm/W)	3500	3700	2000
P_L (W)	4.4	100	40
p/P_L (Pa/W)	10^{-1}	1.1×10^{-1}	6×10^{-2}
Parameters determined by the coherent acoustic background noise (SNR = 1)			
V_N^{ac} (rms)[a] (μV)	2.6	0.5	0.81[b]
S_{cell} (W cm^{-1})	2.6×10^{-8}	4.0×10^{-8} [c]	4.4×10^{-8}
S_{sys} (cm^{-1})	5.9×10^{-9}	5.1×10^{-10} [d]	1.1×10^{-9}
c_{lim} (pptV)	200	17[e]	34
Parameters determined by the coherent photoacoustic background signal (SBR = 1)			
V_N^b (rms) (μV/W)	2.7	1.5	26
V_{min} (rms) (μV)	12	117[d]	1040
p_{min} (Pa)	4.2×10^{-4}	3.3×10^{-2}	1.1×10^{-1}
c_{min} (ppbV)	0.9	3.8	46
α_{min} (cm^{-1})	2.7×10^{-8}	1.2×10^{-7}	1.4×10^{-6}

[a] For a bandwidth of 1 Hz.
[b] The authors claim that they determined a coherent acoustical background noise of 0.1 μV, but their measured S_{sys} and c_{lim} correspond to V_N^{ac} = 0.81 μV.
[c] The value of $S_{cell} = 1.4 \times 10^{-8}$ cm^{-1}, as cited by the authors, was corrected for a peak-to-peak value of the coherent acoustical background noise.
[d] A factor of 0.78 was introduced either in the absorption coefficient of ethylene or in the intracavity laser power to compensate for the influence of saturation; the value of $S_{sys} = 1.8 \times 10^{-10}$ cm^{-1} claimed by the authors was corrected for a peak-to-peak value of the coherent acoustical background noise.
[e] The value of 6 pptV claimed by authors was corrected for a peak-to-peak value of the coherent acoustical background noise

Table 3. Comparison of our results (extracavity PA cell) with the experimental parameters determined with intracavity PA cells.

To keep the flow noise at a sufficiently low level, the flow must be in the laminar regime. Another practical requirement is the time response, which is determined by the gas sample exchange rate in the resonant cell. In addition, a delay occurs because the gas flows from the inlet to the resonant cell first through the acoustic filter. A response time of < 1 s and a delay of < 10 s may occur in the continuous flow mode. Taking into account the largest dimension and the limiting value of the Reynolds number ($Re \leq 2300$ for laminar flow), the flow velocity should not exceed 1.7 m/s. This value is far too large, since it would give a flow rate of 4.8 L/min (large gas consumption). As an operating value a flow rate of up to 0.5 L/min was used. With this value, the maximum flow velocity was about 15 cm/s, the Reynolds number $Re < 200$, and the calculated response time $\tau < 0.7$ s. The response time determined from the rise time of the PA signal is longer due to gas mixing in the buffer volumes of the cell. Nevertheless, the measured response times are below 10 s for nonadsorbing gases. Adsorption on the PA cell may increase the response time significantly. Note that the adsorption effect can be effectively reduced by using appropriate wall materials and higher wall temperatures. The adsorption effects can prevent accurate determination of the vapor pressure, since even with an *in situ* sample the vapor pressure will equilibrate between the rate of vapor emission and rate of plating out.

Let us consider the system to be a cell with constant volume. For an ideal gas, $dT/dE = (nC_v)^{-1}$, so that $dp/dE = R/(C_vV)$. Here T is the temperature, p is the pressure, E is the absorbed energy, n is the number of moles of the gas, C_v is the molar heat capacity at constant volume of the gas mixture, R is the gas constant, and V is the volume of the cell. For a real photoacoustic system the contents of the cell do not completely equilibrate at the modulation frequency; thus the above equations are a simplification. However, they do emphasize the basic constraints for maximizing the photoacoustic signal; (1) the gas mixture in the cell should have the lowest possible molar heat capacity, and (2) the effective volume of the cell should be as small as possible. The latter is also advantageous when the sample amount is limited. The ideal conditions for PA spectroscopy are a gas mixture consisting of a small amount of sample buffered in a large amount of a nonabsorbing gas with a low C_v, such as rare gases, inside a cell with the smallest possible volume. Since the laser beam has cylindrical symmetry, the best way to minimize the volume is to use a cylindrically symmetric cell with an internal diameter barely larger than the diameter of the laser beam.

Cylindrical cells, without additions, such as baffles, gas valves, sample fingers, or microphone tubes, and having a diameter of less than about 1/4 of wavelength, behave as simple pipes. Since the speed of sound is inversely proportional to the mass density of the gas (see Eq. 6, Part I), the frequency is lower for a more massive gas. The speed of sound is effectively independent of the total buffer gas pressure. Some works suggest that using the heavier noble gases as buffers increases the signal-to-noise level in acoustically resonant PA spectroscopy (Thomas III et al., 1978).

Davidson et al. (Davidson et al., 1990) investigated the importance of window noise and the role of acoustic baffles in photoacoustic spectroscopy. Small amounts of dirt or imperfections can cause heating at the windows and thus the production of a photoacoustic background signal. Window absorption is a major problem for intracavity photoacoustic spectroscopy because of the high light intensity inside the cavity. The signature of window noise is that it is laser frequency independent; its intensity tracks the intensity of the exciting radiation. This signal can actually mask the signal of interest (see spectrum D in Fig. 17). It was shown that the window noise should decrease with increasing modulation frequency

(Rosengren, 1975), which suggests that a relatively high modulation frequency is advantageous. The approach to this problem is to keep the windows as clean as possible and place acoustic baffles between the windows and the body of the sample cell where the microphones are mounted.

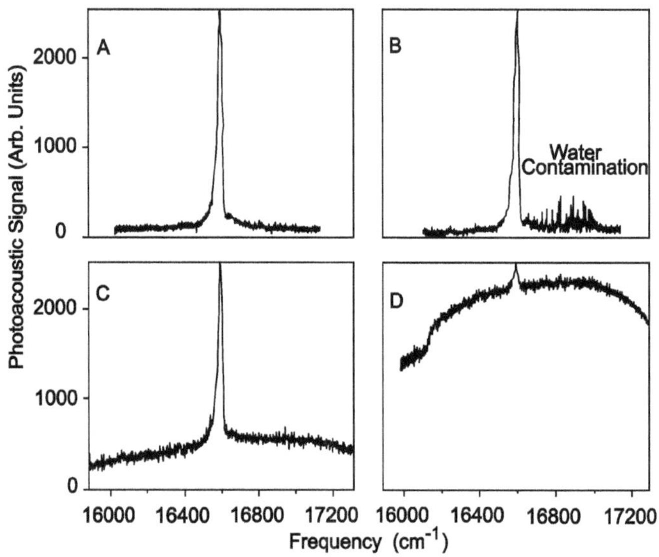

Fig. 17. Photoacoustic spectra: (A) clean windows on baffled cell; (B) clean windows on unbaffled cell; (C) dirty windows on baffled cell; (D) dirty windows on unbaffled cell (Davidson et al., 1990).

The acoustic baffles hinder the propagation of the window noise into the central region of the cell near the microphones. In an attempt to quantify the usefulness of the baffles, four spectra are shown in Fig. 17. These spectra illustrate the microphone signal versus laser frequency. For both the baffled and unbaffled cells, a longitudinal resonance frequency was used. From this figure it is immediately obvious how important it is to have clean windows. If the sample makes it difficult to keep the windows clean, a cell with baffles will perform much better than one without them. Even with clean windows, the baffles give a flatter base line.

The random noise level visible on the base lines of spectra A and B in Fig. 17 is probably due to a combination of the ambient lab noise, noise from the microphones, associated electronics, and the fluctuations of the laser itself. The ultimate limit of a microphone's sensitivity is set by the random thermal fluctuations in the sample and of the microphone diaphragm. In practice, the random fluctuations of the laser do not seem to be critical. Also, the combined electronic noise of the lock-in, preamplifier, and FET amplifier in the microphones totaled about 3 μV. This is at least a factor of 10 less than the noise level observed in the spectra. The noise that is visible in the spectra stems almost exclusively from thermal fluctuations and ambient lab noise. One of the major ambient noise sources is the mechanical chopper. In practice, this seems to set the detection limit. Acoustically isolating the chopper improves the noise, but replacing it with a nonmechanical modulator is better still, as it also speeds up modulation.

In an attempt to reduce the effects of microphone noise, the output of two microphones was summed in the low-noise preamplifier. Two microphones seem better than one when observing the noise level, but the difference is not as obvious in the spectra that were taken. However, the absolute signal level does increase with an increasing number of microphones; when the sample is weakly absorbing, more microphones are an advantage.

High isolation against window absorption may be obtained by introducing acoustic baffles between the windows and the resonator. However, the resonator also introduces additional noise. Pressure fluctuations from turbulence in the gas and acoustic flux reaching the resonator from the surroundings will be amplified along with the signal. Furthermore the amplification provided by a resonator is accompanied by an increased sensitivity to parameters which affect the resonance frequency, such as gas composition, temperature, and pressure.

Besides cleaning the cell windows, a very careful rinsing of the inner walls of the cell is also very important. If the inner of the cell is not properly cleaned before a measurement, a considerable drift of the photoacoustic background signal is observed if the gas flow is interrupted. This fact demonstrates that the desorption of IR-absorbing gases and vapors from the cell walls can make a large contribution to the background signal. That is why no measurement started until after the PA cell had been rinsed with pure nitrogen till the coherent photoacoustical background noise reached the minimum value of 2.7 μV/W.

To distinguish the gas absorption signal from other signals (e.g., from the walls, windows, or interfering gases), one has to switch the CO_2 laser to other laser lines. However, repositioning the laser beam to its original wavelength can change the configuration of the laser cavity (deviation in grating position, thermal drift) and result in irreproducible absorption signals if the operation is not carefully conducted. Using a CO_2 laser stabilized on the top of the gain curve ensures that both the laser frequency and output power are reestablished with high accuracy when the laser operation is changed from one line to another.

2.4 Gas handling system

Whenever monitoring is performed by flowing the gas mixture through a cell, a crucial question is whether the measured signal, which represents the trace gas concentration in the interaction region, also reflects the concentration at the source. On their way to the cell, the different components of a gas mixture may react with one another, form clusters or aerosols, and react with or be adsorbed on particles present, or on the sampling line and cell walls. Adsorption problems are particularly severe for polar molecules with large dipole moments, such as water and ammonia, but they can be reduced by a proper choice of materials.

The vacuum/gas handling system is an important element in these measurements owing to its role in ensuring PA cell and gas purity. The Teflon/stainless steel system can perform several functions without necessitating any disconnections. It can be used to pump out the cell, mix gases in the desired proportions, and monitor the total pressure of gases. Whenever possible, the PA cell was employed in the gas flow mode of operation to minimize any tendency for the vapor to stick to the cell walls and the effects of the subsequent outgassing of contaminants, which would otherwise lead to increasing background signals during an experimental run.

To design an efficient vacuum/gas handling system to be used in LPAS, one must make sure that the following operations can be carried out:

i. evacuation by the vacuum system of the entire gas handling system, including the PA cell, either totally or in different sections;
ii. controlled introduction of a gas or gas mixture either for rinsing the PA cell and the gas handling system with pure nitrogen or for calibrating the PA spectrometer with a certified gas mixture;
iii. pressure measurement in the PA cell and in different sections of the system;
iv. safe insertion in the gas handling system of a sample cuvette (usually made of Pyrex glass) or aluminum-coated plastic bag with the trace gas sample;
v. filtration of certain gases (carbon dioxide and water vapors), which interfere with the trace gas to be measured;
vi. controlled introduction of the trace gas to be measured from the sample cuvette or bag into the PA cell by a nonabsorbing gas (nitrogen or synthetic air) acting as carrier;
vii. controlled change of the sample and carrier gas flow rates within a broad range (10-1000 sccm);
viii. simultaneous measurement of two sample gases (e.g., ethylene and ammonia);
ix. quick monitoring of the trace gas concentration in the sample gas by ensuring a response time on the order of minutes or even seconds.

A vacuum/gas handling system to be used in PA experiments was designed and implemented based on these guidelines. The schematic of the gas handling chain is shown in Fig. 18, while a general view of the valves and distribution system is given in Fig. 19.

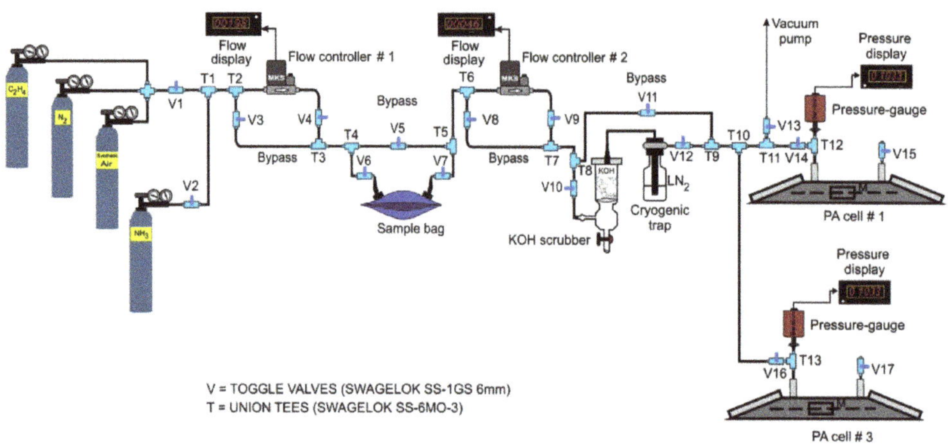

Fig. 18. Gas handling system.

Gas transport lines throughout the gas mixing station were made of Teflon (Swagelok PFA-T6M-1M-30M, 6 mm inner diameter and 1 mm wall thickness) to minimize adsorption and contamination. The toggle valves V1-V17 (Swagelok SS-1GS6mm) and union tees T1-T11 (Swagelok SS-6MO-3) were made of stainless steel. No valve grease was used. The PA cell gas inlet and outlet were connected to the gas handling system with Swagelok fittings (male connectors SS-6MO-1-2RT). Connections to the inlet and outlet valves of the PA cell were made via flexible Teflon tubing so as to minimize the coupling of mechanical vibrations to the PA cell. The flexible lines also make it possible to position the PA cell during optical alignment.

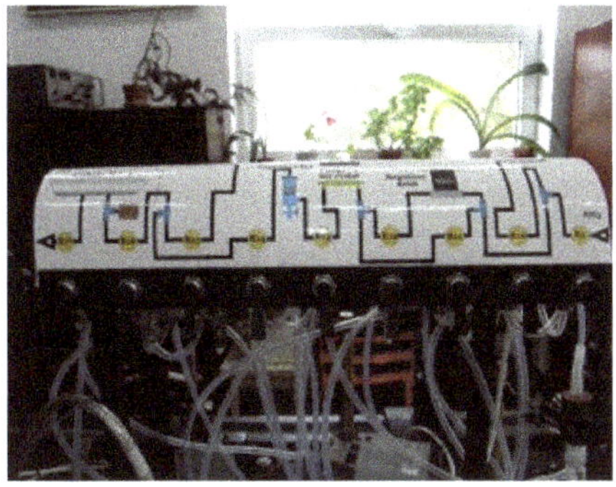

Fig. 19. General view of the gas mixing station.

The pressure of the gases added to the PA cell was determined by means of three Baratron pressure gauges (MKS Instruments, Inc.): 622A (0-1000 mbar), 122A (0-1000 mbar), and 122A (0-100 mbar), connected to a digital two-channel unit PDR-C-2C.

We use thermal mass flowmeters, or mass flow controllers (MFC), to deliver stable and known gas flows to the PA cell. The most critical processes will require flow measurement accuracies of 1% or better in the range 1000 to 10 sccm ($7x10^{-4}$ to $7x10^{-6}$ mol/sec; 1 sccm (at 0^oC) = $7.436x10^{-7}$ mol/sec). The digital MFCs sense the mass flow from the temperature difference between two temperature sensors in thermal contact with the gas stream and then process the information digitally with a microcontroller. The analog sensor output is amplified and digitized before it is sent to a microprocessor to compute the final control valve position. The gas flow in our gas handling system is adjusted by two gas flow controllers, MKS 1179A (0 - 1000 sccm) and MKS 2259CC (0 - 200 sccm), which are connected to a digital four-channel instrument MKS 247C.

By using an adequate scrubber for CO_2 filtration, the CO_2 interference problem can be resolved. The CO_2 trap must neither alter the ethylene concentration level, nor introduce new interfering gases. By using a CO_2 trap with a volume of 120 cm^3 filled with fresh KOH pellets, we succeeded to reduce the CO_2 content in the exhaled breath (mixt expiratory air collection) of a healthy nonsmoking young person from 3.4% (equivalent to an ethylene concentration of 2.35 ppmV) to 156 ppmV (equivalent to an ethylene concentration of 10.8 ppbV), that is a reduction factor of 218 (see Section 2.7).

The water vapors could be additionally filtered by a cryogenic trap filled with liquid nitrogen. Using a simple and small cryogenic trap, we demonstrated the negative influence in our experiments (Dumitras et al., 2008). The liquid nitrogen temperature -196^oC (77 K) is bellow the frozen point of the ethylene gas -169.2^oC (104 K), so the practical effects is just to frozen both water vapors and ethylene. Introducing in the flow a calibrated mixture of 1 ppm C_2H_4 in N_2, we observed after filling the cell at 1 atm pressure that the maximum ethylene concentration is only 51 ppb, diminished by a factor of 20 from the initial ethylene concentration. The level starts to increase suddenly at the point where we stopped the liquid nitrogen admission in the trap. In conclusion, a simple nitrogen trap is not suited for our experiments involving ethylene, but a special thermocontrolled trap can do the job, setting the working temperature below -78.5^oC (194.5K), the freezing point of CO_2, but above -169.2^oC (104 K), the freezing point of ethylene.

The following gases were used throughout the experiments:

- ethylene: Linde Gaz Romania, 0.96 ppmV (±5%) C_2H_4 diluted in nitrogen 5.0 (purity 99.999%) and 9.88 ppmV (±2%) C_2H_4 diluted in nitrogen 6.0 (purity 99.9999%)
- nitrogen: Linde Gaz Romania, nitrogen 5.0 (purity 99.999%) and 6.0 (purity 99.9999%);
- synthetic air: Linde Gaz Romania, 20% oxygen and 80% nitrogen (impurities: hydrocarbons max. 0.1 ppmV, nitrogen oxides max. 0.1 ppmV);
- carbon dioxide: Linde Gaz Romania, purity 99.95% (impurities: oils max. 1 ppmV);
- ammonia: Linde Gaz Romania, purity 99.98% and 9.66 ppmV (±5%) NH_3 diluted in nitrogen 6.0 (purity 99.9999%).

The flow rate was usually set at a low value of 30-100 sccm in all experiments in order to eliminate the acoustic noise of the gas flow, and all measurements were carried out with the PA cell at atmospheric pressure. Flow noise increases upwards of 10 L/h (167 sccm) were found to limit the minimum response time of the detector. The flow velocity minimizes the accumulation of the produced gases in the sampling cell. The carrier gas we used was either nitrogen or synthetic air, and its flow rate through the system was monitored by the calibrated flowmeter. Provision is made for bypassing the flowmeter with the gas mixture flow prior to a measurement to equilibrate the feedline surfaces. This ensures that the measured rise times are an exclusive function of the cell characteristics. A measurement is initiated by diverting the gas flow from the bypass through the flowmeter and PA cell and monitoring the photoacoustic signal rise that follows.

As far as the sampling procedure is concerned, we use an extractive method, based on the collection of trace gas samples by some type of container or collecting medium and subsequent analysis in the laboratory. A problem may arise at this point due to some alterations of the gas composition caused by adsorption and desorption processes on the inner surface of the collecting container. The breath samples we analyzed were obtained from volunteers who agreed to provide such samples at certain time intervals. The volunteers were asked to exhale into a sample bag with a normal exhalation flow rate. The breath samples were collected in 0.75-liter aluminum-coated bags (QuinTron, Milwaukee, Wisconsin, USA) equipped with valves that sealed them after filling (Fig. 20). The bags were inserted into the gas handling system, which ensured a better control by means of two independently adjusted flow controllers of the upstream pressure and the flow rate through the sample bag.

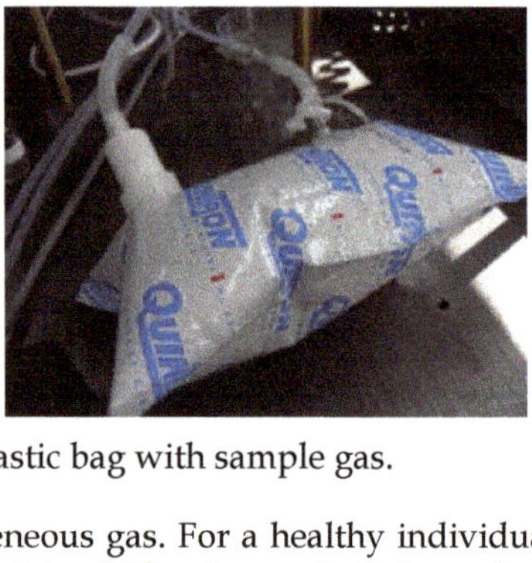

Fig. 20. Aluminum-coated plastic bag with sample gas.

The exhaled air is a heterogeneous gas. For a healthy individual, the first part of a exhaled breath, roughly 150 mL, consists of "dead-space" air from the upper airways (such as the mouth and trachea), where air does not come into contact with the alveoli of the lungs. The following part of a breath, about 350 mL, is "alveolar" breath, which comes from the lungs, where gaseous exchange between the blood and breath air takes place. Dead space air can be interpreted as essential for the detection, and depends on the type of molecule detected from the breath test. For example, the dead-space is used to quantify the amount of the NO molecules. In the case of an asthmatic patient, if the airways are inflamed, a high-level of NO is released into the airways and into the dead-space air. But for volatile organic compounds (VOCs) exchanged between blood and alveolar air, the dead-space air is a "contaminant" diluting the concentrations of VOCs when breath air is collected. In terms of the origin of the collected breath gases, there are three basic collection approaches: 1. *upper airway collection* for NO test; this means that only dead-space gas is collected (it is only for the NO test); 2. *alveolar collection*; this means that pure alveolar gas is collected (for tests of other inorganic gases and VOCs); 3. *mixed expiratory collection*; this means that total breath air, including dead-space air and alveolar gas is collected (appropriate for tests of special gases and VOCs). Because the mixed expiratory collection method is easy to perform in spontaneously breathing subjects requiring no additional equipment, it has been most frequently used in practical applications. However, concentrations of endogenous substances in alveolar air are two to three times higher than those found in mixed expiratory samples, because there is no dilution by dead-space gas. Collection of breath air can be performed for a single breath or for collection of individual breathes over a certain period of time. If the sample is collected through a single breath, one has to be sure that this single breath is representative.

The rate of change in concentration of a species i in a flowing cell is given by:

$$\frac{dC_i(t)}{dt} = \frac{R_{flow}}{V}\left[F_i - C_i(t)\right], \tag{4}$$

where R_{flow} is the gas flow rate (liter/min or sccm), F_i is the feed concentration of species i, and V is the cell volume. This equation assumes that the adsorption rate of i is zero, and that gas mixing inside the cell is instantaneous. Integrating Eq. (4) with the initial condition that $C_i(0) = 0$ gives

$$C_i(t) = F_i\left[1 - exp\left(-R_{flow}t / V\right)\right], \tag{5}$$

and $\tau = V/R_{flow}$ is the renewal time constant.

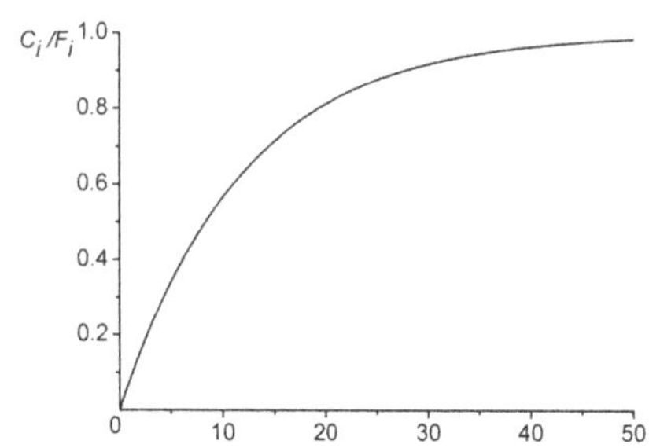

Fig. 21. Time response of the PA spectrometer for a gas flow rate of 100 sccm.

The solid line in Fig. 21 shows the rise time curve predicted by Eq. (5) for the experimental parameter values: V = 1200 cm^3 and R_{flow} = 100 sccm. The total renewal time of the gas content in the system (sampling cell, scrubber, and photoacoustic cell) is τ = 12 min (1/e time). This τ value is small compared to the time response of certain biological samples (e.g., the C_2H_4 production of a single flower, 0.02-0.3 nL/g/h) (Harren et al., 1990b).

2.5 Data acquisition and processing

The acquisition and processing of the recorded data was done with Keithley TestPoint software. TestPoint data acquisition software provides a development environment in which data acquisition applications can be generated. A graphical editor is provided for creating a user interface, or "panel", which the user sees and interacts with as the application executes. A user panel is made of pictorial elements that represent such things as switches, variable controls, numerical, text and selection boxes, bar displays, graphs, and strip charts. In addition, an application editor is provided, which ensures some interactive means of specifying how the visual elements on the user panel interact with the data sources and processing functions to achieve application goals. TestPoint uses an automated textual description of the operations carried out by each user panel element.

We developed a modular software architecture aimed at controlling the experiments, collecting data, and preprocessing information. It helps automate the process of collecting and processing experimental results. The software controls the chopper frequency, transfers powermeter readings, normalizes data, and automatically stores files. It allows the user to set parameters such as the PA cell responsivity (a constant used to normalize raw data), gas absorption coefficient, number of averaged samples at every measurement point, sample acquisition rate, and total number of measurement points. This software interfaces the following instruments:

- lock-in amplifier;
- chopper;
- laser powermeter;
- gas flowmeter.

The software user interface allows the user to set or read input data and instantaneous values for the PA voltage (rms), average laser power after chopper, and trace gas concentration. Users may set experimental parameters for the PA cell responsivity and gas absorption coefficient. They are also provided with a text input to write a description of the experiments or take other notes. The user interface also provides data visualization (Fig. 22).

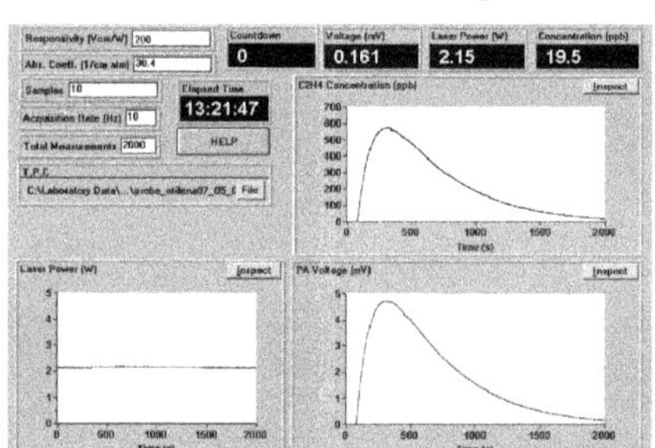

Fig. 22. Software user interface used to record trace gas concentrations.

The software user interface contains three panels which display in real time the following parameters: CO_2 laser power level; PA signal; and trace gas concentration. Another window (countdown) indicates the number of remaining measurement points.

All settings and properties are stored to disk from session to session. In addition, a file may be automatically generated when running an experiment, including: a) **Laser power** stores powermeter readings of power incident on the sample as a function of time; b) **PA signal** stores the instantaneous values of the PA signal measured by the lock-in amplifier as a function of time; c) **Trace gas concentration** stores the time evolution of the trace gas concentration for a given laser wavelength.

2.6 Low power vs. high power lasers

We designed and characterized two experimental set-ups with the PA cell in an external configuration: the first one with a low power CO_2 laser where the saturation effects are negligible, and a second one with a high power CO_2 laser where the saturation effects are important and have to be taken into consideration. We measured all relevant features and determined all quantities used in literature to compare our findings with the best results reported in the previous published papers. All measurements were done in nitrogen and ethylene with the 10P(14) line of a cw CO_2 laser. We succeeded to obtain a minimum detectable concentration better by more than a factor of 10 compared to the best results previously reported in the literature.

To investigate the possibility of using a high power laser in an extracavity configuration, we introduced in the experimental set-up a commercial CO_2 laser (Coherent GEM SELECT 50™ laser) with output power till 50 W and tunable on 73 different lines (Dumitras et al., 2010). When this laser is tuned on 10P(14) line, the maximum power delivered after chopper and focusing lens is 14.5 W.

To change the laser power inside the PA cell we tried either to modify the input power in the laser (RF power supply) or to introduce a beam splitter in the path of the laser beam. Unfortunately, both methods change significantly the beam path inside the PA cell, thus perturbing unacceptably the results of the measurements. The waveguide laser has a poor beam pointing because it has a short optical resonator and the variation of the transverse RF excitation modifies the laser gain profile.

The saturation effects at high laser power were investigated by using the method of truncation of a gaussian laser beam. This approach was possible because the laser beam is very close to a gaussian beam ($M^2 < 1.1$). The method consists in passing the beam through an aperture with known diameter (Fig. 23). To avoid deformations owing to heating, we used water cooled metallic diaphragms with diameters between 1.42 mm and 5.03 mm. All diaphragms were placed at a distance of 450 mm from the beam waist of the laser.

When a gaussian beam of radius w is truncated by an aperture of radius a (Fig. 24), the power transmitted through the aperture is $T = P(a)/P = 1 - \exp(-2a^2/w^2)$. When $2a = 2w$, $T \cong 86\%$, that is 86% of the laser power is transmitted through the aperture (this is known as 86% criterion). When $2a = \pi w$, $T \cong 99\%$, that is 99% of the laser power is transmitted through the aperture and we have the 99% criterion. This formula offers a possibility to measure precisely the diameter of the laser beam at the position of the diaphragm. By knowing the radius of the aperture (a) and by measuring the laser power before and after the aperture (P

and *P*(*a*), respectively), we can determine immediately the radius of the laser beam (w). As it can be seen in Fig. 25, by using five different diaphragms, the resulting average diameter is $2w$ = (7.09 ± 0.2) mm, with an error of less than 3%.

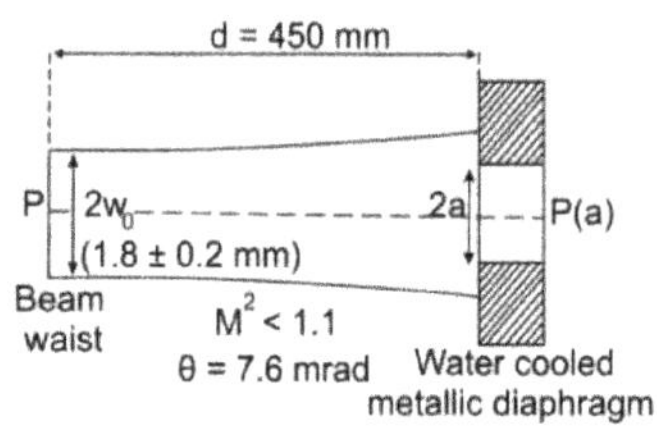

Fig. 23. Attenuation of a laser beam by a diaphragm.

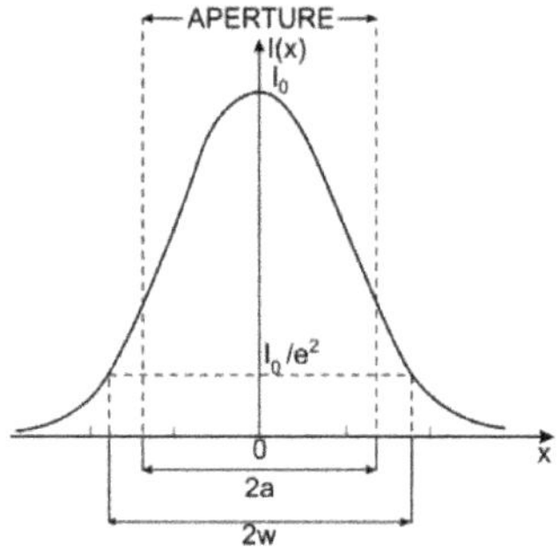

Fig. 24. Truncation of a gaussian laser beam.

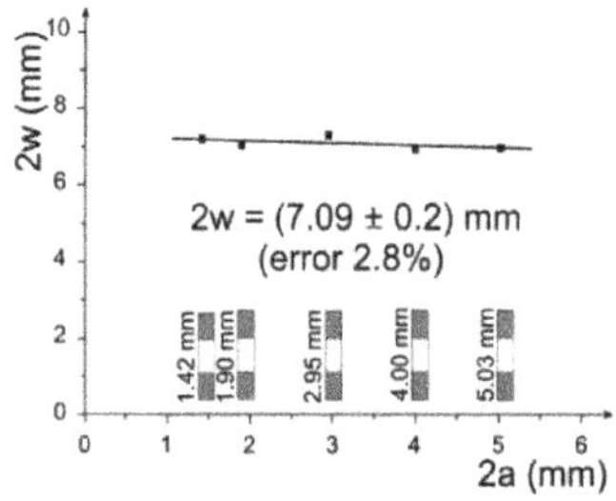

Fig. 25. Measurement of laser beam diameter by the method of truncation.

Figure 26 shows the attenuation of the laser beam when different diaphragms were placed in its path. The solid line is the theoretical curve given by the above equation. By introducing these five diaphragms, the laser power was varied between less than 2 W and near 10 W. In this way, we were able to investigate the laser power range from low power where saturation effects have no significance till high power where saturation effects manifest strongly.

We investigated the influence of saturation by measuring the dependence of the PA cell responsivity function on laser power (Fig. 27). From low laser power regime (under 2 W) where the saturation effects are not important till high power regime (14.5 W, no diaphragm), the PA cell responsivity decreases from 312 V cm/W till 52 V cm/W, that is by a factor of 6. We can observe that the saturation effects manifest immediately as the laser power is increased more than 2 W. The cause of the saturation is that the collisional relaxation to de-excite the molecules cannot keep up with the excitation rate by the laser beam intensity. By increasing laser intensity, the excitation pumping rate of the molecules grows higher and a molecule is more likely to absorb a nearby photon before it relaxes to the ground state. So, as the number of molecules in the excited state increases, the number of molecules which can absorb laser radiation is reduced. That is why we introduced a supplementary scale in Fig. 27, representing the cell responsivity function on the intensity of the focused beam inside the PA cell. A thorough analysis of the laser beam propagation through the focusing lens and in the PA cell was done (Dumitras et al., 2007). For a lens with a focal length of 400 mm, we got in this case a beam diameter at beam waist of 0.89 mm. It can be remarked that saturation starts at laser intensities greater than 0.5 kW/cm^2. This result shows that saturation effects manifest even at low laser intensities. Previously, Harren et al. (Harren et al., 1990a) observed a strong saturation at a much higher intensity (200 kW/cm^2; the laser power was ten times higher and the beam area was ten times smaller than in our case at 13 W), when the PA cell was placed intracavity of a waveguide CO_2 laser. Our conclusion is that high power lasers could be used in PA systems, but saturation effects should be taken into consideration (by making a correlation between the PA cell responsivity and the working laser intensity, as in Fig. 27).

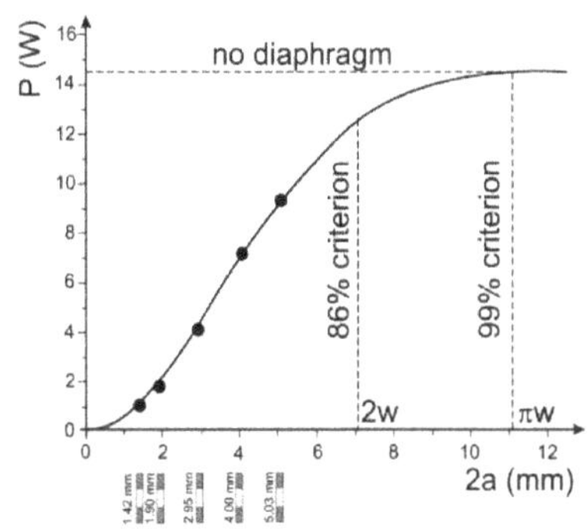

Fig. 26. Variation of laser power function on diaphragm aperture size.

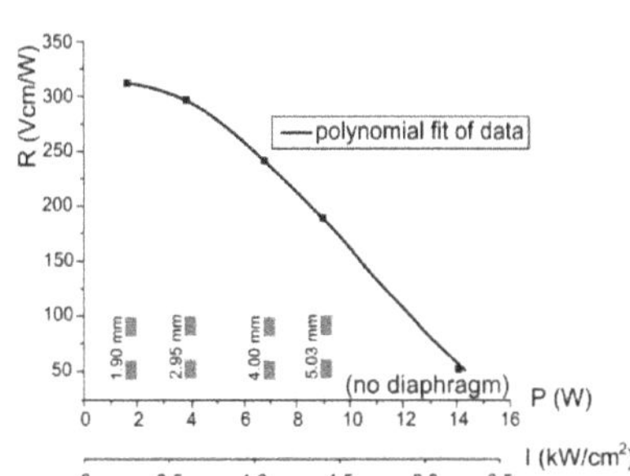

Fig. 27. Saturation effects measured as the dependence of the PA cell responsivity function on laser power and on laser intensity in the focal spot.

An important question is: what happens with the system noises when a diaphragm is introduced in the laser beam path? We proceeded to record the coherent PA background signal on laser power (with and without a diaphragm) and the results are given in Fig. 28. The background signal is huge when a diaphragm is inserted into the system, being of more than 50 times higher than in the case that no diaphragm limits the laser beam. Truncation distorts the intensity pattern of the transmitted beam in both the near-field (Fresnel) and far-field (Fraunhofer) regions. The diffraction effects on an ideal gaussian beam of a sharp-edged circular aperture even as large as $2a = 2w$ (99% criterion) will cause near-field diffraction ripples with an intensity variation $\Delta I/I \cong \pm 17\%$ in the near field, along with a peak intensity reduction of ≅ 17% on axis in the far field (Siegman, 1986). In conclusion, the method of diaphragms used to measure the saturation effects is applicable, but in a laser PA system used in practice an aperture has never to be introduced.

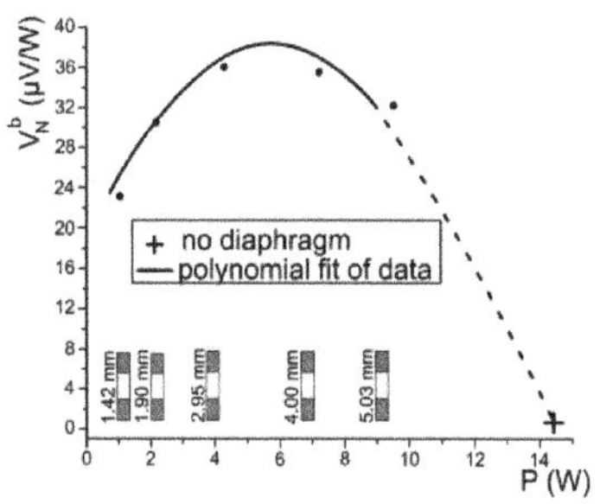

Fig. 28. Dependence of coherent PA background signal on laser power (with and without a diaphragm).

A comparison of low laser power vs. high laser power configurations is presented in Table 4. It seems resonable that high power lasers could be used in PA instruments provided that the saturation is considered and compensated.

Parameter	Low power	High power	Factor
Output laser power (W)	5.5	33	> 6.0
Average laser power (at cell exit) (W)	2.2	14.5	> 6.6
Coherent photoacoustic background signal (μV/W)*	2.7	0.7	< 4.0
Cell responsivity R (V cm/W)**	280	312	> 1.1
Signal saturation**	Small	Very high	> 6.0
Minimum detectable concentration c_{min} (ppbV)**	0.9	0.21	< 4.3
c_{min} (ppbV)***	0.04	0.009	
Minimum detectable absorptivity α_{min} (cm^{-1})**	2.7x10^{-8}	0.64x10^{-8}	< 4.3
Best value previously reported (Harren et al., 1990) (c_{min} = 3.8 ppbV)**	Better 4.2 x	Better 18 x	

Table 4. Comparison of low vs. high laser power configurations: 10P(14) laser line in N_2* ($\alpha = 0$ $cm^{-1}atm^{-1}$), C_2H_4** ($\alpha = 30.4$ $cm^{-1}atm^{-1}$), and SF_6*** ($\alpha = 686$ $cm^{-1}atm^{-1}$).

To this date, the minimum detectable concentrations obtained by us in ethylene (0.9 ppbV with a low power laser and 0.21 ppbV with a high power laser) are the best values reported in the literature, improving this parameter by a factor of 4.2 in the first case and by a factor of 18 in the second case. In a molecular gas with a high absorption coefficient (e.g. SF_6), the minimum detectable concentration could be as low as 9 pptV.

2.7 Removal of interfering gases

Interference of other absorbing substances may impair the theoretical detection limit in a multicomponent analysis of the real samples. Such interference may be caused by other molecular systems present in the environment or substances that are entrained by the carrier flux. If an interfering species is present in the environment, its effect can be minimized by either the introduction of scrubbers and cryogenic traps or the use of dual beam techniques using two photoacoustic (PA) cells.

The CO_2 laser spectral outputs occur in the wavelength region where a large number of compounds possess strong absorption features and where absorptive interferences from water vapors, carbon dioxide, and other major atmospheric gaseous components may influence the measurements.

The breath air is a mixture of nitrogen, oxygen, carbon dioxide, water, inert gases, and traces of VOCs (Table 5). The matrix elements in breath air vary widely from person to person, both qualitatively and quantitatively, particularly for VOCs. More than 1000 trace VOCs have been distinguished in human breath air, at concentrations from ppmV to pptV levels. Only a small number of VOCs are common to everyone, including isoprene, acetone, ethane, and methanol, which are products of core metabolic processes. In addition to these VOCs, exhaled NO, H_2, NH_3, and CO are related to health condition and can reflect a potential disease of the individual or a recent exposure to a drug or an environmental pollutant.

Component	Inhaled air (%)	Exhaled air (%)
Nitrogen	78.0	78.0
Oxygen	21.0	16.0
Carbon dioxide	0.04	3.0-5.0
Argon	0.93	1.0
Water	2.0	5.0-6.0
Other	0.01	
ammonia		250×10^{-9} (250 ppb)
ethylene		6×10^{-9} (6 ppb)

Table 5. Concentration of different components in inhaled and exhaled air.

A healthy adult human has a respiratory rate of 12-15 breaths/min at rest, inspiring and expiring 6-8 L of air per minute. O_2 enters the blood and CO_2 is eliminated through the alveoli. When the end-tidal concentration of CO_2 in healthy persons is measured, a large change of CO_2 concentration is observed between the inhaled air (~ 0.04%) and the exhaled air (~ 4%). The exact amount of exhaled CO_2 varies according to the fitness, energy expenditure and diet of a particular person, with regular values of 3-5%. Due to this high concentration of carbon dioxide in the breath and because CO_2 laser lines are absorbed by this gas, it is necessary to remove most of the carbon dioxide from the exhaled air by introducing a scrubber filled with a chemical active agent, KOH in our case (Bratu et al., 2011).

Due to the exact coincidence of the CO_2 vibrational-rotational transitions with the CO_2 laser lines, carbon dioxide at high concentration in comparison with trace gases like C_2H_4 is inevitably excited by CO_2 laser radiation and the related photoacoustic signal may exceed the trace signal by many orders of magnitude. The absorption coefficient increases strongly with temperature, but it is independent of the CO_2 concentration over a wide range. Ethylene can be excited by the 10P(14) line of the CO_2 laser, where the maximum absorption coefficient $\alpha(C_2H_4)$ has a value of 30.4 cm^{-1} atm^{-1} and ammonia by the 9R(30) line where $\alpha(NH_3) = 56$ cm^{-1} atm^{-1} (Dumitras et al., 2011). A 4% concentration of CO_2 has an absorption strength comparable to 2760 ppbV of C_2H_4 (at the 10P(14) laser line, $\alpha(CO_2) = 2.1\text{x}10^{-3}$ $atm^{-1}cm^{-1}$ and $c(C_2H_4) = c(CO_2)\alpha(CO_2)/\alpha(C_2H_4)$. This equivalent ethylene concentration was found also experimentally (see Fig. 29, measurement without trap). So, the photoacoustic signal is 100 times higher owing to exhaled carbon dioxide in comparison with the usual concentration of ethylene in exhaled air. Similarly, at the 9R(30) line of CO_2 laser, the same concentration of CO_2 has an absorption coefficient equal to that of 1500 ppbV of NH_3. This value is also considerably higher (6 times) compared to the real range of breath concentration which is situated approximately at 250 ppb for ammonia.

Water vapor exhibits a broad continuum with occasional weak lines in the frequency range of the CO_2 laser (for H_2O at the 10P (14) laser line, $\alpha(H_2O) = 2.85\text{x}10^{-5}$ $atm^{-1}cm^{-1}$). The two dominant peaks are the absorption lines on 10R(20) and the most favorable one for ambient air measurement, the 10P(40) laser transition. A 5% concentration of H_2O has an absorption strength comparable to 46.9 ppbV of C_2H_4, that is the normal concentration of water in exhaled air has approximately the same influence in the photoacoustic signal as the normal concentration of ethylene.

Due to the additive character of the photoacoustic signal under normal pressure conditions, the presence of a large amount of water vapor and carbon dioxide impedes C_2H_4 detection in the low-concentration range (ppbV). Consequently, some means of selective spectral discrimination is required if ethylene is to be detected interference free in the matrix of absorbing gases. There are several ways to overcome this problem. One way is to remove CO_2 from the flowing sample by absorption on a KOH-based scrubber inserted between the sampling cell and the PA cell. Taking into account the nature of the specific chemical reactions involved in the CO_2 removal by KOH, a certain amount of water is also absorbed from the sample passing the scrubber. In this way, concentrations below 1 ppmV CO_2 (equivalent to a concentration of 0.07 ppbV of C_2H_4) can be achieved without influencing the C_2H_4 or NH_3 concentration.

Before entering the photoacoustic cell, the gas mixture passes through a KOH scrubber (Fig. 18), which retains most of the interfering carbon dioxide. The removal of CO_2 is limited to the absorbent surface of the pellets. Hence, the larger the surface area or the more porous the granular solid, the larger the capacity of the system to absorb CO_2. At the same time, the flow resistance varies inversely proportional to the particle size. Large particles offer less resistance, but have the disadvantage of providing a smaller total area for reaction. The granules of KOH that we used were typically Merck KOH pellets GR for analysis, with approximate dimensions of 10x7x2 mm. When residence time (time of contact between CO_2 and absorbent) is less than 1 second, CO_2 absorption capacity is greatly reduced, so we introduced flow controllers in order to ensure this pre-requisite.

Potassiun hydroxide is a caustic compound of strong alkaline chemical, dissolving readily in water, giving off much heat and forming a caustic solution. It is a white deliquescent solid in the form of pellets obtained by concentration of purified electrolytic potassium hydroxide solution with very low chloride content. It reacts violently with acid and it is corrosive in moist air toward metals such as zinc, aluminum, tin and lead, forming a combustible, explosive gas. It absorbs rapidly carbon dioxide and water from air. Cautions must be taken when used because the inhaled dust is caustic and irritant, and touching skin or clothes could lead to less or more severe chemical burnings.

We have investigated the efficiency of the KOH scrubber using four recipients with different volumes (13 cm^3, 45 cm^3, 120 cm^3, and 213 cm^3, respectively), and we found out what type has to be used in order to reduce efficiently the amount of CO_2 from the exhaled air sample (Bratu et al., 2011). The KOH scrubber must neither change the ethylene concentration level, nor introduce new interfering gases. The measurements were made each time on the same person (healthy female, 30 years old) and with a new filling of KOH pellets. The gas from the sample bag was transferred into the PA cell at a controlled flow rate of 300 sccm (only for the 13 cm^3 trap) or 600 sccm, in order to ensure a sufficient time of flow in the scrubber column and to minimize any tendency for the vapor to stick to the cell walls or any other effects of internal outgassing of contaminants, which would otherwise lead to increase background signals during an experimental run. The typical resulting final pressure inside the PA cell was around 700 mbar and the corresponding responsivity was 170 cmV/W (see Fig. 3, Part I).

The experimental results without the KOH scrubber showed an equivalent ethylene absorption concentration of 2750 ppbV (with alveolar air collection) and 2350 ppbV (with mixt expiratory air collection), representing mainly the contribution of ethylene, carbon dioxide, water vapors and ammonia to the absorption of 10P(14) CO_2 laser line (Fig. 29). We tested the efficiency of traps filled with KOH and having different volumes (between 13 cm^3 and 213 cm^3) in removing CO_2 from exhaled air.

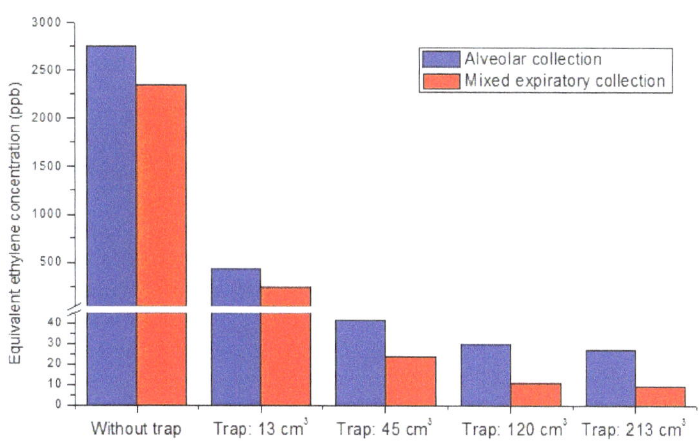

Fig. 29. Efficiency of KOH traps for CO_2 removal from exhaled air.

For the first measurement we used a trap with a small volume of 13 cm^3 of KOH scrubber, and we obtained a decrease of the PA signal down to 1-3 mV. The equivalent ethylene concentration was 435 ppb and 240 ppb, respectively (alveolar air collection vs. mixt expiratory air collection), indicating that the CO_2 concentration was reduced by factors of 6.3

and 9.8, respectively. Only in the case of this trap we observed a peculiar behaviour. Even if the laser power is constant, the PA signal and consequently the equivalent ethylene concentration increases in time after transfering the gas sample in PA cell. The increase of concentration starts from 50 ppbV and continues until it stabilizes at a level of 435/240 ppbV (after 10-15 minutes). It is known that C_2H_4 (28.05 g/mol molar mass) is lighter than CO_2 (44.0099 g/mol molar mass). Because of that, we can say that after passing the KOH scrubber, first C_2H_4 enters in the PA cell and then CO_2 when the trap is no longer effective. So, at the beginning, we measured only the C_2H_4 concentration and then CO_2 starts to strongly interfere in absorption. It is possible that due to the geometry of the cell, a longer time is required in order to attain the total homogeneity of the molecules inside the resonant tube of the cell, but this is not advantageous for repeated measurements.

Larger KOH traps proved to be more efficient in removal of CO_2 from the exhaled air. For the traps with volumes of 45 cm^3, 120 cm^3, and 213 cm^3, respectively, the measured equivalent ethylene concentrations were 41.5/23.6 ppbV, 30/10.8 ppbV, and 26.8/9.1 ppbV, respectively. For larger traps (120 cm^3 and 213 cm^3), approximately same results were obtained, indicating that most of the CO_2 was removed. By using larger traps, a higher transfer rate of the gas mixture in the PA cell is possible, doubling the flow rate to 600 sccm.

For the two largest volumes, we succeeded to reduce the CO_2 content from the exhaled air at a level influencing no more the C_2H_4 and NH_3 concentration values, fact proved by the constant evolution in time of all parameters. Therefore, the trap is effective only for a enough large amount of KOH pellets. We found that a minimum volume of 120 cm^3 of KOH scrubber and a transfer rate of 600 sccm were optimum to insure the required efficiency.

Analyzing the four cases when we inserted the scrubber, the dependence between the removed content of CO_2 and the used KOH quantity proved to be nonlinear, as one could expect. If we consider the content of the sample totally free of CO_2 after passing through the 213 cm^3 and 120 cm^3 KOH traps, we calculated a residual content of CO_2 in alveolar collection of 0.58% (5800 ppm) for the 13 cm^3 trap and of 0.016% (160 ppm) for the 45 cm^3 trap (less than half of the CO_2 concentration in the inhaled air).

We measured also the efficiency of the KOH scrubber when it is used for multiple measurements (Fig. 30). A clear saturation effect is evident: the KOH scrubber is not anymore efficient when the same fill is used for multiple runs (it cannot absorb completely the CO_2 from the gas mixture). In the case of alveolar collection, the equivalent ethylene concentration increases by 2.3 times for the second run, by 2.6 times for the third run and by 3.4 times for the fourth run. When we measured the mixed expiratory collection, this saturation effect is even larger: the equivalent ethylene concentration increases by 2.4 times for the second run, by 8.5 times for the third run and by 20.2 times for the fourth run. The conclusion is that a new fill of KOH scrubber must be introduced after each measurement.

The lungs and airways are always moist, and inspired gas is rapidly saturated with water vapor in the upper segments of the respiratory system. The temperature in the airways and lungs is most identical with deep body temperature (approximately 37°C); at this temperature water vapor has a partial pressure of 47 torr (~6.2%). The increased saturation found at the third and fourth run for mixed expiratory collection is explained by a higher quantity of water vapors in exhaled breath (originating both from lungs and from upper segments of the respiratory system).

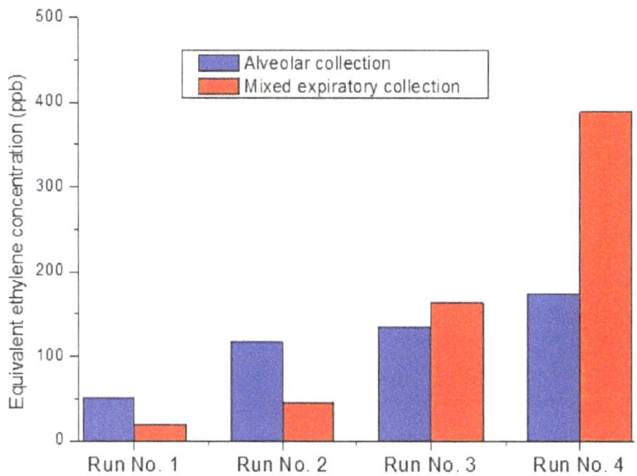

Fig. 30. Decrease of KOH trap efficiency when the same fill was used for multiple measurements.

The nonlinearity of the CO_2 removal could be explained by the mechanism of the chemical reactions. First, the CO_2 combines with the water vapors present in the exhaled air in the form of carbonic acid:

$$CO_2 + H_2O \Rightarrow H_2CO_3. \tag{6}$$

Further, the last one combines with the KOH, creating potassium carbonate and water, and releasing a small amount of heat:

$$H_2CO_3 + 2KOH \Rightarrow K_2CO_3 + 2H_2O + \text{Energy}. \tag{7}$$

In the same time, K_2CO_3 is a highly hygroscopic compound with a retaining capacity of 0.2 g H_2O/1 g K_2CO_3, so the generated water will be only partially returned in the circuit.

The water is of high importance in limiting the rate of CO_2 absorption. High CO_2 concentrations entering the KOH absorber generates large quantities of water, because the reaction (7) is producing water. We know that the absorption rate is greater thanks to the film of moisture coating the pellets, but the same film impedes the access to the active potassium hydroxide pellet volume. More dedicated studies should be made in order to establish the moisture content for an optimum rate of absorption.

In conclusion, we determined experimentally that in the process of CO_2 removal from the breath air samples, a quantity of minimum 120 cm^3 KOH pellets should be used for a sampling bag of 750 mL in order to keep the detection of ethylene and ammonia traces free of CO_2 interference. It should be mentioned that this volume of 120 cm^3 must be reconsidered for sample bags with a greater volume (> 750 mL) or when the gas transfer rate from the bag to the PA cell is larger (> 600 sccm).

2.8 Recipe for an optimum PA system

Based on our experience, we summarize a list of actions to obtain a minimum detectable concentration (c_{min}) as low as possible (Dumitras et al., 2010):

a. Increase the cell constant, C:
- choose resonant operation of the PA cell (design the PA cell as an acoustic resonance chamber and modulate the laser beam at a frequency which coincides with one of the natural resonant acoustic frequencies of the chamber); in this case, the signal is amplified by the quality factor, Q;
- to increase the quality factor, minimize the energy lost over one period by different dissipation processes (surface losses and volume losses); radiation losses through openings can be reduced by terminating the cavity resonator at the openings with acoustic band-stop filters (buffer volumes);
- maintain the chopper frequency close to the top of the resonance curve (a signal reduction less than 2% is obtained if $\Delta T \leq 4^{o}C$);
- in a longitudinally excited resonator use a pipe with a large length ($C \propto L^{1/2}$) and a small diameter ($C \propto r^{-1}$).

b. Increase the microphone responsivity:
- by a higher individual microphone responsivity and/or by an increased number of microphones connected in series.

c. Minimize the electrical noise, V_N^e:
- use state-of-the-art lock-in amplifiers;
- use longer time averaging ($V_N^e \propto \tau^{-1/2}$).

d. Minimize the coherent acoustic background noise, V_N^{ac} (caused by the modulation process):
- use thick detector and tube walls;
- use small flow rates;
- mount the cell and the chopper in separate sound insulating boxes.

e. Minimize the coherent PA background signal, V_N^b:
- use high quality Brewster windows for PA cell (low absorption materials and best possible polished surfaces);
- adjust the windows of the PA cell at Brewster angle and respect the polarization angle of the laser beam;
- place the PA cell windows at the nodes of the mode being excited;
- introduce buffer volumes at both ends of the cell;
- align the laser beam such as to avoid wall interactions;
- do not use tight focused beams to avoid saturation effects.

f. Avoid interference with other absorbing substances:
- to remove CO_2 and H_2O from the flowing sample insert a KOH - based scrubber and a cryogenic trap, respectively, between the sampling cell and the PA cell;
- use multicomponent analysis of the real atmosphere (the minimum number of measurements at different laser wavelengths must be equal to at least the number of unknown trace gases);
- use the phase information of the PA response to suppress the CO_2 signal (a high concentration of CO_2 yields a phase shift of the signal with respect to the acoustic signal of ethylene).

g. Increase the laser power while maintaining the noises at lower values:
- increase the laser power either in extracavity configuration (higher output power) or in intracavity configuration (smaller cavity transmission coefficient);
- maintain all noises and especially the PA background signal at as low as possible values;
- avoid any aperture in the path of the laser beam;
- consider the saturation effects in PA cell responsivity.

The recipe presented above applied on an extracavity laser PA configuration allowed us to achieve one of the most sensitive PA system with a detection limit of $2.7x10^{-8}$ cm^{-1} for a low power laser or even $0.64x10^{-8}$ cm^{-1} for a high power laser. In this way, the method based on laser photoacoustic spectroscopy became a powerful tool for measurement of trace gases (Dumitras et al., 1996a).

3. Applications

3.1 Measurement of absorption coefficients

We have measured precisely the absorption coefficients of ethylene (Dumitras et al., 2007) and ammonia (Dumitras et al., 2011) at CO_2 laser wavelengths.

The ethylene absorption coefficients for various CO_2 laser transitions have been measured in various experiments. Discrepancies as high as ~15% have been found in the absolute IR values observed at many laser transitions. Such discrepancies are typical of many other gases and are partially associated with the difficulty of producing proper gas samples with known concentration levels. Unfortunately, large discrepancies are also found between measurements of the relative spectral signatures (the ratio between absorption coefficients at different wavelengths). Knowledge of the relative spectral signatures rather than absolute ones is sufficient for trace gas identification. We also note that it is rather problematic to obtain highly accurate measurements of the absolute values of the absorption coefficients of gases using the photoacoustic effect. The reason for this is the need of an absolute calibration of the cell. The calibrations cited in the literature are all based on *a priori* knowledge of the absorption coefficient of a gas at some selected wavelength. However, in all cases the absorption coefficient was actually known only to a few percent.

Photoacoustics is emerging as a standard technique for measuring extremely low absorptions independent of the path length and offers a degree of parameter control that cannot be attained by other methods. Radiation absorption by the gas creates a pressure signal which is sensed by the microphone. The resulting signal, processed by a phase sensitive detector, is directly proportional to the absorption coefficient and laser power (or laser power absorbed per unit volume). The sensitivity of the technique is such that absorptions of $<10^{-7}$ cm^{-1} can be measured over path lengths of a few tens of centimeters. The small volume of the chamber makes it possible to accurately control the gas parameters, and the system can be operated with static fills or in continuous gas flow mode.

The set of values of the absorption coefficients α, for all laser wavelengths, for a particular gas or vapor and at a common concentration is called the optoacoustic absorption spectrum or signature and is unique to a combination of vapor and laser. These signatures or "fingerprints" are absolute entities, unique only to the laser frequency and species, which provide the specifics of instrument performance in terms of detection limit and interference rejection (Cristescu et al., 2000b).

To improve the measurement of ethylene absorption coefficients, a special procedure was followed. Prior to each run, the gas mixture was flowed at 100 sccm for several minutes to stabilize the boundary layer on the cell walls, since a certain amount of adsorption would occur and possibly influence background signals; after this conditioning period, the cell was closed off and used in measurement. For every gas fill with 0.96 ppmV ethylene buffered in pure nitrogen, the responsivity of the cell was determined supposing an absorption coefficient of 30.4 $cm^{-1}atm^{-1}$ at 10P(14) laser transition. After measurements at all laser lines, the cell responsivity was checked again, to eliminate any possibility of gas desorption during the measurement. The partial pressure of ethylene was enough to have significant PA signals for all laser lines and low enough to be far away from the saturation regime (observations were only made at a C_2H_4 concentration of 100 ppmV). The α values at each laser line were obtained from Eq. (29, Part I) using the measured PA signal and laser power (the cell responsivity and ethylene concentration were known). An average over several independent measurements at each line was used to improve the overall accuracy of the results. The values to be presented are thought to be the best published to date.

The absolute magnitudes of the absorption coefficients were calculated as mean values of several independent measurements. An absorption coefficient corresponding to each CO_2 laser transition was determined from two sets of 50 different measurements. Every set of measurements was initiated by the frequency stabilization of a given line of the CO_2 laser. From one set of measurements to another, the closed loop of the frequency stabilization circuit was interrupted, the laser was tuned again to the top of the gain curve, and then the frequency stabilization was set and checked by watching the long term stability. Inside one set, 50 independent measurements were made at a rate of one per second to assess reproducibility. From one measurement to the next, the error measurement of the absorption coefficient was calculated as the ratio between the maximum difference (maximum value minus minimum value) and the average value. The final value of the ethylene absorption coefficient is given by the arithmetic mean of the two sets of measurements, while the absorption coefficient error is chosen as the larger value of the two sets. The same procedure was applied for every absorption coefficient of ethylene.

To measure the absorption coefficients of ethylene, the software user interface was modified to allow that the laser power, PA signal, and calculated absorption coefficients function on time (or number of measurements) be recorded on different panels.

The results of our measurements for ethylene are given in Fig. 31. Because of the large spacing between laser transitions (1.2-2 cm^{-1} apart), strong differences of absorption occur. Our results are compared to those of Brewer *et al.* (Brewer et al., 1982) that were also obtained by a photoacoustic method. The difference between the two spectral patterns suggests problems in the measurement techniques (for example, frequency deviation from the laser line center, gas calibration, system purity, linearity, precision) and/or data analysis. The different temperatures and atmospheric pressures at which the measurements were made cannot account for the discrepancies, because Persson *et al.* (Persson et al., 1980) measured a change in absorption coefficient of only 5% at the 10P(14) line for a temperature change of 30°C (negative temperature coefficient), while the changes caused by a pressure difference of 40 Torr are <5% for all CO_2 laser wavelengths.

The random coincidence between the emission and absorption lines will be such that some laser lines will lie close to the centers of the absorbing lines and others will be far away in the wings. The result is a spectral representation unique to that molecule. As a consequence of the superposition of different pressure-broadened C_2H_4 transitions (ν_7 vibration), a strong absorption is obtained at the 10P(14) laser line (absorption coefficient of 30.4 $cm^{-1}atm^{-1}$ at 949.479 cm^{-1}). C_2H_4 has weaker absorption coefficients at the 10P(12) and 10P(16) CO_2 laser transitions (4.36 $cm^{-1}atm^{-1}$ at 951.192 cm^{-1} and 5.10 $cm^{-1}atm^{-1}$ at 947.742 cm^{-1}, respectively). Also, in Fig. 31 ethylene is seen to possess moderately strong absorption profiles within the 9.4-μm band.

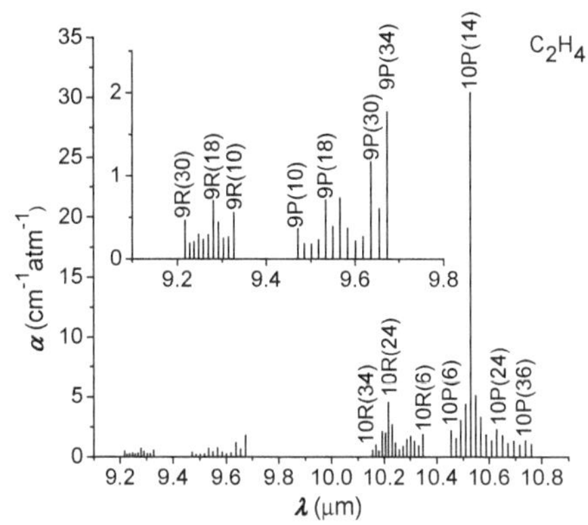

Fig. 31. Absorption coefficients of ethylene at CO_2 laser wavelengths. The inset shows an enlarged view of the measurements for the 9-μm band.

There is general agreement with the results of Brewer et al. (Brewer et al., 1982) for the 00^01-10^00 band. The difference between our results and those obtained by the above mentioned authors is less than 10% for the majority of the investigated lines while only for five lines the discrepancy is higher, between 10% and 20%. By contrast, the values determined in the present work are consistently higher in the 00^01-02^00 band. The difference is larger by 10-50% for the P branch, while our values in the R branch are higher by a factor of 1.5-5.5 (the largest discrepancies are recorded for the 9R(28), 9R(30), and 9R(22) laser lines).

The present work was carried out using a methodology which gave the best possible control over the ethylene partial pressure and background signals. The background levels and calibration of the PA cell were checked before and after every experimental run. The present study is considered reliable, particularly in view of the careful attention that was paid to controlling the gas composition and noise signals. No apparent fault could be found with either the apparatus or methodology that would account for the discrepancy by factors of 2-5 from other reported data in the case of the CO_2 laser 9R lines.

For the measurement of the absorption coefficients of ammonia (Dumitras et al., 2011), the software user interface allows to record the laser power, the PA signal and the calculated absorption coefficients on different panels. The evolution in time of the measurement of the absorption coefficients can also be displayed.

The gas mixture was flowed at 100 sccm for several minutes to stabilize the boundary layer on the cell walls, since a certain amount of adsorption would occur and possibly influence background signals; after this conditioning period, the cell was closed off and used in measurement. For every gas fill with 10 ppm ammonia buffered in pure nitrogen, the responsivity of the cell was determined supposing an absorption coefficient of 57.12 $cm^{-1}atm^{-1}$ at 9R(30) laser transition. This is in accordance both to the measurements reported by Brewer & Bruce (Brewer & Bruce, 1978) and by our tests, when the responsivity of the PA system was checked by measuring the well known absorption coefficient of ethylene at 10P(14) line of the CO_2 laser. After measurements at all laser lines, the cell responsivity was checked again, to eliminate any possibility of gas desorption during the measurement. The α values at each laser line were obtained by using the measured PA signal and laser power and by knowing precisely the ammonia concentration (10.6 ppm) and the responsivity of the

PA cell (312 V cm/W for high power laser). An average over several independent measurements at each line was used to improve the overall accuracy of the results.

The results of our measurements for ammonia are given in Fig. 32. The experimental results show a spectral representation unique to the ammonia molecule. As it can be seen from Fig. 32, ammonia has weaker absorption coefficients at other CO_2 laser transitions; some other significant values for the absorption coefficient were found for 9R and 9P bands: 9R(16) - α = 11.29 cm^{-1}atm^{-1} (error ± 1.4%), 9P(20) - α = 2.10 cm^{-1}atm^{-1} (error ± 2%) and 9P(34) - α = 3.99 cm^{-1}atm^{-1} (error ± 0.62%). In the 10R band the measurements gave: 10R(14) - α = 6.17 cm^{-1}atm^{-1} (error ± 1.5%), 10R(8) - α = 20.08 cm^{-1}atm^{-1} (error ± 1.3%), 10R(6) - α = 26.2 cm^{-1}atm^{-1} (error ± 1.7%), and for the 10P band: 10P(32) - α = 12.45 cm^{-1}atm^{-1} (error ± 2.9%), 10P(34) - α = 14.07 cm^{-1}atm^{-1} (error ± 0.48%) and 10P(36) - α = 7.39 cm^{-1}atm^{-1} (error ± 0.83%). Compared to the other values reported previously in the literature (Brewer & Bruce, 1978), our measurements indicate a general good agreement.

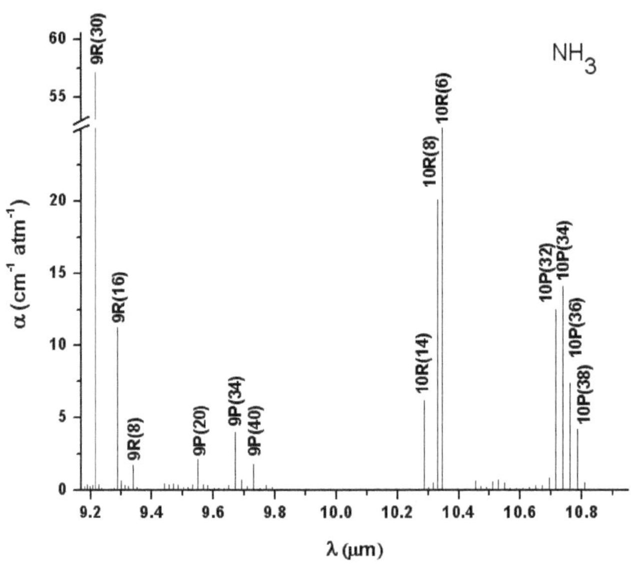

Fig. 32. Absorption coefficients of ammonia at CO_2 laser wavelengths.

3.2 Applications in plant physiology

Ethylene acts as a vegetal hormone produced by all plant tissues. It is transported by diffusion through plant tissues and increases the plasmatic membrane permeability. It has multiple effects on the cell metabolism: increases the oxidative processes, the transport inside the cells and the biodegradation of the organic acids and chlorophyll. Ethylene plays a major role in many metabolic processes: seed and bud dormancy, seed germination promotion, roots induction, development of plantlets (inhibitor of elongation and promotion of lateral shoots), grown promotion, leaf expansion, epinasty (downward curvature of leaves due to the growth of cells on the upper side of the petiole), flowering, wilting of flowers, fruit ripening (ethylene induces some biochemical modifications which produce polyalcohols, hydrocarbons and different oxygenated combinations responsible for the taste, aroma and texture of the fruit), aging and senescence of leaves and flowers and finally, the abscission of leaves and fruits (Cristescu et al., 1998; Cristescu et al., 1999; Dumitras et al., 2004).

The ethylene biosynthesis process in plants follows the MSAE pathway: L-methionine (amino acid) - SAM (S-adenosyl methionine) - ACC (aminocyclopropane-1-carboxylic acid) - C_2H_4. Ethylene, or its precursor ACC, stimulates seed germination of many species at concentrations as low as 0.2 ppb. During germination, a complex cross-talking between several plant hormones exists.

Tomato is an useful model plant for studying ethylene action. Three tomato mutants altered in ripening process affects different steps in ethylene synthesis and perception, resulting in a delay of fruit maturation and pigmentation: Never ripe (*Nr*) is mutated in an ethylene receptor and exhibits delayed and incomplete fruit maturation; ripening inhibitor (*rin*) is a delayed gene that causes the block of ripening before the respiratory burst; and non ripening (*nor*) shows pleiotropic effects analogue to *rin*. The aim of our study was to investigate the ethylene emission during seed germination of these 3 mutants, correlation with their germination ability and analysis of ethylene role on the loss of germinability during seed senescence.

The ethylene production per seed measured during seed germination and seedling elongation is presented in Fig. 33. In these genes, ethylene influences not only fruit ripening, but also the seed germination. The germination index and the percentage of germination of the 5-years-old seeds of the mutants are higher in respect to the control (Ny - New Yorker), in spite of the lower ethylene production of germinating seeds. Conversely to other species, in 5-years-old tomato seeds an inverse correlation between ethylene production and percentage of germination exists. During seed senescence, ethylene accumulation occurs and some processes, triggered during germination, result altered. Further analysis is required to clarify the interaction between ethylene and other hormones like auxin, ABA and cytochinin.

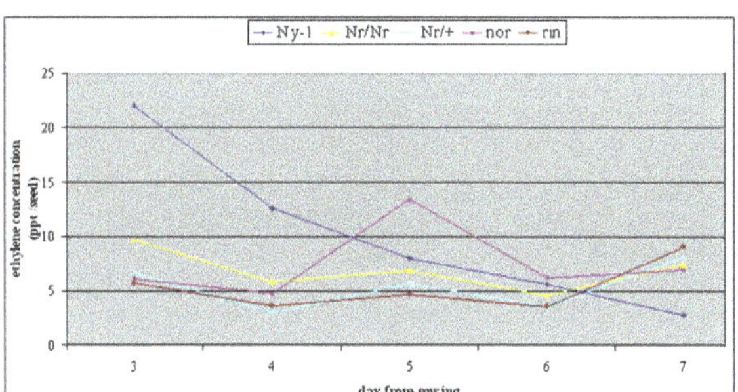

Fig. 33. Ethylene production (ppt/seed) measured during seed germination and seedling elongation.

Climacteric fruits show a respiratory rise during ripening (tomato, pear, fig, mango, banana), while others belong to nonclimacteric fruits (cherry, strawberry, lemon). Fruit ripening (yellowing, softening, respiration, autocatalytic ethylene production) and abscission are regulated by ethylene. During ripening, tomatoes show a strong increase in ethylene production coinciding with the climacteric rise in respiration (CO_2 production). Ethylene is also involved in the postmaturation processes, playing an important role in fruit ripening. Storage and shipping of fruits in terms of wounding effects, temperature, composition of atmospheric gases, postharvest pathogens or seal-packing conditions are important factors to establish the optimal environment necessary for their long term conservation. The aim of our experiments was to monitor the ethylene emission in plants and fruits at low temperature, together with the effect of the temperature at different ripening stages (important for optimization of different stages in agricultural procedures) and to study the effects of mechanical wounding of fruits.

Ethylene emission was monitored from plantlets at different temperatures and it saturates after about 20 min (Lai et al., 2003). The temperature effect is evident in the emission intensity, which increases almost a factor four from 15° to 25°C. There is a remarkable decrease in the lag time for the gas emission at the optimum temperature for biosynthesis of ethylene (25°C). At temperatures lower than 22°C, this lag time is about 6 min, while drops to less than 2 min at 25°C. Temperature does not influence the ethylene emission of immature fruits (0.004 ppb/g and 0.005 ppb/g for 15°C and 25°C, respectively), while it becomes important when the ripening process is triggered in a maturated fruit (0.012 ppb/g and 1.56 ppb/g for 15°C and 25°C, respectively). The same result is obtained for plantlets. Temperature is important for ACC oxidase activity (decreased at low temperatures). Mechanical wounding exerts its effect at the step where SAM is converted to ACC, the direct precursor of ethylene; this step, regulated by the enzyme ACC synthase is rate limiting in the cascade of events leading to an increase of ethylene production.

3.3 Investigation of lipid peroxidation

The oxidative modification of biological molecules is an essential part of the normal biological activity in the human organism. An excess in some oxidant activities does cause injury to cells and tissues. Particular attention is devoted to the oxidant activity of free radicals. An increased free radical formation in the organism is involved in the pathophysiology of several diseases. One of the events generated by free radicals interaction with biomolecules is the oxidative degradation of fatty acids. Oxidative stress is the origin or cause of lipid peroxidation and, as a consequence, of a wide variety of pathophysiological processes. Lipid peroxidation is the free-radical-induced oxidative degradation of polyunsaturated fatty acids. Biomembranes and cells are thereby disrupted, causing cell damage and cell death. As a marker of free-radical-mediated damage in the human body, the measurement of the exhaled volatile hydrocarbons, such as ethylene, is a good noninvasive method to monitor lipid peroxidation.

We have studied lipid peroxidation as a consequence of ionising radiation and heavy metals in living cells (Dumitras et al., 2004). Most heavy metals have a toxic action on human cells and may induce lipid peroxidation. Cadmium is a toxic agent which is supposed to affect the transport of ion through the cell membrane. Cadmium and calcium ionic radii are similar, so Cd can be picked up through the Ca transport mechanism. On the other hand, the Cd permeability through the calcium channel is very poor, so Cd can be considered as a blocker of the calcium channel as well. We tried to determine the extent of the toxic action of Cd *in vitro* by monitoring the ethylene concentration in the breathing air of human cells cultured in a liquid medium to which cadmium chloride was added. Cells of the human leukemic T cell line (Jurkat) were kept in a culture in RPMI 1640 medium containing 10% FBS, 1% L-glutamine and 1% penicillin streptomycin at 37°C in a humidified incubator with 5% CO_2 and 95% air.

The measurement of ethylene before and after treatment of the culture of human cells with $CdCl_2$ has shown that the concentration has increased from 0.5 ppb for control (live cells) to

1.2-1.4 ppb for dead cells, both after 5 hours and 24 hours from the treatment. The increase of ethylene content clearly demonstrates that lipid peroxidation took place owing to the toxic effect of heavy metals (Fig. 34). The measurements were performed on 10P(14) line of the CO_2 laser by means of a nitrogen flow-through system (25 sccm). The CO_2 laser, tuned off resonance on the 10P(20) line from time to time, causes a clear drop in the observed signal.

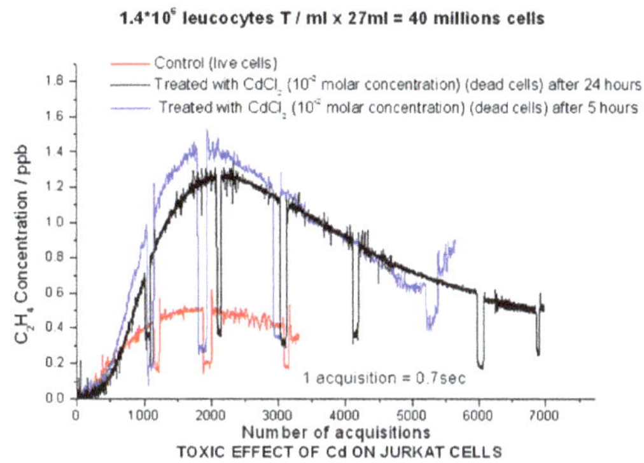

Fig. 34. Toxic effect of Cd on Jurkat cells.

The radiation damage in living matter develops along complex chains of events that follow the absorption of energy: a) physical stage: the energy transfer from the radiation to the matter leads mainly to molecular excitations and ionization; b) chemical stage: the primary reactive species (free atoms and radicals, that are usually extremely unstable), undergo secondary single reactions or a succession of reactions among each other and with their environment, causing damage to molecules of biological importance; c) biological stage: molecular changes occurring in a living organism may cause alterations in the system organization, which become macroscopically observable as biological effects. A substantial part of the total ionising radiation effect concerns water radiolysis, water being a major component of living tissues present in all biological systems. Many water ions and radicals are generated inside tissues as primary reactive species. Aqueous free radicals are very reactive and induce oxidative degradation of phospholipids in cell membranes (lipid peroxidation). The aim of our investigation was to measure the X-ray induced ethylene emission in mice breath and to analyse breath exhaled from patients under external X-ray beam therapy for cancer treatment.

For the purpose of verifying the radioinduced effect, living mice (B6C3F1 and C57B1/6J male mice, between 3 and 6 months old) have been exposed to the total body action of a 250 kV X-ray apparatus GILARDONI model CHF-320-G. At 250 kV voltage and 15 mA current by using a 0.5 mm Cu filter, the measured dose rate was 90.1 cGy/min at 68.4 cm from the source. The value of the X-ray dose given to the treated mice (9 Gy per total body) is comparable, as order of magnitude, to the therapeutic doses given to a human patient in the course of cancer treatment by radiotherapy. The mice were divided in treated and control groups. Each treated mouse received a substantial amount of X-rays in the whole body, while the control mouse received a zero dose. Samples of the breathing air have been collected before and after irradiation. The breathing air has been concentrated on active coal absorbing pellets for a time as long as 1.5 hours, successively expanded into 0.5 liters sample bags, and then transferred into the photoacoustic cell in order to perform the analysis of ethylene content. The PA analysis of ethylene content, by using the above described procedure, takes only few minutes and, after calibration, allows for immediate data release. The radioinduced production of ethylene in the animal appears to be at clearly detectable levels, since the exhaled ethylene increases more than 4 times in the mouse breath after the irradiation (12.4 ppb for control mice before exposure, compared to 55.9 ppb for irradiated mice, after exposure) (Giubileo et al., 2003).

The cigarette smoke contains many toxic components (heavy metals, free radicals, chemicals) that may induce ethylene formation by lipid peroxidation in the lung epithelium (Dumitras et al., 2008; Giubileo et al., 2004). In order to monitor the damages caused by the inhaled smoke, we performed a breath test which gives us information about the volatile compounds under normal and stress circumstances. The exhaled air from the subject being tested was collected inside aluminized bags and then the sample gas was transferred into the measurement PA cell. In all experiments, a high value of ethylene concentration was found immediately after smoking, followed by a slower decrease. A total concentration of 4040 ppb of ethylene was measured in cigarette smoke. In the exhaled breath of a smoker, we found an ethylene concentration of 39 ppb immediately after smoking and even 1.4 ppb at half an hour from smoking a single cigarette, compared to 0.6 ppb as base (before smoking). Ethylene is dangerous for smokers because ethylene oxide is a chemical product that induces cancer in the lungs. For the moment, it is difficult to separate the exogenous and endogenous origin of the ethylene in the smoker's breath.

3.4 Measurement of human biomarkers

The application of laser photoacoustic spectroscopy for fast and precise measurements of breath biomarkers has opened up new promises for monitoring and diagnostics in recent years, especially because breath test is a non-invasive method, safe, rapid and acceptable to patients.

The detection of biomarkers in the human breath for the purpose of diagnosis has a long history. Ancient Greek physicians already knew that the aroma of human breath could provide clues to diagnosis. The perceptive clinician was alert for the sweet, fruity odor of acetone in patients with uncontrolled diabetes; the musty, fishy reek of advanced liver disease; the urine-like smell that accompanies failing kidneys; and the putrid stench of a lung abscess. Modern breath analysis is a noninvasive medical diagnostic method that distinguishes among more than 1000 compounds in exhaled breath.

Human breath includes hundreds of VOCs in low concentrations even though fewer than fifty of these are found in the majority of normal human's breath. Some of these VOCs (ethane, n-pentane, butane, ethanol, acetone) have been identified as biomarkers to some specific pathologies, including lipid peroxidation, heart failure, asthma, cystic fibrosis, diabetic ketoacidosis, alcohol intoxication, renal failure, and others. However, due to the low concentrations and presence of a large number of chemical species in exhaled air, breath analysis requires high sensitive and selective instrumentation to detect and identify the atypical concentrations of specific biomarkers (Cernat et al., 2010; Popa et al., 2011a). In order to assess the physiological meaning and the diagnostic potential of these substances, the biochemical pathways of generation have to be known.

Ethylene from the human breath is a marker of oxidant stress (in patients on hemodialisys, in acute myocardial infarction, in inflammatory diseases and ultraviolet radiation damage of human skin) and can be directly attributed to biochemical events surrounding lipid peroxidation (Dumitras et al., 2005).

Free radicals come from two major sources: endogenous and exogenous. Endogenous free radicals are produced in the body by different mechanisms whereas exogenous sources of free radicals include air pollution, ionizing radiation, cigarette smoke, inflammation etc. The ultimate step in the peroxidative chain reaction is the formation of different hydrocarbons molecules, depending on the molecular arrangement of the fatty acid involved. In the human body, the fatty acids inside the membrane lipids are mainly linoleic acid and arachidonic acid. The peroxidation of these fatty acids produces two volatile alkanes: ethylene and pentane, respectively. Both of them are considered in literature to be good biomarkers of free radical induced lipid peroxidation in humans. The fact that ethylene is highly volatile, not significantly metabolized by the body and not soluble in body fat, means that this diffuses rapidly into bloodstream after generation and it is transported to the lungs. In the lungs, the gas is excreted in the expired breath and then is collected.

Generally speaking, exhaled breath analysis (called breath test) can be represented as follows: production of the biomarker during a particular biochemical reaction or a complex metabolic process; diffusion of biomarker through tissues and input into haematic flow; possible intermediate accumulation (buffering); possible trapping of biomarker by utilization and assimilation systems or natural chemical transformation; transport to the lungs; transmembrane diffusion to the air space of lungs; diffusion of biomarker and their mixing with inhaled air in the alveolar space of lungs; release of biomarker in the breathing air; collection of a breath sample and assessment of the biomarker in the breath sample.

To get an efficient breath air sample, we used aluminized multi-patient collection bags (750 mL aluminum-coated bags-QuinTron), composed of a disposable mouthpiece and a tee-mouthpiece assembly (it includes a plastic tee and a removable one-way flutter valve). Multi-patient collection bags (Fig. 35) are designed to collect multiple samples from patients and hold a sample for maximum 6 hours.

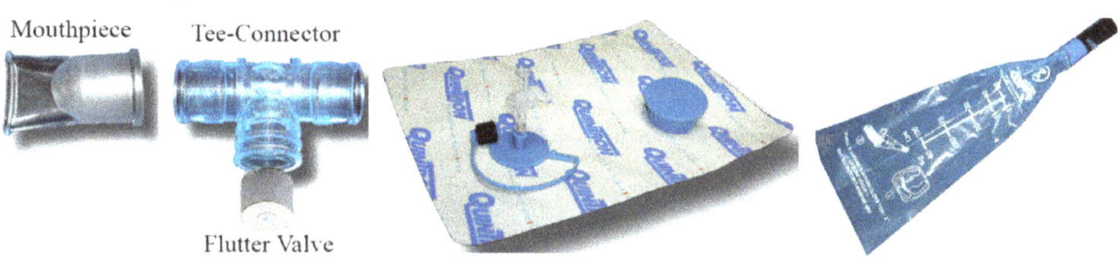

Mouthpiece and tee-connector 0.75 L aluminum-coated bag 0.40 L discard bag

Fig. 35. Breath sample collection system.

How is properly collected a breath sample? After an approximately normal inspiration (avoiding filling the lungs at maximum), the subject places the mouthpiece in his/her mouth, forming a tight seal around it with the lips. A normal expiration is then made through the mouth, in order to empty the lungs of as much air as required to provide the alveolar sample. The first portion of the expired air goes out, after which the valve is opened the tee-piece, the remaining expired air being redirected into the collection bag. When a suitable sample is collected, the patient stops exhaling and removes the mouthpiece.

After the alveolar air sample is collected, the sample gas is transferred into the PA cell and can be analyzed immediately or later. In either case, it is recommendable to seal the large port with the collection bag port cap furnished with the collection bag. The use of the port cap assures that the sample volume will not be lost due to a leak. Its use also avoids the contamination of the sample by gas diffusion through the one-way valve in the large port, if the sample is stored for a long period of time prior to its analysis.

Radiation therapy (also called radiotherapy, X-ray therapy, or irradiation) is the use of certain type of energy (called ionizing radiation) to kill cancer cells and shrink tumors. Radiation therapy injures or destroys cells in the area being treated by damaging their genetic material, making it impossible for these cells to continue grow and divide. Although radiation damages both cancer cells and normal cells, most normal cells can recover from the effects of radiation and function properly. The goal of radiotherapy is to damage as many cancer cells as possible, while limiting harm to nearby healthy tissue.

The effect of ionizing radiation on living cells is supposed to modify the oxidative stress status in the human body through an increase in the peroxidation processes started by the free water radicals generated by indirect radiation effect in living tissue. Important events of the peroxidation take place in the cell membranes determining the release of small linear hydrocarbon molecules through the lipid peroxidation pathways. A fraction of the hydrocarbon molecules generated in the tissue (one among them is the ethylene) will be transported to the lungs by the blood and release in the exhaled breath.

We have analyzed exhaled air from 6 patients between 32 and 77 years old receiving radiation treatment based on X-ray external beam after malign tumor surgery (Dumitras et al., 2008). Breath samples were taken from volunteers at certain time intervals (before, immediately after and at 15 minutes after the X-ray therapy). The patients received fractional doses as high as 2 to 8 Gy depending on the type of cancer. For this experiment patients were asked to exhale into sample bags at a normal exhalation flow rate.

The exhaled air sample was transferred in the PA cell and analyzed in the continuous nitrogen flux. The KOH trap inserted in the gas circuit is used to remove as much as possible the high quantity of CO_2 from the exhaled air. To subtract from the final results any influence of the interfering gases (CO_2, H_2O vapors) we applied the changing lines method, using 3 lines: 10P(14), 10P(16), and 10P(26).

We have measured the following levels of ethylene for a patient (female, 77 years old) with mammary cancer treated by X-ray therapy with a dose of 8 Gy: a) before X-ray therapy: c = 18.6 ppbV; b) Immediately after X-ray therapy: c = 23.17 ppbV; c) 15 min after the X-ray therapy: c = 10.83 ppbV. As a first observation of our measurement we see that, indeed, after the X-ray irradiation the ethylene concentration rises, showing that lipid peroxidation took place. So, it is possible to detect the process in the very first minute after irradiation. The effect of lipid peroxidation is more powerful on the cancer cells, while the healthy cells even affected have higher recovery ability. A surprising decrease in the level of ethylene concentration was observed in the exhaled air after 15 minutes, the level being even lower than the normal level of the patient (e.g. the level measured before any irradiation). This could be explained as a body reaction to the increased level of peroxidic attack: higher the rate of damage, higher the self-defense response of the human organism. Further work is required in order to verify this hypothesis.

In separate studies (Popa et al., 2011b; Popa et al., 2011c), we investigated the breath ethylene and the breath ammonia levels in patients with renal failure receiving haemodialysis (HD) treatment. Human bodies use ammonia in a number of ways, including for the maintenance of the normal pH balance necessary to sustain life. Ammonia is processed in the liver, kidneys and skeletal muscles. Typically, ammonia and ammonium ions (in a healthy individual) are converted into urea in the liver through the urea cycle (Krebs-Henseileit cycle). The urea is then transported through the blood-stream to be excreted into urine by the kidneys. The reversibility of the process requires an equilibrium concentration of ammonia related to the blood urea nitrogen (BUN) loading of the blood. As small molecules, ammonia and ammonium ions can penetrate the blood-lung barrier, and appear in exhaled breath. In the case of kidney dysfunction, urea is unable to be excreted, causing an excessive build up of ammonia in the blood. People with kidney failure have a marked odor of ammonia ("fishy") on their breath, which can be an indicator of this disease.

Volunteer participants (Table 6) were recruited from patients receiving HD treatment at the renal dialysis clinics at the IHS Fundeni (International Healthcare Systems), Bucharest. Subjects were dialyzed 3 times per week, with a 4 h dialysis session. They were instructed to use antiseptic mouthwash before each breath sampling, to avoid oral bacteria (over 700 species of bacteria live in our mouths and can interfere with our molecules of interest). HD was accomplished with BAXTER dialysis machines using DICEA (and XENIUM) high performance cellulose diacetate hollow fibre dialyser-gamma series (DICEA 170G) with following characteristics: surface area of 1.7 m^2, ultrafiltration rate 12.5 mL/hr/mmHg, inner diameter of 200 microns and membrane thickness of 15 microns.

Patient	Gender	Age	HD since	U_{preHD} (mg/dl)	U_{postHD} (mg/dl)	C_2H_4 (ppm)			NH_3 (ppm)		
						before HD	during HD	after HD	before HD	during HD	after HD
P1	Male	67	2005	147	37	0.03	0.13	0.52	4.63	3.58	2.39
P2	Male	80	2004	131	39	0.23	0.51	0.93	4.28	2.82	1.53
P3	Male	79	2008	136	22	0.17	0.31	0.91	2.89	2.06	0.67
P4	Male	22	2010	135	21	0.14	0.19	0.84	5.71	4.08	3.24
P5	Male	54	2010	174	48	0.18	0.43	0.89	4.79	3.07	1.5
P6	Male	66	2005	147	66	-	-	-	2.8	2.01	1.66

Table 6. The particular data of patients and the experimental measurements of breath ethylene and ammonia concentrations (± 10% data error).

A special mention should be made: NH_3 is a highly adsorbing compound and the results of successive measurements are often altered by the molecules previously adsorbed on the pathway and cell walls. To ensure the quality of each measurement, an intensive cycle of N_2 washing was performed between samples, in order to have a maximum increase of 10% for the background photoacoustic signal. It has to be underlined that the measured photoacoustic signal is due mainly to the absorption of ammonia and ethylene, respectively, but some traces of CO_2, H_2O, ethanol, etc., influence the measurements (overall contribution is less than 10%).

Experimental measurements in order to detect traces of ethylene and ammonia were performed for a healthy volunteer (C. A. male, 26 years old) and for 6 patients with renal failure. Particular data of patients are summarized in Table 6. The exhaled air samples were collected before, during (about 1 hour after the start of HD) and immediately after the HD procedure. Analysis of pre-dialysis urea level and post-dialysis urea level (normal limit in the range of 19 - 43 mg/dL) was made at MedCenter, Bucharest (VITROS 51). The results are also presented in Table 6.

Experimental measurements of breath ethylene and ammonia concentrations for the patients (P1-P6) with renal failure and for the healthy subject (P0) were performed and the results are presented in Fig. 36. The control P0 values are 0.006 ppm ethylene and 0.25 ppm ammonia. All measurements were made at 10P(14) CO_2 laser line (10.53 μm), where the ethylene absorption coefficient has the largest value (30.4 $cm^{-1}atm^{-1}$), and at 9R(30) CO_2 laser line (9.22 μm), where the ammonia absorption coefficient has the maximum value of 57 $cm^{-1}atm^{-1}$.

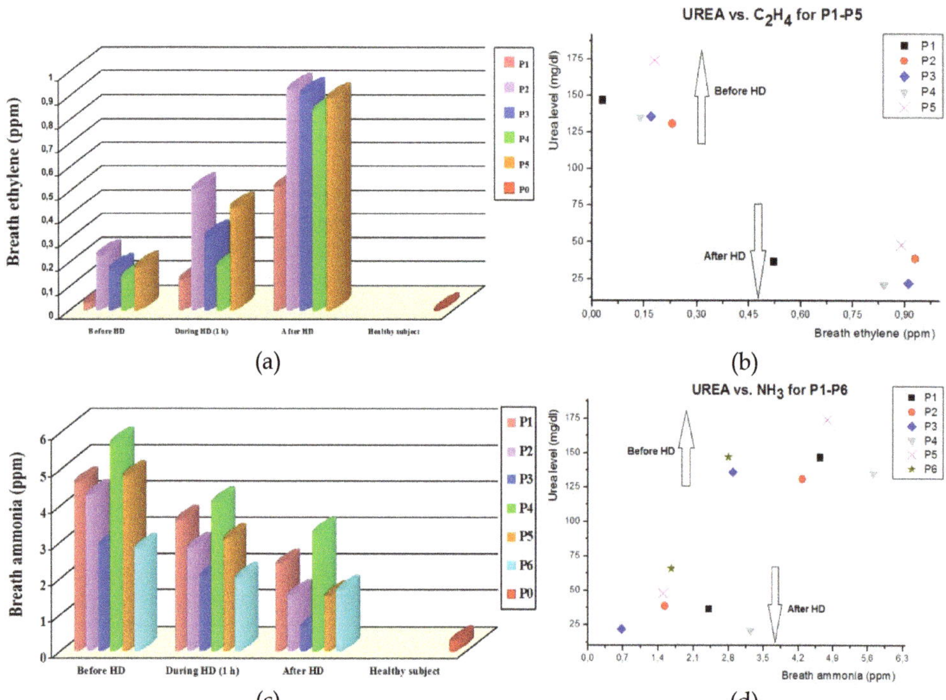

Fig. 36. Ethylene and ammonia concentrations in exhaled breath of patients under HD treatment: (a) ethylene concentrations; (b) breath ethylene concentration correlation with urea level; (c) ammonia concentrations; (d) breath ethylene concentration correlation with urea level.

As a first observation of our measurements (shown in Fig. 36a), we see that, immediately after HD treatment, the ethylene concentration increases, proving the presence of lipid peroxidation. Oxidative stress is a persistent manifestation at patients with renal failure, showing an imbalance between oxidant and antioxidant systems. HD is associated with increased oxidative stress and all treated patients are exposed to this stress. This situation appears to be due to an increased production of free radicals during HD and immediately after HD and a net reduction of many antioxidants.

In Fig. 36 (c) we can observe, as expected, a reduction of ammonia concentration in exhaled breath at patients under HD treatment, which means that ammonia detection in human breath using LPAS system can be used for determining the exact time necessary at every session for the desired state of HD for a patient with end stage renal disease and, in the same time, could serve as a broad noninvasive screen for incipient renal disease.

The most important result is the correlation found between urea data (measured by blood analysis) and the individual breath ammonia and ethylene concentrations (measured by photoacoustic technique), shown in Figs. 36 (b) and 36 (d) for six patients.

This analysis demonstrates that HD determines simultaneously a large increase of the ethylene concentration in the exhaled breath (owing to the oxidative stress) and a reduction of the ammonia concentration, correlated to the level of blood urea nitrogen.

Laser photoacoustic spectroscopy technique demonstrated that it will play an important role in the future of exhaled breath analysis. The key attributes of this technique is sensitivity, selectivity, fast and real time response and ease to use.

4. Conclusions

The applications of resonant PA spectroscopy include concentration measurements and trace gas analysis, accurate determinations of thermophysical properties, detections of dynamic processes such as gas mixing or chemical reactions, relaxation processes (determinations of collisional lifetimes of specified quantum states and routes of energy exchange in polyatomic molecules), spectroscopic experiments, studies of aerosols, etc. Trace-gas sensing is a rapidly developing field, in demand for applications such as process and air-quality measurements, atmospheric monitoring, breath diagnostics, biology and agriculture, chemistry, and security and workplace surveillance.

More than 250 different volatile organic compounds including air pollutants originating from the burning of fossil fuels, traffic, or natural sources can be identified and measured with a CO_2 laser based PA instrument. Such studies are prompted by the growing public concern about serious environmental problems such as acid rain, photochemical smog, stratospheric ozone depletion, and global climatic changes.

Breath analysis is a noninvasive medical diagnostic method that distinguishes among more than 1000 compounds in exhaled breath. Many of these compounds, if measured accurately at very low concentration levels, typically in the range of few ppbV, can be used to identify particular medical conditions. Measuring human biomarkers in exhaled breath is expected to revolutionize diagnosis and management of many diseases and may soon lead to rapid, improved, lower-cost diagnosis, which will in turn ensure expanded life spans and an improved quality of life. For example, ammonia levels in the breath can be used to determine the exact time necessary for an optimal degree of dialysis for a patient with end-stage renal disease at every session.

Trace-gas detection techniques based on PA spectroscopy make it possible to discover and control plant physiology mechanisms such as those responsible for germination, blossoming, senescence, ripening, wounding effects, post anaerobic injury, etc. Many agriculturally interesting gases (ethylene, methane, water vapor concentration, carbon dioxide, ammonia, ozone) can be measured *in situ* and in real time with CO_2 and CO laser based photoacoustic spectrometers.

In chemistry, PA spectroscopy is useful in the monitoring of chemical processes (reaction rates, equilibrium constants, enthalpies), identification of different compounds (even isomers and radicals), and dimerization of fatty acid vapors.

The techniques of PA spectroscopy can be extended to the detection of a wide variety of industrial gases, including benzene, hydrogen cyanide, acetylene, carbon monoxide, and carbon dioxide, as well as a broad range of chemical warfare agents, including nerve gases (Sarin, Soman, Tabun), blistering agents (phosgene, mustard gas), and poisonous gases (hydrogen cyanide), explosives (TNT, PETN), and harmful drugs (heroin, morphine, narcotine).

Our previous research on LPAS (Dumitras et al., 1993; Dutu et al., 1994a; Dutu et al., 1994b; Dumitras et al., 1996a; Dumitras et al., 1996b; Cristescu et al., 1997; Cristescu et al., 2000a) has led to the development of new applications in plant physiology (seed germination, ripening of climacteric fruits, plant response to pathogen infection) (Cristescu et al., 1998; Cristescu et al., 1999; Lai et al., 2003), measurement of gas absorption coefficients (ethylene, ammonia) (Cristescu et al., 2000b; Dumitras et al., 2007; Dumitras et al., 2011), and medicine (cultures of human cells doped with heavy metals, ionizing radiation damage in living organisms, lipid peroxidation in lung epithelium following the inhalation of cigarette smoke, exhaled breath from patients treated by anti-tumor radioisotope therapy and patients under HD treatment) (Giubileo et al., 2003; Dumitras et al., 2004; Giubileo et al., 2004; Dumitras et al., 2005; Cernat et al., 2010; Popa et al., 2011a; Popa et al., 2011b; Popa et al., 2011c; Popa & Matei, 2011).

Extensions of various aspects of this work are currently being pursued in our laboratory.

5. References

Angeli, G.Z.; Solyom, A.M.; Miklos, A. & Bicanic, D.D. (1992). Calibration of a Windowless Photoacoustic Cell for Detection of Trace Gases. *Anal. Chem.*, Vol.64, No.2, (January 1992), pp. 155-158, ISSN 0003-2700

Beck, S.M. (1985). Cell Coatings to Minimize Sample (NH_3 and N_2H_4) Adsorption for Low-Level Photoacoustic Detection. *Appl. Opt.*, Vol.24, No.12, (June 1985), pp. 1761-1763, ISSN 0003-6935

Bernegger, S. & Sigrist, M.W. (1987). Longitudinal Resonant Spectrophone for CO-laser Photoacoustic Spectroscopy. *Appl. Phys. B*, Vol.44, No.2, (October 1987), pp. 125-132, ISSN 0946-2171

Bijnen, F.G.; Reuss, J. & Harren, F.J.M. (1996). Geometrical Optimization of a Longitudinal Resonant Photoacoustic Cell for Sensitive and Fast Trace Gas Detection. *Rev. Sci. Instrum.*, Vol.67, No.8, (August 1996), pp. 2914-2923, ISSN 0034-6748

Bratu, A.M.; Popa, C. ; Matei, C. ; Banita, S.; Dutu, D.C.A. & Dumitras, D.C. (2011). Removal of Interfering Gases in Breath Biomarker Measurements. *J. Optoelectron. Adv. Mater.*, Vol.13, No. 8, (August 2011), pp. 1045-1050, ISSN 1454-4164

Brewer, J. & Bruce, C.W. (1978). Photoacoustic Spectroscopy of NH_3 at the 9-μm and 10-μm $^{12}C^{16}O_2$ Laser Wavelengths. *Appl. Opt.*, Vol.17, No.23, (December 1978), pp. 3746-3749, ISSN 0003-6935

Brewer, R.J.; Bruce, C.W. & Mater, J.L. (1982). Optoacoustic Spectroscopy of C_2H_4 at the 9- and 10-μm $C^{12}O_2^{16}$ Laser Wavelengths. *Appl. Opt.*, Vol.21, No.22, (November 1982), pp. 4092-4100, ISSN 0003-6935

Cernat, R.; Matei, C.; Bratu, A.M.; Popa, C.; Dutu, D.C.A.; Patachia, M.; Petrus, M.; Banita, S. & Dumitras, D.C. (2010). Laser Photoacoustic Spectroscopy Method for Measurements of Trace Gas Concentrations from Human Breath. *Rom. Rep. Phys.*, Vol.62, No.3, pp. 610-616, ISSN 1221-1451

Crane, R.A. (1978). Laser Optoacoustic Absorption Spectra for Various Explosive Vapors. *Appl. Opt.*, Vol.17, No.13, (July 1978), pp. 2097-2102, ISSN 0003-6935

Cristescu, S.; Dumitras D.C. & Godeanu, A. (1999). Role of Ethylene in the Germination Metabolism of Ricinus Communis. *Proc. Tenth International Conference "Photoacoustic and Photothermal Phenomena"*, Vol.463, F. Scudieri, M. Bertolotti (Eds), pp. 652-654, American Institute of Physics, ISBN 978-1-563-96805-3, Melville, NY, USA

Cristescu, S.; Dumitras, D.C. & Duţu, D.C.A. (1998). Photoacoustic Detection of Ethylene Released by Biological Samples under Stress Conditions. *Proc. SPIE ROMOPTO '97: Fifth Conference on Optics*, Vol.3405, V.I. Vlad, D.C. Dumitras (Eds.), pp. 627-631, SPIE, ISBN 978-0-819-42857-8, Bellingham, WA, USA

Cristescu, S.; Dumitras, D.C. & Dutu, D.C.A. (2000a). Characterization of a Resonant Cell Using the Acoustic Transmission Line Model. *Proc. SPIE SIOEL '99: Sixth Symposium on Optoelectronics*, Vol.4068, T. Necsoiu, M. Robu, D.C. Dumitras (Eds.), pp. 263-272, SPIE, ISBN 978-0-819-43705-1, Bellingham, WA, USA

Cristescu, S.; Dumitras, D.C. & Dutu, D.C.A. (2000b). Ammonia and Ethene Absorption Measurements with a Tunable CO_2 Laser-Based Photoacoustic Trace Gas Detector. *Proc. SPIE ALT '99 International Conference on Advanced Laser Technologies*, Vol.4070, V.I. Pustovoy, V.I. Konov (Eds.), pp. 457-464, SPIE, ISBN 978-0-819-43707-5, Bellingham, WA, USA

Cristescu, S.; Dumitras, D.C.; Dutu, D.C.A. & Mujat, C. (1997). Real-Time Detection System for Laser Photoacoustic Applications. *Rom. Rep. Phys.*, Vol.49, No.8-9-10, pp. 757-768, ISSN 1221-1451

Davidson, J.; Gutow, J.H. & Zare, R.N. (1990). Experimental Improvements in Recording Gas-Phase Photoacoustic Spectra. *J. Phys. Chem.*, Vol.94, No.10, (May 1990), pp. 4069-4073, ISSN 0022-3654

Dewey, C.F. (1977). Design of Optoacoustic Systems. In *Optoacoustic Spectroscopy and Detection*, Ch. 3, Y.-H. Pao (Ed.), 47-77, Academic, ISBN 978-0-125-44159-9, New York, NY, USA

Dumitras, D. C.; Alexandrescu, R. & Morjan, I. (1993). Laser Absorption Measurements Using Photoacoustic Detection. In *Proc. 10-th Int. FASE Symposium*, D. Rucinski (Ed.), 189-192, ICPE Publishing House, Bucharest, Romania

Dumitras, D.C.; Banita, S.; Bratu, A.M.; Cernat, R.; Dutu, D.C.A.; Matei, C.; Patachia, M.; Petrus, M. & Popa, C. (2010). Ultrasensitive CO_2 Laser Photoacoustic System. *Infrared Phys. Technol.*, Vol.53, No.5, (September 2010), pp. 308-314, ISSN 1350-4495

Dumitras, D.C.; Dutu, D.C.; Comaniciu, N. & Draganescu, V. (1976). Sealed-Off CO_2 Lasers. *Rev. Roum. Phys.*, Vol.21, No.6, pp. 559-568, ISSN 0035-4090

Dumitras, D.C.; Dutu, D.C.; Comaniciu, N.; Draganescu, V.; Alexandrescu, R. & Morjan, I. (1981). Frequency Stabilized CO_2 Laser Design. *Rev. Roum. Phys.*, Vol.26, No.5, pp. 485-498, ISSN 1221-1451

Dumitras, D.C.; Dutu, D.C.; Draganescu, V. & Comaniciu, N. (1985). *Frequency Stabilization of CO_2 Lasers*. Preprint LOP-55, CIP Press, Bucharest, Romania

Dumitras, D.C.; Dutu, D.C.; Matei, C.; Magureanu, A.M.; Petrus, M. & Popa, C. (2007). Laser Photoacoustic Spectroscopy: Principles, Instrumentation, and Characterization. *J. Optoelectr. Adv. Mater.*, Vol.9, No.12, (December 2007), pp. 3655-3701, ISSN 1454-4164

Dumitras, D.C.; Dutu, D.C.; Matei, C.; Magureanu, A.M.; Petrus, M.; Popa C. & Patachia, M. (2008). Measurements of Ethylene Concentration by Laser Photoacoustic Techniques with Applications at Breath Analysis. *Rom. Rep. Phys.*, Vol.60, No.3, pp. 593-602, ISSN 1221-1451

Dumitras, D.C.; Dutu, D.C.A.; Cristescu, S. & Mujat, C. (1996a). Laser Photoacoustic Spectroscopy: A Powerful Tool for Trace Gas Measurements. *Proc. SPIE 17th Congress of the International Commission for Optics: Optics for Science and New Technology*, Vol.2778, J.-S. Chang, J.-H. Lee, C.-H. Nam (Eds.), pp. 670-671, SPIE, ISBN 978-0-819-42164-7, Bellingham, WA, USA

Dumitras, D.C.; Dutu, D.C.A.; Cristescu, S. & Mujat, C. (1996b). Trace Gas Analysis Using Laser Photoacoustic Spectroscopy Methods. *Optoelectronica,* Vol.4, pp. 21-27

Dumitras, D.C.; Dutu, D.C.A.; Matei, C.; Cernat, R.; Banita, S.; Patachia, M.; Bratu, A.M.; Petrus, M. & Popa, C. (2011). Evaluation of Ammonia Absorption Coefficients by Photoacoustic Spectroscopy for Detection of Ammonia Levels in Human Breath. *Laser Phys.,* Vol.21, No.4, (April 2011), pp. 796-800, ISSN 1555-6611

Dumitras, D.C.; Giubileo, G. & Puiu, A. (2005). Investigation of Human Biomarkers in Exhaled Breath by Laser Photoacoustic Spectroscopy. *Proc. SPIE Advanced Laser Technologies 2004,* Vol.5850, I.A. Shcherbakov, A. Giardini, V.I. Konov, V.I. Pustovoy (Eds.), pp. 111-121, SPIE, ISBN 978-0-819-45847-6, Bellingham, WA, USA

Dumitras, D.C.; Puiu, A.; Cernat, R.; Giubileo, G. & Lai, A. (2004). Laser Photoacoustic Spectroscopy: A Powerful Tool for Measurement of Trace Gases of Biological Interest at Sub-ppb Level. *Molec. Cryst. Liquid Cryst. J.,* Vol.418, No.1, (January 2004), pp. 217/[945]-227/[955], ISSN 1542-1406

Dumitras, D.C.; Sporea, D. & Dutu, D.C.A. (2006). Set-Up for Power Calibration Transfer at CO_2 Laser Wavelength. *Proc. SPIE Advanced Laser Technologies 2005,* Vol.6344, I.A. Shcherbakov, K. Xu, Q. Wang, A.V. Priezzhev, V.I. Pustovoy (Eds.), pp. 643-653, SPIE, ISBN 978-0-819-46420-0, Bellingham, WA, USA

Dutu, D.C.; Draganescu, V.; Comaniciu, N. & Dumitras, D.C. (1985). Plasma Impedance and Optovoltaic Effect in Sealed-Off CO_2 Lasers. *Rev. Roum. Phys.,* Vol.30, No.2, pp. 127-130, ISSN 0035-4090

Dutu, D.C.A.; Cristescu, S. & Dumitras, D.C. (1994b). Measurement of Photoacoustic Signal and Noises in a Sensitive Spectrophone with a Frequency Stabilized CO_2 Laser. In *Proc. SPIE ROMOPTO'94-Fourth Conference in Optics,* Vol.2461, V.I. Vlad (Ed.), pp. 308-316, SPIE, ISBN 978-0-819-41813-5, Bellingham, WA, USA

Dutu, D.C.A.; Dumitras, D.C.; Cristescu, S. & Sarkozy, L. (1994a). Evaluation of Photoacoustic Signal and Noises in a Sensitive Spectrophone Irradiated by a CO_2 Laser Beam. *Rom. Rep. Phys.,* Vol.46, No.7-8, pp. 639-644, ISSN 1221-1451

Fink, T.; Büscher, S.; Gäbler, R.; Yu, Q.; Dax, A. & Urban, W. (1996). An Improved CO_2 Laser Intracavity Photoacoustic Spectrometer for Trace Gas Analysis. *Rev. Sci. Instrum.,* Vol.67, No.11, (November 1996), pp. 4000-4004, ISSN 0034-6748

Fung, K. H. & Lin, H.-B. (1986). Trace Gas Detection by Laser Intracavity Photothermal Spectroscopy. *Appl. Opt.,* Vol.25, No.5, (March 1986), pp. 749-752, ISSN 0003-6935

Gandurin, A. L.; Gerasimov, S.B.; Zheltukhin, A.A.; Konovalov, I.P.; Kornilov, S.T.; Mel'nik, G.F.; Mikhalevich, Yu.Yu.; Ogurok, D.D.; Petrishchev, V.A. & Chirikov, S.N. (1986). Optoacoustic Gas Analyzer for NO, NO_2, NH_3, C_2H_4, and Saturated Hydrocarbon Pollutants. *Zh. Prikl. Spektrosk.,* Vol.45, No.2, (August 1986), pp. 337-343, ISSN 0514-7506

Gerlach, R. & Amer, N.M. (1980). Brewster Window and Windowless Resonant Spectrophones for Intracavity Operation. *Appl. Phys.A,* Vol.23, No.3, (November 1980), pp. 319-326, ISSN 0947-8396

Giubileo, G.; DeDominicis, L.; Lombardi, C.C.; Mancini, C.; Antonini, A. & Dumitras, D.C. (2003). Detectability by Photoacoustic Spectroscopy of X-Ray Induced Ethylene in Mice Breath. *Proc. SPIE ALT'02 International Conference on Advanced Laser Technologies* , Vol.5147, H.P. Weber, V.I. Konov, T. Graf (Eds.), pp. 219-225, SPIE, ISBN 978-0-819-45017-3, Bellingham, WA, USA

Giubileo, G.; Puiu, A. & Dumitras, D.C. (2004). Detection of Ethylene in Smokers Breath by Laser Photoacoustic Spectroscopy. *Proc. SPIE ALT'03 International Conference on Advanced*

Laser Technologies: Biomedical Optics, Vol.5486, R.K. Wang, J.C. Hebden, A.V. Priezzhev, V.V. Tuchin (Eds.), pp. 280-286, SPIE, ISBN 978-0-819-45418-8, Bellingham, WA, USA

Harren, F.J.M. & Reuss, J. (1997). Spectroscopy, Photoacoustic, In *Encyclopedia of Applied Physics,* Vol.19, G.L. Trigg (Ed.), 413-435, VCH Publishers, ISBN 978-3-527-40478-0, New York, USA

Harren, F.J.M.; Bijnen, F.G.C.; Reuss, J.; Voesenek, L.A.C.J. & Blom, C.W.P.M. (1990a). Sensitive Intracavity Photoacoustic Measurements with a CO_2 Waveguide Laser. *Appl. Phys. B,* Vol.50, No. 2, (February 1990), pp. 137-144, ISSN 0946-2171

Harren, F.J.M.; Reuss, J.; Woltering, E.J. & Bicanic, D.D. (1990b). Photoacoustic Measurement of Agriculturally Interesting Gases and Detection of C_2H_4 Below the ppb Level. *Appl. Spectrosc.,* Vol.44, No.8, (September 1990), pp. 1360-1368, ISSN 0003-7028

Henningsen, J. & Melander, N. (1997). Sensitive Measurement of Adsorption Dynamics with Nonresonant Gas Phase Photoacoustics. *Appl. Opt.,* Vol.36, No. 27, (September 1997), pp. 7037-7045, ISSN 0003-6935

Hubert, M.H. (1983). *Laser Optoacustic Detector Measurement of Signatures of a Selection of Environmental Contaminants.* Rep. No. 83-715-1, Ultra Lasertech Inc., Mississauga, Canada

Karbach, A. & Hess, P. (1985). High Precision Acoustic Spectroscopy by Laser Excitation of Resonant Modes. *J. Chem. Phys.,* Vol.83, No.3, (August 1985), pp. 1075-1084, ISSN 0021-9606

Kavaya, M.J.; Margolis, J.S. & Shumate, M.S. (1979). Optoacoustic Detection Using Stark Modulation. *Appl. Opt.,* Vol.18, No.15, (August 1979), pp. 2602-2606, ISSN 0003-6935

Koch, K.P. & Lahmann, W. (1978). Optoacoustic Detection of Sulphur Dioxide Below the Parts per Billion Level. *Appl. Phys. Lett.,* Vol.32, No.5, (March 1978), pp. 289-291, ISSN 0003-6951

Kosterev, A.A.; Bakhirkin, Y.A. & Tittel, F.K. (2005). Ultrasensitive Gas Detection by Quartz-Enhanced Photoacoustic Spectroscopy in the Fundamental Molecular Absorption Bands Region. *Appl. Phys. B,* Vol.80, No.1, (January 2005), pp. 133-138, ISSN 0946-2171

Kritchman, E.; Shtrikman, S. & Slatkine, M. (1978). Resonant Optoacoustic Cells for Trace Gas Analysis. *J. Opt. Soc. Am.,* Vol.68, No.9, (September 1978), pp. 1257-1271, ISSN 1084-7529

Lai, A.; DeDominicis, L.; Francucci, M.; Giorgi, M.; Giubileo, G. & Dumitras, D.C. (2003). Detection of Ethylene Traces by Photoacoustic Spectroscopy. *Proc. SPIE Third GR-I International Conference on New Laser Technologies and Applications,* Vol.5131, A. Carabelas, G. Baldacchini, P. Di Lazzaro, D. Zevgolis (Eds.), pp. 295-299, SPIE, ISBN 978-0-819-44997-9, Bellingham, WA, USA

Meyer, P.L. & Sigrist, M.W. (1990). Atmospheric Pollution Monitoring Using CO_2-Laser Photoacoustic Spectroscopy and Other Techniques. *Rev. Sci. Instrum.,* Vol.61, No.7, (July 1990), pp. 1779-1807, ISSN 0034-6748

Miklós, A. & Lörincz, A. (1989). Windowless Resonant Acoustic Chamber for Laser-Photoacoustic Applications. *Appl. Phys. B,* Vol.48, No.3, (April 1989), pp. 213-218, ISSN 0721-7269

Nägele, M. & Sigrist, M.W. (2000). Mobile Laser Spectrometer with Novel Resonant Multipass Photoacoustic Cell for Trace-Gas Sensing. *Appl. Phys. B,* Vol.70, No.6, (June 2000), pp. 895-901, ISSN 0946-2171

Nodov, E. (1978). Optimization of Resonant Cell Design for Optoacoustic Gas Spectroscopy (H-Type). *Appl. Opt.,* Vol.17, No.7, (April 1978), pp. 1110-1119, ISSN 0003-6935

Olaffson, A.; Hammerich, M.; Bülow, J. & Henningsen, J. (1989). Photoacoustic Detection of NH_3 in Power Plant Emission with a CO_2 Laser. *Appl. Phys. B,* Vol.49, No.2, (August 1989), pp. 91-97, ISSN 0946-2171

Olafsson, A.; Hammerich, M. & Henningsen, J. (1992). Photoacoustic Spectroscopy of C_2H_4 with a Tunable CO_2 Laser. *Appl. Opt.*, Vol.31, No.15, (May 1992), pp. 2657-2668, ISSN 0003-6935

Persson, U.; Marthinsson, B.; Johansson, J. & Eng, S.T. (1980). Temperature and Pressure Dependence of NH_3 and C_2H_4 Absorption Cross Sections at CO_2 Laser Wavelengths. *Appl. Opt.*, Vol.19, No.10, (May 1980), pp.1711-1715, ISSN 0003-6935

Popa, C. & Matei, C. (2011). Photoacoustic Assessment of Oxidative Stress in Dialysis and Radiotherapy by LPAS System. *Optoelectron. Adv. Mater. – Rapid Commun.*, Vol. 5, No.11, (November 2011), pp. 1237-1242, ISSN 1842-6573

Popa, C., Bratu, A.M.; Cernat, R.; Dutu, D.C.A. & Dumitras, D.C. (2011a). Spectroscopic Studies of Ethylene and Ammonia as Biomarkers at Patients with Different Medical Disorders. *U. P. B. Sci. Bull., Series A,* Vol.73, No.2, pp. 167-174, ISSN 1223-7027

Popa, C.; Bratu, A.M.; Matei, C.; Cernat, R.; Popescu, A. & Dumitras, D.C. (2011b). Qualitative and Quantitative Determination of Human Biomarkers by Laser Photoacoustic Spectroscopy Methods. *Laser Phys.*, Vol.21, No.7, (July 2011), pp. 1336-1342, ISSN 1555-6611

Popa, C.; Cernat, R.; Dutu, D.C.A. & Dumitras, D.C. (2011c). Ethylene and Ammonia Traces Measurements from the Patients Breath with Renal Failure via LPAS Method. *Appl. Phys. B,* Vol.105, No.3, (November 2011), pp. 669-674, ISSN 0946-2171

Pushkarsky, M.B.; Weber, M.E.; Baghdassarian, O.; Narasimhan, L.R. & Patel, C.K.N. (2002). Laser-Based Photoacoustic Ammonia Sensors for Industrial Applications. *Appl. Phys. B,* Vol.75, No.4-5, (April 2002), pp. 391-396, ISSN 0946-2171

Rooth, R.A.; Verhage, A.J.L. & Wouters, L.W. (1990). Photoacoustic Measurement of Ammonia in the Atmosphere: Influence of Water Vapor and Carbon Dioxide. *Appl. Opt.*, Vol.29, No. 25, (September 1990), pp. 3643-3653, ISSN 0003-6935

Rosengren, L.-G. (1975). Optimal Optoacoustic Detector Design. *Appl. Opt.*, Vol.14, No.8, (August 1975), pp. 1960-1976, ISSN 0003-6935

Ryan, J.S.; Hubert, M.H. & Crane, R.A. (1983). Water Vapor Absorption at Isotopic CO_2 Laser Wavelengths. *Appl. Opt.*, Vol.22, No.5, (March 1975), pp. 711-717, ISSN 0003-6935

Sauren, H.; Bicanic, D.; Jelink, H. & Reuss, J. (1989). High-Sensitivity, Interference-Free, Stark-Tuned CO_2 Laser Photoacoustic Sensing of Urban Ammonia. *J. Appl. Phys.*, Vol.66, No.10, (November 1989), pp. 5085-5087, ISSN 0021-4922

Siegman, A.E. (1986). *Lasers*, University Science Books, ISBN 978-0-935-70211-3, Sausalito, CA, USA

Sigrist, M.W.; Bernegger, S. & Meyer, P.L. (1989). Atmospheric and Exhaust Air Monitoring by Laser Photoacoustic Spectroscopy, In *Topics in Current Physics "Photoacoustic, Photothermal and Photochemical Processes in Gases"*, Ch.7, Vol.46, P. Hess (Ed.), 173-211, Springer, ISBN 978-3-540-51392-2, Berlin, Germany

Thomas III, L.J.; Kelly, M.J. & Amer, N.M. (1978). The Role of Buffer Gases in Optoacoustic Spectroscopy. *Appl. Phys. Lett.*, Vol.32, No.11, (June 1978), pp. 736-738, ISSN 0003-6951

Thöny, A. & Sigrist, M.W. (1995). New Developments in CO_2-Laser Photoacoustic Monitoring of Trace Gases. *Infrared Phys. Technol.*, Vol.36, No.2, (February 1995), pp. 585-615, ISSN 1350-4495

Tonelli, M.; Minguzzi, P. & Di Lieto, A. (1983). Intermodulated Optoacoustic Spectroscopy. *J. Physique (Colloque C6),* Vol.44, No.10, (October 1983), pp. 553-557, ISSN 0449-1947

Zharov, V.P. & Letokhov, V.S. (1986). *Laser Optoacoustic Spectroscopy*, Vol.37, Springer, ISBN 978-3-540-11795-4, Berlin, Germany

Part 2

New Systems

4

Diffusion Cooled V-Fold CO_2 Laser

Rakesh Kumar Soni
*Raja Ramanna Centre for Advanced Technology, Indore (M.P.)
India*

1. Introduction

A laser is light amplifier. The acronym LASER stands for Light Amplification by Stimulated Emission of Radiation. It is an electromagnetic radiation with wavelength ranging from ultraviolet to infrared. The fundamental concept of laser operation was first introduced by Einstein in 1917 in one of his three papers on the quantum theory of radiation (Einstein 1917). Almost half a century later, in 1960, T.H. Maiman was the first person to demonstrate the laser by using a ruby crystal. It is a coherent, convergent and monochromatic beam of light. Lasers have various applications in various fields and to appreciate the competency of a laser radiation it is essential to comprehend the basic operation mechanism and properties of laser radiation. The fundamental concept of laser operation is stimulated emission. The three processes required to produce the high energy laser beam are: (a) population inversion, (b) stimulated emission and (c) amplification. Population inversion is a necessary condition for stimulated emission and corresponds to a non-equilibrium distribution of electrons such that the higher energy states have a larger number of electrons than the lower energy states. The process of achieving the population inversion by exciting the electrons to the higher energy states is referred to as pumping (Svelto and Hanna 1989). In general, population inversion is achieved by optical pumping and electrical pumping. In optical pumping, gas-filled flash lamps are most popular. Flash lamps are essentially glass or quartz tubes filled with gases such as xenon and krypton. Some wavelength of the flash (emission spectrum of flash lamp) matches with the absorption characteristics of the active laser medium facilitating population inversion. This is used in solid-state lasers like ruby and Nd:YAG (yttrium–aluminum–garnet). The basic differences between lasers and other light sources are the characteristics often used to describe a laser: (i) the output beam is narrow (ii) the light is monochromatic and (iii) the emission is coherent. The laser light is categorized by different properties and many applications of lasers use these properties. These properties are: (a) mono-chromaticity (b) collimation (c) coherence (d) brightness or radiance (e) focal spot size (f) low divergence (g) transverse modes and (g) temporal modes.

2. Gas lasers

After the demonstration of the first ruby laser, the laser action has been demonstrated in many materials. Lasers are generally classified depending on the physical nature of the active medium used: (I) solid-state lasers (II) gas lasers (III) semiconductor lasers and (IV) dye lasers. It is beyond the purview of this chapter to describe the principles of operation of all these lasers. Here only gas laser systems and typically V-fold CO_2 laser is explained.

The first gas laser, a helium-neon type, conceived and developed by Ali Javan. It was demonstrated for the first time on December 12, 1960, at Bell Telephone Laboratories in Murray Hill, New Jersey. Gas lasers have certain advantages such as homogeneous medium, easy transportation for replenishment, cooling and relatively inexpensive. However, due to physical nature of the gases (low densities), a large volume of gas is required to achieve the significant population inversion for laser action. Hence, gas lasers are usually relatively larger than the solid-state lasers. Gas lasers can be classified into atomic, ionic, and molecular lasers depending on whether the laser transitions are taking place between the energy levels of atoms, ions, and molecules respectively. There are several laser systems in each class. Only some of the typical gas lasers and their wavelengths are shown below in Table-1.

Laser Type	Wavelength (nm)
ArF	191
KrF	249
XeCl	308
HeCd	325, 441.5
XeF	351
Argon	488, 514.5
Copper vapor	510.6, 578.2
Krypton	520-676
Gold vapor	628
HeNe	632.8
CO_2	10,600

Table 1. Gas lasers and Their Wavelengths

2.1 Carbon dioxide lasers

C.K.N. Patel in 1964 working at Bell laboratories made the most efficient gas laser, known as carbon dioxide (CO_2) laser. The carbon dioxide laser is one of the most versatile type laser on the market today and most widely used materials processing laser. Also, they are efficient and inexpensive in terms of cost per unit power. It emits infrared radiation between 9 and 11 micro-meters (μm), either at a single line selected by the user or on the strongest lines in un-tuned cavities. It can produce continuous output powers ranging from well under 1 watt (W) for scientific applications to many kilowatts (kW) for material processing. It can generate pulses from the nanosecond to millisecond regimes. Custom-made CO_2 lasers have produced continuous beams of hundreds of kilowatts for military laser weapon research (Hecht, 1984) or nanosecond-long pulses of 40 kilojoules (kJ) for research in laser-induced nuclear fusion (Los Alamos National Laboratory, 1982). This versatility comes from the fact that there are several distinct types of carbon dioxide lasers. Thus users see several distinct types, such as waveguide, low-power sealed-tube, high-power flowing-gas, and pulsed transversely excited CO_2 lasers. The great interest in carbon dioxide lasers stems from their continuous power capability, high efficiency and ease of construction. Table-2 illustrates their advantages over other gas lasers.

Laser Type	Linear Power Density (W/m)	Max. Power (W)	Efficiency (%)
HeNe	0.1	1	0.1
Argon	1-10	50	0.1
CO_2	60-80	1200	15-20

Table 2. Comparison of Gas Lasers

2.2 Excitation mechanism of CO_2 lasers

The CO_2 laser is a gas discharge device which operates by electric excitation. The active medium in a CO_2 laser is a mixture of carbon dioxide, nitrogen, and helium. Each gas plays a distinct role. Carbon dioxide is the light emitter. The CO_2 molecules are excited so they vibrate in three different types such as symmetric stretching, bending, and asymmetric stretching (Fig. 1). The molecules then lose part of the excitation energy by dropping to one of two other, lower energy vibrational states as shown in Fig.2. Once the molecules have emitted their laser photons, they continue to drop down the energy-level ladder until they reach the ground state. The nitrogen molecules help to excite CO_2 to the upper laser level. Nitrogen molecules are excited first. This is most often done with high voltage direct current, but may also be accomplished by radio frequency excitation. Energy level of the nitrogen molecule is nearly resembles to the (001) vibrational levels of CO_2 molecule. Laser transition takes place between initial level (001) and final levels (100) and (020), resulting in 10.6 and 9.6 μm laser radiations, respectively. The nitrogen molecules mechanically transfer energy to CO_2 molecules via collisions. In practice, the presence of N_2 significantly enhances laser operation, and that gas is almost always present in CO_2 lasers. Helium plays a dual role. It serves as a buffer gas to aid in heat transfer and helps the CO_2 molecules drop from the lower laser levels to the ground state, thus maintaining the population inversion needed for laser operation. However, the laser radiation at 10.6 μm is the strongest and forms the most usual mode of operation. This process is efficient only if the carbon dioxide is cold, so that its energy levels match that of the nitrogen. High-power systems use elaborate heat exchangers to keep the gas cool. The type of CO_2 lasers as slow flow, transverse or cross flow and fast axial flow determines the properties of a CO_2 laser. CO_2 lasers are capable of both continuous wave (CW) and pulsed operation (Wilson and Hawkes 1987) and in most systems; the electric excitation is controlled to do this.

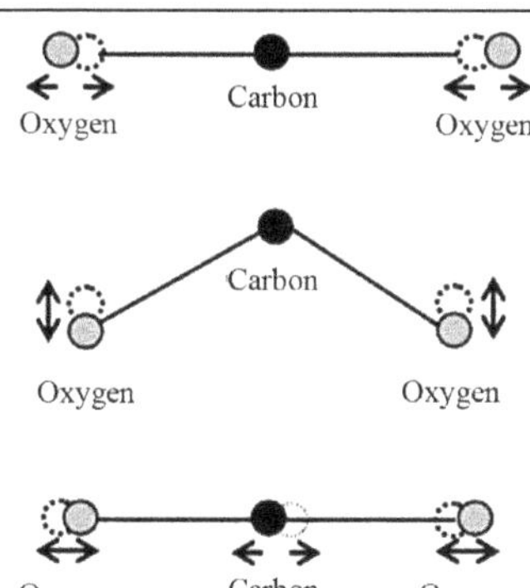

Fig. 1. Vibrational Modes of CO_2 Molecule

The energy level diagram for the operation of CO_2 laser is shown in Fig.2.

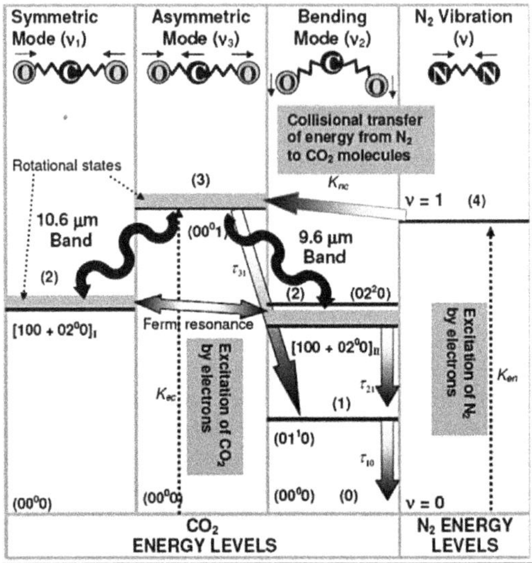

Fig. 2. Energy Level Diagram of CO_2 Laser

2.3 Types of CO_2 lasers

2.3.1 Sealed-tube lasers

The sealed-tube CO_2 laser is a glass tube filled with CO_2, He, and N_2, with mirrors forming a resonant cavity, as shown in Fig.3.

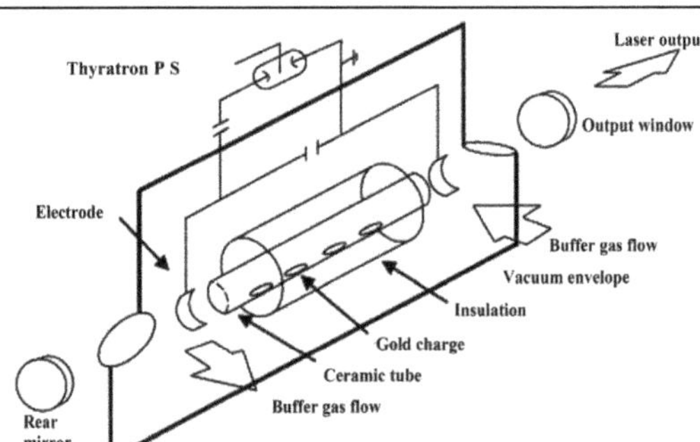

Fig. 3. Sealed Tube Laser

Electrodes are placed near the two ends of the tube. Proper gas mixtures are filled in the tube and seal it. A high voltage is applied to the electrodes to pass a discharge through the gas. A sealed CO_2 laser with an ordinary gas mixture would stop operating within a few minutes. The electric discharge in the tube breaks down the CO_2 in CO and O_2. Catalyst is added in the path to regenerate CO_2. Nickel cathode (at 300°C) can catalyze the recombination reaction. Such measures can be used to produce sealed CO_2 lasers which can operate for up to several thousand hours before their output seriously degrades. Sometimes hydrogen or water to the gas mixture is added so that it can regenerate CO_2 by the carbon monoxide produced by the discharge. In traditional sealed CO_2 lasers, the maximum output power possible with this longitudinal discharge is about 50 W per meter of cavity length, and maximum continuous-wave output is about 100 W. A new methodology is radio-frequency (RF) discharge transverse to the tube axis. This design does not require high-voltage electrodes and offers some other advantages, including the ability to electronically control output at rates to 10 kilohertz (kHz), lower operating voltage and potentially lower tube cost. On the other hand, RF power supplies are more complex and less efficient than DC supplies. RF excitation has been growing in popularity for sealed-tube CO_2 lasers. It can generate more power because it can excite a broader area than a DC discharge, but it also works well at low powers. All sealed-tube CO_2 lasers are limited in output by the difficulty in removing heat.

2.3.2 Waveguide lasers

This type of laser structure is efficient way to produce a compact CW CO_2 laser. It consists of two transverse radio-frequency (RF) electrodes separated by insulating sections. An RF power supply is connected to the electrodes to provide a high-frequency alternating field across the electrodes within the bore region. The waveguide modes access the entire gain volume since the modes reflect off the discharge walls in a zigzag fashion. The waveguide itself traverses the laser length in a zigzag. Waveguide lasers are a type of sealed CO_2 laser in which the inner diameter of a sealed CO_2 laser is shrunk to a few millimeters and the tube is constructed in the form of a waveguide, as shown in Fig. 4.

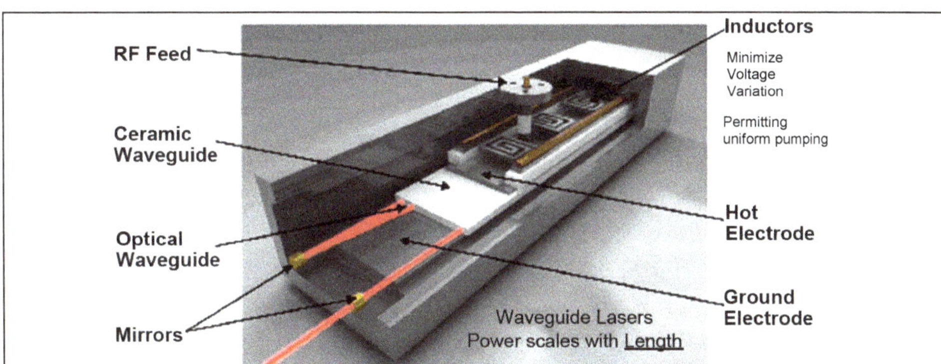

Fig. 4. Waveguide Laser

The waveguide design limits diffraction losses that would otherwise impair operation of a narrow-tube laser. The tube normally is sealed with a gas reservoir separate from the waveguide itself. Waveguide lasers may be excited by DC discharges or intense RF fields. Waveguides may be made of metal, dielectric or combinations of the two. The waveguide laser is very attractive for low powers, particularly under about 50 W. It provides a good beam quality. It can operate continuously or pulsed and can be readily tuned to many discrete lines in the CO_2 spectrum. Its size is comparable to the size of a helium-neon laser but able to generate power in watts.

2.3.3 Longitudinal (axial) slow flow laser

These lasers are operated as conventional gas discharge lasers in the form of long, narrow, cylindrically shaped glass enclosures with electrodes at opposite ends from which the discharge excitation current is introduced as shown in Fig.5. These lasers can be either pulsed or continuous wave and can have lengths of up to several meters. In some versions the discharge enclosure is sealed off and in other versions the gas flows through the tube longitudinally and can be re-circulated to conserve the gases. A water coolant jacket usually surrounds the discharge region. Electric discharge is applied along the tube's axis.

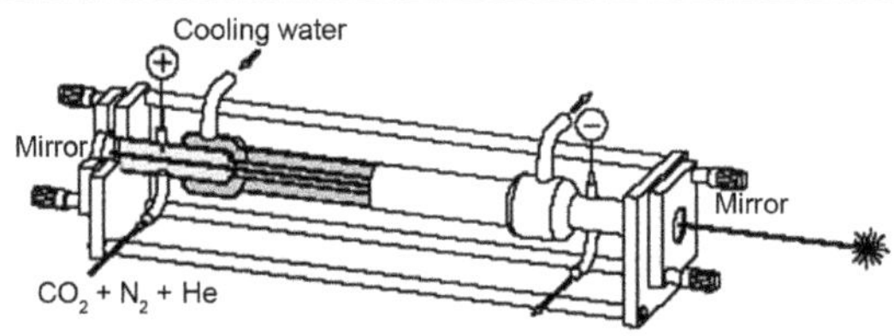

Fig. 5. Longitudinal (Axial) Slow Flow Laser

Low gas pressure and low consumption of gas by recycling methods are some of the salient features of this laser. Slow axial-flow CO_2 lasers produce continuous-wave output proportional to the tube length. Average or continuous power of about 500 W can be produced by folding the laser beam with mirrors through multiple tube segments. This also makes the system compact and the design is simple enough. Heat is removed by conduction mode of heat transfer. Laser gases transfer its heat to the walls of the tube and ultimately that heat can be removed by water circulation or other coolant around the tube.

2.3.4 Fast axial flow laser

The efficiency of axial flow lasers can be increased dramatically by using a pump or turbine to move the gas rapidly through the discharge area as shown in Fig.6.

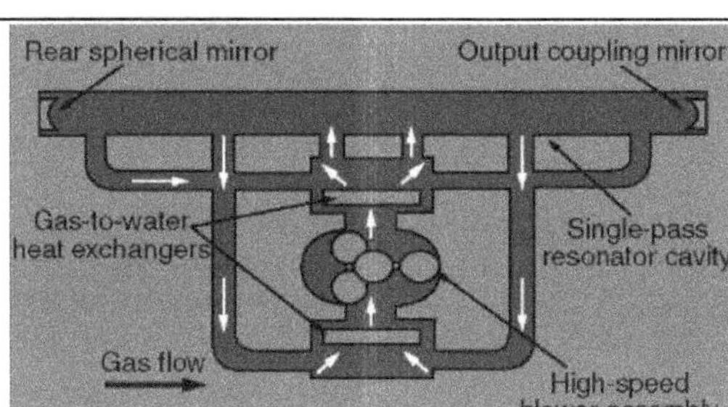

Fig. 6. Fast Axial Flow Laser

This design allows short resonators to produce relatively high powers; 800 W/m is a typical value of power per unit length. Excitation usually is with a longitudinal discharge, as in slow axial-flow lasers, but some fast axial-flow lasers are powered by radio-frequency discharges. The main advantage of the fast flow is that it cools the laser gas better than slow-flow lasers because the gas moves very quickly through the discharge zone. After leaving the discharge zone, the gas is cooled by heat exchanger. The fast axial-flow laser has become the most common industrial CO_2 laser in the power range of 500 W to 5 kW, because of short resonator and small floor space required. Besides the advantages, these lasers have some limitations of complex system design and poor mode quality.

2.3.5 Transverse flow laser

In transverse flow lasers, gas flow direction, electric discharge and direction of laser cavity axis are in three mutually perpendicular directions as shown in Fig.7. It can produce very high power of the order of 10 kW per meter.

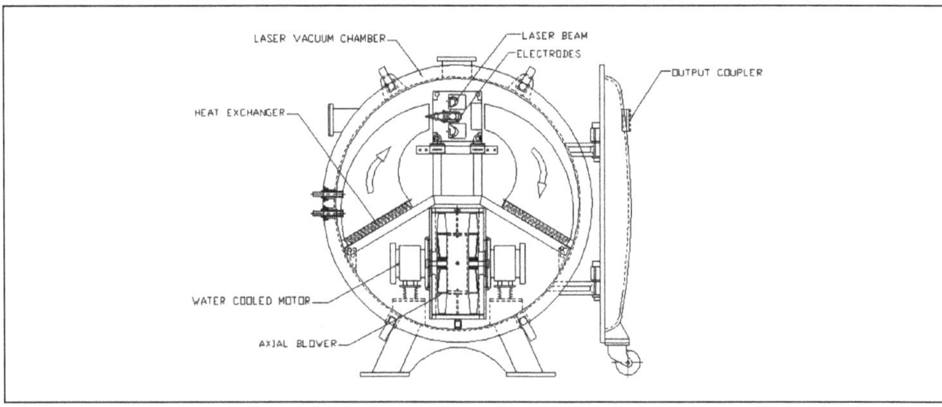

Fig. 7. Transverse Flow Laser

The gas flows across a much wider region and recycled by passing it through a system which regenerates CO_2 and adds some fresh gas to the mixture. In this laser, beam mode structure and beam symmetry are considerably poorer than in fast or slow axial-flow lasers.

2.3.6 Gas dynamic laser

At the end of the 1960s, the gas-dynamic laser was an important breakthrough that made it possible for the first time to reach power levels of 100 kW or more. Basic structure of gas dynamic laser is shown in Fig.8. In gas dynamic lasers the gas is flowed in the transverse direction to the laser axis. Laser gas which is initially at a pressure of several atmospheres is heated electrically or thermally to excite the molecules and population inversion takes place. The high speed pumps are used to rapidly flow the gas. It is then allowed to expand supersonically through an expansion nozzle into a low-pressure region. This expansion causes the gas to supercool and thereby provide rapid relaxation of the lower laser level from the highest rotational states to the lowest rotational states, leaving a population inversion of those empty higher lying rotational states with respect to the upper laser level. A laser beam is extracted from the gas by placing a pair of mirrors on opposite sides of the expansion chamber. Lasers of this design have produced CW output powers greater than 100 kW. This type of excitation was developed primarily for military applications, but lower-power versions have found applications in materials processing.

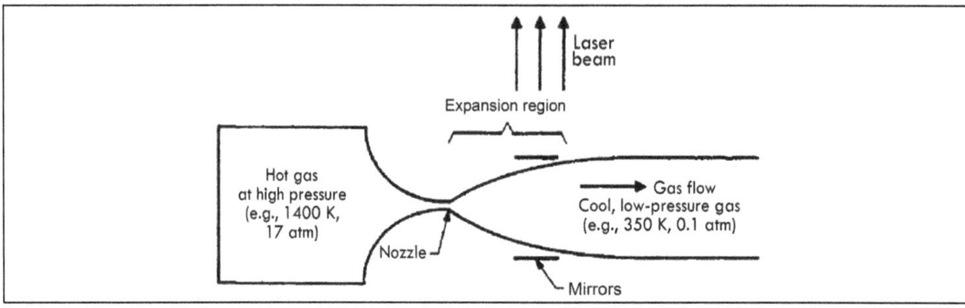

Fig. 8. Gas Dynamic Laser

2.3.7 Transversely Excited Atmospheric (TEA) flow laser

These lasers operate at high total gas pressures of 1 atmosphere or more in order to benefit from obtaining a much higher energy output per unit volume of gas. A schematic of TEA CO_2 laser is shown in Fig.9.

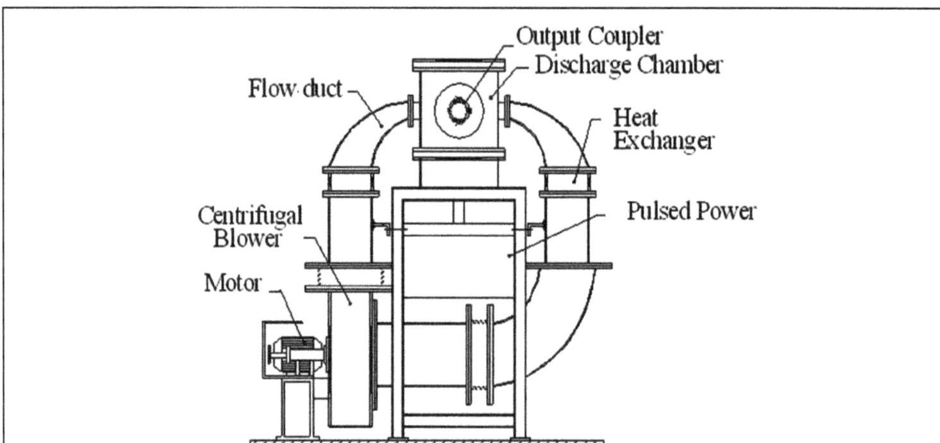

Fig. 9. Transversely Excited Atmospheric (TEA) Flow Laser

Extremely high voltages are required initially to ionize the gas and thereby initiate the discharge process to operate the laser at high pressure. Due to the high gas pressure, arcing tends to form within the discharge. In a transverse discharge, the two electrodes are placed parallel to each other over the length of the discharge and a high voltage is applied across the electrodes. Pre-ionization is used to ionize the space between the electrodes uniformly before applying the high voltage. With this pre-ionization, the discharge can then proceed in a uniform fashion over the entire electrode assembly rather than forming a narrow high-current arc at just one location. The pre-ionization is produced by flashes of ultraviolet light from a row of pre-ionizing UV spark discharges. Such lasers can produce many joules of energy for unit discharge volume. Tens of nanoseconds to microseconds pulse can be produced by passing electric pulses through the gas in a direction transverse to the laser cavity axis. TEA lasers are available in versions with sealed tubes, slow or fast axial flow, or transverse gas flow, depending on power levels. The prime attractions of TEA lasers are their generation of short, intense pulses and the extraction of high power per unit volume of laser gas. High-pressure operation also broadens emission lines, permitting the use of mode locking techniques to generate pulses lasting about 1 nanosecond. It allows tuning over most of the CO_2 wavelength range.

Table-3 illustrate a comparison among details of attainable laser power per cubic cm of active volume in the different types of CO_2 lasers.

CO_2 Laser System	Power Scaling (W/m)
Sealed-off systems	70
Slow flow systems	100
Fast flow systems	800
Pulsed system (TEA Laser)	1.2 TW pulse

Table 3. Comparison of Power Scaling of Different Types of CO_2 Lasers

3. V-fold diffusion cooled CO_2 laser

In the previous paragraphs, we studied about a brief history of lasers and some details about the CO_2 lasers. Here we are going to study about the topic of this chapter i.e. "V-fold diffusion cooled laser" in detail. Fig. 10 is a real photograph of 500 W diffusion cooled CO_2 laser indigenously developed at Department of Atomic Energy, Raja Ramanna Centre for Advanced Technology, Indore, MP, India.

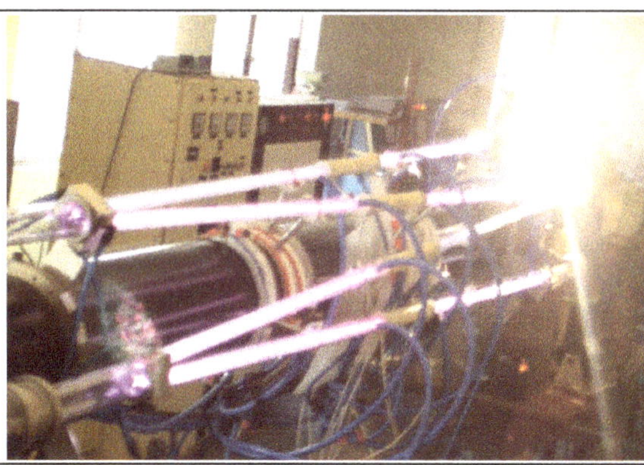

Fig. 10. Photograph of Diffusion Cooled V-fold CO_2 Laser

V-fold laser is also a type of CO_2 laser with some salient features. The name V-fold is given to this laser because of its resonator geometry which is V-folded resonator. Basically this laser is slow flow diffusion cooled CO_2 laser. Convection accompanied by conduction is the mode of heat transfer of this laser. Compare to convective cooled lasers, diffusion cooled laser devoid of bulky heat exchangers and blowers. It makes laser head more attractive, compact & simple in the power range of 300-500 W. In the diffusion cooled laser the laser power can be scaled up by increasing the discharge length at the rate of 50 W/m. We adopted symmetric concave resonator geometry to reduce diffraction loss. V-folding over a cylindrical surface minimizes the astigmatism effect. We obtained more than 380 W laser power in a 7.5 meter discharge length.

3.1 Design considerations

In order to design a V-fold CO_2 laser, the physical dimensions of the active volume, gas flow velocity, output coupling of optical resonator are to be decided. The desired output power P_o can be calculated for the required volume of the active medium, if we know the typical input power density P_{in} that can be dissipated in the homogeneous and stable discharge. P_{in} depends on several factors such as electrode design, gas mixture, its pressure, excitation method, gas flow velocity and its uniformity. Following considerations are taken into account in determining the design parameters such as the discharge length, discharge aperture, optimum reflectivity and gas pressure.

i. The maximum laser power density should be less than the damage threshold of optical elements, however, it should be more than the saturation intensity I_s which is proportional to $p \times n$, where p is gas pressure in mbar and n = 2 in slow flow laser. The damage threshold intensity of the ZnSe mirror, usually used as output coupler in CO_2 lasers is about 2 kW/cm^2. Considering this the incident intensity I_c on the output coupler should be maintained at around 1.0 kW/cm^2.
ii. The optimum output coupling or transmissivity (T) of the resonator can be estimated with the knowledge of the discharge length 'L' small signal gain 'g_o' and the intra-cavity losses (a) by the following relation:

$$T_{opt} = 1 - exp\left[-2L\left\{(g_0 \times a)^{1/2} - a\right\}\right] \tag{1}$$

iii. The small signal gain is usually experimentally measured and it is in the range of 0.5 to 1% per cm. In optimum laser design it can be seen that the transmissivity (T) is almost constant, independent of small signal gain and the laser power. We can write for the intra-cavity intensity I_c incident on the output coupler as:

$$I_c = I_s \times g_0 \times L \tag{2}$$

$$\frac{I_c}{I_s} = g_0 \times L \tag{3}$$

iv. The damage threshold of the output coupler limits the maximum value of I_c and thus the maximum value of $g_0 \times L$ is also limited. In the optimum laser design the intra-cavity losses $a \times L$ is kept minimum and this is also independent of laser power. Usually the total intra-cavity loss should not be more than 5% of total gain. Thus, $g_0 \times L$ and $a \times L$ being constant, the optimum transmittivity T_{opt} is also constant. For the typical values of $I_s$and I_c are about 300-500 W/cm^2 and 1 kW/cm^2 respectively. $g_0 \times L$ is in the range of 2-3 in high power lasers. For these conditions:

$$T_{opt} \approx 50 - 60\% \tag{4}$$

v. Minimum diffraction loss in the resonator criterion should also be considered in designing the V-fold resonator.

In a convective cooled laser, the laser power can be scaled up with the following equation:

$$P_L = \left[\frac{\eta}{\eta - 1}\right] . \rho . C_p . \Delta T . V_f . L . d \approx 120\,\dot{M} \tag{5}$$

Where

η = electro-optic efficiency,
ρ = laser gas density,
C_p = specific heat of laser gas,
ΔT = rise in laser gas temperature,
V_f = flow velocity of laser gas,
L = discharge length and
d = discharge height or electrode separation,
$\dot{M}$ = mass flow rate of gas through discharge zone

The above relation is valid only when the rise in laser gas temperature ΔT ~250°C, without bottlenecking at the lower laser level and maintaining a stable and uniform discharge. The temperature above 250°C populates the lower laser level and destroys population inversion. From Eq.(5), a larger mass flow rate is required for higher laser power. Mass flow rate depends upon area of discharge zone, gas flow velocity or gas mixture density. Since the density for a gas mixture is constant at a particular pressure. So increasing either area of discharge zone or gas flow velocity can only increase power. Discharge Area (A) is the function of electrode separation or discharge height (d) and discharge length (L). So the laser power would increase with the increase of d or L. But it is observed that the maximum discharge current, discharge voltage and the laser power remained almost constant for different electrode separations (d). This is because of the electric field would remain constant to maintain the same discharge current. Laser power may also increase with the discharge length (L) but we found that on increasing the length after a certain optimum value, power decreases due to saturation and due to predominance of cavity losses. Also there are limitations of space and alignment on increasing the discharge length. Therefore length cannot be increased after a certain optimum value to increase the power. Thus, after certain value, increasing either discharge length (L) or electrode separation (d) cannot increase laser power i.e. the discharge area cannot be increased too much. Thus to increase the power gas flow velocity may be increased. So to achieve more gas flow velocity, higher capacity pumps/blowers with high discharge and high pressure are required. An effective heat exchanger is needed to dissipate the heat and to keep the gas temperature below 250°C in discharge zone.

CO_2 Laser Systems	Power (kW/m^3)
Diffusion Cooled (length scaling)	500
Transverse Flow	1500
Fast Axial Flow	3000
Slab Laser	3300

Table 4. Power per Unit Volume of Laser Gas

CO_2 Laser Systems	Typical volume of discharge region compared to total volume (%)
Fast Axial Flow	10
Transverse Flow	14
Diffusion Cooled	20
Slab Laser	27

Table 5. Discharge Volume to Total Volume Ratio of Different Types of CO_2 Lasers

As we go from diffusion-cooled lasers to convective cooled lasers, the power-scaling move from length to volume. From calculation, slab lasers give more power per cubic meter of laser gas compared to various types of CO_2 lasers. Following tables shows the laser power output for unit volume of laser gas (Table-4) and typical volume of discharge region to total volume in percentage (Table-5) for various types of CO_2 laser.

From above two tables, it is concluded that the maximum power could be achieved in slab laser and power is moderate in transverse flow laser. In all other laser, except transverse flow lasers, the power scaling up to multi-kilowatt is not easy. The laser power depends on length of active medium (diffusion cooled) or area of discharge electrode (in slab laser) but in transverse flow lasers, power is scaled-up by volume so it is relatively easy. From the above data, it is clear that the power per unit laser gas and discharge volume to total volume ratio is maximum for slab laser. So, if we somehow move from transverse configuration to Slab (area) or diffusion cooled configuration then we can definitely enhance the power of our laser.

3.2 Construction of V-fold laser

The complete laser assembly is mounted on a 3 meter long aluminum pipe (Fig.11). Outer diameter of aluminum pipe is 200 mm. Since the whole laser assembly is mounted on this pipe only therefore best possible straightness of pipe was required. It is very difficult to get single pipe of 3 meter length and straightness 1 mm therefore the whole pipe is casted in 3 segments, each of 1 meter. All the three segments are welded with straightness in 1 mm. To maintain the straightness and rigidity, both the ends are joined with a flange and tie rod. The aluminum pipe is supported at the ends by a support system made of stainless steel plate of 10 mm thick. Bottom of support system is bolted with the support table. There is no middle support for the pipe due to assembly constraints of glass tube. Since the straightness of tube is very important we calculated the deflection at the mid-point of pipe and it is found that the deflection is insignificant. Five rings of stainless steel 304 (SS304) are inserted in the pipe.

Anode support ring is supporting the anode part of this laser at the center. Additional rings of nylon are also placed near to this central ring to give extra support to the joint of glass tube and anode block. Anode block is made of metalon-6 which acts as insulator (Fig.12). Anode pins made of SS304 are placed at the center separated/isolated by metalon-6 tube. The anode block contains two anodes at each end. Anodes are made of stainless steel. Viton® O-rings are used in between glass tube and anode for sealing. Gas inlet ports are also provided on the anode block. Gas flows from the anode block to cathode through the glass tube. Low thermal expansion borosilicate glass tubes are used. These tubes have jacketed construction. Inner tubes have outer diameter 12 mm, inner diameter 9 mm and length 750 mm. Outer (jacket) tubes are having 22 mm OD (Fig.13). Outer tubes also have ports for water inlet and outlet. Water flows through the annular space between inner and outer tube. A chiller unit supplies water at a total flow rate 12 lit/min and 15°C in water jackets for cooling of gases.

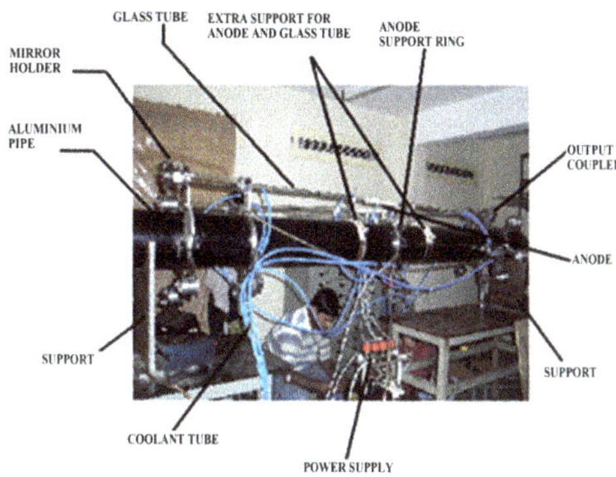

Fig. 11. Components of V-fold CO_2 Laser

Fig. 12. Photograph of Anode Block

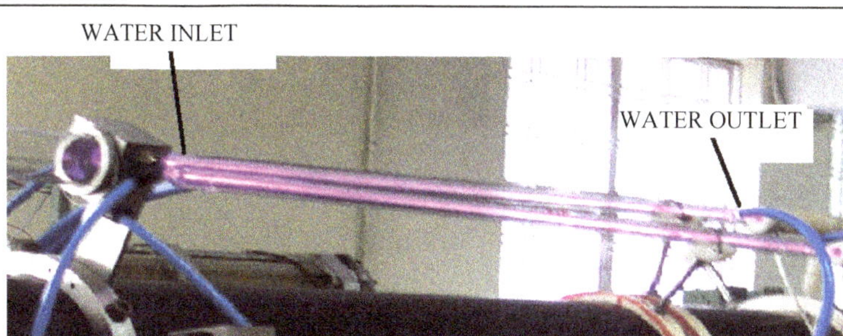

Fig. 13. Schematic of Water Jacket

The jacketed glass tube is supported by cathode block which is ultimately supported by a plate and ring over the aluminum pipe (Fig.14.). Two glass tubes in V-shape are supported by a cathode block on one side. Cathode block is made of SS304 have the advantage of low scaling problem caused by electrical discharge. A mirror holder is connected on the other side of the cathode block through a glass tube of 45 mm OD (Fig.14 & 15). Each mirror holder consists of one mirror and they are placed at the extreme ends on both sides. Mirror holder assembly is also supported on pipe through a ring and plate. Rear mirrors and folding mirrors are made of OFHC Copper substrate of 25 mm diameter and radius of curvature (ROC) 5 meter. Mirrors are gold coated with ~99% reflectivity. Two micrometer screws are fitted on the back side of the each mirror holder to align the laser beam. Alignment is the most critical part of this laser. The alignment accuracy of 0.5 mrad mirror tilt was targeted and achieved by the micrometer screw. Output power is obtained through a ZnSe output coupler having concave geometry of ROC 5 meter and 17% reflectivity.

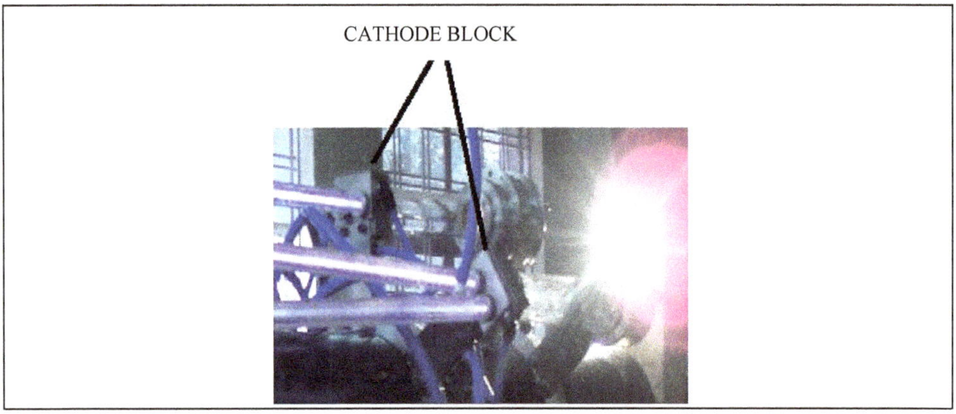

Fig. 14. Cathode Block

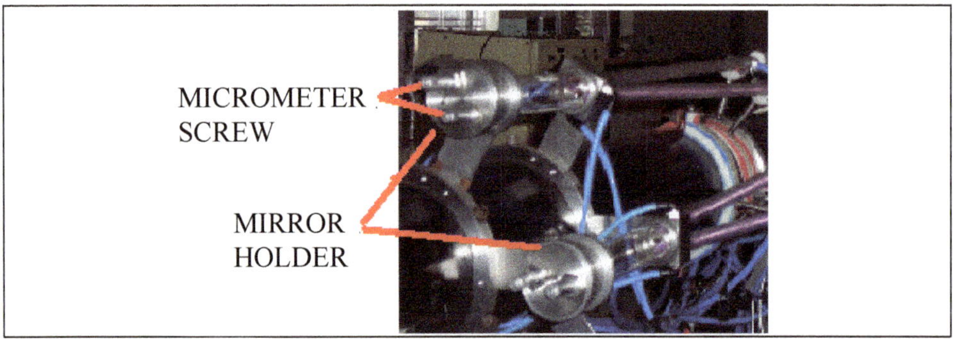

Fig. 15. Mirror Holder

3.3 Working of V-fold laser

The working principle of the laser is similar to other CO_2 lasers. The gas mixture of CO_2, N_2, and He enters in each discharge tube at its center and flows symmetrically towards the cathode blocks, which are connected to a rotary vane vacuum pump of pumping speed of 500 lit/min. Pressure, temperature and gas mixture have been optimized for the maximum output power. Optimum gas pressure is 30 mbar. In the diffusion cooled laser the laser power can be scaled up by increasing the discharge length at the rate of 50 W/m. With the increase of discharge length and therefore optical resonator length, the Fresnel number $N_F = r^2 / \lambda.l$ where r, l and λ are the radius of mirror clear aperture, resonator length and laser wavelength respectively. N_F reduces and with this the diffraction loss increases. Due to this the input power in a laser with plano-concave resonator did not scale up with discharge length beyond 3-4 meters. We adopted the symmetric concave resonator geometry to reduce diffraction loss and V-folding over a cylindrical surface instead of a flat surface for laying the discharge tubes to minimize the astigmatism effect. Each section of V-fold laser has about 1.5 meter discharge length, distance between two mirrors is 2.5 meter. All resonator mirrors i.e. rear reflector, ZnSe output coupler and all folding mirrors are having concave surface of 5 meter ROC. Since, the laser mode formed in any section are sustained in all the other sections therefore the length of one section determines the Fresnel number. Corresponding to the resultant Fresnel number the diffraction loss is low. Introduction of curved folding mirrors through a small folding angle of 5° could introduce considerable aberration due to astigmatism after large number of folding. In order to minimize the overall effect of astigmatism, the tubes were mounted on a cylindrical surface instead of a flat surface to have ~2π folding. The central supporting aluminum pipe due to high moment of inertia have minimum deflection thus minimizes the misalignment. With a fully reflecting mirror on the left and a partially transmitting mirror on the right, the device becomes a V-fold laser which radiates in the far infrared at 10.6 microns. Till date, 420 W power in 10.5 meter discharge length is obtained from this laser system.

3.4 Electrical characteristics of V-fold laser

All gas discharges operated in the glow discharge region have electrical characteristics similar to those indicated in Fig.16. The voltage and current values and the exact shape of the curve depend on the type of gases, gas pressure and the length & diameter of the discharge tube.

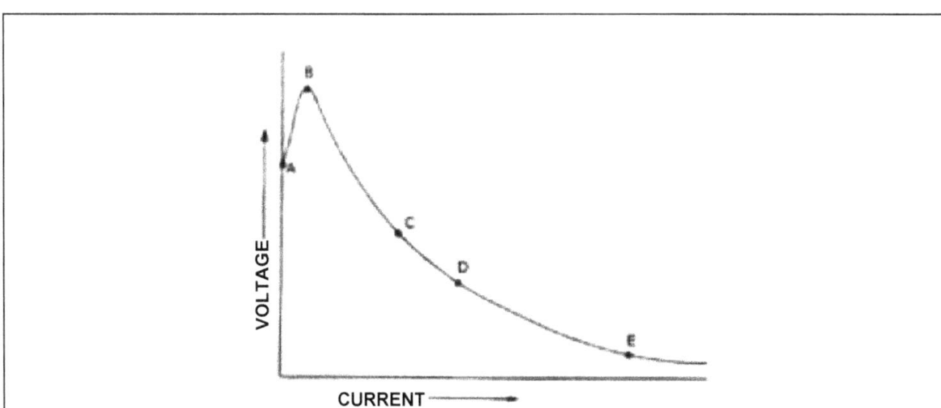

Fig. 16. Voltage-Current Curve of a Gaseous Discharge

Before ionization, the current through the gas is essentially zero. Increasing the voltage on the gas results in a small pre-breakdown current due to a very small amount of easily ionized matter, which is always present in a gas near room temperature (point A). Increasing the applied voltage further will increase this current slightly until the breakdown voltage is reached (point B). At this voltage level, a significant number of atoms become ionized because of the high electric field present in the gas. The free negative electrons are attracted toward the anode and the heavier positive ions toward the cathode. This increases conductivity of the gas and lowers the electrical resistance of the discharge. The electrons are sufficiently accelerated by the electric field to free other electrons through collisions with gas atoms or molecules. Thus, as current increases (from point C to point D), ionization increases and voltage across the discharge tube decreases. This means that an increase in current results in a decrease in resistance. This property of gas discharges is called negative dynamic resistance. This does not mean that the resistance of the tube is a negative value, but that the slope of the voltage-current curve has a negative value. Current through the gas will increase until it is limited by some other electrical component in the circuit or until the power supply can no longer sustain the current. In the case of low-current CW devices such as He-Ne laser tubes, the current is limited at a lower level (point C). In the flashlamps of pulsed solid-state lasers, current is allowed to increase to a value of many kilo-amps (point E) before energy stored in the capacitors is exhausted.

3.5 Power supply of V-fold laser

The Pulser/Sustainer technique is utilized for the production of uniform electrical discharge in the glow discharge regime. The Pulser/Sustainer concept produces pressure and volume scalable plasmas by essentially applying two successive discharges to the gas. The first fast high-voltage pulse creates the electron density uniformly between its electrodes using only a small amount of energy. However a second discharge applies the proper voltage to this plasma to tune the electrons to a temperature sufficiently high for efficient laser pumping but not high enough to generate any appreciable further increase in electron density. Thus, the dominant amount of energy is put into the gas (by the sustainer) exactly where it is desired (vibration excitation of N_2 and CO_2) without triggering. Such plasma instabilities as arcs and sparking are usually associated with substantial ionization rates. The plasma is then with two "knobs"- one controlling electron density, the other electron temperature. The result is a stable uniform tuned high-power-density plasma that is not wall controlled and, hence a high power efficient N_2/CO_2 laser. To realize this concept we have used a 25 kV DC Power supply, 500 mA of current and a pulser with 9 kV of peak voltage, 2 μsec pulse and 5 kHz frequency. The schematic circuit diagram of laser power supply is shown in the Fig.17. An experiment was also performed to know the minimum pulse energy required per pulse to create the uniform discharge. This was studied by the use of another pulser which was available to us with peak voltage of 6 kV, 5 kHz frequency and with variable pulse width. By changing the pulse width we got the situation where we got the uniform smooth discharge. To initiate the discharge in all tubes simultaneously, pre-ionization technique has been adopted. For pre-ionization, a high frequency pulser of peak voltage 6 kV and repetition rate 2-5 kHz has been developed. Pulse width can be varied from 2 to 8 μsec. Pulser is connected to the anode pins by a DC power supply of 30 kV / 750 mA rating through a capacitor of 1.7 nF to block the high voltage DC excitation current. Thick film non-inductive resistors of 191 kΩ are used between DC Supply and anode pins as ballast resistance. Pre-ionization initiates discharge in all the tubes simultaneously and maintain it stable at even low currents. Ballast resistor is required to control the current flowing in the circuit, as discharge has a negative dynamic resistance; hence ballast resistance is an important parameter in getting a uniform stable discharge. If ballast resistance is not proper, it may result in large flow of current, which may result in formation of arcs and no laser action take place. Moreover we require discharge intensity equal in all zones, if instability creeps into one zone, it will affect the other zone and we will not get uniformity in the discharge. If the ballast resistance is of high value, there will be much of power losses in the ballast resistors. We experimented with four different values of ballast resistors. They are 140, 249, 300 and 191 kΩ. With 140 kΩ we could not get the required current density for maximum output optical power. The other three gave us stable discharge and the optimum current in each discharge zone is found to be 26 mA. We finally used the 191 kΩ resistor in our circuit considering the maximum overall efficiency of 10.6%.

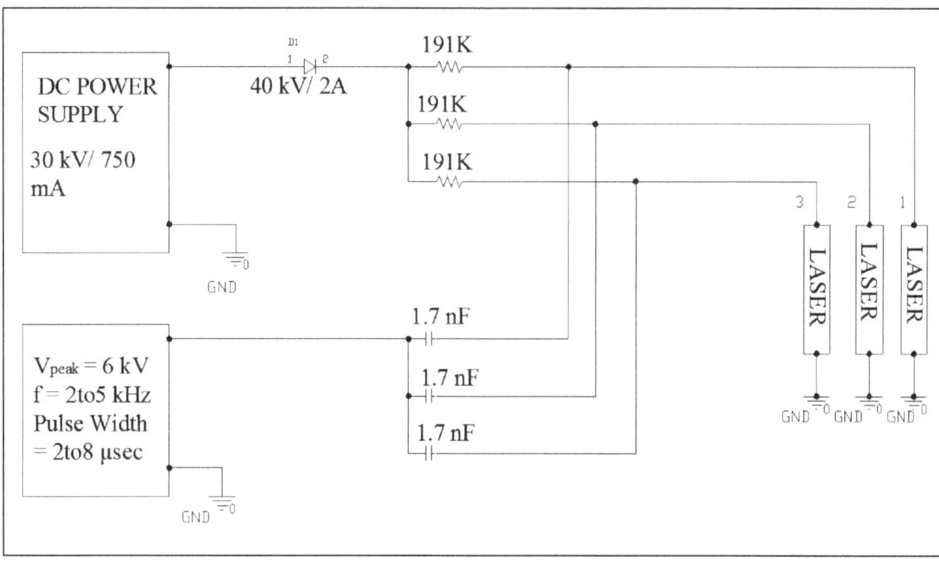

Fig. 17. Schematic of Power Supply of V-fold Laser

3.6 Laser resonator of V-fold laser

Design of a suitable optical resonator is needed to extract the laser power from the annular discharge region and also to provide the feedback to the laser. Resonators are classified depending on beam stability inside the resonator and named as follows:

I. Stable
II. Unstable

The simplest optical resonator (The Fabry-Perot resonator or confocal) consists of a pair of plane or spherical mirrors located opposite one another. They are cantered to a common optical axis and are aligned perpendicular to this axis. For lasers in the low to medium power range (1 mW - 200 W), the hemispherical resonator is mainly used and for high power laser both stable and unstable types of resonator are used. There are many combinations depending on their stability criteria given below:

$$0 \le g_1 g_2 \le 1 \quad \text{stability condition} \tag{6}$$

$$g_i = 1 - \frac{L}{R_i} \quad \text{g-parameter} \tag{7}$$

Where

L = Length of resonator,
R_i = Radius of curvature of resonator

We use the resonator mostly which satisfy this condition. The stability curve shown below represents that which resonator is preferred in stability criteria.

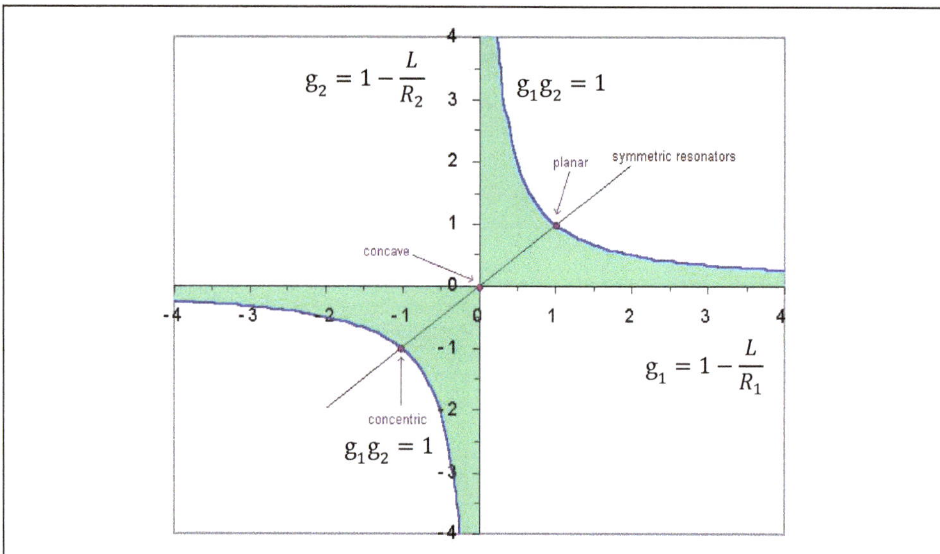

Fig. 18. Resonator Stability Curve

In our present laser we are using a concave-concave type resonator (where 2L=R) in a V-fold manner. Resonator mirrors for visible laser are generally made of glass but in CO_2 laser the radiation is of 10.6 μm which comes in infrared region and this wavelength is absorbed by glass. So a special type of output coupler made up of ZnSe is generally used. The V-Fold laser resonator is a stable resonator comprising of concave mirrors of radius of curvature of 5 meter. The distance between the mirrors is 2.5 meter. Concave mirrors keep the beam bound inside the cavity and tends to reduce the diffraction losses. For a Gaussian beam to exist in a resonator, its wave fronts must fit exactly into the curvature of the mirrors. Thus beam radius at the waist and at the mirrors can be found out using the following equation:

$$\omega = \omega_0 \left[1 + \left(\frac{\lambda z}{\pi \omega_0^2} \right)^2 \right]^{1/2} \tag{8}$$

$$R = z \left[1 + \left(\frac{\pi \omega_0^2}{\lambda z} \right)^2 \right]^{1/2} \tag{9}$$

where

ω = beam radius at the mirrors

ω_0 = minimum spot size

z = distance from the waist

λ = wavelength of CO_2 laser

R = radius of curvature of the mirror

Equation 1 and 2 gives the value of ω_0 and ω is 2.72 mm and 3.153 mm respectively.

3.7 Optimization of V-fold laser

Performance of CO_2 lasers may be optimized in several ways: maximize multimode power; maximize single- mode power; maximize efficiency; and/or minimize size and complexity. The parameters that affect such optimization for flowing gas systems are:

- Tube length, diameter and wall temperature
- Gas mixture, pressure, and flow speed
- Optical mode control, wavelength control, and output coupling
- Electrical discharge control and current density

Optimization is by no means simple, because the various parameters are strongly interrelated. All results, therefore, should be viewed only as indicative of performance trends. The engineer should be prepared to perform experimental exploration of his own system.

3.7.1 Alignment procedure of V-fold laser

Aligning this laser was very challenging job for us. Since the inner diameter of the discharge tube is 9 mm, we require alignment accuracy in microns. Since small amount of misalignment can lead to appreciable loss in output power, a great deal of work was done in making the system rigid. Height or position of the glass CO_2 laser tube should never change because any small movement throws it out of alignment and this could take days to realign. Instead, change the laser system by varying the mirror orientations, grating orientation and He-Ne laser orientation. The idea is to make the two mirrors at the ends of the laser cavity reflect a beam back-and forth many times without striking the walls of the tube. There are a few tricks in aligning this particular laser. Step by step, they are as follows:

i. Make sure that there is no high voltage at the electrodes of the laser tube by checking that the power supply is turned off.

ii. Set up a He-Ne laser alongside the cavity with a pin hole exiting the He-Ne Laser. Use two mirrors and direct the beam down the center bore of the CO_2 laser tube. The He-Ne laser beam should be positioned on the center of the mirrors for adjustment purposes. In the beginning, blank off the back mirror with a piece of paper so that reflections don't confuse matters the set-up.
iii. Direct the He-Ne beam through the middle of the output mirror (the first mirror it passes through). You will see more than one dot reflecting back.
iv. Adjust the mirrors until the He-Ne laser beam goes through the middle of the bore without reflecting off the walls of the tube. It may not look as if it goes through the middle of the Brewster windows, and it may not go exactly through the middle of the output mirror. Going down the center of the bore is the most important.
v. Remove the paper blocking the back mirror and adjust the mirror so that the reflection is centered on the output port of the He-Ne laser (it is easier to align if you place a card with a small hole punched in it at the output port of the Helium- Neon laser).
vi. Now adjust the output mirror so that the inner surface reflection of that mirror (the bigger, dimmer one of the two) is centered on the back mirror reflection spot at the Helium- Neon laser. Fringes can usually be seen on the reflections when the two are aligned (Fabry-Perot interferometer). Alignment is pretty much complete. It may take you a day or two to get to this point.
vii. Blank off the output port of the He-Ne laser with a fire brick to protect it from the CO_2 beam. Place the power detector in front of the CO_2 output port and place a fire brick behind the detector. Whenever you change scales on the power meter, you should reset it to zero.

3.7.2 Power scaling of V-fold laser

The output power of the laser scales up with the input power and input electrical power is limited by two factors. First is the rise in laser gas temperature and second is discharge instability. The most common being the ionization thermal instability. For efficient and reliable laser operation the input power density should be smaller and determined by the cooling and the discharge stabilization processes. In V-fold laser, the maximum input power density is limited by the heating effect and not by the discharge instability. Also, laser power in a V-fold diffusion cooled laser is directly proportional to the discharge length and is independent of the tube diameter and gas pressure. Thus, the laser power in V-fold diffusion cooled CO_2 laser can be scaled up by increasing the active length only and it has been incorporated by introducing several discharge tubes arranged optically in series.

3.8 Losses in optical cavities of V-fold laser

The following factors contribute to losses within the optical cavities of the lasers:

a. Misalignment of the mirrors - If the mirrors of the cavity are not aligned properly with the optical axis, the beam will not be contained within the cavity, but will move farther toward one edge of the cavity after each reflection.
b. Dirty optics - Dust, dirt, fingerprints and scratches on optical surfaces scatter the laser light and cause permanent damage to the optical surfaces.

c. Reflection losses - Whenever light is incident on a transparent surface, some portion of it always is reflected. Brewster windows and antireflection coatings greatly reduce this loss of light but cannot eliminate it entirely.

d. Diffraction loss - Part of the laser light may pass over the edges of the mirror or strike the edges of the aperture and be removed from the beam. This is the largest loss factor in many lasers. When a light beam passes through a limiting aperture, the waves at the edge of the beam bend outward slightly, causing the beam to diverge. This phenomenon is termed "diffraction". When laser light moves, diffraction occurs at the aperture and the beam diverges. When the beam returns to the aperture after reflection from the mirror, its diameter is larger than the diameter of the aperture and the edges of the beam are blocked. The portion of the beam that does pass through the aperture is diffracted again and experiences additional loss on the next pass.

e. Absorption Loss – This loss occurs due to the mirrors either fully or partially reflecting. No mirror is considered to be the 100% reflecting mirror and some part of incident laser get absorbed in the mirror. So as the number of mirrors will increase, the loss will also increase.

3.9 Misalignment sensitivity of V-fold laser

In order to ensure the high-power and stable CO_2 laser operation, misalignment sensitivity has to be known. The power and stability of the laser greatly depends on the misalignment of the optical resonator. In such type of resonator in which a V-fold resonator is used, misalignment is the main cause of reduction in power. So the effect of mirror misalignment of folded resonators is investigated experimentally and compared to first-order perturbation theory. An expression D is derived, which characterizes the misalignment sensitivity of any folded resonator. It is proved experimentally that this misalignment sensitivity depends on the effective resonator length L^* and the g_i parameters only.

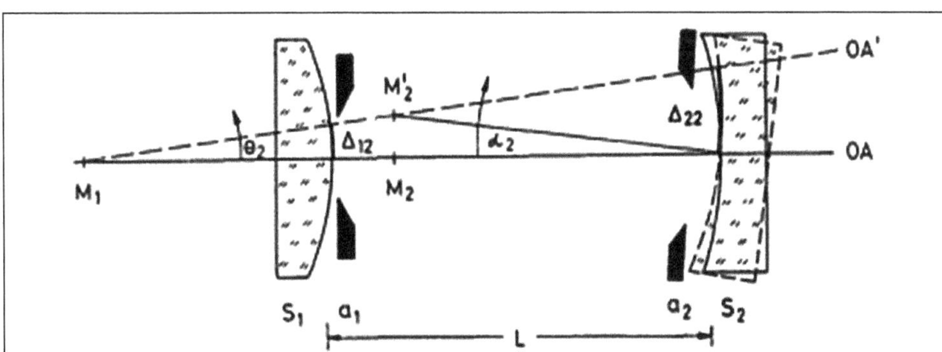

Fig. 19. Misaligned Spherical Resonator

The misalignment sensitivity of a resonator is defined as the sensitivity with which the diffraction losses or the output power are changed due to mirror tilt. By adapting the diameter of the TEM_{00} mode to the diameter of the active medium, the efficiency of a laser oscillator can be increased considerably. This requires either a large mirror distance L or an optical resonator operating near the limit of stability. In either case the resonator becomes very sensitive to a misalignment of the mirrors. From symmetry we may deduce that the increase of diffraction loss due to misalignment is proportional to the square of the mirror-tilting angle α_{oi}. Therefore, a suitable expression for the loss factor V_i per resonator bounce is:

$$V_i = V_o[1 - (\alpha_i/\alpha_{oi})^2] \tag{10}$$

Where i indicates mirror S_i, which is tilted by an angle α_i with respect to the resonator axis (see Fig.19). The misalignment sensitivity of the resonator is characterized by α_{oi}. In the following sections the relation between α_{oi} and the resonator parameters is investigated experimentally and theoretically.

3.9.1 Background

There are few papers dealing with the influence of misalignment on diffraction losses. Numerical calculations were carried out for special systems such as symmetric or confocal resonators and plane-plane resonators using first-order perturbation theory. But they assume that the aperture of the system does not disturb the field distribution of the infinite mirror. The laser oscillator consists of two spherical mirrors, radii of curvature R_1 and R_2 in a distance *(L)* and refractive index. It is assumed to be homogeneous. The mode properties of the resonator are characterized by the effective length L^* and the g_i parameters. For infinite mirrors, the spot size of the TEM_{00} mode is given by:

$$W_i^2 = \frac{\lambda L^*}{\pi}\left(\frac{g_j}{g_i(1-g_1g_2)}\right)^{1/2} \tag{11}$$

The resonator axis is defined by the two centers of mirror curvature M_1 and M_2. If mirror S_i is tilted by an angle α_i, the resonator axis is rotated by an angle θ_i, and the centers of the field intensity patterns are shifted. A simple geometric consideration delivers the relations:

$$\theta_i = \alpha_i \frac{1-g_i}{1-g_1g_2} \tag{12}$$

$$\Delta_{ii} = \alpha_i g_i L^* / (1-g_1g_2) \tag{13}$$

$$\Delta_{ij} = \alpha_j L^* / (1-g_1g_2) \qquad i \neq j \tag{14}$$

Δ_{ij} means the displacement of the intensity pattern at mirror S_i, if mirror S_j is tilted by α_j. Near the limit of stability $(g_1g_2 \to 1)$, the beam steering angle θ_i and the displacement Δ_{ij} may become considerably large. Nevertheless, as long as infinite mirrors are considered, the resonator remains aligned, and there are no diffraction losses. But if a limiting aperture is inserted into the resonator, e.g., the active medium or a mode selecting pinhole, diffraction losses occur and increase rapidly with increasing mirror tilt angle. Tilting a mirror is equivalent to a displacement of the pinhole. For a system with only one pinhole, Berger et al calculated the dependence of diffraction loss factor V on the pinhole displacement (Δ). A first-order perturbation theory for the TEM_{00} mode delivers:

$$V = 1 - \left[1 + 2(\Delta/w)^2 (a/w)^2\right] \exp\left[-2(a/w)^2\right] \tag{15}$$

Where

a = pinhole radius,
w = beam diameter of the TEM_{00} field pattern at the pinhole, and
V = loss factor per resonator bounce.

Generally a resonator has limiting apertures on both mirrors. Then the loss factor by tilting mirror S_i is given by:

$$V_i = \left(V_{ii} . V_{ji}\right)^{1/2} \qquad i \neq j \tag{16}$$

$$V_{ji} = 1 - \left[1 + 2\left(\frac{\Delta_{ji}}{w_j}\right)^2 \left(\frac{a_j}{w_j}\right)^2\right] . \exp\left[-2\left(\frac{a_j}{w_j}\right)^2\right] \tag{17}$$

For small losses ($1\text{-}V_{ji}$, $1\text{-}V_{ii} << 1$), Eq. (16) combined with Eq. (17) can be approximated by

$$V_i \cong V_o - \left\{\left[\left(\frac{\Delta_{ji}}{w_j}\right)\left(\frac{a_j}{w_j}\right)\right]^2 \exp\left[-2\left(\frac{a_j}{w_j}\right)^2\right] + \left[\left(\frac{\Delta_{ii}}{w_i}\right)\left(\frac{a_i}{w_i}\right)\right]^2 \exp\left[-2\left(\frac{a_i}{w_i}\right)^2\right]\right\} \tag{18}$$

V_o is the loss factor of the aligned system with

$$V_o = 1 - \frac{1}{2}\left\{\exp\left[-2\left(\frac{a_i}{w_i}\right)^2\right] + \exp\left[-2\left(\frac{a_j}{w_j}\right)^2\right]\right\} \tag{19}$$

For minimizing diffraction losses on the one hand and preventing multimode oscillation on the other hand, it is reasonable to use pinhole radii a bit larger than the beam radii. Combining the above equations, we finally get

$$V_i = V_o\left(1 - \alpha_i^2 \frac{S^2}{\exp 2S^2 - 1} D_i^2\right) \tag{20}$$

$$V_o = 1 - \exp\left(-2S^2\right) \tag{21}$$

$$D_i^2 = \frac{\pi L^*}{\lambda}\left(\frac{g_j}{g_i}\right)^{1/2} \frac{1 + g_1 g_2}{\left(1 - g_1 g_2\right)^{3/2}} \tag{22}$$

Equation (20) represents the diffraction loss factor V_i per resonator bounce, if mirror S_i is tilted by α_i. Misalignment sensitivity D_i, which according to Eq. (22) depends on the resonator length L^* and the g_i parameters. If a mirror is tilted by an angle $\alpha = 1/D_i$, additional losses of 10% are caused. This gives a clear idea of the meaning of the misalignment sensitivity D_i. However, the low-gain lasers are affected much more by an additional loss of 10% than the high-gain lasers. Thus, misalignment sensitivities of different resonator configurations may be compared if their gains are the same. If both mirrors are misaligned, the losses proportional to D_i^2 are summed up. Therefore, the misalignment of the complete system is defined as $D=\left(D_1^2+D_2^2\right)^{1/2}$ and is given by:

$$D=\left[\left(\frac{\pi L^*}{\lambda}\right)\frac{1+g_1g_2}{\left(1-g_1g_2\right)^{3/2}}\frac{\left|g_1+g_2\right|}{\left(g_1g_2\right)^{1/2}}\right]^{1/2} \tag{23}$$

Where, 'D' is a number characterizing any spherical resonator with respect to its sensitivity against mirror tilting. High value of D means high misalignment sensitivity. The most insensitive resonator is the symmetric con-focal one with $g_1=g_2=0$.

$$D_0=\left(\frac{2\pi L^*}{\lambda}\right)^{1/2} \tag{24}$$

But, from the stability diagram, we learn that $g_1=g_2=0$ represents a discontinuity. Small deviations from symmetry may cause high losses and high misalignment sensitivity.

3.9.2 Experimental investigation

The power and stability of a laser system is mainly governed by the misalignment sensitivity of optical resonator. To ensure stable and high power from laser system misalignment sensitivity has to be known. The effect of reflector and output coupler misalignment for concave -concave & Plano-concave resonators in single and double limbs of V-fold laser are investigated experimentally and compared to first-order perturbation theory. Eq.23 is used to quantify the misalignment sensitivity of the V-fold laser resonator. It is proved experimentally that this misalignment sensitivity depends on the effective resonator length L^* and the gi parameters only. High value of D means high misalignment sensitivity. The influence of mirror misalignment on laser output and field distribution was investigated by various authors. Experiment was carried out for four different arrangements.

a. Single limb with concave-concave resonator
b. Single limb with Plano-concave resonator
c. Double limbs with concave-concave resonator and
d. Double limbs with Plano-concave resonator.

Laser was operated with all these arrangements and then misaligned with the help of micrometer screw fitted on the backside of the optics. These four arrangements gave the misalignment characteristics for the single and double limb as well as Plano-concave and concave-concave resonator. Power was measured in the best-aligned condition then graphs were plotted for laser power v/s misalignment (Fig.20). The experimental results are verified by theoretical calculation of the misalignment sensitivity parameter 'D' (Table-6).

Misalignment sensitivity increases with L* i.e. no. of limbs. It is also observed that the plano-concave resonator is more sensitive to misalignment then the concave-concave resonator (Fig.20 & 22). It is also interesting to observe that the output coupler is less sensitive to misalignment compare to the rear concave reflector (Fig.21 & 23). This is due to very high

S.No.	Type of Resonator	Active Medium Length (cm)	D (mrad)
1	Concave-Concave	150	1.687
2	Concave-Concave	300	2.378
3	Plano-Concave	150	2.286
4	Plano-Concave	300	3.223

Table 6. Theoretical Value of Misalignment Sensitivity Parameter '*D*'

coupling loss of resonator & long gain length. The radiation, which begins from the output coupler-end, sees the round trip gain while the radiation which begins from the rear mirror; sees only single trip, and the starting intensity of radiation in the first case is relatively smaller than that in the second case. Therefore the misalignment in first case (output coupler) has relatively less effect on the laser power build up compared to the misalignment of the second case (rear reflector). Furthermore, the experimental results indicate that sensitivity parameter '*D*' is a suitable parameter to describe the alignment stability of a resonator.

SL - Single limb, DL - Double limb, CC - Concave-Concave resonator, PC - Plano-Concave resonator, M1 - Micrometer1, M2 - Micrometer2
Note: Micro-meters are numbered 1 & 2 in anticlockwise direction.

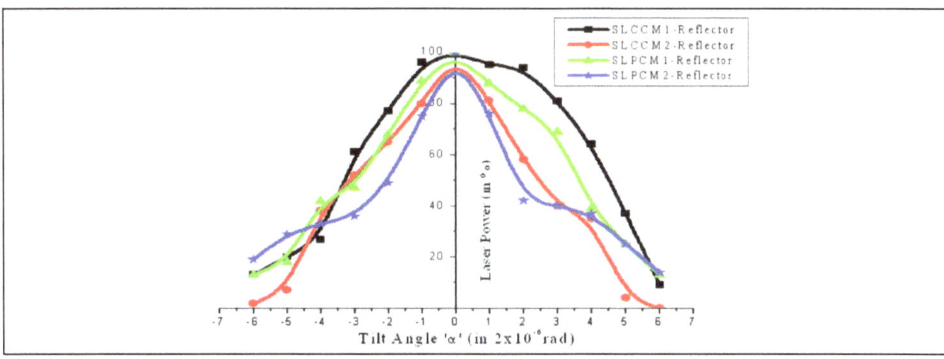

Fig. 20. Misalignment in Single Limb for Reflector

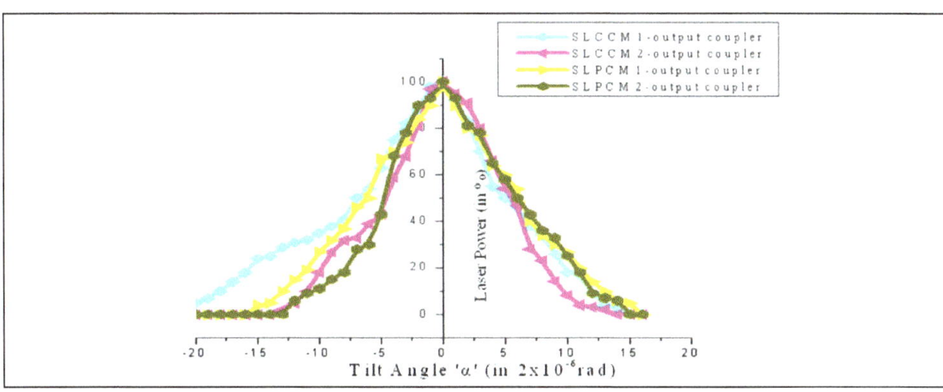

Fig. 21. Misalignment in Single Limb for Output Coupler

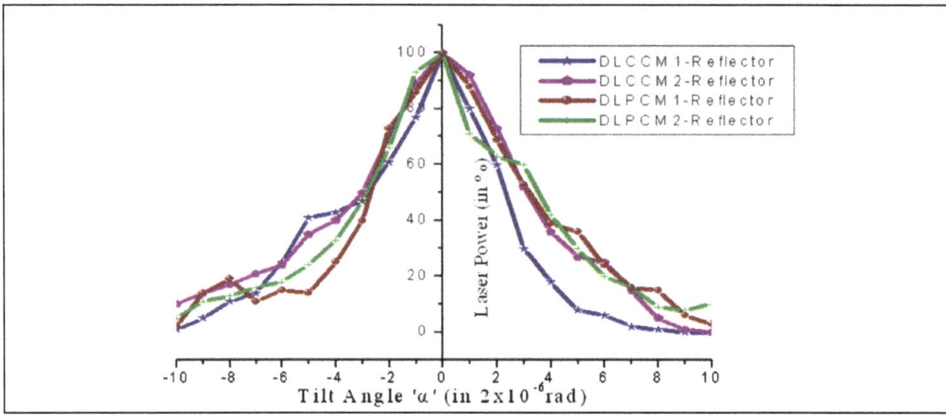

Fig. 22. Misalignment in Double Limb for Reflector

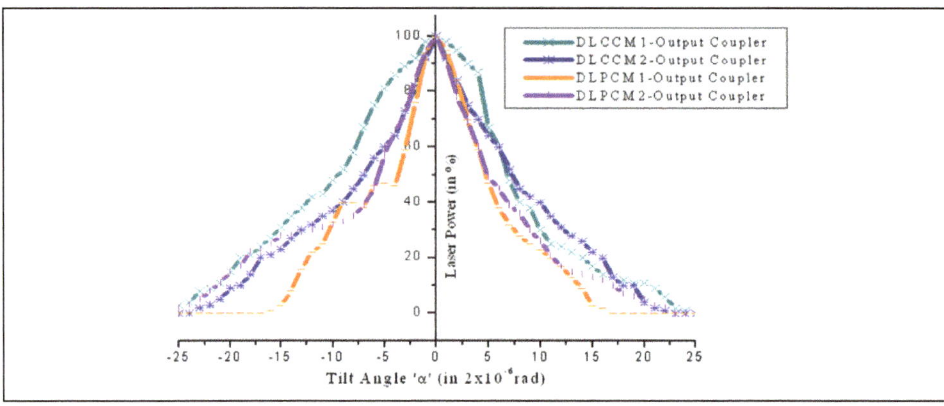

Fig. 23. Misalignment in Double Limb for Output Coupler

3.10 Optimum reflectivity of output coupler of V-fold laser

In V-fold type of resonator since there are more number of limbs and each limb has different output coupling reflectivity. So the output power of a laser that can be extracted depends on the reflectivity/transmission of the output coupler (Eq.25).

$$P_{out} = A_b I_s \frac{1-R}{1-R+\sqrt{R}\left(1/a-a\right)}\left[g_0 l - \ln\sqrt{R\times a^2}\right] \quad (25)$$

Where

A_b = Cross section area of medium,
R = Reflectivity,
a = cavity losses,
I_s = Saturation Intensity and
$g_0 l$ = Small signal gain.

Decreasing reflectivity to extract more power increases the overall loss of the system, requiring greater pumping power to reach threshold. Increasing the output coupler reflectivity increases the cavity photon life time, thereby increasing the photon loss and resulting in decrease of laser output power (Fig.24).

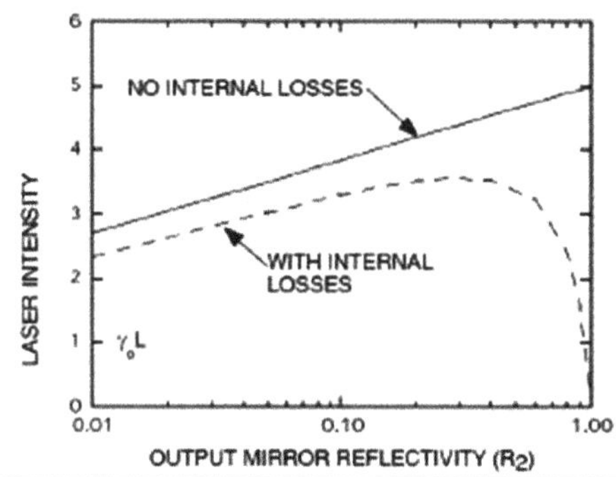

Fig. 24. Theoretical Curve for Output Coupling Reflectance

There must be an optimum reflectivity of an output coupler at which the radiant output power will be a maximum. This part reports the variation of output power as a function of output coupler reflectivity and active medium length for a V-fold diffusion cooled CO_2 gas laser. A relationship (Eq.26) is used for optimum transmission coefficient of the output couplers to verify experimental measurements.

$$T_{opt} = (g_0 La)^{1/2}\left[1-\left(\frac{a}{g_0 L}\right)^{1/2}\right] \tag{26}$$

In the development of a high-power CW CO_2 laser; it is a design challenge to reach high output power simultaneously with good beam quality. The problem becomes stringent in multi-fold diffusion cooled CO_2 lasers that uses a stable resonator configuration, where many meters of resonator length are required to generate a few kilowatts of energy, owing to the low aspect ratio between the discharge diameter and the discharge length necessary to obtain a mono mode beam. A laser will operate satisfactorily with many possible combinations of output coupler reflectivity, provided that the gain in a single pass through the amplifier is sufficiently large to equal or exceed the mirror transmission losses (or other losses).

Experiment is carried out to test the performance of the laser for different reflectivity of output couplers and different active medium length. We used a concave-concave resonator; consist of gold coated copper mirror and a concave ZnSe output coupler of 5 meter radius of curvature each. In our experimental set-up, we have taken five different output couplers of reflectivity 5, 10, 17, 50 and 60% and corresponding output power was measured for 1.5, 3.0, 4.5 and 6.0 meter active medium length. These results are plotted for active medium length v/s output power for different output couplers (Fig.25) & reflectivity v/s output power for above stated active medium lengths (Fig.26). Output power of diffusion cooled laser is proportional to active medium length but we can see (Fig.25) that as the length increases power increases but the rate of increase of output power decreases. This is because of diffraction losses increases with increase of length. For theoretical calculation, in order to estimate g_0 and a in our laser, we have used the Eq.25 of laser power in a V-fold CO_2 laser. Substituting the value of laser power for three different reflectivity of the output coupler, the three unknowns i.e. $g_0, a \& I_s$ are calculated theoretically. Thus using these values in expression (Eq.26), the T_{opt} is estimated to be 66% for 6 meter active medium length theoretically. Experimentally also we have observed that laser output power is 209 watts for 83% transmissivity and 150 watts for 50% transmissivity. From the above data we can predict that the optimum value of transmissivity lies somewhere between 50 & 83%.

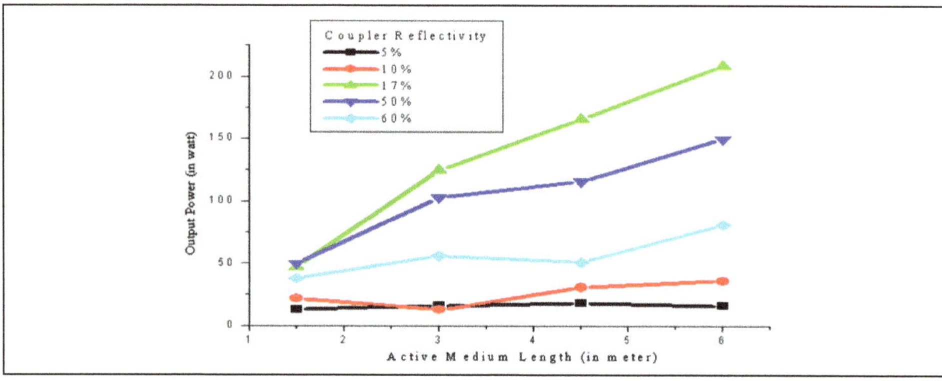

Fig. 25. Experimental Curve: Output Power v/s Active Medium Length

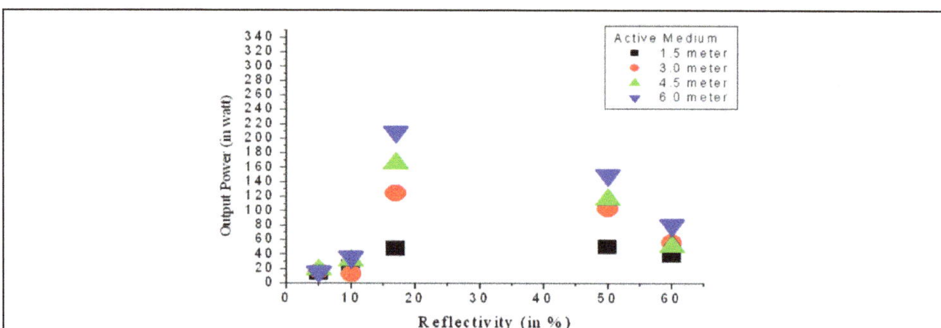

Fig. 26. Experimental Curve: Output Power v/s Reflectivity

4. Safety precautions

Some general considerations when working with V-fold CO_2 lasers are as follows:

- Provide a beam stop capable of safely absorbing this power on a continuous basis.
- Clearly mark and if possible, block off access to the path of the beam.
- Reflected beams may have nearly as much power as the original and are just as dangerous. Although many common materials will block 10.6 µm, specular surfaces will reflect it quite well.

- Make sure that everyone in the vicinity of the laser or anywhere the beam (or its reflection) may be is fully aware of the safety issues and has proper eyewear.
- Provide visible and unambiguous indications that the laser is powered and the beam is on.
- A kill switch is essential and should be located far enough from the laser tube so that it is accessible in an emergency even if a total meltdown is in progress.
- For flowing gas lasers, provide adequate ventilation. While the lasing gasses (helium, nitrogen, and carbon dioxide) are not toxic, and not very much is involved for laser operation, a leak in the gas delivery system could go undetected. CO_2 in particular is heavier than air so it will displace air in an enclosed space which may result in various symptoms from nausea to asphyxiation.
- Where maintenance or repair is involved, be aware of the properties of the specific materials used for the optics and elsewhere. For example, the biohazards of zinc selenide and beryllia.

5. Conclusion

In the present laser, power of 380 Watts from 7.5 m discharge length and maximum 420 W from seven limbs (10.5 meter discharge length) has been achieved. The maximum average power of 50 W/m is obtained from this laser, which is comparable to other diffusion-cooled laser developed till now. Studies have shown that dissociation of CO_2 molecules increases with the increase of no of tube or discharge length. Care has been taken to have a low gas residence time to reduce the deleterious effect of CO_2 dissociation. The electro-optic efficiency of the laser is about 13%.

The power and stability of a laser system is mainly governed by the misalignment sensitivity of the optical resonator. To ensure stable and high power from a laser system misalignment sensitivity has to be known. The experimental results indicate that sensitivity parameter D is a suitable parameter to describe the alignment stability of a resonator.

The output power of a laser that can be extracted depends on the reflectivity/transmission of the output coupler. There must be an optimum reflectivity of an output coupler at which the radiant output power will be a maximum.

According to Rigrod's formula if length increases power reduces, as there are many other parameters, which are not optimized. So power goes on decreases when length increases. Beam size also affects the output power.

6. Acknowledgement

Author is thankful to Sh. Mukesh Jewariya, Sh. Firoz Koser, Sh. D.D. Saha, Sh. M.B. Pote, Sh. S.V. Deshmukh, Sh. A.K. Nath, Sh. Dinesh Nagpure, Sh. Abrat Verma and all other colleagues of Laser and Material Processing Division, RRCAT, who directly or indirectly involve in design and development of this laser. Author is also thankful to Sh. Abhay Kumar (IMA Section) and Sh. Arup Ratan Jana (Accelerator and Beam Physics Lab.).

7. References

Dahotre, Narendra B. & Harimkar, Sandip P. (2008). *Laser Fabrication and Machining of Materials*, Springer Science + Business Media, ISBN 978-0-387-72343-3, USA

Endo, Masamori & Walter, Robert F. (2007). *Gas Lasers*, CRC Press, ISBN 0-8493-3553-1, USA

Freiberg, R.J. and Halsted, A. S. (1969). Properties of Low Order Transverse Modes in Argon Ion Lasers. *Applied Optics*, Vol. 8, No. 2, (1969), pp. (355-362), ISSN 1559-128X

Hitz, B.; Ewing, J.J. and Hetch, J. (2005). *Introduction to laser technology* (Third Edition), John Wiley & Sons Inc., USA

Hoag, Ethan; Pease, Henry; Staal, John and Zar, Jacob. (1974). Performance Characteristics of a 10-kW Industrial CO2 Laser System. *Applied Optics*, Vol. 13, No. 8, (August 1974), pp. (1959-1964), ISSN 1559-128X

Hodgson, N. (1997). Optical resonators: fundamentals, advanced concepts and applications, Springer, ISBN 3-540-76137-3, Berlin

Koechner, W. (1988). Solid State Laser Engineering, Springer-Verlag, ISBN 3-540-18747-2, Berlin

Migliore, Leonard. (1996). *Laser Materials Processing*, Marcel Dekker Inc., ISBN 0-8247-9714-0, New York, USA

Nath, A.K. and Golubev, V.S. (1998). Design considerations and scaling laws for high power convective cooled CW CO_2 lasers. *Pramana*, Vol. 51, No. 3-4, (September-October 1998), pp. (463-479), 0304-4289

Scott, Marion W. and Myers, Gary D. (1984). Steady-state CO2 laser model. *Applied Optics*, Vol. 23, No. 17, (September 1984), pp. (2874-2878), ISSN 1559-128X

Seigman, A.E. (1986). *Lasers*, University Science Books, Mill Valley, California, USA, ISBN 0-935702-11-5, USA

Soni, R.K. et al. (2005). Diffusion Cooled V-Fold CO2 Laser, *Proceedings of Fourth DAE-BRNS National Laser Symposium, 185 January 10-13, 2005*, ISBN 8177647342, 9788177647341, Mumbai (India), January 2005

Soni, R.K. et al. (2005). Misalignment Sensitivity of a V-Fold Optical Resonator, *Proceedings of Fifth DAE-BRNS National Laser Symposium (NLS-5), December 7-10, 2005, page 96*, ISBN , Tamil Nadu (India), December 7-10, 2005

Soni, R.K. et al. (2005). Optimum Reflectivity for a Diffusion Cooled CO2 Laser, *Proceedings of Fifth DAE-BRNS National Laser Symposium (NLS-5), December 7-10, 2005, page 98*, ISBN , Tamil Nadu (India), December 2005

Svelto, Orazio. (2010). *Principles of Lasers* (Fifth Edition), Springer Science+Business Media, www.springer.com

Thyagarajan, K. & Ghatak, Ajoy. (2010). *Lasers: Fundamentals and Applications* (Second Editiion), Springer Science+Business Media, New York, USA

Weber, H., Herziger, G. & Poprawe, R. (2006). *Laser Physics and Applications Sub volume A: Laser Fundamentals Part 2*, Springer, ISBN 978-3-540-28824-4, Germany

Xinju, Lan. (2010). *Laser Technology* (Second Edition), CRC Press, ISBN 978-1-4200-9171-7, USA

5

Ultrashort Pulses

Mikhail N. Polyanskiy and Marcus Babzien
Brookhaven National Laboratory
USA

1. Introduction

1.1 Niche for ultrashort-pulse CO_2 lasers

Ultrashort pulses usually are defined as those lasting less than a nanosecond. Low-energetic ultrashort pulses can be used, for instance, as a probe for studying the dynamics of ultrafast processes. Amplifying these pulses can deliver extremely high peak power, allowing applications such as laser particle acceleration, or γ-ray generation via Compton scattering on relativistic electrons. Existing laser systems can provide pulses as brief as hundreds of attoseconds (10^{-18} s) and as powerful as tens of petawatts (10^{15} W); more advanced ones are being constructed or are planned (Corkum & Krausz, 2007, Korzhimanov et al., 2011).

Virtually all modern ultrashort-pulse lasers are based on solid-state technology and usually operate at ~1 μm wavelength. Nevertheless, these applications relying not only on the ultrashort duration or the extreme power of the laser pulse, but also on its wavelength, leave a niche for laser systems utilizing different types of active medium. The 10-micron wavelength of the CO_2 laser particularly is beneficial for some usages.

To demonstrate the potential of ultrashort, mid-IR pulses, we consider their employment for laser ion acceleration, presently one of the main drivers for developing ultrashort-pulse CO_2 lasers (Palmer et al., 2011, Norreys, 2011). The motivation underlying the search for alternatives to conventional accelerators, wherein a system of electrodes and magnets creates the accelerating field, is the need to reduce the size and the cost of these devices. Acceleration in the electromagnetic field of a laser beam is a promising alternative for traditional acceleration schemes. In laser ion acceleration, an intense laser pulse focused on a target first ionizes it and then accelerates the charged particles from the resulting plasma. Usually this target is a metal foil, and a ~1 μm, multi-TW peak power, solid-state laser provides the ionizing/accelerating field. A very efficient acceleration regime called *Radiation Pressure Acceleration* (RPA) featuring a narrow energy-spectrum of the accelerated ions is reached when laser pulse interacts with plasma having a near-critical density (Esirkepov et al., 2004). The following formula defines critical plasma density, N_c (the density at which plasma becomes opaque):

$$N_c[\text{cm}^{-3}] \approx 1.115\times10^{21}\left(\frac{n}{\lambda[\mu\text{m}]}\right)^2, \tag{1}$$

where n is the refractive index, and λ the laser wavelength. According to the Eq. (1), N_c is ~10^{21} cm^{-3} for the 1-μm lasers, and ~10^{19} cm^{-3} for the 10-μm lasers. These densities are much lower than that of solid materials. The critical density at 10 μm is comparable with the density of gases (~2.7×10^{19} cm^{-3} at 1 bar); thus, it readily is achievable for the CO_2 laser's wavelength when the gas jet is used as a target. On the other hand, realizing the critical-density jet for 1-μm solid-state lasers is challenging. Another advantage of the longer wavelength for ion acceleration in the RPA regime is the λ^2 scaling of the ponderomotive force, implying that a 100x lower intensity of the CO_2 laser field will suffice for reaching a given ion energy compared with a solid-state laser.

Yet another simple consideration favors CO_2 lasers for certain applications: A 10-μm laser pulse carries 10x more photons than a 1-μm pulse of the same energy. Photon density is important for applications such as γ-ray generation by Compton scattering on relativistic electrons (Yakimenko & Pogorelsky, 2006).

1.2 Challenge of high peak power

The minimum achievable duration of the pulse is defined by the bandwidth of the gain spectrum. In the simplest case of the absence of chirping (frequency variation with time), the spectrum of a pulse is the Fourier transform of its temporal profile. The pulse then is called transform-limited, and its duration, τ_p, is inversely proportional to its spectral width, $\Delta\nu_p$. The following expression is valid for a transform-limited Gaussian pulse (Paschotta, 2008):

$$\tau_p \approx \frac{0.44}{\Delta\nu_p}, \tag{2}$$

where τ_p and $\Delta\nu_p$ are defined as full-width at half-maximum (FWHM). The duration of a chirped pulse is longer than that of a transform-limited pulse of the same shape and spectral width. Fig. 1 shows the spectra of several femto- and pico-second transform-limited Gaussian pulses. Fig. 2 compares them with the gain spectrum of a typical CO_2 laser mixture, clearly revealing the challenge of generating and amplifying ultrashort pulses in a CO_2 laser.

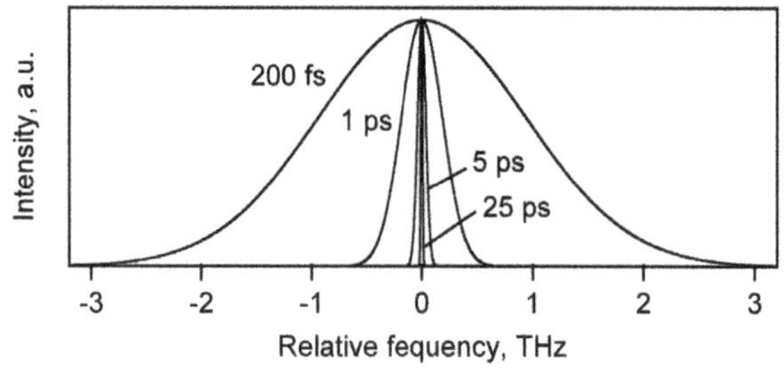

Fig. 1. Spectra of transform-limited Gaussian pulses. Durations are full-width at half-maximum (FWHM).

At low pressures (less than a few bars), the gain spectrum consists of individual rotational lines whose bandwidth (FWHM) is given by the following expression (Brimacombe & Reid, 1983):

$$\Delta\nu_{pressure}[\text{GHz}] \approx P[\text{bar}]\cdot(5.79\Psi_{CO2} + 4.25\Psi_{N2} + 3.55\Psi_{He}), \quad (3)$$

where P is the total gas pressure, and Ψ_x is the relative concentration of the component x. Eq. (3) yields $\Delta\nu_{pressure}\approx3.8$ GHz at 1 bar of a typical mixture $[CO_2]:[N_2]:[He]=1:1:8$. Although direct substituting this bandwidth in Eq. (2) gives $\tau_p\approx120$ ps, in reality the pulse's spectrum is narrower than the gain bandwidth, and, correspondingly, the pulse's duration is longer. The latter fact is comprehensible by realizing that amplification is the strongest in the center of the laser line, and thus, upon amplification, the ratio between the intensity at the central frequency and that at the wing of the gain-spectrum line increases, thus narrowing the pulse's spectrum. Consequently, the minimum achievable duration of the pulse for an atmospheric-pressure CO_2 laser is about 1 ns (Abrams & Wood, 1971). Pressure can be increased to broaden the gain bandwidth, and thus somewhat reduce the pulse's duration. However, the complete overlap of rotational lines separated by 55 GHz (P-) or 40 GHz (R-branches) that would allow using an entire rotational branch, does not occur below 20-25 bar; this is not feasible in discharge-pumped lasers having a ~10 bar practical pressure limit.

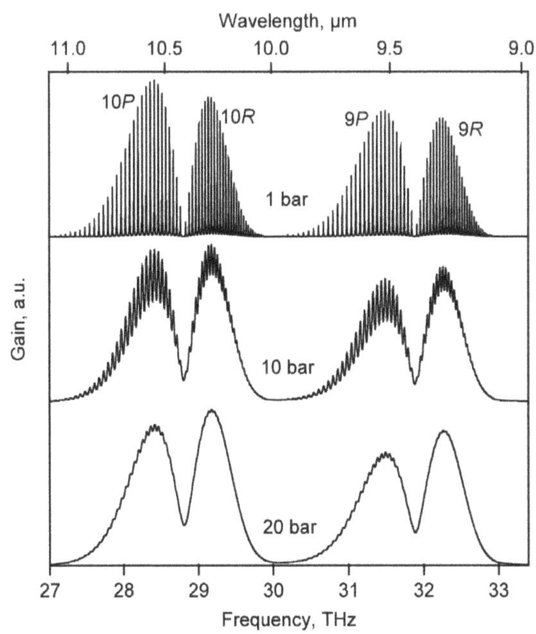

Fig. 2. Simulated normalized gain spectra at different gas pressures. 10P, 10R, 9P, and 9R: Rotational P- and R-branches of the two vibrational transitions supporting lasing at ~10 μm and ~9 μm respectively.

Another important consequence of the rotational structure of the CO_2 gain spectrum relates to the amplification of the ultrashort pulses. The spectrum of a few-picosecond-long pulse, according to Eq. (2), overlaps several rotational lines of the gain spectrum. Hence, upon amplification, the pulse's spectrum acquires the corresponding periodic structure. In the time-domain (inverse Fourier-transform of the spectrum), this is equivalent to a pulse train with pulse-to-pulse distance equal to the inverse separation of the spectral lines (18 ps in the P-, and 25 ps in the R-branches). Fig. 3 demonstrates this effect for a 5-ps pulse amplified in a 10-bar active medium.

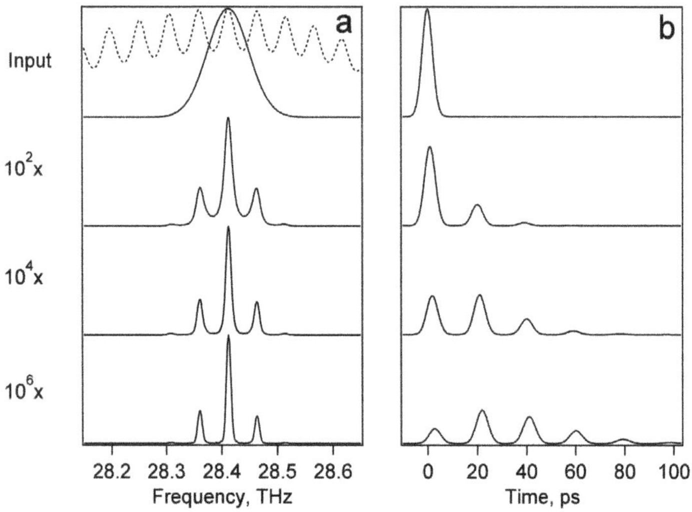

Fig. 3. Simulated dynamics of a 5-ps (FWHM) pulse spectrum (a), and its temporal structure normalized by the total energy in the train (b) amplified in a 10-bar CO_2 amplifier. Dashed line: normalized gain spectrum.

We discuss below different approaches to overcome these difficulties.

2. Generating the seed pulse

In all modern schemes for producing ultrahigh-power laser pulses there are at least two major stages: (1) Generation of a low-power, ultrashort seed pulse, and, (2) amplification of the pulse. In this section, we consider different methods of creating ultrashort 10-μm pulses. Section 3, following, discusses the amplification stage.

2.1 Mode-locking

Mode-locking is the primary technique of generating ultrashort pulses in solid-state systems. In this approach, cavity modes present in the spectrum of the laser field are synchronized (locked) by modulating the cavity's Q-factor at a frequency that is a multiple of the inverse of the resonator's round-trip time, either actively (e.g., with an intra-cavity acousto-optical modulator), or passively (e.g., with a saturable absorber). Without synchronization, each mode is independent and defines its own pulse; thus, the pulse's duration is only limited by the relatively narrow bandwidth of an individual mode. However, after mode-locking, the time structure of the laser field is defined by the entire spectrum comprising *all* active modes. For the periodic spectrum of longitudinal modes, this structure is a periodic train of pulses; the pulse's repetition rate is equal to the lock-in frequency (i.e., a multiple of the inter-mode distance), and individual pulse's duration is inversely proportional to the total extent of the spectrum.

The minimum pulse duration achievable via mode-locking is related to the bandwidth, $\Delta\nu$, (FWHM) of the gain spectrum according the following expression (Siegman & Kuizenga, 1969):

$$\tau_p \approx 0.45 \cdot \left(\frac{g}{M}\right)^{1/4} (f \cdot \Delta\nu)^{-1/2}, \tag{4}$$

where $g \equiv 2aL$ is the saturated round-trip excess gain (a - excess gain per unit length, L - resonator length), M is the modulation, and $f=c/2L$ is its frequency. For a typical transversely excited atmospheric pressure (TEA) CO_2 laser (g=1; M=1; f=150 MHz; $\Delta\nu$=3.8 GHz) the duration is $\tau_p \approx 600$ ps. Wood et al. describe their technical realization of the mode-locked TEA CO_2 laser (Wood et al., 1970, Abrams & Wood, 1971). A 10-bar mode-locked system was reported by Houtman and Meyer (Houtman & Meyer, 1987). Application of mode-locking is limited in 10-micron systems by relatively long (hundreds of picoseconds) pulse duration achievable via this technique.

Passive self-mode-locking often occurs in TEA CO_2 lasers producing a comb pulse-structure unless special countermeasures are taken. This effect is believed to be due to gain saturation (Kovalev, 1996). When a smooth pulse is required, the laser must be forced to operate at a single cavity mode that usually is achieved by using a low-pressure intra-cavity CW discharge cell ("smoothing tube"). Such a laser, combining two gain sections in a single resonator; viz., main atmospheric-pressure and a low-pressure one for spectrum narrowing, is called *hybrid* laser. Fig. 4 is an example of a hybrid TEA CO_2 lasers' pulse profile with and without discharge in the smoothing tube. As Fig. 4a shows, self-mode-locking occurring when the smoothing tube not activated results in modulation at a frequency that is a multiple (here, double) of the inter-mode spacing. Activating the smoothing tube eliminates the modulation, Fig. 4b.

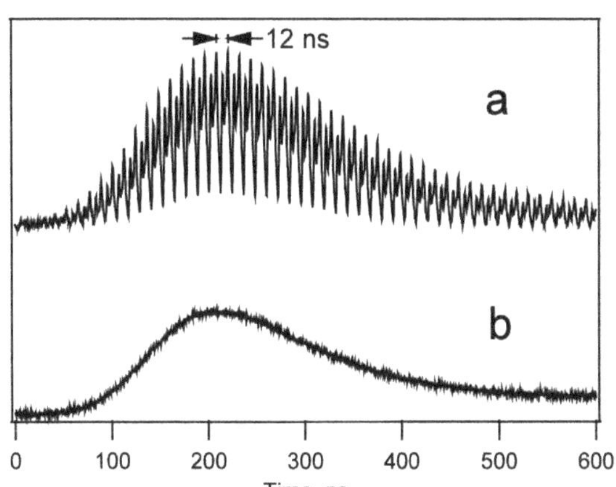

Fig. 4. Temporal structure of the output of a hybrid TEA CO_2 laser with smoothing tube discharge OFF (a), and ON (b). Resonator round-trip time is 12 ns; self-mode-locking occurs at the doubled round-trip frequency.

2.2 Plasma shutter and optical free-induction decay

Highly intense laser pulses focused in gas media can initiate avalanche ionization (laser breakdown). If gas density is high enough, overcritical plasma forms, blocking the trailing part of the laser pulse, partially absorbing and partially reflecting it. This effect itself, usually termed *plasma shutter*, can be used to cut the tail of the pulse thus reducing its overall duration. However, it is not sufficient for producing an ultrashort pulse because the front part of the initial pulse passes the plasma shutter unchanged. The possibility of generating a pulse as short as a few optical cycles lies in the fact that the fast switching-out of the laser field by plasma entails a very rapid variation in the field spectra. Essentially, we can approximate the spectrum of the transmitted pulse by the Fourier-transform of a step function. Frequencies different from those present in the original pulse briefly appear in the spectrum. If we now filter-out the frequency of the initial pulse from the resulting spectrum, we end up with a very short pulse (Yablonovich, 1973, 1974a, 1974b). Free-induction decay technique is relatively simple to realize and can allow achieving ~20 ps pulse duration at the expense of large losses and strong alteration of the spectrum.

2.3 Pulse-slicing techniques

Another possibility for producing a low-energy ultrashort pulse is to slice a small fraction out of a longer one (for instance, a hundred-nanosecond output of a hybrid TEA CO_2 laser similar to that in Fig. 4b) using a fast optical switch. Here, the switching speed limits the duration of the resulting pulse. Two techniques often employed for this purpose are a semiconductor optical switch (Alcock & Corkum, 1979), or a Kerr cell (Filip et al., 2002). Both rely on an ultrashort pulse of another laser (usually a solid-state one) to trigger the switch by inducing a short-living "plasma mirror" in the case of a semiconductor switch, or birefringence in that of the Kerr cell. Fig. 5 illustrates the principles of these two techniques.

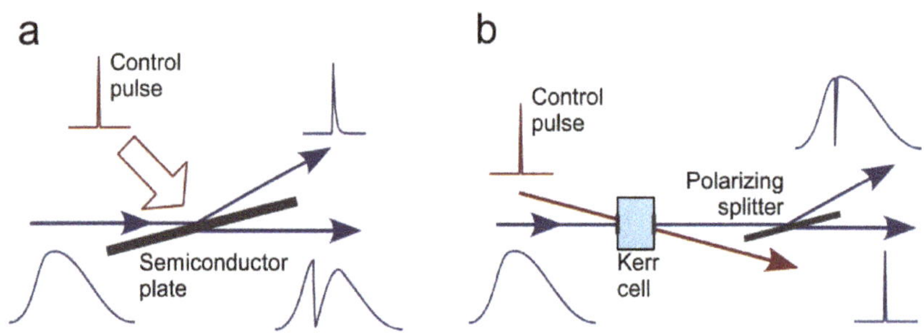

Fig. 5. Simplified schematics of the techniques for pulse slicing by (a) the semiconductor switch, and, (b) the Kerr cell.

A powerful laser pulse irradiating the surface of a semiconductor partially ionizes it creating a "plasma" of free charge carriers. If the plasma's density exceeds the critical one (N_c), the semiconductor surface turns into a mirror, reflecting the entering laser beam. At below-critical densities, the beam is attenuated mostly by free-carrier absorption. The duration of the reflection is determined mainly by that of the control pulse, and the speed of free carrier diffusion (typically hundreds of picoseconds). Absorption generally lasts longer (hundreds of nanoseconds), and its persistence is defined by the free carrier's recombination time. The inverse-quadratic relationship between N_c and the laser's wavelength (Eq. (1)) allows our realization of a scheme wherein a relatively low-energy pulse from a solid-state laser controls a high-power CO_2-laser pulse. Combining two semiconductor switches, the first of which operates in a reflection- and the second in the transmission- configuration (Fig. 6) enables us to slice a CO_2 pulse on both edges, thus producing one whose duration is limited only by the rise-time of the control pulse.

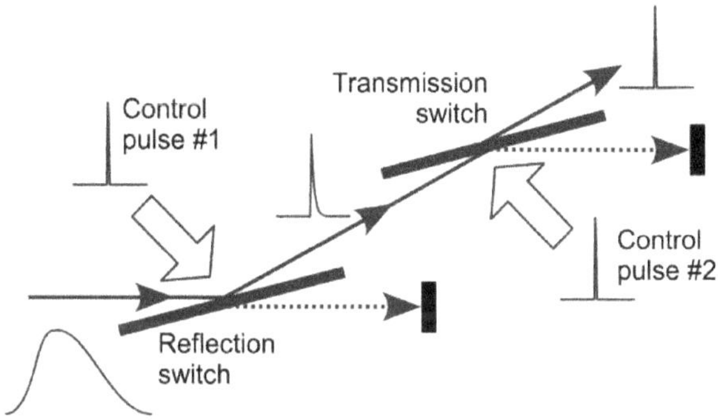

Fig. 6. Combination of reflection- and transmission- semiconductor switches to generate an ultrashort pulse.

Reportedly, Rolland & Corkum, (1986) achieved a 130-fs pulse via this technique. Germanium is the commonest material used in semiconductor switches; silicon and cadmium telluride also proved usable in this application.

The laser-stimulated birefringence employed in the Kerr-cell technique has the advantage of low inertness, so supporting the production of an ultrashort pulse in a single step. Its minimum achievable duration is limited by that of the control pulse, and the relaxation speed of the induced birefringence. The Kerr cell depicted in Fig. 5b rotates the polarization of the CO_2 laser-beam while it is being irradiated by the control pulse. A polarization filter on the cell output separates the part of the pulse with rotated polarization from the main pulse. Liquid carbon disulphide (CS_2) featuring ~2 ps relaxation time usually serves as the active medium in the optical Kerr cell for slicing the CO_2 laser pulses. For optimum switching efficiency, the phase angle between the control- and the CO_2 laser- pulses must be equal to $\pi/4$.

Currently, pulse-slicing is the main method of producing low-energy ultrashort (few-picosecond) 10-micron seed pulses.

2.4 All-solid-state systems

Solid-state ultrafast oscillators producing several-cycle and longer optical pulses in the near-infrared spectral region are a well-established technology. Using frequency conversion via *optical parametric amplification* (OPA) one can generate an ultrashort mid-IR pulse. One of the earliest theoretical treatments of parametric amplification was by Armstrong, Bloembergen, Ducuing, and Pershan (Armstrong et al., 1962). Using a resonant cavity to enhance output, Giordmaine and Miller demonstrated the principle (Giordmaine and Miller, 1965). Okorogu et al. demonstrated efficient, single-stage difference frequency downconversion from near-to mid-IR (Okorogu et al., 1998). The combination of compact size, various free-space or fiber-based configurations, and efficient pump sources provide an advantageous starting point for CO_2 laser seed sources. One method used for frequency conversion is covered below with attention toward stable and reliable operation as a sub-component of a larger CO_2 laser system.

Titanium-doped sapphire is now the dominant laser system in many industrial and research fields such as physical chemistry, materials science and processing, and strong-field physics. Therefore, a large commercial infrastructure exists for producing reliable amplifiers delivering high energy pulses suitable for nonlinear conversion. Diode-pumped Neodymium lasers which are frequency-doubled for pumping the broadband Ti:Al_2O_3 gain medium are turn-key and have lifetimes on the order of 10 000 hours. Many amplifier systems are available producing pulse energies above 5 mJ near 1 kHz repetition rates. In such a configuration, the pump pulses have energy stability better than 1% because the energy is removed on a time scale comparable to the upper-state lifetime, and a quasi-steady-state exists between pumping and energy extraction.

After amplification to high energy, the use of OPA provides a path for the generation of significant seed energies at 10 μm with the full bandwidth and tunability to cover the entire gain spectrum of CO_2. One such commercial approach is shown in Fig. 7.

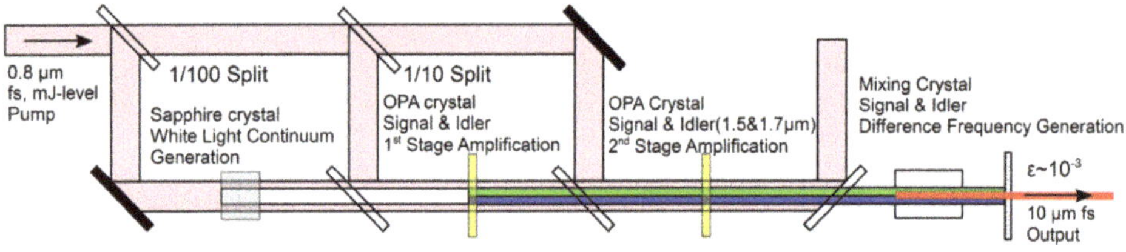

Fig. 7. Commercially-available frequency conversion system from 0.8 to 10 microns

In this configuration, the pulses from the Ti:Al_2O_3 amplifier are used in three separate nonlinear processes. The first is white light continuum generation in sapphire. This process creates an ultra-broadband spectrum spanning the entire visible and near-IR region while preserving the phase and temporal structure of the original pulses. This broadband radiation makes an ideal seed source for the following sections as it allows free tuning to the desired wavelengths. Because the next section uses three-wave mixing, this seed power also eliminates instabilities that would be encountered from quantum fluctuations in a pure parametric generator.

The next two stages rely on standard nonlinear crystals and act as simple parametric amplifiers that are angle-tuned to achieve gain at the desired signal and idler wavelengths. For conversion from 0.8 to 10 micron, these are approximately 1.5 and 1.7 micron, respectively. By utilizing two stages, the gain of each section can be optimized while preserving bandwidth that would be limited by a longer single crystal. The second parametric amplification stage therefore utilizes most of the pulse energy delivered from the Ti:Al_2O_3 amplifier.

In the final section, the signal and idler from the previous stages generate a difference frequency pulse in another nonlinear material that has transparency in the mid-IR region.

These cascaded nonlinear processes allow stable, repeatable conversion of the ultrafast pump pulses from the near- to the mid-IR region, while providing broad bandwidth, wavelength tunability, and ultrashort duration. The theoretical maximum energy conversion efficiency from 0.8 to 10 micron via this cascaded three-photon mixing is near 3.5%, however when real beams which are non-uniform in time and space are considered, as well as losses on the large number of optical elements required, the realized efficiency is approximately an order of magnitude lower. Pulsewidths under 500 fs are easily achievable, as well as bandwidth covering any single branch of the CO_2 gain spectrum.

All-solid-state systems provide good control of pulse synchronization and shape, but are much more elaborate than the other techniques used for ultrashort mid-IR pulse generation.

3. Amplification

3.1 "Smoothing" the gain spectrum

A major problem in seamlessly amplifying picosecond pulses is the discrete rotational-line structure of the gain spectrum, causing modulation of the pulse spectrum and pulse splitting. The gain spectrum's modulation can be smeared either by broadening individual rotational lines, thus assuring their better overlap, or by increasing their density. In the first case, we can use pressure- (collision-) and/or field- (power-) broadening effects. Several approaches increase line density: (1) Using an R-branch of a laser transition having a 1.4 times denser line structure than a conventional P-branch; (2) isotopic enrichment of the CO_2 molecules, wherein the superposition of the slightly shifted spectra of different isotopic species (*isotopologues*) generates a denser effective spectrum; and, (3) combining the *sequence* bands of laser transition with regular ones. We briefly overview these approaches next.

Pressure broadening. As discussed in the Section 1.2, increasing the pressure lowers modulation in the gain spectrum. Complete suppression occurs when collisionally broadened bandwidths of the rotational lines become about twice the interline distance. However, the 20-25 bar pressure required for this, according to the Eq. (3), is impractical due to difficulties in arranging the uniform electric-discharge pumping the large volume of active gas required for building a high-power CO_2 laser amplifier. Replacing discharge- with optical- pumping may afford using a pure-CO_2 active medium and increased working pressure, thus considerably extending the pressure broadening effect. Rapid progress in the solid-state laser technology might well lead to the availability of a reliable source for optical excitation of CO_2 active medium in the near future. Gordienko & Platonenko, (2010) consider that here the ErCr:YSGG (2.79 μm) laser is a promising candidate.

Field broadening. We can approximate the magnitude of line broadening (FWHM) appearing in the intense laser field due to the Autler-Townes (or dynamic Stark) effect (Autler & Townes, 1955) by the doubled Rabi frequency Ω: $\Delta\nu_{field} = 2\Omega = 2\frac{d \cdot E}{h}$, where d is the transition dipole momentum, E is the laser field, and h is Plank's constant. Substituting the laser field with its expression through the intensity, I, and using the numerical values of the involved constants, we get the following equation:

$$\Delta\nu_{field}[\text{GHz}] \approx 0.02764\, d[\text{D}]\sqrt{I[\text{W/cm}^2]} \tag{5}$$

With Eq. (5), and the known value of the laser's transition dipole momentum d=0.0275 D (the value for the 10P(20) line from HITRAN database (Rothman et al., 2009)), we find that field broadening is sufficient to completely suppress modulation at a laser intensity 15-20 GW/cm^2; this is reachable in the modern high-power picosecond CO_2 laser systems. Capitalizing on this approach supported the attainment of 15 TW peak power in the CO_2 laser system of Neptune Laboratory of the University of California, Loss Angeles (Section 6.2).

R-branch. The R-branches of the CO_2 laser transitions have a rotational structure 1.4 times denser than the more often used P-branches (Witteman, 1987); thus, they offer better overlap between collisionally broadened lines, and, hence, yield a smoother gain spectrum (Fig. 2). Interestingly, under high-pressure conditions, such as 10 bar or higher, the overlap between rotational lines increases the peak intensity of the R-branch compared to that of the P-branch that otherwise prevails in conventional low-pressure lasers.

Isotopic CO_2. By partially substituting the ^{16}O atoms in CO_2 gas with another stable ^{18}O isotope, we obtain almost perfectly smooth combined spectrum after superimposing the spectra of three CO_2 isotopologues (molecules with different isotopic composition): ^{16}O-^{12}C-^{16}O, ^{16}O-^{12}C-^{18}O, and ^{18}O-^{12}C-^{18}O (Fig. 8). They often are denoted as 626, 628, and 828 wherein 2, 6, and 8, respectively, represent ^{12}C, ^{16}O and ^{18}O.

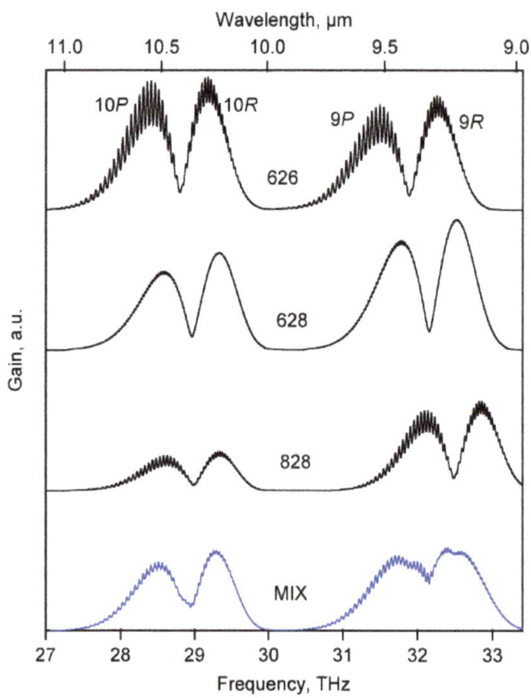

Fig. 8. Simulated gain spectra of three CO_2 isotopologues with different combinations of oxygen-16 and oxygen-18 atoms (no enrichment in carbon isotopes), and the effective spectrum of their mixture in the proportion [626]:[628]:[828]=0.16:0.48:0.36 (statistical equilibrium in the case of [^{16}O]:[^{18}O]=0.4:0.6). Total gas pressure is 10 bar.

For a given proportion $[^{16}O]:[^{18}O]$, independent of the initial distribution of ^{16}O and ^{18}O between the CO_2 molecules, statistical equilibration via intermolecular isotope-exchange leads to $[626]:[628]:[828]= [^{16}O]^2:2[^{16}O]\,[^{18}O]:[^{18}O]^2$. We note that due to the broken symmetry of the 628 molecule, it has twice as many rotational lines in each rotational branch compared to more symmetric 626- and 828- isotopologues. The combination of three CO_2 isotopologues, as depicted in Fig. 8, results in a smooth spectrum already apparent at 10 bar. The gain of the isotopic mixture in the 10-micron branches at this pressure is 1.4 times lower than that of the regular gas, mainly reflecting the relatively low gain of the 828 CO_2 isotopologue. Thus, a longer path through an active medium or a higher CO_2 concentration is needed to maintain the same net amplification. The isotope-based approach is practically implemented in the CO_2 laser of Accelerator Test Facility at Brookhaven National Laboratory (Section 6.1).

Sequence bands. Transitions between high-lying vibrational overtones of the CO_2 molecule can contribute considerably to the high-pressure amplifier gain. In this case, the rotational spectra of the regular- and sequence- bands overlap, so providing a denser effective spectrum. Exploiting the sequence band for smoothing the gain spectrum seems especially promising for the 10R branch wherein the rotational lines belonging to the sequence band $00^02\text{-}[10^01,02^01]_I$ fall very close to the centers of the gaps between the lines of the regular band $00^01\text{-}[10^00,02^00]_I$. Simple estimation of the ratio of gains of the sequence- and the regular- band, assuming the Boltzmann energy distribution within a vibrational mode (Reid & Siemsen, 1976):

$$G_{seq} / G_{reg} = 2\exp(-h\nu_3 / kT_3)\,, \tag{6}$$

where T_3 and ν_3 are the vibrational temperature and frequency of the asymmetric stretch-mode of the CO_2 molecule, and h and k respectively are the Plank's and Boltzmann's constants, show that sequence band's gain reaches 50% of the regular band's gain at T_3=2500 K, viz., comparable to the conditions of high-pressure CO_2 amplifiers.

3.2 Effects in optical materials

To assure highly intense laser fields, special attention must be given to properly selecting and utilizing the optical elements, and to accounting for their influence on the laser field. This especially is challenging in the 10-μm spectral region because of the dearth of optical materials compared to the visible or near-IR diapasons, and lack of data on the materials' behavior under ultrashort mid-IR pulses. Below, we summarize the properties of optical materials most important for using in the high-peak-power 10-μm laser field. Table 1 gives numerical data on the refractive indices and dispersion of some popular IR materials used in CO_2 lasers.

Chromatic dispersion plays an important role due to the hundreds of GHz-wide spectrum of (sub-) picosecond 10-μm pulses that may entail considerable pulse stretching. For example, a pulse of 1 ps (FWHM) spreads to 1.27 ps after a single pass through a 10-cm NaCl window. Accordingly, the amount and thickness of optical elements should be minimized, and/or a grating compensator added for recompressing the pulse.

Nonlinear index, B-integral. A high-power laser pulse propagating through a medium changes its refractive index n (the *Kerr effect*):

$$n = n_0 + n_2 I , \tag{7}$$

where n_0 is the linear (low-intensity) refractive index, n_2 the nonlinear index, and I the optical intensity. Variation of refractive index across the beam's cross-section degrades its quality. For instance, lensing (*Kerr lensing*) occurs when phase-shift in the center of the beam is considerably larger than on its edge. The additional phase-retardation introduced to the beam after propagating through an optical element due to the nonlinear index (*B-integral*) is the accepted parameter for quantifying this effect (Paschotta):

$$B = \frac{2\pi}{\lambda} \int n_2 I(z) dz , \tag{8}$$

where λ is the wavelength, $I(z)$ the optical intensity along the beam's axis, and, z the position in the beam's direction. Usually, a noticeable self-focusing occurs if the B-integral exceeds 3-5. Strong wave-front distortion and eventually catastrophic filamentation may occur at larger values of the B-integral. An estimate of the B-integral using the n_2 index from the Table 1 yields B = 13.4 for a 1 ps (FWHM), 500 mJ/cm^2 pulse passing through a 10 cm of NaCl. Expectedly, the effect will be much stronger for other materials, implying that special care must be taken in selecting materials and controlling the fluence through the optical elements. The Kerr effect also is responsible for self-chirping due to temporal variation of the phase shift defined by the pulse's temporal structure (see Section 4.1).

	n_0 @ 10.6 μm [a]	$dn_0/d\nu$ @10.6 μm (10^{-3} THz^{-1}) [a]	n_2 (10^{-16} cm^2/W) [b]
KCl	1.45	1.51	5.7 @ 1.06 μm
NaCl	1.49	2.64	4.4 @ 1.06 μm
ZnSe	2.40	2.44	290 @ 1.06 μm
CdTe	2.67	1.04	-3000 @ 1.06 μm
Si	3.42	0.0914	1000 @ 2.2 μm[c]
Ge	4.00	0.293	2800 @ 10.6 μm

Table 1. Linear refractive index n_0, chromatic dispersion $dn_0/d\nu$, and nonlinear index n_2 of some IR materials. (a) RefractiveIndex.INFO; (b) Sheik-Bahae et al., 1991; (c) Bristow et al., 2007.

Optical breakdown threshold. We know far less about the ultrashort-pulse breakdown thresholds for mid-IR wavelengths and the materials used at these wavelengths than we do for visible light and near-IR. Reportedly, the values are ~0.5 J/cm^2 for NaCl, and 1-2 J/cm^2 for gold-coated stainless-steel mirrors for 2-ps, 10-μm pulse (Corkum, 1983). Variations in the breakdown threshold with pulse duration are best studied for fused silica at a wavelength of ~800 nm (Du et al., 1994, Stuart et al., 1996, Tien et al., 1999). Jia et al. (2006) also explored the wavelength dependence of the damage threshold over 250-2000 nm for 150 fs pulses; they concluded that there was a relatively small variation in the threshold at wavelengths above 800 nm. At pulses <10 ps, the damage threshold decreases slow with declining duration of the pulse, rather than displaying the $\tau_p^{1/2}$ dependence valid for longer pulses; this difference is explained by the gradual transition from a thermally dominated damage regime to one dominated by collisional- and multi-photon- ionization and plasma formation. Assuming a similar behavior in mid-IR, we conclude there is relatively small variation of the breakdown threshold fluence as a function of pulse duration for pulses of a few picoseconds or shorter; thus, as a guideline in system design, in most cases we can adopt 0.5 J/cm^2 for transparent optics, and 1 J/cm^2 for mirrors.

3.3 Chirped-pulse amplification

We can minimize the high-power effects in optical materials on the pulse by employing the method of chirped-pulse amplification (CPA), the principle of which is illustrated in Fig. 9.

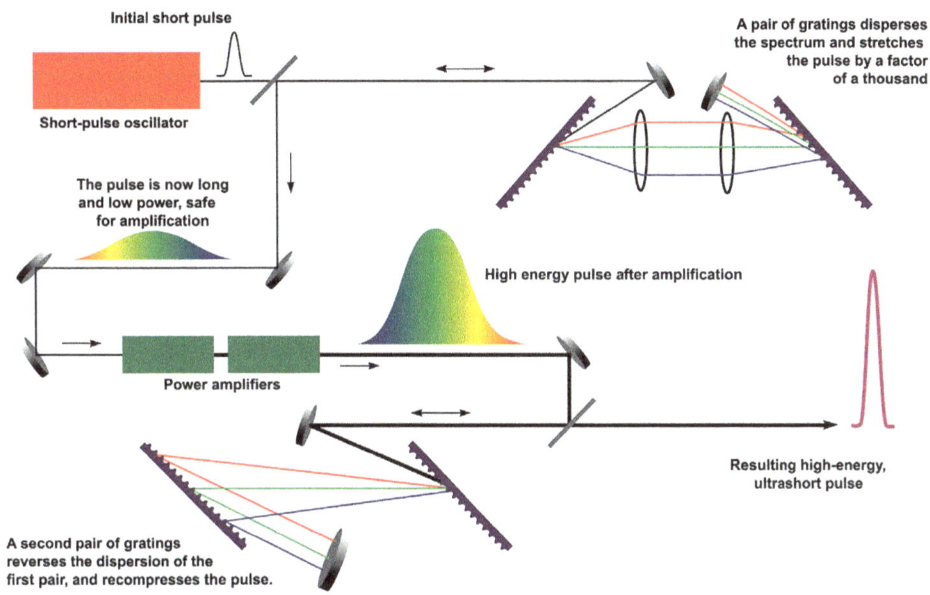

Fig. 9. Chirped-pulse amplification schematic (Perry et al., 1995). Reproduced with permission, courtesy of the Lawrence Livermore National Laboratory.

Before being amplified, an ultrashort pulse is *chirped*: A wavelength-dependent delay is introduced in the pulse using a *stretcher* consisting of a couple of gratings and lenses, as shown in the Fig. 9. Because ultrashort pulses have broad spectral bandwidth, inversely proportional to their duration (Eq. (2)), it is relatively straightforward to stretch them by a few orders-of-magnitude. The chirped pulse carries virtually the same energy as the original one, but its peak power is reduced in inverse proportion to the stretching factor. The chirped pulse can be amplified to energies much higher than that achievable by directly amplifying ultrashort pulses wherein high-power effects in optical materials and active medium appear earlier. After amplification, the pulse is re-compressed in another two-grating device (the *compressor*).

The invention of CPA for reducing light intensity during ultrashort pulse amplification was a dramatic breakthrough in solid-state laser technology. Although implementing the CPA technique in a (sub-)picosecond CO_2 laser remains to be done (the corresponding project is planned at the Accelerator Test Facility of the Brookhaven National Laboratory), it is reasonable to expect that the outcome will comparably be valuable.

3.4 Energy extraction efficiency

Optimal use of the energy stored in the active medium is a key characteristic of a mature laser design. The difficulty in efficiently extracting excitation energy from gaseous active media via an ultrashort pulse arises from the fact that the pulse's duration is either shorter than, or comparable to the characteristic times of the involved excitation- and relaxation-processes. Below, we briefly consider these processes and their influence on the efficiency of extracting energy.

Excitation and vibrational relaxation. In a typical CO_2 laser, an electric discharge is used to create population inversion. Fast electrons accelerated by electric field collide with CO_2 and N_2 molecules, so exciting their upper vibrational states. Nitrogen serves as reservoir for storing the excitation energy which is then transferred to CO_2 via vibrational relaxation. The typical duration of discharge in a pulsed CO_2 laser is about a microsecond, and vibrational relaxation times range from 1-10 ns·bar. This implies that a (sub-)picosecond laser pulse only extracts energy already available at the upper vibrational laser-level, with negligible energy deposition from the discharge or from redistribution from other vibrational levels during pulse propagation. To maximize extraction, wherever possible a regenerative amplification scheme is used, wherein the pulse passes the amplifier medium many times allowing repopulation of the upper vibrational level of the laser transition between the passes. However, realizing this scheme practically is problematic for high-energy pulses when beam must be wide to increase the active volume and avoid damaging optical elements. Thus, two-stage amplification usually is adopted (see Section 6); a regenerative amplifier providing amplification up to millijoules level is followed by a final high-energy amplifier arranged either in a single-pass configuration, or several passes are only partially overlapped. Slow pumping and vibrational relaxation limit energy extraction in the final amplification stage. A possible solution is replacing pumping by electric discharge, with optical pumping by a short laser pulse that quickly and directly excites the upper laser level, and eliminates the need for redistributing vibrational energy.

Rotational relaxation. Rotational relaxation processes limit the fraction of energy extractable from the upper vibrational laser level in a single pass through the active medium. The laser pulse interacts only with a limited number of rotational transitions that is defined by the overlap between the pulse's spectrum and the amplification band. At high pressure, when collisionally broadened rotational lines overlap, or for a very short pulse, when pulse's spectrum covers several rotational lines, energy is extracted from several rotational sub-levels. Otherwise, energy is extracted only from a single sub-level containing about $1/15^{th}$ of the entire energy stored in that vibrational level. Three scenarios support the complete emptying of the vibrational level: (1) The pulse is long enough to provide time for repopulation of the active rotational levels. Typical rotational relaxation times are ~100 ps·bar; about 15 collisions are required effectively to empty all rotational sub-levels through a single active rotational transition. Thus, the minimum required pulse duration is ~1.5 ns·bar (e.g., 150 ps at 10 bar). (2) Pressure broadening allows all rotational transitions to interact with the laser field. Using Eq. (3) we find that ~100 bar is needed to broaden the rotational lines to the ~0.5 THz necessary for this regime. (3) The spectrum of the pulse covers the entire rotational branch, as happens when the pulse duration is ≤1 ps (Eq. (2)). The last regime is preferable if the aim is to keep pulse's span as brief as possible.

4. Pulse compression

4.1 Self-chirping in nonlinear media

Despite the usual attempts researchers take in trying to avoid the undesirable effects of the nonlinear response on the high-power pulses, it might be possible to employ these phenomena to further compress ultrashort pulses. When an optical pulse propagates through a media, the refractive index of the latter changes according to Eq. (7), following the intensity of the optical field (the Kerr effect). Phase velocity, $v_p=c/n$, varies accordingly. The field frequency of the pulse, whose phase velocity continuously changes, shifts proportionally to its derivative dv_p/dt, so generating pulse chirping. Fig. 10 illustrates this process.

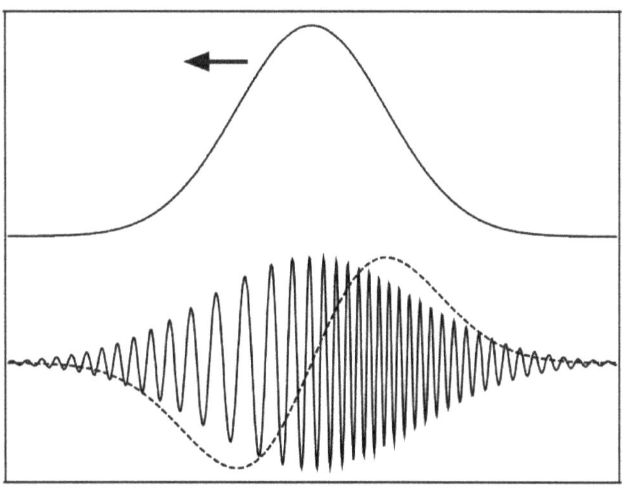

Fig. 10. Self-chirping of a pulse propagating through a nonlinear media. Top: Pulse intensity profile (leading edge is on the left); bottom dashed: Phase velocity derivative; bottom solid: Wave-packet of the chirped pulse.

It is important to realize the difference between the chirping, as in the CPA, and chirping, as in self-chirping. In the first case, the pulse's spectrum is unchanged whereas its duration increases; in the second, the spectrum broadens while pulse duration does not change. A dispersive compressor similar to that utilized in the CPA can be arranged to shrink the self-chirped pulse well below its original duration.

4.2 Self-chirping in plasma

Self-chirping in plasma essentially is a variation of the self-chirping in nonlinear media. The distinctive feature of this case is that the variation in refractive index is caused by laser-induced gas-ionization rather than the Kerr effect. Refractive index of the ionized gas is determined by the linear refractive index of the media, n_0, and the plasma density, N_e:

$$n = n_0 \left(1 - \frac{N_e}{N_c}\right)^{1/2}, \tag{9}$$

where N_c is the critical density (Eq. (1)). Unlike the Kerr effect, where refractive index usually *increases* with laser intensity, strong ionization in high-intensity fields *decreases* the refractive index. Therefore, the direction of chirping is reversed: The field frequency of the leading edge of the pulse is higher than that of the trailing one (*blue chirp*), allowing the compression of the chirped pulse via the linear dispersion in an optical material. By sending the pulse through a window of a properly selected thickness, made of a material with negative group velocity dispersion (e.g., NaCl), we can delay the blue-shifted leading edge of the pulse more than the trailing edge, thus shrinking the pulse.

Inert gases (e.g., xenon) are considered promising candidates for the role of nonlinear media for CO_2 laser pulse self-chirping (Gordienko et al., 2009). A complication arises because the field intensity is not constant across the beam; thus, self-chirping that is pronounced in the center of the beam becomes negligible at its edges. The beam can be homogenized using a hollow waveguide (Nisoli et al., 1997, Voronin et al., 2010), or a filamentation regime (Couairon et al., 2006, Gordienko et al., 2009). Fig. 11 shows the results of simulations of an 1.2-ps pulse compression via self-chirping in xenon in filamentation regime followed by a NaCl compressor.

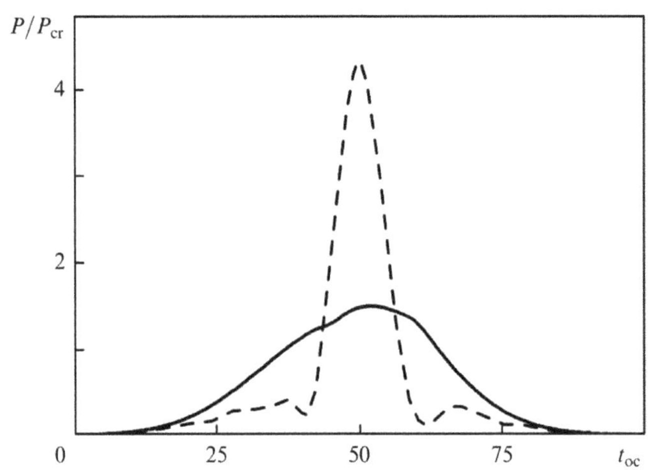

Fig. 11. Simulated 1.2-ps (FWHM) pulse before (solid) and after (dashed) compression via self-chirping in xenon plasma and dispersive compression in NaCl (Gordienko et al., 2009). Time is measured in optical cycles (1 o.c.≈35 fs); P is power and P_{cr} is critical self-focusing power in xenon: $P_{cr} \approx \lambda^2/4\pi n_0 n_2$. Reprinted by permission of Turpion Ltd.

Corcum observed unintentional pulse shortening due to plasma chirping in a picosecond CO_2 amplifier (Corcum, 1985). The suggested explanation capitalized on the pulse's self-chirping in the partially ionized active media, and compression in the material of the cavity windows. Controlled shortening was realized using an external gas cell by Tochitsky et al. (Tochitsky et al., 2001).

5. Pulse diagnostics

Measuring the temporal structure of ultrashort 10-μm pulses is not fundamentally different from measuring visible or near-IR pulses. However, due to low demand, there are very few commercial diagnostic instruments (e.g., Frequency Resolved Optical Gating, FROG) suitable for direct use in the mid-IR region. Several techniques used for diagnosing CO_2 laser pulses are schematically represented in Fig. 12.

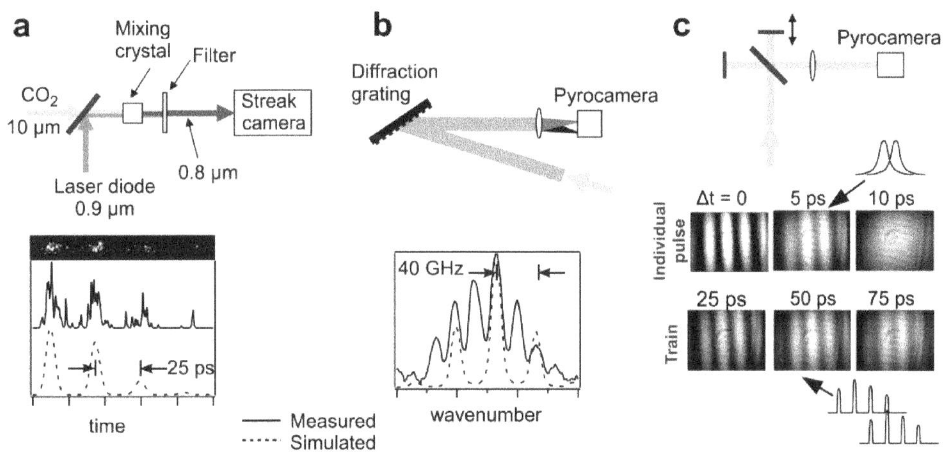

Fig. 12. Apparatuses for ultrashort mid-IR pulse diagnostics. (a) Streak camera; (b) Spectrometer (additional peaks in the measured spectrum are attributed to a sequence band not included in the simulations); and, (c) Autocorrelator.

A *streak camera* is a convenient tool for monitoring the pulse structure, providing a resolution of 1-2 ps. Because photocathodes used in streak cameras are insensitive to the mid-IR wavelengths, a frequency conversion technique must be used to shift the pulse wavelength to the visible- or near-IR- diapason. For this purpose, either a difference-frequency mixing in a nonlinear crystal (Fig. 12a), or a Kerr-cell-based optical switch where the CO_2 laser's pulse controls a visible or mid-IR beam is suitable. The resolution of the streak camera is sufficient for measuring pulse splitting due to modulation on the rotational gain spectrum. However, the accuracy of measuring the duration of individual pulses is limited by 1-2 ps.

A *spectrometer* can be used for indirect measurement of the pulse's duration. In the absence of chirping, the total bandwidth of the spectrum is inversely proportional to the duration of the individual pulses (Eq. (2)). If the beam's quality is good enough, a simple grating spectrometer can be arranged similar to that shown in Fig. 12b.

An *autocorrelator* technique must be established to attain the most reliable results. For instance, it can be used to periodically validate the measurements from the streak camera and the spectrometer. An autocorrelator splits the measured beam into two, and recombines it on an active element where the pulses interact, providing a measurable signal that is a function of the temporal overlap. By recording the interaction signal as a function of the time-delay between pulses, we obtain information about the temporal profile of the pulse. Usually, a nonlinear crystal is used as the active element, generating a harmonic frequency when the two pulses overlap in time. In a simpler design, the pulses' temporal overlap is evaluated by measuring the modulation of the interference pattern resulting from the interaction of the two beams. In the autocorrelator design shown in the Fig. 12c an intentionally induced slight misalignment of the interferometer's arms generates an interference pattern on the pyroelectric camera's sensor. The interference contrast serves as a measure of the temporal overlap between the pulses; the maximum modulation corresponds to zero delay, whereas the complete separation of the pulses in time entails the disappearance of interference. With this technique, we can study both the duration of individual pulses and the train's structure, which results in the periodic appearance of the interference pattern with gradually reduced modulation at delays that are multiples of the pulse-splitting period.

6. Existing terawatt CO_2 lasers

Two systems worldwide now can generate terawatt peak-power, 10-μm pulses: One at the Accelerator Test Facility of the Brookhaven National Laboratory (BNL) (Polyanskiy et al., 2011); and, the other at the Neptune Laboratory of the University of California, Loss Angeles (UCLA) (Haberberger et al., 2010). These systems are described briefly below.

6.1 1-TW system at BNL's Accelerator Test Facility (BNL-ATF)

The CO_2 laser system depicted in the Fig. 13 consists of a picosecond pulse-generator that produces a linear-polarized 0.1-μJ, 5-ps pulse, along with two high-pressure amplifiers that ultimately boost the laser pulses' energy to the ~5 J level.

In this system, a 5-ps pulse is sliced by the sequence of optical switches from the 200-ns, 20-mJ output of a hybrid TEA CO_2 laser tuned to the 10R(14) line (10.3 μm). First, a 10-ns pulse is cut off the initial pulse with a Pockels cell, and intensified in a 3-bar UV-pre-ionized electric-discharge pre-amplifier. Then, a semiconductor optical switch, which is controlled by a 14-ps YAG laser, slices off a ~200 ps part of the pulse. Finally, the 200-ps pulse is sent through a CS_2 Kerr cell controlled by a co-propagating, 5-ps, frequency-doubled YAG laser-pulse. A polarization filter placed after the Kerr cell selects a 0.1-μJ, 5-ps seed pulse that then is raised to 10 mJ in multiple round-trip passes through a regenerative amplifier filled with a gas mixture featuring *isotopically enriched* carbon dioxide to prevent pulse splitting upon amplification. The amplifier is energized with UV-pre-ionized transverse electric discharge. Further amplification to 5 J is attained in 6 passes through a large-aperture (8x10 cm^2), x-ray pre-ionized final amplifier. Field broadening prevents pulse splitting in the final amplifier despite using regular CO_2 gas therein. The output is a single 5-ps pulse implying ~1 TW peak power. Table 2 summarizes the technical details of these two amplification stages.

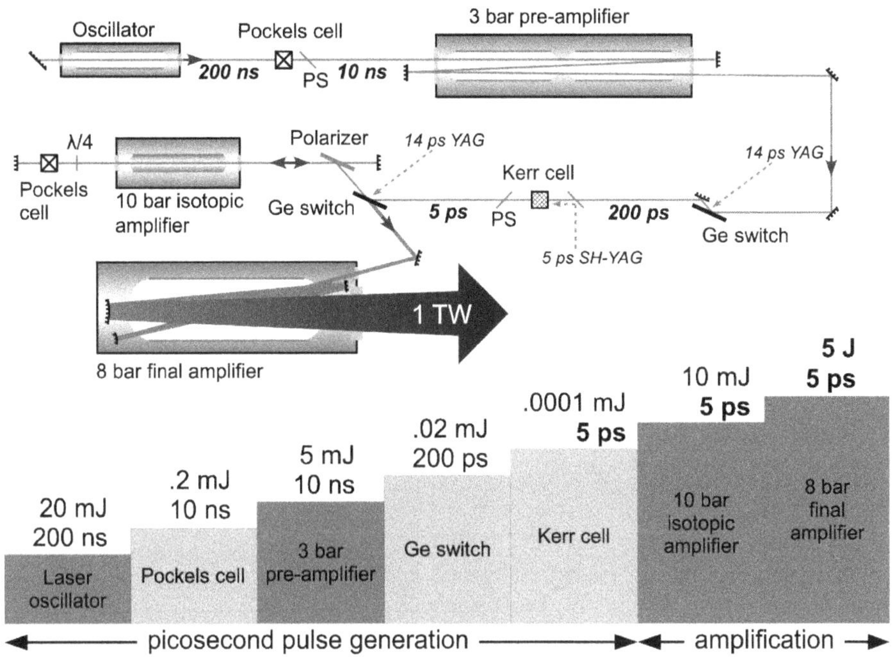

Fig. 13. Layout and pulse dynamics in the BNL-ATF CO_2 laser system; PS: Polarizing splitter.

	Regenerative amplifier	Final amplifier
Working pressure	10 bar	8 bar
Gas mixture: $[CO_2]$:$[N_2]$:[He]	1:1:18 (isotopic CO_2)	2:1:28
Active volume	1×1×80 cm^3	8×10×100 cm^3
Small-signal gain	1-2 %/cm	1.5-2 %/cm
Number of passes	8-12 round-trips	6 passes
Net amplification	10^5	10^3

Table 2. Parameters of BNL-ATF laser amplifiers.

6.2 15-TW system at UCLA's Neptune Laboratory

Fig. 14 is a scheme of the CO_2 laser system of the UCLA's Neptune Laboratory operating at 10.6 μm wavelength (10P branch). The 3-ps injection pulse is produced by slicing a portion of the output of a hybrid TEA CO_2 laser (comprising a low-pressure smoothing tube to suppress energy modulation caused by self-mode-locking). The slicing is realized as a single-step process, using a CS_2-filled Kerr cell controlled by a 3-ps pulse of a solid-state laser. The nanojoule injection pulse first is amplified in an 8-bar regenerative amplifier, reaching milijoules energy and splitting into a train of ~7 sub-pulses separated by 18-ps intervals. The final 2.5-bar amplifier boosts the pulse energy up to 100 J, and simultaneously mostly suppresses splitting via the field-broadening effect. The output pulse consists of 2-3 sub-pulses with ~45% energy in the first of them, implying ~15 TW peak power.

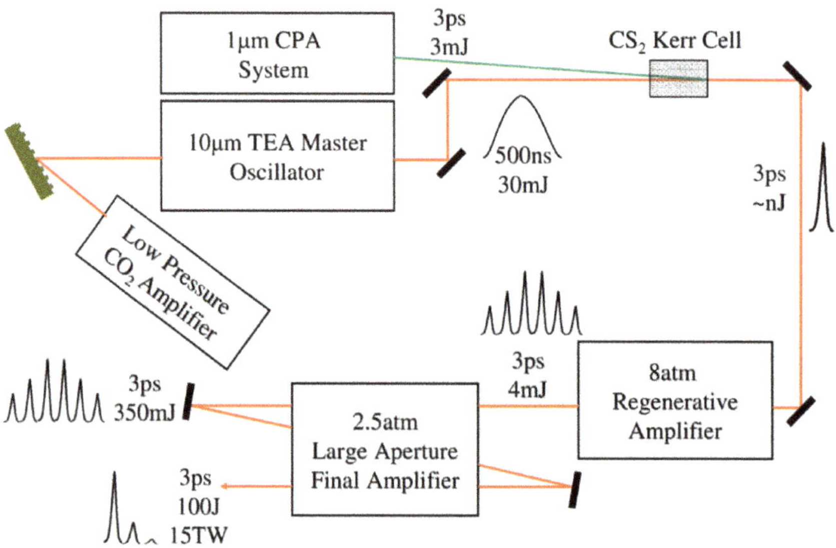

Fig. 14. Layout and pulse dynamics of the UCLA's Neptune Laboratory laser system (Haberberger et al., 2010). Reproduced with permission.

Table 3 summarizes the parameters of this system's amplifiers.

	Regenerative amplifier	Final amplifier
Working pressure	8 bar	2.5 bar
Gas mixture: $[CO_2]$:$[N_2]$:[He]	1:1:14	4:1:0
Active volume	1×1×60 cm^3	20×35×250 cm^3
Small-signal gain	-	2.6 %/cm
Number of passes	-	3 passes
Net amplification	10^7	10^5

Table 3. Parameters of UCLA's Neptune Laboratory laser amplifiers.

7. Conclusion

We overviewed the underlying physics and technical approaches to generating and amplifying ultrashort 10-μm pulses. Modern CO_2 laser systems can generate pulses as brief as few picoseconds and as powerful as several terawatt. Potential applications, among which are the high-energy physics experiments and the proton acceleration for cancer therapy, call for even higher peak power.

Achievements in solid-state laser technology can help the further development of ultrashort-pulse, high-peak-power CO_2 laser systems. Modern solid-state lasers can be directly used in mid-IR systems, e.g., for controlling optical switches, pumping CO_2 laser transition, or generating the ultrashort 10-μm seed pulses via nonlinear frequency conversion and parametric amplification. Apart from that, the advanced techniques initially developed for solid-state lasers (e.g. chirped pulse amplification) can be adopted in CO_2 laser systems.

8. Acknowledgements

This work is supported by the US DOE contract DE-AC02-98CH10886 and by the BNL Laboratory Directed R&D (LDRD) grant #07-004. The authors are thankful to Sergei Tochitsky from UCLA's Neptune Laboratory for providing information on the Neptune's CO_2 laser.

9. References

Abrams, R. L. & Wood, O. R. (1971). Characteristics of a mode-locked TEA CO_2 laser. *Appl. Phys. Lett.*, Vol.19, No.12, (December 1971), pp. 518-520, ISSN 0003-6951

Alcock, A. J. & Corkum, P. B. (1979). Ultra-fast switching of infrared radiation by laser-produced carriers in semiconductors. *Can. J. Phys.*, Vol.57, No.9, (September 1979), pp. 1280-1290, ISSN 0008-4204

Armstrong, J. A., Bloembergen, N., Ducuing, J. & Pershan, P. S. (1962). Interactions between light waves in a nonlinear dielectric. *Phys. Rev.*, Vol.127,No.6, (September 1962), pp. 1918-1939, ISSN 1943-2879

Autler, S. H. & Townes, C. H. (1955). Stark effect in rapidly varying fields. *Phys. Rev.*, Vol.100,No.2, (October 1955), pp. 703-722, ISSN 1943-2879

Brimacombe, R. K. & Reid, J. (1983). Accurate measurements of pressure-broadened linewidths in a transversely excited CO_2 discharge. *IEEE J. Quantum Electron.*, Vol.19, No.11, (November 1983), pp. 1668-1673, ISSN 0018-9197

Bristow, A. D., Rotenberg, N., & van Driel, H. M. (2007). Two-photon absorption and Kerr coefficients of silicon for 850–2200 nm. *Appl. Phys. Lett.*, Vol.90, No.19 (May 2007), p. 191104, ISSN 0003-6951

Corkum, P. B. & Krausz, F. (2007). Attosecond science. *Nature Physics.*, Vol.3, No.6, (June 2007), pp. 381-387, ISSN 1745-2473

Corkum, P. B. (1983). High-power, subpicosecond 10-μm pulse generation. *Opt. Lett.*, Vol.8, No.10, (October 1983), pp. 514-516, ISSN 0146-9592

Corkum, P. B. (1985). Amplification of picosecond 10 μm pulses in multiatmosphere CO_2 lasers. *IEEE J. Quantum Electron.*, Vol.21, No.3, (March 1985), pp. 216-232, ISSN 0018-9197

Couairon, A., Biegert, J., Hauri, C. P., Kornelis, W., Helbing, F. W., Keller, U. & Mysyrowicz, A., F. (2006). Self-compression of ultra-short laser pulses down to one optical cycle by filamentation. *J. Modern Opt.*, Vol.53, No.1-2, (January 2006), pp. 75-85, ISSN 0950–0340

Du, D., Liu, X., Korn, G., Squier, J., & Mourou, G. (1994). Laser-induced breakdown by impact ionization in SiO_2 with pulse widths from 7 ns to 150 fs. *Appl. Phys.Lett.*, Vol.64, No.26 (June 1994), pp. 3071-3073, ISSN 0003-6951

Esirkepov, T., Borghesi, M., Bulanov, S. V., Mourou, G. & Tajima, T. (2004). Highly efficient relativistic-ion generation in the laser-piston regime. *Phys. Rev. Lett.*, Vol.92, No.17 (April 2004), p. 175003, ISSN 0031-9007

Filip, C. V., Narang R., Tochitsky, S. Ya., Clayton, C. E. &. Joshi, C. (2002). Optical Kerr switching technique for the production of a picosecond, multiwavelength CO_2 laser pulse. *Appl. Opt.*, Vol.41, No.18, (June 2002), pp. 3743-3747, ISSN 0003-6935

Giordmaine, J. A. & Miller, R. C. (1965). Tunable coherent parametric oscillation in $LiNbO_3$ at optical frequencies. *Phys. Rev. Lett.*, Vol.14, No.24 (June 1965), p. 973-976, ISSN 0031-9007

Gordienko, V. M., & Platonenko, V. T. (2010). Regenerative amplification of picosecond 10-μm pulses in a high-pressure optically pumped CO_2 laser. *Quant. Electron.*, Vol.40, No.12, (December 2009), pp. 1118-1122, ISSN 1063-7818

Gordienko, V. M., Platonenko, V. T. & Sterzhantov, A. F. (2009). Self-action of a high-power 10-μm laser radiation in gases: control of the pulse duration and generation of hot electrons. *Quant. Electron.*, Vol.39, No.7, (July 2009), pp. 663-668, ISSN 1063-7818

Haberberger, D., Tochitsky, S. & Joshi, C. (2010). Fifteen terawatt picosecond CO_2 laser system. *Opt. Express*, Vol.18, No.17, (August 2010), pp. 17865-17875, ISSN 1094-4087

Houfman, H. & Meyer, J. (1987). Ultrashort CO_2 laser pulse generation by square-wave mode locking and cavity dumping. *Opt. Lett.*, Vol.12, No.2, (February 1987), pp. 87-89, ISSN 0146-9592

Jia, T. Q., Chen, H. X., Huang, M., Zhao, F. L., Li, X. X., Xu, S. Z., Sun, H. Y., Feng, D. H., Li, C. B., Wang, X. F., Li, R. X., Xu, Z. Z., He, X. K. & Kuroda, H. (2006). Ultraviolet-infrared femtosecond laser-induced damage in fused silica and CaF_2 crystals. *Phys. Rev. B*, Vol.73, No.5 (February, 2006) p.054105, ISSN 0163-1829

Korzhimanov, A. V., Gonoskov, A. A., Khazanov, E. A. & Sergeev, A. M. (2011). Horizons of petawatt laser technology. *Physics - Uspekhi*, Vol.54, No.1, (January 2011), pp. 9-28, ISSN 1063-7869

Kovalev, V. I. (1996). Mechanism of self-mode-locking in the active medium of CO_2 lasers. *Quant. Electron.*, Vol.26, No.2, (February 1996), pp. 131-132, ISSN 1063-7818

Nisoli, M., Stagira, S., De Silvestri, S., Svelto, O., Sartania, S., Cheng, Z., Lenzner, M., Spielmann, C. & Krausz, F. (1997). A novel high-energy pulse compression system: generation of multigigawatt sub-5-fs pulses. *Appl. Phys. B*, Vol.65, No.2, (August 1997), pp. 189-196, ISSN 0946-2171

Norreys, P. (2011). Particle acceleration: Pushing protons with photons. *Nature Photonics*, Vol.5, No.3, (March 2011), pp. 134-135, ISSN 1749-4885

Okorogu, A. O., Mirov, S. B., Lee, W., Crouthamel, D. I., Jenkins, N., Dergachev, A. Yu., Vodopyanov, K. L. & Badikov, V. V. (1998). Tunable middle infrared downconversion in GaSe and $AgGaS_2$. Opt. Commun., Vol.155, No.4-6, (October 1998), pp. 307-312, ISSN 0030-4018

Palmer, C. A. J., Dover, N. P., Pogorelsky, I., Babzien, M., Dudnikova, G. I., Ispiriyan, M., Polyanskiy, M. N., Schreiber, J., Shkolnikov, P., Yakimenko, V. & Najmudin, Z.

(2011). Monoenergetic proton beams accelerated by a radiation pressure driven shock. *Phys. Rev. Lett.*, Vol.106, No.1 (January 2011), p. 014801, ISSN 0031-9007

Paschotta, R. (2008). *Field Guide to Laser Pulse Generation,* SPIE Press, ISBN 978-0-8194-7248-9, Bellingham, Washington, USA

Paschotta, R.. B-Integral. *Encyclopedia of laser physics and technology.* Available from: http://www.rp-photonics.com/b_integral.html. Accessed October 6, 2011

Perry, M., Shore, B., Boyd, R. & Britten, J. (1995). Multilayer dielectric gratings: Increasing the power of light. LLNL Sci. Technol. Rev., (September 1995), pp. 24-33

Polyanskiy, M. N., Pogorelsky, I. V. & Yakimenko, V. (2011). Picosecond pulse amplification in isotopic CO_2 active medium. *Opt. Express*, Vol.19, No.8, (April 2011), pp. 7717-7725, ISSN 1094-4087

RefractiveIndex.INFO. *Refractive index database.* Available from http://refractiveindex.info. Accessed September 30, 2011

Reid, J. & Siemsen, K. (1976). New CO_2 laser bands in the 9–11-μm wavelength region. *Appl. Phys. Lett.*, Vol.29, No.4 (August 1976), pp. 250-251, ISSN 0003-6951

Rolland, C. & Corkum, P. B. (1986). Generation of 130-fsec midinfrared pulses. *J. Opt. Soc. Am. B,* Vol.3, No.12, (December 1986), pp. 1625-1629, ISSN 0740-3224

Rothman, L. S., Gordon, I. E., Barbe, A. et al. (2009) The HITRAN 2008 molecular spectroscopic database. *J. Quant. Spectr. Rad. Transfer,* Vol.110, No.9-10 (June-July 2009), pp. 533-572, ISSN: 0022-4073

Sheik-Bahae, M., Hutchings, D. C., Hagan, D. J. & Van Stryland, E. W. (1991). Dispersion of bound electron nonlinear refraction in solids. *IEEE J. Quantum Electron.,* Vol.27, No.6, (June 1991), pp. 1296-1309, ISSN 0018-9197

Siegman, A. E. & Kuizenga, D. J. (1969). Simple analytical expressions for AM and FM mode-locked pulses in homogeneous lasers. *Appl. Phys. Lett.,* Vol.14, No.6, (March 1969), pp. 518-520, ISSN 0003-6951

Stuart, B. C., Feit, M. D., Rubenchik, A. M., Shore, B. W., and Perry, M. D. (1996). Nanosecond-to-femtosecond laser-induced breakdown in dielectrics. *Phys. Rev. B,* Vol.53, No.4, (January 1996), pp. 1749-1761, ISSN 0163-1829

Tien, A.-C., Backus, S., Kapteyn, H., Murnane M., & Mourou G. (1999). Short-pulse laser damage in transparent materials as a function of pulse duration. *Phys. Rev. Lett.,* Vol.82, No.19 (May 1999), pp. 3883-3886, ISSN 0031-9007

Tochitsky, S. Ya., Filip, C., Narang, R., Clayton, C. E., Marsh, K. A. & Joshi, C. (2001). Efficient shortening of self-chirped picosecond pulses in a high-power CO_2 amplifier. *Opt. Lett.,* Vol.26, No.11, (June 2001), pp. 813-815, ISSN 0146-9592

Voronin, A. A., Gordienko, V. M., Platonenko, V. T., Panchenko, V. Ya. & Zheltikov, A. M. (2010). Ionization-assisted guided-wave pulse compression to extreme peak powers and single-cycle pulse widths in the mid-infrared. *Opt. Lett.,* Vol.35, No.21, (November 2010), pp. 3640-3642, ISSN 0146-9592

Witteman, W. J. (1987). *The CO_2 laser.* Springer-Verlag, ISBN 0-387-17657-8, Berlin Heidelberg New York London Paris Tokyo

Wood, O. R., Abrams R. L. & Bridges, T. J. (1970). Mode locking of a transversively excited atmospheric pressure CO_2 Laseer. *Appl. Phys. Lett.,* Vol.17, No.9, (November 1970), pp. 376-378, ISSN 0003-6951

Yablonovich, E. (1973). Spectral broadening in the light transmitted through a rapidly growing plasma. *Phys. Rev. Lett.*, Vol.31, No.14, (October 1973), pp. 877–879, ISSN 0031-9007

Yablonovich, E. (1974a). Short CO_2 laser pulse generation by optical free induction decay. *Appl. Phys. Lett.*, Vol.25, No.10, (November 1974), pp. 580-582, ISSN 0003-6951

Yablonovich, E. (1974b). Self-phase modulation and short-pulse generation from laser-breakdown plasmas. *Phys. Rev. A*, Vol.10, No.5, (November 1974), pp. 1888-1895, ISSN 1050-2947

Yakimenlo, V. & Pogorelsky, I. (2006). Polarized γ source based on Compton backscattering in a laser cavity. *Phys. Rev. ST Accel. Beams*, Vol.9, No.9, (September 2006), p. 091001, ISSN 1098-4402

High Average Power Pulsed CO_2 Laser for Short Wavelength Light Sources

Akira Endo
Research Institute for Science and Engineering, Waseda University, Tokyo Japan

1. Introduction

1.1 Background of the emerging technology

Increase of average power of pulsed CO_2 laser was required by strong demand of the semiconductor industry, in pursue of the next generation of lithography light source at 13.5nm (Endo, et.al, 2006). The target average EUV power was increased from 10W level in the beginning to several 100W levels in the recent maturing period. No existing solid state laser technology satisfies the demand of the average power as the laser driver, by counting the laser-EUV conversion efficiency around 1%. Intensive research of one decade also showed that opacity, namely self absorption of the generated EUV light is less significant in high Z plasma, driven by a longer wavelength laser. CO_2 laser produced Tin plasma showed more than 4% conversion efficiency in practical target geometry. Details are recently reviewed by A.Endo (Endo, 2010) and V.Y.Banine (Banine et.al, 2011). Interested readers are advised to refer to these articles. The established architecture is shown in Fig.1 as the laser produced Tin plasma which is generated from mist target of 300μm diameter, irradiated by 15ns CO_2 laser pulse. The mist target is produced from a 10μm Tin droplet after irradiation by a solid state laser with smaller pulse energy.

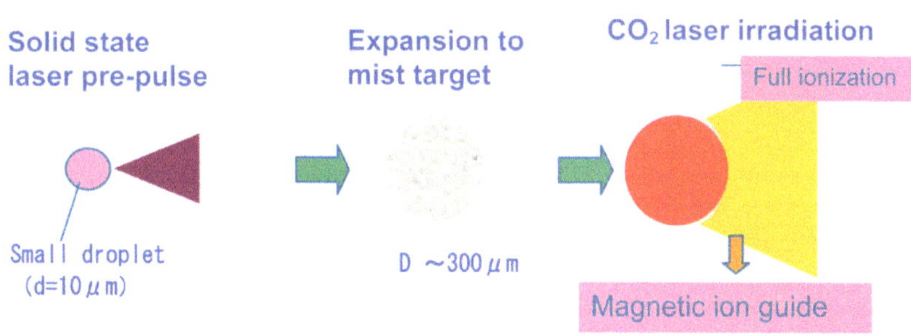

Fig. 1. Schematic of Tin plasma by double pulse method

The conversion efficiency (CE), from the input laser pulse energy to the generated EUV pulse energy at 13.5nm (2% bandwidth, 2π sr), is the major parameter for improvement in high average power EUV light source for better economy. Low repetition rate pulsed CO_2 laser was composed of transverse discharge modules, and often employed in laser plasma experiments until 90's, but gradually disappeared from laboratories after improvements of solid state pulsed lasers. It was once employed as a driver of a plasma X-ray laser from a carbon target (Suckewer, et.al., 1983).

Medium average power pulsed CO_2 laser systems are very successful tools for various applications ranging from material processing of metals, glass, ceramics and epoxy, paint removal and medical or spectroscopic applications, to the generation of laser produced plasmas as UV, EUV and soft X-ray sources. One drawback is the limited repetition rate of TEA CO_2 laser based source, another drawback is limited controllability of the pulse width in low pressure microwave excited lasers. Attempts were reported in early 90's to operate microwave excited CO_2 laser modules in a Q-switched oscillator mode of CW 2kW device (Sakai et.al., 1994) and an oscillator-amplifier mode of CW 7kW system (Bielesch et.al., 1992). Typical performances were at the repetition rate of 4 kHz with output average power of 680 W with pulse energy of 170 mJ and pulse width in full width half maximum (FWHM) of 250 ns, and at the repetition rate of 10 kHz with average power of 800W, with pulse energy of 70 mJ, and 35 ns pulse width, respectively. Laser extraction efficiencies, however, were not very high in both cases in the short pulse mode. Commercially available short pulse CO_2 laser oscillator was known typically as EOM-10 from De Maria Electro Optics Systems, Inc (now Coherent Inc). The specification was average power of 10W at 100 kHz repetition rate with 15ns pulse width. The design guideline of a multi kW short pulse CO_2 laser system is characterized by high repetition rate, high pulse energy, high amplification efficiency and high beam quality. The system is based on commercial high average power CW CO_2 laser modules as amplifiers.

A short pulse oscillator was installed in our laboratory as the seeder for the amplifiers. The laser was an EO Q-switched, pulse length of 15~30 ns, single P(20) line, RF pumped waveguide CO_2 laser with 60 W output at a repetition rate of 100 kHz. The repetition rate was tunable as 10~140 kHz. Commercial 5 kW and 15 kW CW CO_2 lasers were employed as amplifiers. Every unit was 13.56 MHz RF-excited, fast axial flow lasers from Trumpf Inc. Lasers were modified as amplifiers by replacing both cavity mirror with ZnSe windows. The 5 kW laser used a standard gas composition of CO_2:N_2:He=5:29:66 at 120 Torr gas pressure. The axial gas flow speed was sufficiently high to keep the laser gas temperature low inside the operational condition. The length of a single gain region was 15 cm, and 16 cylindrical gain regions were connected in series in one laser unit; the tube inner diameter was 17mm. The total length of the optical pass inside the laser was 590 cm. The laser operated at 5 kW CW output power with a M^2 =1.8 beam quality. The electrical input power was 36 kW. The 15 kW laser as the main amplifier, used a standard gas composition of CO_2:N_2:He=2:10:48 at 150 Torr gas pressure. The length of a single gain region was 28 cm, and 16 active cylindrical gain regions were connected in series; the tube inner diameter was 30 mm. The total length of the optical pass inside the laser was 890 cm. The maximum electrical input power was 88 kW. The key parameters of the amplifier are the extraction efficiency and beam quality. A series of experiments were performed to clarify these parameters to estimate the final possible values (Hoshino et.al., 2008).

The experimental setup is shown as Fig.2 with expected output power of 10kW. The maximum average output power of 8 kW was experimentally obtained at a repetition rate of 100 kHz with 3kW input power to the main amplifier. Parasitic oscillations and/or optical coupling between amplifier modules were not significant in burst mode. It was successful to extract 5kW power in pulsed mode from CW 15kW laser. The extraction efficiency (output power-input power/ CW output power) was over 30%. Initial estimation of extractable power from the main amplifier was 7kW, but the experimentally obtained power of 5kW indicated many factors for further improvement.

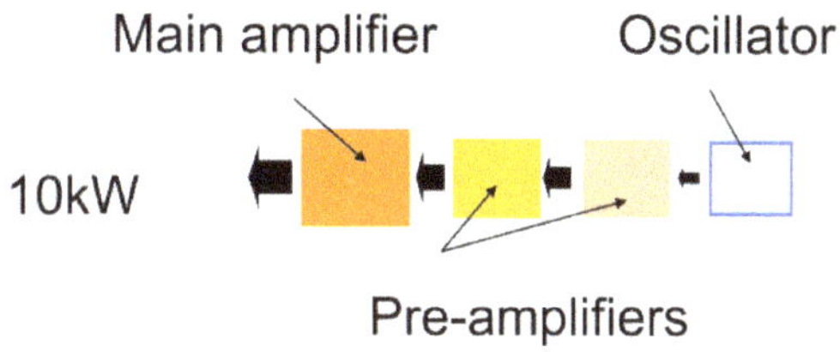

Fig. 2. Schematic diagram of multi stage MOPA, short pulse CO_2 laser

The laser beam quality was measured with a ZnSe lens of 508mm focal length and a slit-scan type beam profiler (Photon Inc., NanoScan). The laser beam size at the lens focus was measured for the oscillator and amplifier, resulting in a beam quality factor M^2 as 1.1. Especially, the laser beam size was identical before and after amplification, i.e. the amplification did not cause any phase distortion. Fig.3 shows a typical spatial beam profile. Fig.4 shows the temporal laser pulse profile of the amplified laser output. The pulse duration was 20 ns (FWHM) and the pedestal was below 10% of the total pulse energy. A pedestal and/or tail of the seed laser pulse could be amplified and reduce the laser gain. Back scattering light from Tin mist target is experimentally less than 10% of the input laser energy, and backward amplification must be carefully avoided by full depletion of residual laser gain.

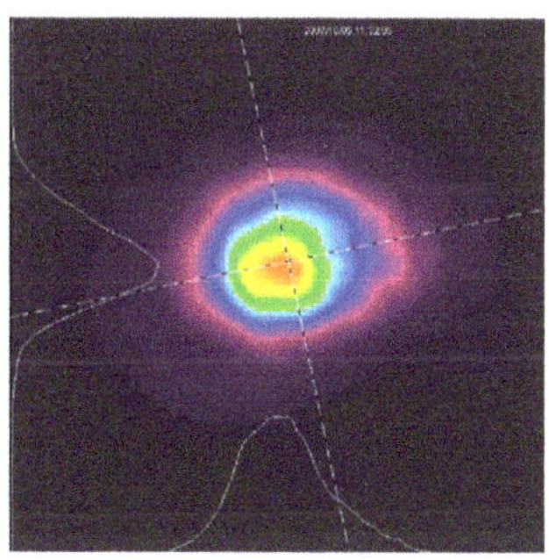

Fig. 3. CO_2 laser beam of M^2=1.1

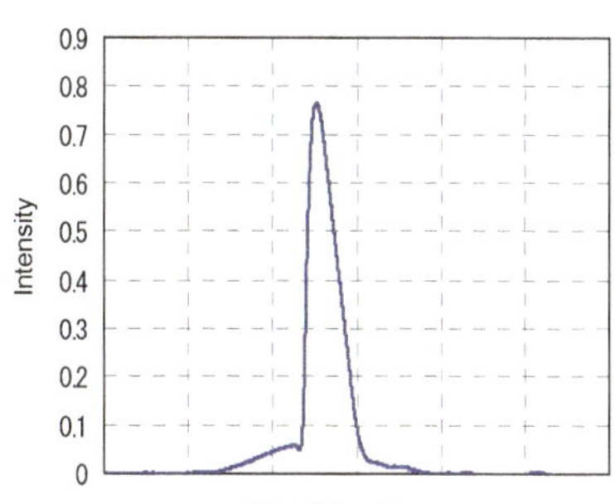

Fig. 4. CO_2 laser pulse with low pedestal

2. 10ns, 100 kHz operation of CO_2 laser at 10kW average power

CO_2 molecular dynamics is the fundamental subject to understand the operational parameters of pulsed CO_2 laser. Figure 5 shows a typical energy diagram of CO_2 laser active medium.

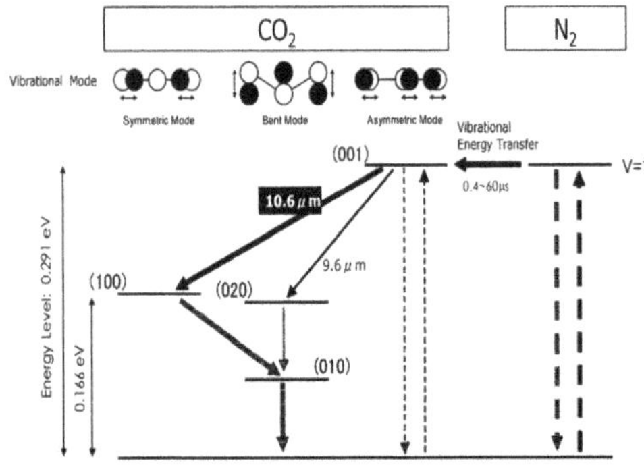

Fig. 5. Energy level diagram of active CO_2 laser medium

Short pulse, high repetition operational CO_2 laser is the most suitable laser technology to meet the requirements of HVM EUV source. Free electron lasers (FEL) are also capable of realizing high average power, short pulse coherent beam, based on the emerging superconducting energy recovery linac (sERL) technology (Krafft, 2006). However, the pulse energy is in sub mJ range with a few ps pulse width at 75MHz repetition rate, which does not match to the plasma production specifications for EUV sources.

Short pulse CO_2 laser technology was well studied by TEA discharge oscillator-amplifier configuration up until the 1990's (Decker, et.al, 1991). There were two major limitations on this scheme, namely, a low repetition rate up to 10Hz and backscatter amplification in non-saturated amplifier medium. The gain medium in the RF pumped CO_2 laser is a $CO_2/N_2/He$ mixture of typically 100 Torr pressure. The CO_2 molecule stores energy in the rotational-vibration mode from electric collision excitation in the 00^01 band, and the typical relaxation time for the vibration is 0.5μsec, which is shorter than the pulse interval time of 10μsec for 100kHz repetition rate. The amplification that is described in this case is short pulse amplification and expressed by the Frantz-Nodvik equation (Rheaut, et.al, 1973), where E_{in} is the input fluence in mJ/cm^2, E_{out} is the output fluence, E_s is the saturation fluence, g_0 is the small signal gain coefficient in cm^{-1}, L is the gain medium length in cm. Maximum available fluence after single pass amplification E_m is given by $g_0 \cdot L \cdot E_s$ in mJ/cm^2. g_0 and E_s are functions of medium parameters proportional to the upper state molecule numbers as, σN^*, and inversely proportional to the radiative cross section as $h\nu/2\sigma$, each other. The Frantz-Nodvik equation describes the short pulse amplification as,

$$E_{out} = E_s \cdot \ln[1 + \exp(g_0 \cdot L)[\exp(\frac{E_{in}}{E_s}) - 1]] \tag{1}$$

and the equation is valid provided that the pulse duration τ_p is long compared to the rotational relaxation time τ_r , or short, namely in the case of $\tau_p \gg \tau_r$ or $\tau_p \ll \tau_r$. CO_2 laser amplification in the intermediate region, namely in the case $\tau_p \approx \tau_r$, is treated by rotational reservoir model calculations, which requires non practical numerical solutions (Harrach,

1975). The rotational relaxation time is calculated as τ_r = 2.3ns for 100 Torr and a 450°C medium, while the typical laser pulse width in the example is τ_p = 15ns. The characteristic number τ_p / τ_r is 6.5 for 100 Torr medium, and the work of F.Rheaut deals with the case of almost the same ratio for 1 atmosphere.

As a parameter study, small signal gain coefficient g_0 is tentatively assumed as 1%/cm, and the saturation fluence is as 10mJ/cm² for a 100 Torr RF pumped CO_2 laser amplifier. The maximum fluence E_m is then given as 10mJ/cm² for L=1m gain length. Figure 6 shows one early experimental result of 15ns pulse amplification at 100 kHz of an axial flow CW 15kW laser. The average beam diameter was ~ 15mm, and amplification parameters were calculated as g_0=0.43%/cm, and E_s=8.0mJ/cm², by numerical fitting of the experimental data to the equation (1). The fitting gave reasonable results in spite of the $\tau_p \gtrsim \tau_r$ condition (Ariga, 2007). It was concluded that the Frantz-Nodvik equation was practically usable in this semi-intermediate region for characterization of the amplifier parameters. It seems reasonable by accounting the measured E_s, together with an effort to increase g_0 to 1%/cm,to obtain 100mJ pulse energy from the amplifier once the gain cross section A is 10cm², or the gain length L is 10m with A=1cm². The cross section A is depending on the amplifier design, and the gain length L is limited by self oscillation of the amplifier.

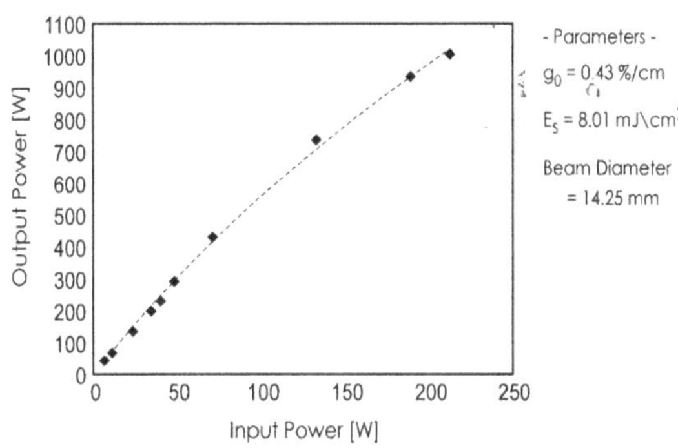

Fig. 6. Short pulse amplification data of 15kW CW output power module with 15ns pulse width at 100 kHz. The results were best fitted by the Frantz-Nodvik equation to give the effective value of g_0 and Es.

A short pulse amplifier is characterized by its medium gain diameter, D and gain length, L. The most common limitation of the available gain is set by the amplified spontaneous emission (ASE) on the optical axis, which deteriorates the pulse signal to noise ratio (SN), and depletes the available gain. A general mathematical formula on this phenomenon has been reported by Lowenthal et.al. (Lowental et.al,1986). On axis ASE flux is given by equation (11) from this work, for the purpose of numerical calculation. Figure 16 of the paper shows the achieved g_0L with parameters L/D. It is generally advised to keep g_0L less than 3 for a single pass amplifier, especially for storage lasers which have no coherent flux to extract gain simultaneously with pumping (short pulse amplification in cw pumped medium). A short pulse CO_2 laser amplifier has a typical gain diameter of 1cm, an optical gain length of 1m and an aspect ratio L/D large enough to consider the configuration as one dimensional. The gain depletion effect is less significant compared to the cubic gain medium previously reported (Lowenthal, et.al. 1986), but the on axis ASE effect is similar. Parasitic oscillation is experienced in actual experiments and this is often the practical limit for full amplification. The phenomena are strictly dependant on the device design, namely partial reflection from laser wall, optics holders or leakage through isolators. It is an issue to be treated for each laser system, but the physical fundamental is the same as the double pass case by Lowenthal et.al.

It is preferable to employ gaseous saturable absorbers like SF_6 in the multi stage amplifiers, to avoid damage to solid-state saturable absorbers like p-doped Germanium (Haglund, et.al. 1981). The typical switching threshold in the gaseous saturable absorbers is 10mJ/cm^2. Also, the optical beam delivery system requires optimization to efficiently depress pulse pedestal. An electro optical switch, which uses solid-state material like CdTe, is employed in the lower average power stage to realize better noise depression (Slattery, et.al. 1975). The damage threshold is dependant on the laser parameters such as pulse width, fluence, average power and beam uniformity. Laser beam containing hot spots can cause damage to the material surface even at much lower fluence. Solid-state materials suffer stronger thermal lens effects as shown by the following formula (Koechner, 1999).

$$f = \frac{KA}{P_a}\left(\frac{1}{2}\frac{dn}{dT} + \alpha C_{r,\phi} n_0^3 + \frac{\alpha r_0 (n_0 - 1)}{L}\right)^{-1} \tag{2}$$

The first term is the temperature dependant refractive index change, the second term is stress induced refractive index change, and the final term is temperature dependant surface modification. Thermal lensing is the most significant phenomenon in the power amplifier stage with optical components, especially for ZnSe or Diamond windows. The effective focus depends on the beam power, and active feedback is necessary to realize stable beam propagation throughout the whole laser optical path. Figure 7 shows a model calculation of the thermal loading effect to the beam propagation through 2 stage amplifiers of axial flow type of 5kW CW output, with an input of pulses in 20ns at 100 kHz repetition rate input with 60W average power (Nowak, et.al, 2008).

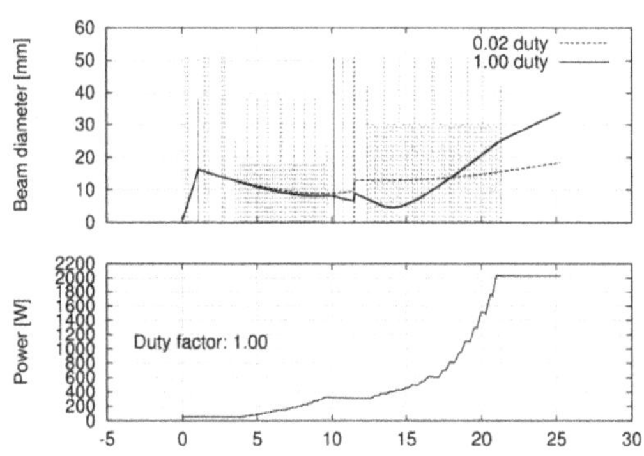

Fig. 7. Calculated results of the amplified beam diameter behavior depending on the operational duty. Input power is 60W at a distance of 0m, and the output optics of the first 5kW amplifier is at a distance of 10m. The input optics of the second module is at a distance of 11m. These windows suffer from strong thermal lensing effects which lead to beam diameter fluctuation.

It is indicated that the thermal lens effect is not deterministic at low duty operation (2%), but significant at 100% duty operation. Filling factor Φ, is the parameter used to measure the usability of the gain region by the propagating beam,

$$\Phi = \text{beam volume}/\text{gain volume} \quad (3)$$

This is reduced at higher duty operation due to a shorter thermal lens focus length, and amplification saturation is lower at higher duty operation with reduced Φ. Specifically designed active feedback control is necessary to stabilize the beam propagation at high duty operation. Figure 8 shows an experimental example of the beam diameter at the amplifier exit, with and without active beam control (Nowak, et.al, 2008). Lower spatial quality oscillator beam causes micro lens effects in the transparent optical components, and leads to chaotic beam amplification with a higher M^2 number. Design of a controlled spectrum oscillator with high quality spatial profile is important for better amplified beam quality, and efficient amplification.

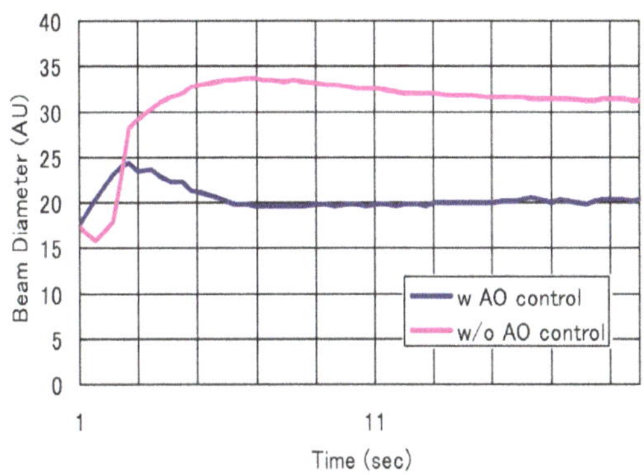

Fig. 8. Measured beam diameter after output window with and without active optical feedback.

3. New generation oscillator technologies

Precise control of the amplification depends on the quality of the oscillator pulse. Recent development of two different IR source technologies, namely OPA(Optical Parametric Amplifier) and QCL(Quantum Cascade Laser) are reviewed in this paragraph. Compact slab CO_2 laser is employed as the gain medium of regenerative amplifier to boost the weak initial IR beam. OPA and QCL are potentially controllable in its pulse width even in the pico second range.

Laser beam is characterized by its spectrum, and the low pressure CO_2 laser medium is composed of many vibration-rotational lines. The P(20) line is the single laser line in normal oscillation conditions at 10.6μm wavelength. The rotational relaxation time is calculated as τ_r = 2.3ns for a 100 Torr and 450°C medium, which is not negligibly small compared to the typical laser pulse width τ_p = 15ns. Electrically excited energy distributes in many other rotational modes of CO_2 molecules, and collisional relaxation to the P(20) line is limited during the amplified pulse period. Figure 9 shows the spectrum structure of the laser lines, with a broad continuous spectrum from a solid state seeder overlapped, to fully extract the stored energy. Recent advanced nonlinear laser technology is at the stage of operating a broad band optical parametric oscillator (OPA) at the center wavelength of 10.6µm, with more than 10mW at 150kHz (Light Conversion, "ORPHEUS", 2011). This specification is enough to seed a single transverse mode CO_2 laser oscillator.

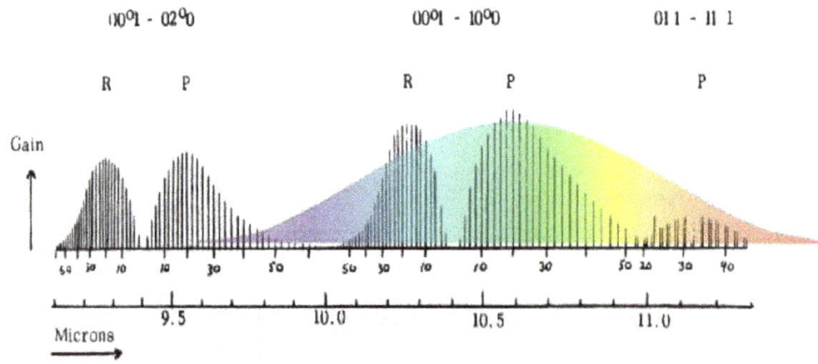

Fig. 9. CO_2 laser spectrum structure overlapped with a broad OPA seeder spectrum

Multiline amplification was calculated to evaluate the effectiveness of the low pressure CO_2 laser at 15kW CW output power. The beam diameter was assumed to be 18mm. Numerical result shows the amplification enhancement, with 4 lines composed of P(16,18,20,22) as 1.3 times higher than the single P(20) amplification case. Figure 10 shows the calculation result of the gain Γ. Emerging quantum cascade lasers (QCL) are available, which can generate specific lines of the P band of the CO_2 laser (Cascade Technologies, LS03D, 2009) and possibly seed a CO_2 laser oscillator with a discrete spectrum. QCL lasers are thus ideal as compact and robust seed sources. They can be accurately tuned to particular gain lines of the CO_2 medium with sufficient accuracy. QCL lasers are capable of tens of mW output power at typical pulse durations of 10ns, providing good bandwidth matching to a lasing line in a typical CO2 medium. A QCL can provide at least 3 orders of magnitude higher power per self oscillation lines of a small oscillator, thus relaxing the requirement of the roundtrip gain and the number of roundtrips in the seeded oscillator, thereby improving power output and stability. The theoretical prediction of Fig.10 was recently confirmed by QCL multiline seeded pulses in a large slab amplifier (Nowak, 2011).

Multiline amplification effectively enhances small signal gain g_0 compared to CW gain, and saturation fluence E_s, by improving the spectrum factor, and this leads to the enhancement of maximum available flux E_m. The final optical limit, typically 1J/cm², is given by the optical damage of the output window, which is more than one magnitude higher than the available E_m. The small signal gain g_0 and saturation fluence E_s are the two basic parameters to characterize for any amplifiers (DeaAutels, et.al. 2003). It is reasonable to expect double enhancement of E_m to 20mJ/cm² after optimization of amplifier parameters, and the available beam energy as 200mJ with the effective gain volume as LA = 1000 cm³ (1 litter). Typical repetition rate of 100 kHz gives the average output power as 20kW, after meeting all requirements described in this article.

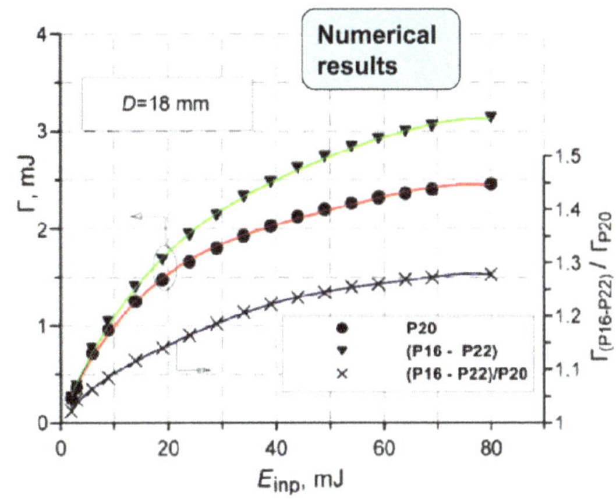

Fig. 10. Absolute (●,▼) and relative (X) energy extraction from a main amplifier tube pumped by a CW RF discharge with discharge power 88 kW. The input beam shape is Gaussian with diameter D=18mm. The energy extraction is plotted to the input pulse energy. The symbol -X- is the relative gain ratio of single P20 line to 4 lines (P16-P22) amplification

4. Further developments

It is understood that the high conversion efficiency of EUV light comes from the UTA (unresolved transition arrays) of highly ionized high Z plasma. The peak wavelength of the UTA is depending on the Z number, and further shorter wavelength plasma has a similar physical behavior (Li, et.al. 2011). The first generation EUV source at 13.5nm wavelength works well up to 11nm node of semiconductor mass production within the next one decade. A possibility to switch to a shorter wavelength (BEUV: beyond EUV) is required to be studied in advance (Banine, et.al, 2010). Availability of high reflectivity mirror at 6.x nm region initiated basic research to find candidate heavy elements with high UTA emission in this wavelength region (Churilov, et.al, 2009). Intensive research has revealed that the UTA emission has peak intensity from Gadolinium at 6.7nm wavelength comparably efficient to that of Tin at 13.5nm. Lower density plasma is favorable for less opacity effect, and short pulse CO_2 laser is again the best driver (Higashiguchi et.al, 2011). The optimum plasma temperature increases from 40eV for Sn at 13.5nm, to 150eV for Gd at 6.x nm. The laser plasma temperature is expressed as

$$T \propto (I_0 \lambda^2)^{2/3} \quad (4)$$

where I_0 is the laser peak intensity, and λ is the laser wavelength (Ramis et.al, 1983). It is understood that a short pulse CO_2 laser is better fitted to higher plasma temperature due to its longer wavelength. Optimum laser pulse width for highest CE for Sn at 13.5nm is typically 15nsec, but higher temperature Gd plasma dissipates faster, and shorter pulse length may be required for highest CE. Minimum sustainable pulse width in low pressure CO_2 laser amplifier is estimated from rotational gain bandwidth Δv(Abrams, 1974) as,

$$\Delta v = 7.58\,(\varphi_{CO2} + 0.73\varphi_{N2} + 0.64\varphi_{He}) \times P(300/T)^{1/2} \quad (5)$$

where φ is the partial ratio of each component gas, P is the total pressure in Torr, and T is the gas temperature in K. Δv is given as

$$\Delta v = 424 \text{ MHz} \tag{6}$$

for typical gas parameters as

$$CO_2 : N_2 : He = 1:1:8$$

$$P = 100 \text{ Torr} \tag{7}$$

$$T = 450 \text{ K}$$

The minimum pulse width is estimated from the Fourier transform limit of a Gaussian pulse as

$$\Delta v \cdot \Delta t = 0.44 \tag{8}$$

and the resulting Δt is around 1 nsec. The present oscillator is a QCL seeded Q-switched, cavity dumped laser based on a RF pumped low pressure CO_2 laser. Typical out put pulse width is 15nsec at 100 kHz repetition rate with 5W average power. Shorter pulse width is available by various methods like electro-optical or laser pulse slicing, depending on the requirement from future plasma experiments. Careful optical design of amplifiers can sustain the amplified pulse width for the requirement by dispersion compensation.

Another important field where 10μm wavelength is effective for short wavelength light generation, is the laser Compton X-ray generation. It is already well studied on the optimization of the laser-Compton hard X-ray source by single shot base (John, 1998, Endo, 2001). Experimental results agreed well with theoretical predictions. Highest peak brightness is obtained in the case of counter propagating laser pulse and electron beam bunch, in the minimum focusing before nonlinear threshold. The new short wavelength light source is now well matured to demonstrate single-shot phase contrast bio imaging in hard X-ray region (Oliva, et.al, 2010).

The major challenge of the laser Compton source in the EUV/SXR region is the lower electron beam voltage, which in turn results in a larger interaction cross section. Figure 11 describes the schematic of the laser-Compton interaction between electron beam and laser.

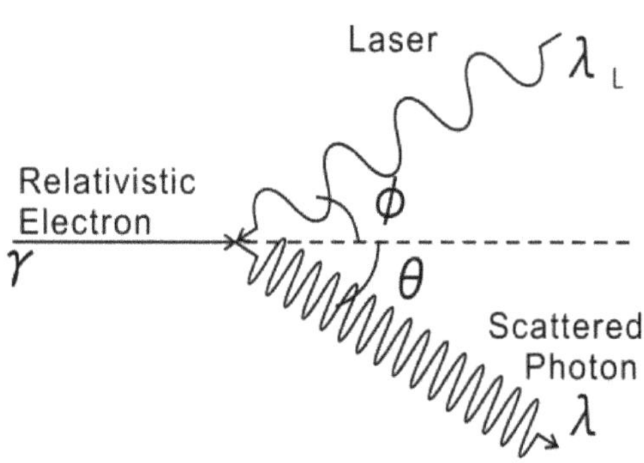

Fig. 11. Schematic of laser-Compton scattering process

Laser-Compton scattering photon spectrum has a peak in the forward direction at a wavelength;

$$\lambda_p = \frac{\lambda_L (1 + \frac{K^2}{2})}{2\gamma^2 (1 + \beta cos\phi)} \tag{9}$$

where γ and β are Lorentz factors, λ_L the laser undulation period (laser wavelength), K the K parameter of the undulator which is equivalent to the laser intensity parameter, and Φ the colliding angle. The spectrum depends on the angular distribution; the wavelength λ is emitted at

$$\theta = \frac{1}{\gamma}\sqrt{\frac{\lambda - \lambda_p}{\lambda_p}} \tag{10}$$

It is seen that higher γ electron beam produces less divergent light. Figure 12 shows the relationship between electron beam energy and maximum (forward) photon energy for both laser wavelengths by Nd:YAG laser (1.06μm) and CO_2 laser (10.6μm). As a result of Fig. 12, the required electron beam energy is 3.2MeV (1.06μm) and 10.2MeV (10.6μm) in order to produce 6.7nm SXR. It makes large difference to treat such a low energy electron beam. The Lorentz factor γ is 6.3 and 20.0, respectively, which means 3 times better directivity with 10.6μm laser than that of 1.06μm.

The general formula of obtainable photon flux N_0 is calculated in the normal collision by the following expression.

$$N_0 \propto \frac{\sigma_c N_e N_p}{4\pi r^2} \tag{11}$$

where σ_c is the Compton cross section (6.7×10^{-25} cm^2), N_e the total electron number, N_p the total photon number, and r the interaction area radius.

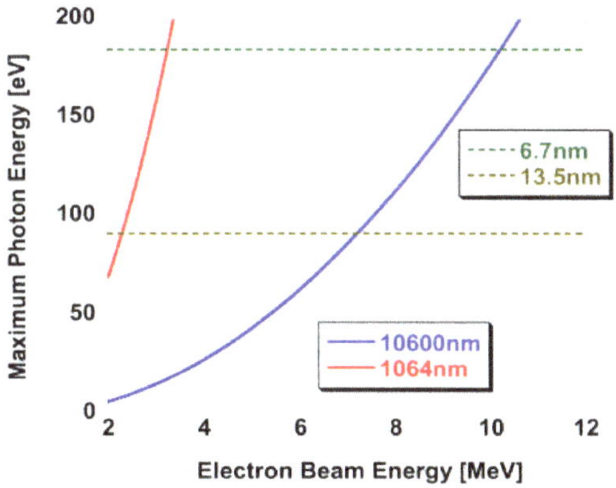

Fig. 12. Laser-Compton photon energy vs electron beam energy. Comparison 1064nm (Nd:YAG) with 10600nm (CO2) wavelength as the photon target.

It is useful to calculate standard SXR photon numbers obtainable in ideal parameters in both cases. As described in Eq. (11), the SXR number is proportional to the laser photon number. The approach to increase the photon average flux is to increase N_e, N_p and decrease r, but there are instrumental limitations to realize these simultaneously. The practical limitation of laser average power is determined by a damage on optical components, which is determined by average and peak intensity (W/cm^2). It is suggested that the usage of 10.6μm CO_2 laser has an advantage to produce one order larger number of SXR photons by the same intensity compared with 1μm solid state lasers. Another limitation is the onset of the nonlinear threshold of the higher harmonics generation, which is evident over 10^{17}W/cm^2 laser irradiation intensity (Kumita, 2008).

Usual approach is to increase the repetition rate of the event, and the obtainable photon average flux is expressed as;

$$N = f \times N_0 \tag{12}$$

where f is the repetition frequency. Characterization of the laser-Compton X-ray source has been undertaken with f as 1-10 Hz typically. High flux mode requires f in the range from kHz to MHz region.

It is under development of pulsed solid state laser storage in an optical super-cavity for laser-Compton X-ray sources (Sakaue, 2010, 2011). The enhancement inside the optical cavity was 600, in which the finess was more than 2000, and the waist of 60μm (2σ) was stably achieved using a 1μm wavelength Nd:Vanadium mode-locked laser with repetition rate 357MHz, pulse width 7ps, and average power 7W. Schematic of super-cavity is shown in Figure 13.

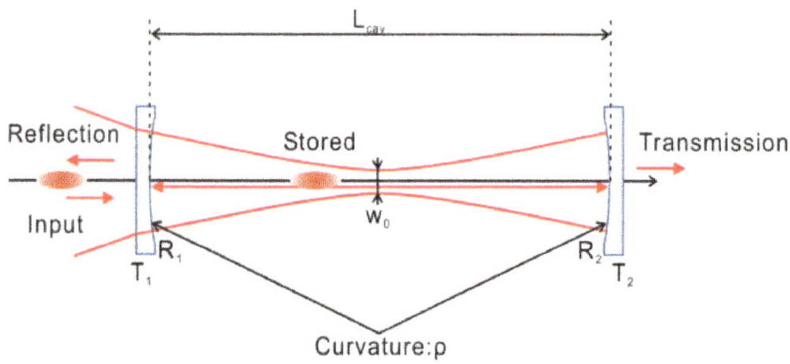

Fig. 13. Schematic of laser storage super-cavity.

Design and optimization is described here on a super-cavity for storing 10.6μm CO_2 laser pulses. Super-cavity requires high reflectivity and high transmittance mirror i.e. ultra-low loss mirror as an input and high reflectivity mirror as an output for high enhancement. The enhancement is presented by using cavity finesse (F) as (Hodgson, et.al, 2005);

$$S_{cav} = \frac{F}{\pi} \tag{13}$$

It is noted that the assumed cavity length is perfectly matched with input laser light. Finesse (F) is given by;

$$F = \frac{\pi\sqrt{R_{eff}}}{1 - R_{eff}} \tag{14}$$

where R_{eff} is $\sqrt{R_1 R_2}$. As described above, higher reflectivity provides a higher enhancement super-cavity. Particularly, the loss, which includes both absorption and scattering, on the reflection coating is critical issue for storing a high power laser beam as described above. Such a high quality optical mirror was difficult for far infrared wavelength, however, there are now some products usable for super-cavity mirrors (Ophir Optics, 2009). Super-cavity for 10.6μm laser pulse can be achieved with the enhancement of about 600 by using best mirrors available. Figure 14 shows the calculated transmission, reflection and stored power of the super-cavity as a function of the phase advance in one cavity circulation i.e. super-cavity length. The dotted line shows the reflection/transmission from the cavity and the solid line is stored power inside the super-cavity, assuming as input power is 1. There is no transmission light because Mirror 2 transmission is 0%. The enhancement of 600 is achieved by this super-cavity. The precision of cavity length adjustment is one issue for stable operation. The requirement is one-order relaxed due to the wavelength in case of CO_2 laser, thus the stable operation with enhancement of 600 and more can be easily achieved from our experiences in 1μm laser storage. Critical issue for higher enhancement is to obtain the extremely low loss and high reflectivity mirrors, which is the key R&D of multi-layer coatings with high resistance substrates.

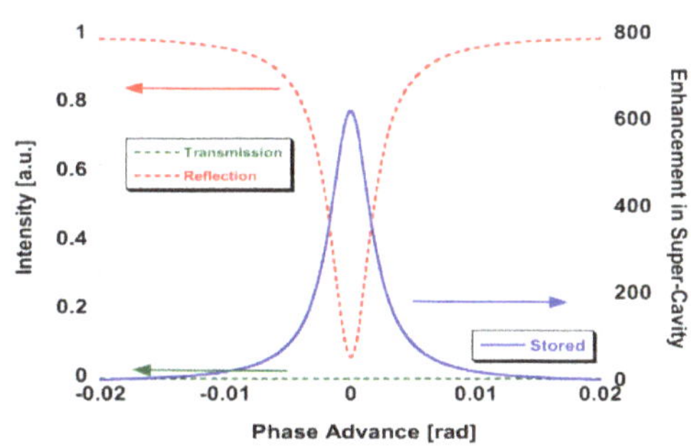

Fig. 14. Calculated results of CO_2 laser super-cavity as a function of phase advance in one revolution of cavity. 2π phase advance corresponds to $\lambda/2$ cavity length mismatch with the laser frequency.

Concerning a small waist achievement, our source requires 40μm waist (2σ). The waist of super-cavity is described as;

$$w_0^2 = \frac{\lambda}{\pi} \frac{\sqrt{L_{cav}(2\rho - L_{cav})}}{2} \tag{15}$$

where λ is wavelength of laser, L_{cav} cavity length, ρ curvature of cavity mirror. While high enhancement is easier, small waist cavity is difficult for 10.6μm laser as described in Eq. (15). However the two-mirror, Fabry-Perot cavity would be difficult to achieve small waist due to the cavity structure is confocal, we are developing a concentric, four-mirror super-cavity (Y.Honda, et.al. 2009). This technique can reduce the mirror alignment requirements as two order magnitude. We estimate that 40μm waist can be achieved using concentric super-cavity.

The first preliminary experimental results were obtained by using a single transverse mode, CW 10W CO_2 laser in a two mirror super cavity (Sakaue, et.al. 2011). The reflectivity of the input mirror was 99.5% with 15m curvature, and the CW CO_2 laser was operated with 10W maximum power of single longitudinal mode. The obtained transmitted light is shown on the oscilloscope with a sweeping voltage signal to the Piezo driver. The highest peak signal corresponds to the fundamental transverse mode, followed by higher spatial modes. The measured Finesse was around 300, which is almost half of the calculated value 624. Measured beam waist was 2.1mm, compared to the calculated value 1.8mm. The experiment showed a relatively stable result of the optical storage cavity in the CO_2 laser wavelength. Next step is planned as a demonstration of the optical storage with picosecond pulses.

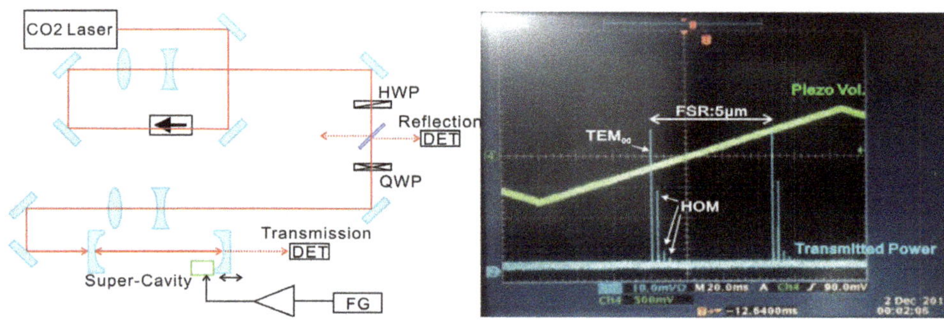

Fig. 15. Experimental setup and first result of CO_2 laser storage. Signal train corresponds to the transmission light with cavity length interval of 5 μm.

5. Conclusion

High average power, short pulse width CO_2 laser is originated in the EUV light source research in the beginning, but expanding its application to universal short wavelength plasma and non plasma sources. Reliable gas laser amplifiers with various geometrical structures are now employed with advanced solid state, semiconductor seeders to control its wavelength more precisely, and with advanced optics to enhance its pulsed average power to unprecedented level.

The author deeply expresses his thanks to his colleagues in the trials of research and development in EUV program. Early study on the CO_2 laser technology was successfully driven by Dr. T. Miura. He contributed also in the application of OPA technology as the broadband seeder for CO_2 laser oscillators. QCL was successfully studied as a seeder for precision control of the lasing lines by Dr. K. M. Nowak. Dr. H. Mizoguchi of Gigaphoton Inc. kindly gave me a chance to write this overview. The author deeply appreciates co-workers in Waseda University, especially Dr. K. Sakaue and Professor M. Washio in the CO_2 laser super-cavity program.

6. References

Abrams, R.L 1974 Broadening coefficients for the P(20) laser transition; Appl.Phys.Lett. 25, pp609-611

Ariga,T 2007 Development of a short pulse and high average power CO_2 laser for EUV lithography, Proc.SPIE 6346 634604 "XVI International Symposium on Gas Flow, Chemical Lasers and High Power Lasers", 4-8 September , 2006, Gmunden Austria

Banine,V Y, Koshele K N, Swinkels G.H.P.M. 2011. Physical processes in EUV sources for microlithography J.Phys.D.App.Phys. 44 253001

Bielesch U Budde M Fischbach M Freisinger B Schaefer J H Uhlenbusch J and Viol W 1992 Q-switched multi kilowatt CO2 laser system excited by microwaves, *Proceedings of SPIE 9th International Symposium on Gas Flow and Chemical Lasers,* pp57- 60 SPIE1810 ISBN: 9780819410108, August 1992, Heraklion, Greece

Churilov S S, Kildiyarova R R, Ryabtsev A N and Sadovsky S V 2009 EUV spectra of Gd and Tb ions excited in laser-produced and vacuum spark plasmas Phys.Scr. 80 045303

DeaAutels L.G. Daniels D. Bagford J.O. Lander M 2003 High power large bore CO_2 laser small signal gain coefficient and saturation intensity measurements J.Opt.A:Pure Appl.Opt.5 pp96-101

Decker,J.E. Lagace,S. Berube,J. Beaudoin,Y. Lin,S.L. (1991); Stable operation of a powerful 3-Hz line tunable TEA CO_2 oscillator-amplifiers system, Appl.Optics 30 pp1888-1890

Endo,A. (2010); CO_2 laser produced Tin plasma light source as the solution for EUV lithography, InTech

Endo,A.; Hoshino, H.; Ariga, T. & Miura, T. (2006). High power pulsed CO_2 laser for EUV lithography, *EUV Source Workshop*, May 2006, Vancouver, B.C. Canada, International Sematech

Haglund,R.F. Nowak,A.V. Czuchlewski,S.J. (1981); Gaseous saturable absorbers for the helios CO_2 laser system, IEEE Quantum Electron. QE17, pp1799-1808

Harrach, R.J. (1975); Effect of rotational and intramode vibrational coupling on short pulse amplification in CO_2, IEEE J.Quantum Electron. QE-11, pp349-357

Higashiguchi,T Otsuka,T Yugami,N Jiang,W Endo,A Li,B Kilbanr,D Dunne,P O'Sullivan,G (2011) Extreme ultraviolet source at 6.7nm based on a low-density plasma, Appl.Phys.Lett. 99, 191502, 2011

Hodgson,N & Weber,H (2005). *Laser resonators and beam propagation: Fundamentals, Advanced Concepts and Applications 2nd edition*, ISBN-10: 0387400788, Springer, Berlin

Honda,Y Shimizu,H Fukuda,M Omori,T Urakawa,J Sakaue,K Sakai,H Sasao,N (2009) Stabilization of a non-planar optical cavity using its polarization property, Opt.Commun. 282 pp3108-3112

Hoshino, H.; Suganuma, T.; Asayama, T.; Nowak, K.; Moriya, M.; Abe, T.; Endo, A. & Sumitani, A. (2008). LPP EUV light source employing high power CO_2 laser, *Proceedings of SPIE Emerging Lithography*, vol.6921, ISBN: 9780819471062, San Jose, CA, February, 2008, SPIE

Koechner,W. (1999); *Solid-State Laser Engineering,* Springer, Berlin

Koechner,W. (1999); *Solid-State Laser Engineering,* Springer, Berlin

Kraft, G.A. (2006) Performance achievements and challenges for FELs based on ERLs FEL2006, TUAAU01, Aug 29, Berlin Germany

Kumita,T Kamiya,Y Babzien,M Ben-Zvi,I Kusche,K Pavlishin,I V Pogorelsky, I V Siddons, D P Yakimenko, V Hirose,T Omori,T Urakawa,J Yokoya,K Cline, D and Zhou,F (2008) Observation of the Nonlinear Effect in Relativistic Thomson

Li,B Endo,A Otsuka,T O'gorman,C Cummins,T Donnely,T Kilbane,D Jiang,W Higashiguchi,T Yugami,N Dunne,P O'Sullivan,G (2011) Scaling of laser produced plasma UTA emission down to 3 nm for next generation lithography and short wavelength imaging, *Proceeding of SPIE Optics+Photonics* vol.8139, San Diego,CA,August 2011

Lowenthal, D.D. Egglestone,J.M. (1986); ASE effects in small aspect ratio laser oscillators and amplifiers with nonsaturable absorption, IEEE J.Quantum Electron. QE-22, pp1165-1173

Nowak, K.M.; Suganuma, T.; Endo, A.; Sumitani, A.; Goryachkin, D.A.; Romanov, N.A.; Sherstobitov, V.E.; Kovalchuk, L.V.; Rodionov, A.Y. (2008). Efficient and compact short pulse MOPA system for laser-produced-plasma extreme-UV sources employing RF-discharge slab-waveguide CO_2 amplifiers, *Proceedings of SPIE High-Power Laser Ablation,* vol.7005, ISBN: 9780819472069, Taos, NM, April 2008, SPIE

Nowak,K.M. (2011). Towards 20kW CO_2 laser system for Sn-LPP EUV source, *2011 International Workshop on EUV and Soft X-Ray Sources*, Dublin, Ireland November 7-9, 2011

Oliva,P. Carpinelli,M. Golosio,B. Delogu,P. Endrizzi,M. Park,J. Pogorelsky,I. Yakimenko,V. Williams,O. Rosenzweig,J (2010), Quantitative evaluation of single-shot inline phase contrast imaging using an inverse Compton x-ray source, Appl.Phys.Lett. 97, 134104

Ramis,R (1983) Electron temperature versus laser intensity times wavelength squared: a comparison of theory and experiments, Nucl.Fusion 23739

Rheault,F. Lachambre,J.L. Gilbert,J. Fortin,R. Blanchard,M (1973); Saturation properties of TEA-CO_2 amplifiers in the nanosecond pulse regime, Opt.Commun. 8, pp132-135

Sakai,T. & Hamada,N. (1994). Q-switched CO2 laser using intense pulsed RF discharge and high speed rotating chopper, *Proceedings of SPIE Gas Flow and Chemical Lasers: Tenth International Symposium*, vol.2502, pp. 25-30, ISBN: 9780819418609, Friedrichshafen, Germany, September 1994, SPIE

Sakaue, K Araki, S Fukuda, M Higashi, Y Honda, Y Sasao, N Shimizu, H Taniguchi, T Urakawa, J and Washio, M (2011) Development of a laser pulse storage technique in an optical super-cavity for a compact X-ray source based on laser-Compton scattering, Nucl.Instrum.Meth. A637 S107-S111

Sakaue, K Endo,A and Washio,M (2011) Development of a 10μm optical storage cavity, *2011 International Workshop on EUV and Soft X-Ray Sources*, Dublin, Ireland November 7-9, 2011

Sakaue, K Washio, M Araki, S K Fukuda, M Higashi, Y Honda, Y Omori, T Taniguchi, T Terunuma, N Urakawa, J and Sasao, N (2009) Observation of pulsed x-ray trains produced by laser-electron Compton scatterings, Rev.Sci.Instrum. 80 123304 1-7

Scattering of Electron and Laser Beams, Laser Phys. 16 pp267-271

Slattery, J.E. Thompson,J.S. Schroeder,J.B. (1975); Thermal pulse damage thresholds in cadmium telluride, Appl.Opt. 14, pp2234-2237

Suckewer, S.; Skinner, C.; Voorhees, D.; Milchberg, D.; Keane, C. & Semet, A. (1983). Population inversion and gain measurements for soft X-ray laser development in a magnetically confined plasma column, *IEEE-QE* 19 pp1855-1860

Yorozu, M Yang, J Okada, Y Yanagida, T Sakai, F Ito, S and Endo, A (2003) Spatial beam profile of the femtosecond X-ray pulses generated by a laser-Compton scheme, Appl.Phys. B76 pp293-297

7

Transmission of CO_2 Laser Radiation Through Glass Hollow Core Microstructured Fibers

A. D. Pryamikov, A. F. Kosolapov, V. G. Plotnichenko and E. M. Dianov
Fiber Optics Research Center of Russian Academy of Sciences
Russian Federation

1. Introduction

In this chapter we would like to highlight and analyze the main problems of transmission and propagation of CO_2 laser radiation in the hollow core microstructured fibers (HC MFs). It is well known that there is a strong need for the fiber delivery systems for 10.6 μm CO_2 lasers due to a wide range of CO_2 laser applications in medicine, spectrometry, industry, military applications and in other fields of science and technology. Research into the possibility of the mid IR laser radiation transmission (especially, CO and CO_2 lasers) with the help of optical fibers as well as with crystalline or glass cores made of different materials has been going hand with hand with the technological development. However, until recently these fibers haven't been used for industrial applications due to a relatively high level of optical losses at the lasers wavelengths and certain physicochemical properties of the fiber materials. These problems mainly arise from a low laser damage threshold, low melting temperatures of most IR transmitting materials and their high nonlinearity.

Hollow waveguides are exempt from many problems that are common to all types of the solid waveguides in this spectral region and thus can serve as much more reliable delivery systems. Glass hollow waveguides, crystalline hollow waveguides, dielectric - coated cylindrical hollow waveguides, polycrystalline fibers are well known examples of such systems. Here we consider only the glass HC MFs with their characteristics and physical phenomena laying the basis for their waveguide mechanisms. In particular, we propose a new type of HC MF for CO_2 laser radiation delivery with the cladding consisting of one row of the glass capillaries. We show that due to complicated boundary conditions and optical properties of an individual capillary it is possible to obtain low loss waveguide regimes for CO_2 laser radiation. Moreover, we show that the HC MFs with a determined symmetry type of a capillary arrangement in the cladding exhibit low bend losses when such low loss waveguide regimes occur.

The chapter is organized as follows. In Section 2 we consider all types of proposed hollow waveguides for CO_2 laser radiation transmission and give a short historical overview to highlight the problem of CO_2 laser beam delivery in the hollow core waveguides. In Section 3 we consider physical mechanisms of the light guiding in the glass HC MFs with cladding consisting of capillaries due to which, in our opinion, it becomes possible to guide the light in the mid IR including CO_2 laser radiation. In Section 4 we offer some numerical analyses and show a possibility to achieve low loss waveguide regimes in such HC MFs by careful selection of geometry parameters characterizing the fibers and a glass refractive index. Section 5 contains conclusions.

2. CO_2 laser radiation transmission through hollow core waveguides

A variety of waveguides has been studied for the delivery of CO_2 laser energy. In this section we consider briefly all types and properties of hollow waveguides and HC MFs for the CO_2 laser radiation transmission known up to now.

2.1 Historical overview of CO_2 laser radiation transmission through hollow waveguides

Rectangular core hollow waveguide structures were the first suggested for the delivery of CO_2 laser radiation [Nishihara et. al., 1974]. The first publication on the IR spectral transmission measurements of a rectangular hollow waveguide dates back to [Garmire, 1976]. The hollow waveguide described was made of aluminium strips with heights greater than 0.5 mm demonstrating a delivery over 200 W of continuous CO_2 laser radiation with no damage to the structure [Garmire et. al., 1979]. However, such metallic rectangular waveguides are not suitable for many practical applications due to their relatively big outer dimensions ~ 1 mm*10mm in cross section while smaller dimensions made the loss was too high for any practical use. Thus the search for materials and waveguides more suitable for practical applications continued and, a few years later, in [Laakmann, 1987] there was proposed a way to decrease the outside dimensions of HC waveguides to under 2 mm in diameter and to increase the reflectivity of the bore inside surface. Silver was used as a substrate metallic material of the rectangular core on which several dielectric coatings were deposited. By doing so, the author succeeded in maintaining a practical transmission level for the hollow rectangular waveguides. However, as imperfections of the inside geometry and surfaces affected the transmission, the design of the rectangular hollow waveguide proposed in [Laakmann, 1987] had to be improved [Mashida et. al., 1991]. The authors proposed a hollow waveguide having the same cross section design as in [Laakmann, 1987] but with multiple dielectric coatings on the inside surfaces to increase its reflectivity. As a result, 1 - mm - core straight rectangular hollow waveguide of such construction had a loss ~ 0.1 dB/m for circularly polarized CO_2 laser radiation. Moreover, the waveguide demonstrated the low loss for a bend. The next attempt to decrease loss for the rectangular core hollow waveguides was described in [Karasawa et. al., 1990]. The authors proposed and fabricated a germanium coated rectangular hollow waveguide with a cross section of 2 mm^2, a length of 80 cm and a loss of less than 0.1 dB/m was fabricated. The resulting waveguide had a relatively low loss even for a bend.

However, the main disadvantage of the rectangular hollow waveguides is their relatively large outer dimensions and low flexibility which has led to a greater popularity of circular hollow waveguides. These waveguides made of glass, metal or plastic are those most commonly used today. Along general lines, circular cross section hollow waveguides for CO_2 laser radiation transmission can be divided into two groups: attenuating total reflecting and leaky waveguides. The metallic or dielectric films are deposited on the inside of metallic, plastic or glass tubing. Attenuating total reflecting hollow waveguides have inner core materials with a refractive index less than 1. Due to this fact, the angle of incident radiation in the core is greater than the critical angle and the light experiences the total internal reflection. An example of such fiber operating at the CO_2 laser wavelength is the sapphire fiber with a refractive index of $n = 0.67$ [Harrington & Gregory, 1990].

For our part, in the following subsections we will consider the second group of the circular cross section hollow waveguides, previously classified as leaky waveguides, in particular, HC MFs. This group of hollow waveguides has inner wall surface with refractive indices greater than 1. Leaky hollow waveguides have metallic or dielectric layers deposited on the inside metallic, plastic or glass tubing to enhance their surface reflectivity. Creation of such dielectric coated cylindrical hollow waveguides presented a complicated technological problem since high quality reflective coatings are not compatible with circular cross section geometry. Traditional vapor deposition techniques don't produce good quality coatings on the inside of a capillary. The theoretical calculated attenuation of a dielectric coated hollow waveguide for the IR region was obtained in [Miyagi & Kawakami, 1984]. The authors have shown that the attenuation is very sensitive to the material and geometry of a dielectric film. Also, the attenuation is very sensitive to the properties of the metal under the dielectric film. A metal should have a low refractive index and a very high extinction coefficient. For example, these can be silver, nickel, copper. A dielectric should be selected with a maximum refractive index, for example, KCl, ZnSe, ZnS etc. The first demonstration of a dielectric coated cylindrical hollow waveguide was perfomed by Prof. Miyagi's group [Miyagi et. al., 1983] in 1983. The 1.2 m - long mandrel of polished aluminum tubing was coated with approximately 0.45 μm of germanium. Then, a layer of nickel up to 200 μm was deposited on the top of germanium before the aluminum mandrel was removed by leaching. The final structure was a nickel tube with optically thick dielectric layers on the inner wall. The fabricated waveguide had a core diameter of 1.5 mm and length of about 1 m. The measured attenuation was ~ 0.4 dB/m in the straight waveguide but the bend loss was very high.

Authors in [Croitoru et. al., U.S. Patent, 1990] have used polyethylene and Teflon tubing as a substrate in which thin and flexible metallic layers of Al followed by AgI were deposited. They reported attenuation in a straight waveguide of about 0.6 dB/m at the bore size value of 4 mm [Croitoru et. al., 1990]. Authors of [Morrow & Gu, 1994] reported a cylindrical hollow waveguide in which Ag and Ag halide coatings were deposited inside Ag tubes. The waveguide with 1 - mm - bore size had attenuation below 0.1 dB/m at $\lambda = 10.6$ μm which was still below 0.8 dB/m at a bend radius up to 25 cm.

At the end of this subsection, we will turn to the currently most popular hollow circular cross section waveguides for CO_2 laser radiation transmission. These are hollow glass waveguides developed initially by Prof. Harrington's group [Abel et. al., 1994]. There are two main advantages of the glass tubing substrate. First, it is easier to make a long, uniform tubing from glass having considerably smoother wall surfaces than metal or plastic tubings. As a result, the scattered losses are less. The second advantage is that the technology of making glass capillary tubings is common and inexpensive. The authors fabricated a hollow glass waveguide using wet chemistry methods. First, an Ag layer was deposited on the inside of a silica glass tubing. Then, an AgI layer was formed over the metallic film. The thickness of the layer was optimized to obtain a high reflectivity at the required wavelength. A straight waveguide with a bore of 530 μm demonstrated a loss of 0.3 dB/m at CO_2 laser wavelength. This fiber could maintain a loss level under 2 dB/m at a bend radius as small as

5 cm. Hollow glass waveguides have been used successfully for a modest CO_2 laser power delivery below ~ 80 W. For the higher power delivery it is necessary to place a water - cooled jacket around the guides. The highest CO_2 laser power delivered through the water - cooled hollow glass waveguide with 700 μm bore was 1040 W [Nubling & Harrington, 1996]. This is comparable to CO_2 laser power delivered through the water - cooled hollow metallic waveguide with 1800 μm bore which was 2700 W [Hongo et. al., 1992].

2.2 CO_2 laser radiation transmission through hollow core microstructured fibers

In this subsection we will consider a new approach to solving the problem of the mid IR transmission (in particular, CO_2 laser radiation) through the glass hollow core microstructured fibers (HC MFs). The possibility of the light confinement in the air core of HC MFs with the cladding consisting of two dimensional periodic array of air holes was predicted by Russell at the beginning of 1990s and theoretically demonstrated by Birks et. al. [Birks et. al., 1995]. The most advanced HC MFs are hollow core photonic crystal fibers (HC PCFs). HC PCFs in turn can be divided into two main groups. The HC PCFs from the first group guide the light by virtue of photonic band gap (BG HC PCFs). The HC PCFs from the second group have no band gaps and guide the light due to an inhibited coupling between the core guided modes and modes associated with a cladding [Benabid et. al., 2002]. They are called inhibited coupling HC PCFs (IC HC PCFs). Both types of HC PCFs have the claddings with very little solid material, usually, with a filling fraction less than 10%.

The guidance mechanism for BG HC PCFs is based on the concept of 'out of plane' band gap. The microstructure of BG HC PCF cladding consists of air holes packed in a triangular arrangement. It gives rise to a full two dimensional photonic band gap [Birks et. al., 1995]. As a result, forbidden frequencies occur for optical waves whose wave vector (axial) component is not equal to zero. Such frequency ranges constitute bands. The first experimental demonstration of light transmission in the BG HC PCF was made in 1999 [Gregan et. al., 1999]. Up to now, considerable efforts have been put forth in experimental and theoretical studies of BG HC PCFs made of silica glass [Humbert et. al., 2004; Benabid et. al., 2004]. This special interest can be partly explained by a need to find a way of yielding a loss level less than 0.2 dB/km for telecommunication spectral region. So far, the BG HC PCFs loss was reduced only to 1.2 dB/km due to intrinsic roughness of the air - glass interfaces in the structure [Roberts, Couny et. al., 2005].

As it was mentioned above, BG HC PCFs made of silica glass have claddings with very little solid material. The bandgap located between 4th and 5th bands is used for guiding in HC PCFs with such high air - filling fraction ($\geq 80\%$) [Humbert et. al., 2004]. The number of each band is counted from the band with the largest value of the propagation constant of the air core mode. However, there is an important need for BG HC PCFs which can be used in the mid and far IR. BG HC PCF made of silica glass with a core diameter of 40 μm demonstrated single mode waveguide regime in a narrow transmission window near the wavelength of $\lambda = 3.14$ μm with an attenuation of ~ 2.6 dB/m [Shephard et. al., 2005]. But silica glass BG HC PCFs cannot be used for CO_2 laser radiation transmission due to a very high material loss of silica. Transmission of light in the mid IR region becomes possible with BG HC PCFs made of glasses which are transparent in this spectral region such as chalcogenide glasses. These glasses are composed of the chalcogen elements Se and Te with an addition of such elements as Ge, As, Sb. The transparency windows of these glasses correspond approximately to the mid IR region 2 - 25 μm.

Modeling of BG HC PCFs made of nonsilica glasses was performed by a number of authors [Shaw et. al., 2003; Pottage et. al., 2003; Pearce et. al., 2005]. In this paper [Shaw et. al., 2003] BG HC PCFs made of As - S (refractive index ~ 2.4) and As - Se (refractive index ~ 2.8) were analyzed. It was shown that there exist several spectral regions with bandgaps for an air filling fraction > 40% in both As - S and As - Se BG HC PCFs. These BG HC PCFs have large bandgap widths at the air filling fraction of 45% to 60%. BG HC PCFs with high air filling fractions > 80% also exhibited large bandgap widths. In the author' opinion, all these results show a possibility of a light transmission in the mid IR using halcogenide BG HC PCFs. The authors of [Pottage et. al., 2003] have carried out a numerical analyses of BG HC PCFs for a wide range of refractive indices from n = 1.5 to n =3.6 and for different values of air filling fractions from 33% to 87%. They discovered a new type of the bandgap which was called type 2 bandgap at an air filling fraction ~ 60% for any glass index beyond 2. The results showed a possibility of obtaining a satisfactory guidance in such BG HC PCFs.

Another important aspect of the problem of the mid IR radiation transmission was discussed in [Pearce et. al., 2005]. Apart from a limitation to attaining a low loss guidance in BG HC PCFs connected with the intrinsic roughness of the air glass interfaces, there is another problem connected with an existence of surface guided modes that are trapped in the core surroundings. Experimental and theoretical studies [Smith et. al., 2003; Humbert et. al., 2004; West et. al., 2004; Saitoh et. al., 2004] carried out for silica BG HC PCFs have shown that the anticrossing between dispersion curves of the surface modes and the air core modes is the main factor leading to a transmission loss in BG HC PCFs. Several methods were proposed to suppress the surface modes. The first method is used to reduce the distortion of the core by including 'fingers' of glass [West et. al., 2004]. The second method is to use thin core walls [Saitoh et. al., 2004] and the third one is to use 'antiresonance' walls [Roberts, Williams et. al., 2005]. The authors of [Pearce et. al., 2005] modeled a realistic design of distorted cores for BG HC PCFs which can guide the light in the type 2 bandgap. They have demonstrated that BG HC PCFs made of high index glass can guide a fundamental air core mode with a fraction of power in the air of up to 98%.

In their paper [Hu & Menyuk, 2007] the authors analyzed BG HC PCFs for refractive indices between 1.4 and 2.8. They found two maxima of the relative bandgap as a function of the air filling fraction and refractive index. The authors also found that the relative bandgap and the level of loss are interrelated. When the relative bandgap increases the loss decreases and vice versa.

Despite the promising results of modeling obtained in the above listed works a practical realization of BG HC PCFs made completely of chalcogenide glass for the mid IR spectral region has not been reported. The only successful realization of a photonic band gap hollow core fiber for the CO_2 laser radiation transmission was 'Omniguide' fiber where the cladding is a Bragg reflector (hollow core Bragg fiber) made of soft glass and polymer [Temelkuran et. al., 2002]. The authors of [Bowden & Harrington, 2009] have studied low and high index chalcogenide glasses for their potential use in the fabrication of all glass hollow core Bragg fiber.

The first work devoted to the fabrication and experimental investigation of BG HC PCF for CO_2 laser radiation transmission has appeared only in 2010 [Deseveday et. al., 2010]. The authors designed BG HC PCF made of chalcogenide glass to guide the light in the air core at λ = 9.3 μm. They also fabricated two BG HC PCFs which could potentially guide CO_2 laser radiation but no guidance was observed. The authors explained this fact by technological difficulties in the fabrication process. They hope to improve the process by avoiding air tightness anomalies and by decreasing the core's wall thickness.

In the next section we will represent our approach to solving the problem of CO_2 laser radiation transmission through the glass HC MFs.

3. Mechanisms of CO_2 laser radiation transmission through the glass hollow core microstructured fiber with the cladding consisting of capillaries

In this section, we will consider physical mechanisms and principles which enable, in our opinion, to obtain a loss level much lower than the material loss of the glass of HC MFs with a negative curvature of the core boundary [Pryamikov et. al., 2011]. The negative curvature of the core boundary is obtained by the cladding consisting of one or several rows of glass capillaries. Such microstructure design leads to a significant complication of the boundary conditions for the air - core modes. To justify our assumptions it will be necessary to consider a plane wave scattering on a cylindrical surface to show an analogy between this phenomenon and the light scattering on the plane optical diffraction grating. An analogy between discrete rotational symmetry of the capillary arrangement in the cladding and the plane diffraction grating will also be outlined. We will also consider the second main factor leading to a loss reduction of the air core modes of the HC MFs and to an increase in the width of transmission regions. It is connected with the geometry parameters of an individual capillary and the glass refractive index. In the end, we will try to justify the statement how these factors can result in the loss reduction of the air core mode in HC MF with the cladding consisting of capillaries with respect to the BG HC PCFs and kagome lattice IC HC PCFs.

3.1 The cylindrical surface as a diffraction grating

In this subsection, we will offer a reason which, in our opinion, lies behind the low loss waveguide regimes for the glass HC MFs with negative curvature of the core boundary. For the first time, an effect of the loss level decrease resulting from the negative curvature of the core boundary was observed for a large pitch kagome - lattice IC HC PCF with a hypocycloid - shaped core structure (the second group of PCFs) [Wang et. al., 2011]. We have used a simple cladding structure of the HC MF consisting of eight silica capillaries (Fig. 1(left)). Such HC MF guided light in the mid IR up to 4 μm despite of very high material losses of silica in this spectral region. In this case, the negative curvature of the core boundary was created by the capillary surfaces. Of course, such long a wavelength guiding is determined by not only the negative curvature of the core boundary but (may be to a greater extent) also by the optical properties of an individual capillary of the cladding. For example, the simple cladding structure consisting of one row of the capillaries has a lower density of eigenstates with respect to the cladding consisting of the solid rods (Fig. 1(right)).

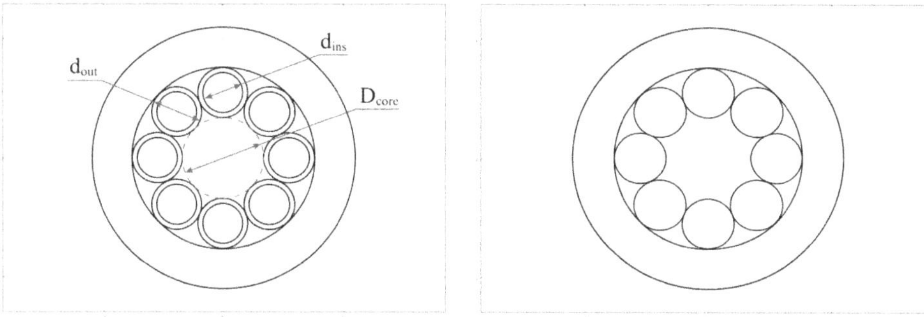

Fig. 1. (left) HC MF with the cladding consisting of eight capillaries (right) HC MF with the cladding consisting of eight solid rods.

In [Pryamikov et. al., 2011] the authors have made a calculation mistake when they were trying to justify their assumptions on the role of the negative curvature of the core boundary. The mistake was made when calculating the waveguide loss for a dielectric tube using an analytical formula from [Marcatili & Schmeltzer, 1964]. Indeed, the loss levels for the dielectric tube and silica HC MF with the cladding consisting of solid rods with equal air core diameters are approximately the same. Despite this fact, the main conclusion of the paper remains accurate, i.e. that to obtain a long wavelength waveguide regime in the mid IR with silica HC MF it is necessary to combine two factors, namely, the negative curvature of the core boundary and the low density of eigenstates of the individual elements of the cladding.

To clarify a role of the negative curvature of the core boundary we should consider a plane wave scattering on the curved surface, for example, of a solid rod. This problem was solved by many authors [Wait, 1955; Lind & Greenberg, 1966]. Depending on the polarization state of the incident plane wave z - component of the electric (TM polarization) or magnetic field (TE polarization) is parallel to the incident plane. Suppose the refractive index of the solid rod is n_1 and the outer space is n_2. In the following, only TE polarized plane wave will be considered. Its z - component of the magnetic field can be expanded according to addition theorem for Bessel functions and taking into account the temporal dependence as $e^{i\omega t}$ one can obtain:

$$H_z^i = H_0 \sin\theta [\sum_{n=-\infty}^{\infty} i^n J_n(\lambda_2 r) e^{-in\varphi}] e^{-i\beta z}$$

$$E_z^i = 0 , \tag{1}$$

where θ is an angle of incidence, φ and r are an azimuthal and radial cylindrical coordinates, H_0 is an amplitude of the incident plane wave and $J_n(\lambda_2 r)$ are the Bessel functions of first kind . If $k_2 = \frac{2\pi}{\lambda} n_2$ is a wavevector in the outer space $\lambda_2 = \sqrt{k_2^2 - \beta^2}$ is its transverse component and $\beta = k_2 \cos\theta$ is its axial component. In such a way, the incident plane wave is expanded into an infinite number of cylindrical harmonics due to the curvature of the cylinder surface. The z - components of the scattered field and field inside the cylinder can be expressed in the same manner:

$$H_z^s = [\sum_{n=-\infty}^{\infty} b_n^s H_n^{(2)}(\lambda_2 r)e^{-in\varphi}]e^{-i\beta z}$$

$$E_z^s = [\sum_{n=-\infty}^{\infty} a_n^s H_n^{(2)}(\lambda_2 r)e^{-in\varphi}]e^{-i\beta z} \tag{2a}$$

$$H_z^{ins} = [\sum_{n=-\infty}^{\infty} b_n J_n(\lambda_1 r)e^{-in\varphi}]e^{-i\beta z}$$

$$E_z^{ins} = [\sum_{n=-\infty}^{\infty} a_n J_n(\lambda_1 r)e^{-in\varphi}]e^{-i\beta z}, \tag{2b}$$

where $H_n^{(2)}(\lambda_2 r)$ is Hankel function, $\lambda_1 = \sqrt{k_1^2 - \beta^2}$ is a transverse components of the wavevector $k_1 = \frac{2\pi}{\lambda} n_1$ for the field inside the rod. On the basis of these expressions for z - components of the incident, scattered and inside fields it is possible to calculate φ and r - components of the fields [Adler, 1952]. In other words, the field of the incident wave is represented by an infinite number of space 'channels' (harmonics) through which the energy of the incident wave is transferred to the scattered fields and the fields inside the dielectric rod. It is seen, that such sets of cylindrical harmonics (1) - (2) have a mode structure and can be considered as radiation or continuous modes of ITE (incident transverse electric) type of the individual dielectric rod [Snyder, 1971].

To calculate the coefficients b_n^s, a_n^s, b_n, a_n it is necessary to apply boundary conditions for z and φ - components of the incident, scattered fields and the fields inside the cylinder. Because of the mode structure of the total field it is not necessary to solve an infinite set of simultaneous linear equations. To obtain the nth order coefficients one needs to solve 4*4 inhomogeneous system of linear equations. For a solid cylinder rod it is possible to obtain analytical expression for the coefficients which includes such terms as $J_n'(\lambda_1 a) / J_n(\lambda_1 a)$ [Wait, 1955]. These terms have resonances (poles) corresponding to zeros of $J_n(\lambda_1 a)$. It is necessary to point out that the resonances of the nth order coefficients of the scattered field are determined not only by nth order functions $J_n(\lambda_1 a)$ but also by $J_{n+1}(\lambda_1 a)$ and $J_{n-1}(\lambda_1 a)$ due to recurrent relations for the derivatives of the Bessel functions. In other words, the different diffraction orders are coupled between each other. Due to this fact, it is possible to observe not only Lorentzian - like resonances for the spectral dependencies of absolute values of the amplitudes $|b_n^s|, |a_n^s|$ but also Fano type resonances [Fano, 1961]. For example, Fano type resonances were analysed in the case of all solid band gap fibers [Steinvurzel et. al., 2006] with the cladding consisting of solid dielectric cylinders with a refractive index higher than the background.

A similar phenomenon occurs when the plane wave is scattered on optical diffraction grating. A short analysis can be carried out based on the work [Hessel & Oliner, 1965].

Suppose the diffraction grating extends infinitely in the y and z directions, with the period d in the z direction only. The refractive index of the outer medium is again n_2. The magnetic field of this incident wave of S polarization is represented as:

$$H_y^i = H_0 e^{ik_x x} e^{-i\beta z} \tag{3}$$

Then, due to periodicity the scattered magnetic field can be represented as [Hessel & Oliner, 1965]:

$$H_y^s = [\sum_{n=-\infty}^{\infty} A_n(\beta) e^{ik_x^n x} e^{-i\frac{2\pi n}{d}z}] e^{-i\beta z}, \tag{4}$$

where $A_n(\beta)$ are the amplitudes of the various spectral orders and $k_x^n = \sqrt{k_2^2 - (\beta + \frac{2\pi n}{d})^2}$ are transverse components of the wavevector of the scattered field. Applying boundary conditions one obtains an infinite inhomogeneous set of simultaneous linear equations [Hessel & Oliner, 1965] for the scattered field amplitudes $A_n(\beta)$. These amplitudes are coupled between each other as in the case of the plane wave scattering on the rod and a set of resonances Lorentzian and Fano types known as Wood anomalies [Wood, 1902; Lord Rayleigh, 1907; Hessel & Oliner, 1965] also occur. In the case of optical grating the conservation of light momentum for the scattered light in the z - direction looks like:

$$k_n^s = \beta + \frac{2\pi n}{d}.$$

In the case of the plane wave scattering on a dielectric rod there is also a conservation of the angular momentum of the light. The fields of the source plane wave are expanded into a set of the 'space' channels with an angular momentum determined by the number of n in $e^{-in\phi}$ (1) and each continuous (radiation) mode of the dielectric rod has the same angular momentum (1, 2). Note, that this is (4) of the same form as an expression (2a) for the scattered field H_z^s in the case of the dielectric rod. Comparing expressions (2a) and (4) one can state that the process of the plane wave scattering on the rod can be represented as scattering on the diffraction quasi - grating. The surface of the dielectric rod can be considered as an azimuthal diffraction grating with periodicity in φ - direction and conservation of the angular momentum of the light. But instead of one incident plane wave as in the case of the plane diffraction grating (3) there is an infinite set of incident cylindrical harmonics with their own angular momentums of the light. It can also be shown that the spectral dependencies of b_n^s, a_n^s in expansions (2a) demonstrate the resonance behaviour analogous to Wood anomalies as in the case of the plane diffraction grating. This problem and its application to the waveguide mechanism in all solid band gap fibers will be considered by us thoroughly in our future publication. In the same way, it is possible to consider a cylindrical surface of the capillary as a diffraction quasi - grating in the φ direction instead of z – direction in the case of the plane diffraction grating.

The main conclusion to draw is that the plane wave incidence on a curved cylindrical surface leads to the appearance of an infinite set of cylindrical harmonics with different angular momentums of the light. As the source of the incident wave is outside of a solid cylinder or a capillary one obtains an inhomogeneous set of linear equations for determining amplitudes of the scattered field harmonics. These scattered field harmonics have the same values of the angular momentums of the light as the harmonics of the incident field. Different orders of these harmonics (diffraction orders) are coupled between each other due to the properties of the Bessel functions. It leads to appearance of Wood anomalies in the spectral dependencies of their amplitudes. As a consequence, the curved cylindrical surface can be considered as an azimuthal diffraction grating with modulation in φ - direction [Pryamikov, to be prepared]. This fact leads to a loss of simplicity of the boundary conditions for the air core modes of the HC MF with the cladding consisting of capillaries compared with BG HC PCF, for example. Similar situation occurs if one considers the boundary conditions for the core modes of all solid photonic band gap fibers [White et. al., 2002] and Bragg fibers [Yeh et. al., 1978].

3.2 Discrete rotational symmetry of the core boundary

Another aspect of the problem of complicated boundary conditions for HC MF with the cladding consisting of N capillaries is that the capillary location in the cladding is also periodic in the azimuthal coordinate φ (Fig.1). As a result, the boundary conditions for the air core modes are also periodic. It is known that if the system transforms into itself for a set of discrete rotations $\delta\varphi = 2\pi k / N$, $k \in (0, N-1)$ around axial vector $\vec{z}$ the eigenfunctions of the system (in other words, the air core modes) can be represented as [Skorobogatiy & Yang, 2009]:

$$\psi_n(r,\varphi,z) = e^{-in\varphi} U_n(r) e^{-i\beta z} \; , \tag{5}$$

and their eigenvalues are:

$$X = e^{in\frac{2\pi k}{N}} \; .$$

These eigenvalues are the same for any $n = n + Nm$, where m is an integer and the eigenfunctions characterized by an integer n are degenerate ones [Skorobogatiy & Yang, 2009]. As a consequence, such eigenstates of the system with a discrete rotational symmetry can be expressed as a superposition of all the degenerate states:

$$\psi_n(r,\varphi,z) = [\sum_{m=-\infty}^{\infty} A(m) U_{n+Nm}(r) e^{-i(n+Nm)\varphi}] e^{-i\beta z} \; , \tag{6}$$

where an expression under the sum sign is a periodic function in φ with a period $\frac{2\pi}{N}$ and $n \in [0, N-1]$. Note that this is (6) of the same form as an expression (2a) and (4) for the scattered field H_z^s in the case of the dielectric rod and the plane diffraction grating. The discrete symmetry of the core boundary gives one more type of azimuthal diffraction grating in the considered HC MFs.

In such a way, the boundary conditions for the air core modes of the HC MF with the cladding consisting of capillaries are complicated not only by the curvature of the cylindrical surface of an individual capillary but also by the discrete rotational symmetry of the core boundary.

3.3 Complicated boundary conditions for the air core mode of HC MF with the cladding consisting of capillaries and low loss guidance

Summarizing the conclusions of the two previous subsections one can state that the air core modes of HC MFs with the cladding consisting of capillaries are formed by a superposition (interference) of several space cylindrical harmonics originated from the azimuthal diffraction quasi - gratings occurred in the cladding. The main factors affecting the air core mode formation are geometry parameters of an individual capillary and the type of a periodic arrangement of the capillaries in the cladding. All these space harmonics interact with the capillary walls in stronger or weaker ways depending on their radial and azimuthal distribution. In this case, the material loss of the capillary walls doesn't have the same affect on the attenuation of each harmonic. In such a way, the energy of the air core mode at the core boundary for HC MF with the cladding consisting of capillaries is distributed in much more complicated way than in the case of BG HC PCFs or kagome lattice IC HC PCFs.

BG HC PCFs and kagome lattice IC HC PCFs do not have such complicated mechanism of the air core modes formation because of a quasi continuous rotational symmetry of the core boundary and the boundaries of succeeding layers in the cladding. Kagome lattices IC HC PCFs sometimes have the discrete symmetry of the core boundary but just with a polygonal shape of the core without a negative curvature of the core boundary. The authors in [Pearce et al., 2007] have shown that the loss behavior of a hollow core fiber with the cladding consisting of concentric glass rings or hexagons explain the qualitative features in the loss curves associated with kagome lattice IC HC PCF. The hollow core photonic band gap fiber with the cladding consisting of concentric glass rings has a continuous rotational symmetry of the core modes and is analogous to solid Bragg fibers [Fevrier et. al., 2006]. In this case, the air core is formed by the boundaries of the concentric glass rings which cannot play the role of the azimuthal diffraction gratings. Applying boundary conditions to each concentric ring of the cladding one obtains a homogeneous system of linear equations determinant of which is a dispersion relation for the air core modes. Each air core mode is formed and described by only one space cylindrical harmonic with the propagation constant β determined from the corresponding dispersion relation. These space harmonics don't interact with each other's as in the case of all solid band gap fibers and Fano type resonances cannot be observed (an exception can be kagome lattice IC HC PCF with a polygonal shape of the air core). As a consequence, all energy of each air core mode is concentrated in one 'space' channel (cylindrical harmonic) which interacts with glass rings of the cladding in the same way according to its azimuthal angular dependence of $e^{-in\varphi}$. This air core mode is leaky and its imaginary part of β is determined from the dispersion relation mentioned above. All resonances in high index layers for this mode are radial and can be described by the ARROW model [Litchinitser et. al., 2002].

The other very important moment is that the geometry parameters of the capillaries d_{ins}, d_{out} and the glass refractive index at the determined value of D_{core} (Fig .1(left)) should be chosen in such a way so as to excite leaky modes with high quality factor. These modes usually have high radial and azimuthal indices. To this end, the capillary wall thickness must be thin enough and comparable with the wavelength. In this case, it is possible to obtain large widths of the transmission regions for the air core modes. Moreover, it seems possible to choose all parameters of the considered HC MF including glass refractive index in such a way so as to obtain a very weak interaction (coupling) of each cylindrical harmonic constituting the air core mode with capillary walls inside the transmission regions. All these factors can give rise to a low loss waveguide regime for CO_2 laser radiation. Several examples of such waveguide regimes will be given in the next section.

4. Numerical modelling of HC MFs with different types of discrete rotational symmetry of the core boundary and glass composition of the capillaries

In the following subsections four types of HC MFs with the claddings consisting of 6, 8, 10 and 12 capillaries will be considered. The calculations will be carried out for two values of the glass refractive indices n = 2.4, 2.8 and with three values of d_{ins} / d_{out} = 0.8, 0.85, 0.9. All calculations will be made in the narrow spectral region near λ = 10.6 µm. This fact is connected with high density of the individual capillary eigenstates (leaky modes with high azimuthal indices [Vincetti & Setti, 2010]) which occur at such high values of the glass refractive indices and individual capillary dimensions with respect to the wavelength. We will show that the way of obtaining a low loss waveguide regime in the glass HC MFs with the cladding consisting of capillaries presents a complicated multiparameter task. Unusual behaviour of the bend loss depending on the bend radii for high index glass HC MFs with the cladding consisting of capillaries will be demonstrated.

4.1 Loss dependencies for HC MF with different number of capillaries in the cladding

First, we will consider the loss dependencies for HC MF with six capillaries in the cladding. As was mentioned above, all loss dependencies are calculated in the narrow spectral range from 10.59 µm to 10.61 µm with a wavelength step equals to 1 nm. The calculations will be made for two values of the ratio of d_{ins} / d_{out} = 0.85 and 0.9 because the losses are very high at lower values of the one. In Fig. 2 these dependencies are shown for two values of the air core diameter D_{core} = 220 µm and 320 µm.

As one can see from Fig. 2(a) HC MF with n = 2.4 and d_{ins} / d_{out} = 0.9 has the minimal loss level in the considered spectral range. In our opinion, it can be explained by the minimal value of density of individual capillary states. In this case, the capillary has a minimal capillary wall thickness with respect to the wavelength and the lowest refractive index. The loss dependence for HC MF with n = 2.8 and d_{ins} / d_{out} = 0.9 is relatively inhomogeneous and has a strong peak at λ = 10.601 µm caused by the excitation of a capillary leaky mode with high azimuthal and radial indices. This mode is shown in Fig. 3(left).

Other curves in Fig.2 (a) for d_{ins} / d_{out} = 0.85 have a higher loss level due to thicker capillary walls compared to the previous case. In Fig. 2(b) HC MF with n = 2.8 and d_{ins} / d_{out} = 0.85 has the maximal loss due to the excitation of the second type of the capillary leaky modes.

These capillary modes have a lower azimuthal index and a different radial dependence (Fig. 3(right)). It is worth pointing out that these two types of the individual capillary leaky modes are the main reason for occurring high loss regions for the considered HC MFs made of high index glasses.

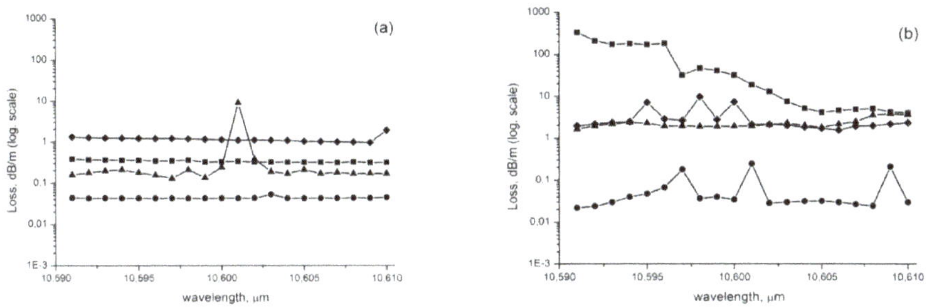

Fig. 2. (a) loss dependence for HC MF with D_{core} = 220 μm and 6 capillaries in the cladding, d_{ins} / d_{out} = 0.9 (n = 2.4, circles), d_{ins} / d_{out} = 0.9 (n = 2.8, triangles), d_{ins} / d_{out} = 0.85 (n = 2.4, rhombuses), d_{ins} / d_{out} = 0.85 (n = 2.8, squares); (b) loss dependence for HC MF with D_{core} = 320 μm the other notations are the same as in (a).

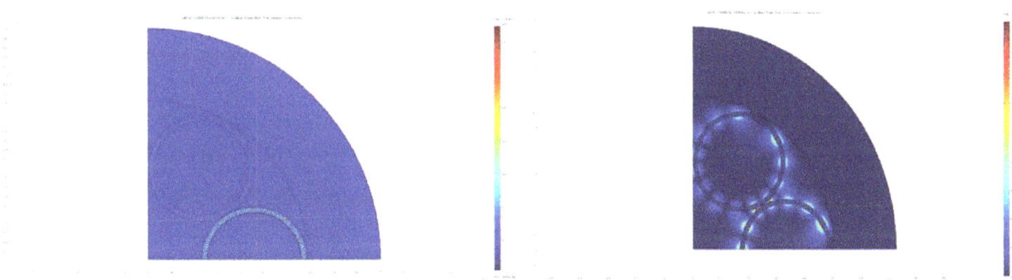

Fig. 3. (left) a typical Pointing vector distribution for a capillary leaky mode in the cladding with a high azimuthal index; (right) the Pointing vector distribution for the second type of the capillary leaky modes with a lower azimuthal index.

Further, we will consider HC MF with eight capillaries in the cladding. The air core diameters and the glass refractive indices will be the same as in the case of HC MFs with six capillaries in the cladding. In Fig. 4 the loss dependencies for this HC MF are shown.

As one can see from Fig. 4(a) HC MF with d_{ins} / d_{out} = 0.85 and n = 2.8 has the highest loss due to the excitation of the leaky modes with a lower azimuthal index in this spectral region (Fig. 3(b)). The HC MF with d_{ins} / d_{out} = 0.85 and n = 2.4 has a high loss for the same reason. Losses for HC MFs with other parameters in Fig. 4(a) are relatively low, especially, in the case of HC MF with d_{ins} / d_{out} = 0.8 and n = 2.4. The losses for HC MFs with D_{core} = 320 μm are very high due to the strong coupling of the air core modes with the capillary modes having a lower azimuthal index (Fig.3(right)) in this spectral range with the exception of the HC MF with d_{ins} / d_{out} = 0.85 and n = 2.8.

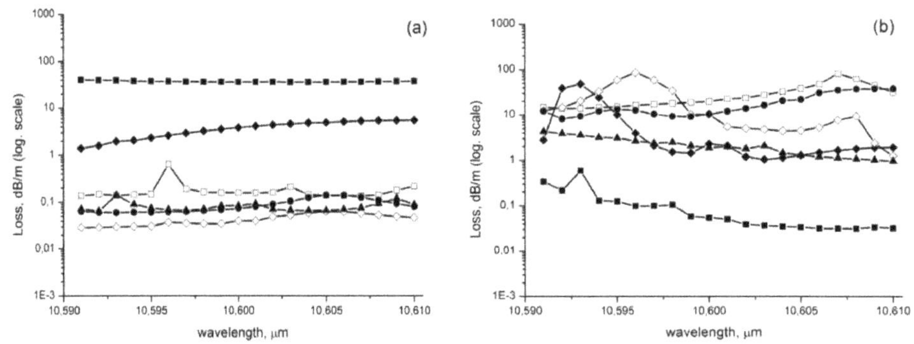

Fig. 4. (a) loss dependence for HC MF with D_{core} = 220 µm and 8 capillaries in the cladding, d_{ins} / d_{out} = 0.9 (n = 2.4, circles), d_{ins} / d_{out} = 0.9 (n = 2.8, triangles), d_{ins} / d_{out} = 0.85 (n = 2.4, rhombuses), d_{ins} / d_{out} = 0.85 (n = 2.8, squares), d_{ins} / d_{out} = 0.8 (n = 2.4, white rhombuses), d_{ins} / d_{out} = 0.8 (n = 2.8, white squares); (b) loss dependence for HC MF with D_{core} = 320 µm the other notations are the same as in (a).

The loss dependencies for HC MFs with ten capillaries in the cladding are shown in Fig. 5. The values of d_{ins}, d_{out} and consequently the capillary wall thicknesses are lower compared to the cases considered above. The loss level for all HC MFs (Fig. 5(a)) is very high due to a strong coupling to the individual capillary leaky modes of both types. Low loss curves correspond only to HC MFs with d_{ins} / d_{out} = 0.8, 0.85 and n = 2.8. The other picture is observed in the case of HC MFs with D_{core} = 320 µm. All loss dependencies have low losses with the exception of HC MF with d_{ins} / d_{out} = 0.8 and n = 2.8 which has strong coupling with the cladding and, consequently, a non propagating regime in this spectral range.

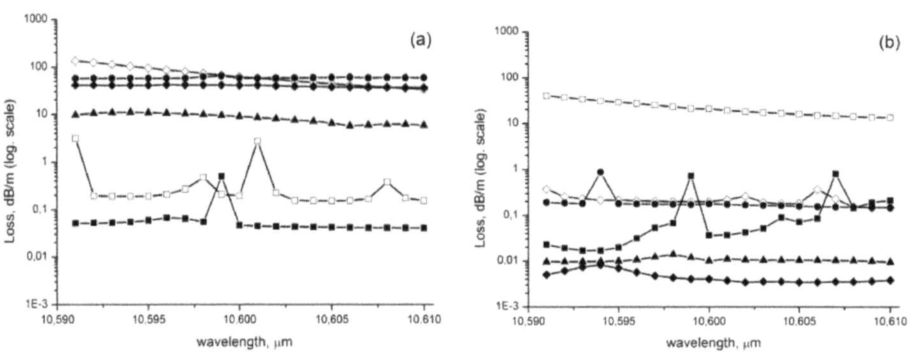

Fig. 5. (a) loss dependence for HC MF with D_{core} = 220 µm and 10 capillaries in the cladding, d_{ins} / d_{out} = 0.9 (n = 2.4, circles), d_{ins} / d_{out} = 0.9 (n = 2.8, triangles), d_{ins} / d_{out} = 0.85 (n = 2.4, rhombuses), d_{ins} / d_{out} = 0.85 (n = 2.8, squares), d_{ins} / d_{out} = 0.8 (n = 2.4, white rhombuses), d_{ins} / d_{out} = 0.8 (n = 2.8, white squares); (b) loss dependence for HC MF with D_{core} = 320 µm the other notations are the same as in (a).

The loss dependencies in Fig. 5 show that a low loss regime for the high index glass HC MFs with the cladding consisting of capillaries can be obtained by the right selection of many parameters characterizing HC MF including D_{core}.

At the end of this subsection, the loss dependencies for HC MFs with the cladding consisting of 12 capillaries will be shown (Fig. 6).

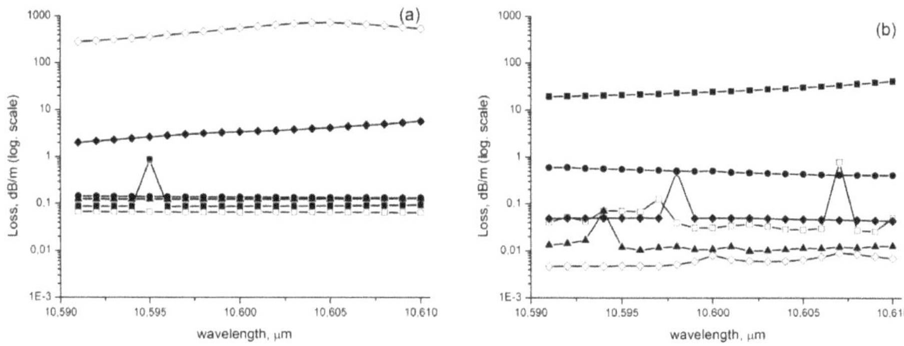

Fig. 6. (a) loss dependence for HC MF with D_{core} = 220 μm and 12 capillaries in the cladding, d_{ins} / d_{out} = 0.9 (n = 2.4, circles), d_{ins} / d_{out} = 0.9 (n = 2.8, triangles), d_{ins} / d_{out} = 0.85 (n = 2.4, rhombuses), d_{ins} / d_{out} = 0.85 (n = 2.8, squares), d_{ins} / d_{out} = 0.8 (n = 2.4, white rhombuses), d_{ins} / d_{out} = 0.8 (n = 2.8, white squares); (b) loss dependence for HC MF with D_{core} = 320 μm the other notations are the same as in (a).

It has thus been shown that the achievement of a low loss waveguide regime for HC MFs with the cladding consisting of capillaries is complicated multi parameter task. All the parameters characterizing the HC MFs such as D_{core}, d_{ins}, d_{out}, n, N (number of the capillaries in the cladding) have an effect on the waveguide regime in the considered spectral range. In this way, two main factors affect the loss level of the HC MFs. The first is the density of eigenstates of the individual capillary and the second is the discrete rotational symmetry of the core boundary. The density of eigenstates of the individual capillary is determined by geometry parameters of a capillary and the value of a glass refractive index. The second factor is connected to the symmetry of the capillary arrangement in the cladding. By comparing the figures in this subsection one can make a conclusion that by decreasing the number of capillaries in the cladding one obtains a stronger dependence on the D_{core}. It seems possible to find a balance between the number of capillaries and the air core diameter. With the increase in the capillary number the role of the discrete rotational symmetry weakens.

4.2 Bend loss dependencies on a bend radius for HC MF with a different number of capillaries in the cladding

In this subsection we will consider characteristics of the bend loss behaviour for HC MFs with a different number of capillaries in the cladding. To reveal the special features of the bend loss one analyses the bend loss behaviour for HC MFs with optimal waveguide regimes found in the previous subsection and, correspondingly, with minimal waveguide losses.

In Fig.7 the bend loss dependencies for two low loss waveguide regimes in the case of HC MF with the cladding consisting of 6 capillaries and 8 capillaries are shown.

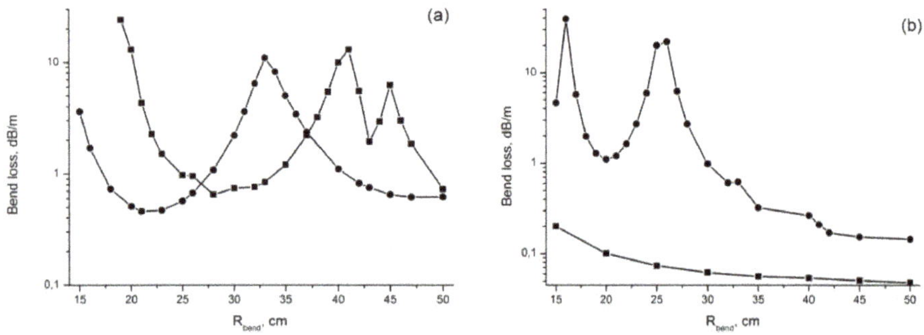

Fig. 7. (a) bend loss dependence on the bend radius for HC MFs with 6 capillaries in the cladding: D_{core} = 220 μm, d_{ins} / d_{out} = 0.9 (n = 2.4, squares) and D_{core} = 320 μm, d_{ins} / d_{out} = 0.9 (n = 2.4, circles); (b) bend loss dependence on the bend radius for HC MFs with 8 capillaries in the cladding: D_{core} = 220 μm, d_{ins} / d_{out} = 0.8 (n = 2.4, squares) and D_{core} = 320 μm, d_{ins} / d_{out} = 0.85 (n = 2.8, circles).

As one can see from Fig. 7(a), the bend loss dependencies have resonance peaks in the case of both values of D_{core}. Just as in the case of HC MF with 8 capillaries in the cladding this resonance behaviour exists only for HC MF with D_{core} = 320 μm (Fig. 7(b)). These resonances are connected with the excitation of capillary eigenstates modes called 'airy' modes [Vincetti&Setti, 2010]. An example of such a resonance tunnelling with the excitation of the 'airy' mode of the individual capillary under bending is shown in Fig. 8 for HC MF with 8 capillaries in the cladding.

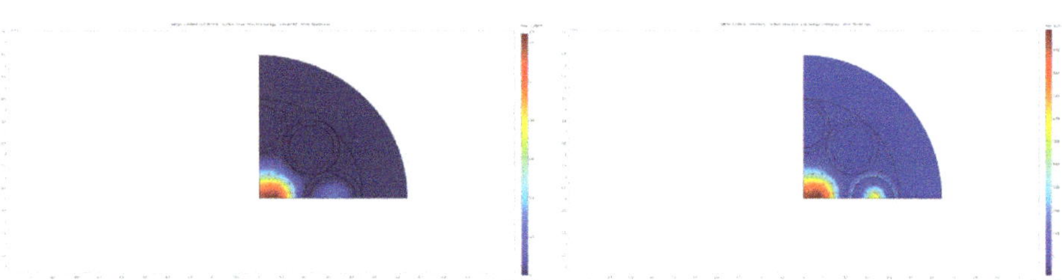

Fig. 8. (left) the air core mode of HC MF with 8 capillaries in the cladding begin to couple with an 'airy' mode of an individual capillary of the cladding at a certain value of the bend radius; (right) the resonance excitation of the 'airy' mode of an individual capillary occurs at a lower value of the bend radius.

Such resonance behaviour of the bend loss occurs due to a very high density of dielectric modes (eigenstates) of an individual capillary at such values of n, d_{ins}, d_{out}. The energy of the air core mode of the HC MFs is tunnelled by these dielectric modes into the capillary 'airy' modes. The higher the values of n, d_{ins}, d_{out} with respect to the wavelength the more effective tunnelling is observed and the excited 'airy' modes of the individual capillary have higher quality factor. For example, the bend loss dependence for HC MF with the cladding consisting of 8 capillaries and D_{core} = 220 μm has no resonance peaks due to suppressing the tunnelling through the capillary walls due to a decrease in the values of d_{ins}, d_{out} or n.

To confirm the above conclusions, bend losses for HC MFs with the cladding consisting of 10 and 12 capillaries were calculated (Fig. 9). As in the case of Fig. 7, HC MFs with the lowest waveguide losses were taken for the bend loss calculations.

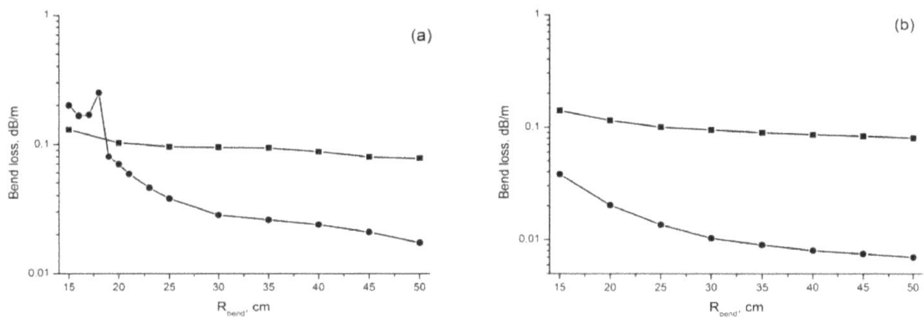

Fig. 9. (a) bend loss dependence on the bend radius for HC MFs with 10 capillaries in the cladding: D_{core} = 220 μm, d_{ins} / d_{out} = 0.85 (n = 2.8, squares) and D_{core} = 320 μm, d_{ins} / d_{out} = 0.85 (n = 2.4, circles); (b) bend loss dependence on the bend radius for HC MFs with 12 capillaries in the cladding: D_{core} = 220 μm, d_{ins} / d_{out} = 0.8 (n = 2.8, squares) and D_{core} = 320 μm, d_{ins} / d_{out} = 0.8 (n = 2.4, circles).

All curves in Fig. 9 have no resonance peaks except for the HC MF with D_{core} = 320 μm, d_{ins} / d_{out} = 0.85 and n = 2.4. In this case, the dielectric capillary mode with a high azimuthal index (Fig. 3(left)) was excited at the bend radius R = 18 cm and a weak tunnelling process into the 'airy' mode occurred. The level of the bend losses for all considered bent HC MFs (Fig. 9) is very close to that of the straight HC MFs.

In conclusion, one can state that the optimal waveguide regime in the spectral region near λ = 10.6 μm for HC MFs made of high index glass (n > 2) is possible at N > 8, where N is a number of capillaries in the cladding. In this case, the process of tunnelling of the air core modes of HC MF into the 'airy' modes of an individual capillary is suppressed due to low quality factor of the 'airy' modes and thus a low loss waveguide regime for a bend becomes possible.

5. Conclusion

The guidance of CO_2 laser radiation in HC MFs with the cladding consisting of capillaries was analysed. Two main factors affecting the waveguide mechanism in these waveguide structures were proposed. The first factor is connected with the representation of the curved air core boundary of the HC MFs as azimuthal diffraction quasi - gratings. These azimuthal diffraction gratings occur due to the cylindrical surface of an individual capillary of the cladding as well as to a discrete rotational symmetry of their arrangement in the cladding. In this way, the air core mode of the HC MF is formed by the interference of space cylindrical harmonics originated from the light scattering on these quasi - diffraction gratings. The process of the air core mode formation is much more complicated compared to the one for the HC MFs with continuous rotational symmetry of the air core boundary. The interaction of cylindrical harmonics forming the air core mode of the HC MF with the capillary walls is different from each other. It leads to a weakened material loss effect on the waveguide regime in comparison with the HC MF with a continuous rotational symmetry of the air core boundary, for example, IC HC PCFs. The second factor is connected with the optical properties of an individual capillary, in particular, with the density of its eigenstates (leaky modes). This factor determines the widths of the transmission regions and the level of waveguide losses. Numerical analyses have shown that a low loss waveguide regime in the spectral region near λ = 10.6 μm for the HC MFs with the cladding consisting of high index glass capillaries becomes possible. The optimisation of the HC MF structure to achieve these regimes is a complicated multiparameter task depending on all geometry parameters characterizing the HC MFs and the value of a glass refractive index. The bend loss of the HC MFs with the cladding consisting of capillaries made of high index glasses has a resonance character depending on the bend radius. To suppress such resonances it is necessary to increase the number of capillaries in the cladding. It leads to a decrease in the capillary sizes and to a decrease in the quality factor of the 'airy' modes of the capillaries.

In the end, we would like to outline the prospects of future investigations in this field. In our opinion, to improve the waveguide properties of the HC MFs it is necessary to study the process of the air core mode formation more carefully. An effect of different types of symmetries of the capillaries arrangements in the cladding and the value of D_{core} on the level of waveguide loss should be investigated. The optical properties of an individual capillary made of high index glass, in particular, its optical eigenstates and the density of these eigenstates depending on the geometry parameters of the capillary and glass refractive indices should be studied. To achieve a low loss guidance it is necessary to perform an optimisation of the HC MFs structure. This complicated optimisation task can be performed by powerful numerical algorithms which have already been applied, for example, to optimisation of the Bragg fibers structures [Biriukov et. al., 2008]. Also, it is necessary to improve the technology of making high index glasses with a low material loss and a technology of the HC MFs fabrication. It is worth mentioning that the early experiments demonstrated the possibility of obtaining waveguide regimes for such HC MFs made of high index chalcogenide glasses at CO_2 laser wavelengths [Kosolapov et. al., 2011]. As the light was well localized in the core, such fibers hold much promise for the delivery of CO_2 laser radiation.

6. Acknowledgment

The authors thank Alexandra Nikolskaya for her assistance in translating this chapter into English.

7. References

Abel, T., Hirsch, J., Harrington, J. (1994). Hollow glass waveguides for broadband infrared transmission. *Optics Letters*, Vol.19, No.14, pp. 1034-1036.

Adler R. B. (1952). Proceedings Of the I. R. E., Vol. 40, p. 339 - 348.

Benabid F., Knight J. C., Antonopoulos G., Russell P. St. J. (2002). Stimulated Raman scattering in hydrogen - filled hollow core photonic crystal fibers. *Science*, Vol. 298, pp. 399 - 402.

Benabid, F., Bowmans, G., Knight, J.C., Russell, P.St.J., Couny, F. (2004). Ultrahigh efficiency laser wavelength conversion in a gas - filled hollow core photonic crystal fiber by pure stimulated rotational Raman scattering in molecular hydrogen. *Physical Review Letters*, Vol.93, No.12, pp. 123903-4.

Biriukov A. S.; Bogdanovich D. V.; Gaponov D. A.; Pryamikov A. D. (2008). Optical properties of Bragg fibers. *Quantum Electronics*, Vol.38, pp. 620 - 633.

Birks T. A., Roberts P. J., Russell P.S., Atkin D. M., and Shepherd T. J. (1995). Full 2- D photonic band gaps in silica/air structures. *Electron. Letters*, Vol. 31, p. 1941 - 1943.

Bowden B. F.&Harrington J. A. (2009). Fabrication and characterization of chalcogenide glass for hollow Bragg fibers. *Applied Optics*, Vol.48, pp. 3050 - 3054.

Croitoru N ., Saba K., Dror J., Goldenberg E., Mendelovic D., and Israel G. (1990). U.S. Patent 4.390.863.

Croitoru N., Dror J., and Gannot I. (1990). Characterization of hollow fibers for the transmission of infrared radiation. *Appl. Optics*, Vol. 29, p. 1805 - 1809.

Deseveday F.; Renversez G.; Troles J.; Houizot P.; Brilland L.; Vasilief I.; Coulombier Q.; Traynor N.; Smektala F.; Adam J -L. (2010). Chalcogenide glass hollow core photonic crystal fibers. *Optical Materials*, Vol.32, pp. 1532 - 1539.

Fano U. (1961). Effects of configuration interaction on intensities and phase shifts. *Phys. Review*, Vol.124, pp. 1866 - 1878.

Fevrier S.; Jamier R.; Blondy J -M .; Semjonov S. L.; Likhachev M. E.; Bubnov M. M.; Dianov E. M.; Khopin V. F., Salganskii M. Y., Guryanov A. N. (2006). Low loss singlemode large mode area all silica photonic band gap fiber. *Optics Express*, Vol.14, pp. 562 - 569.

Garmire, E., McMahon, T., Bass, M. (1976). Propagation of infrared light in flexible hollow waveguides. *Applied Optics*, Vol.15, No.1, pp.145 -150.

Garmire, E., McMahon, T., Bass, M. (1979). Low-loss propagation and polarization rotation in twisted infrared metal waveguides. *Appl. Phys. Letters*, Vol.34, No.1, pp. 35-37.

Gregan R. F., Mangan B. J., Knight J. C., Birks T.A., Russell P. St. J., Roberts P. J., and Allan D.C. (1999). Sinle - mode photonic band gap guidance of light in air. *Science*, v. 285, pp. 1537 - 1539.

Harrington J.A., Gregory, C.C. (1990). Hollow sapphire fibers for the delivery of CO_2 laser energy. *Optics Letters*, Vol.15, No.10, pp.541-543.

Hessel A.&Oliner A. A. (1965). A new theory of Wood's anomalies on optical gratings. *Applied Optics*, Vol.4, pp. 1275 - 1297.

Hongo, A., Morosawa, K., Matsumoto, K., Shiota, T., Hashimoto, T. (1992). Transmission of kilowatt - class CO_2 laser light through dielectric - coated metallic hollow waveguides for material processing. *Applied Optics*, Vol.31, No.24, pp. 5114-5120.

Hu J.&Menyuk C. R. (2007). Leakage loss and bandgap analysis in air - core photonic bandgap fiber for nonsilica glasses. *Optics Express,* Vol.15, pp. 339 - 349.

Humbert G., Knight J. C., Bowmans G., Russell P. St. J., Williams D. P., Roberts P. J., and Mangan B. J. (2004), Hollow core photonic crystal fibers for beam delivery. *Opt. Express,* Vol. 12, pp. 1477 - 1484.

Karasawa S., Miyagi M., Nakamura T., Ishikawa H. (1990). Fabrication of dielectric - coated waveguides for CO_2 laser light transmission. *Electronics and Communications in Japan (Part 2: Electronics),* Vol. 73, Issue 6, pp. 65 – 69.

Kosolapov A.; Pryamikov A.; Biriukov A.; Astapovich M.; Shiryaev V.; Snopatin G.; Plotnichenko V.; Churbanov M.; Dianov E. (2011). Demonstration of CO_2 - laser power delivery through chalcogenide - glass fiber with negative - curvature hollow core. *Optics Express,* Vol. 19, pp. 25723 - 25728.

Laakmann K. (1987). U.S. Patent 4.688.893.

Lind A. C.&Greenberg J. M. (1966). Electromagnetic scattering by obliquely oriented cylinders. *Journal of applied physics,* Vol.37, pp. 3195 - 3203.

Litchinitser N. M.; Abeeluck A. K.; Headley C.; Eggleton B. J. (2002). Antiresonant reflecting photonic crystal optical waveguides. *Optics Letters,* Vol.27, pp. 1592 - 1594.

Lord Rayleigh (1907). Note on the remarkable case of diffraction spectra described by Prof. Wood. *Philosophical magazine Series 6,* Vol.14, pp. 60 – 65.

Marcatili E. A.&Schmeltzer R. A. (1964). Hollow metallic and dielectric waveguides for long distance optical transmission and lasers. *The Bell system technical journal,* Vol. 43, pp. 1783 - 1809.

Mashida H., Nishimura A., Ishikawa M., and Miyagi H. (1991). Low loss lead fluoride - coated square waveguides for CO_2 laser light transmission. *Electr. Letters,* Vol. 27, pp. 2068 - 2070.

Miyagi, M., Kawakami, S. (1984). Design Theory of Dielectric-Coated Circular Metallic Waveguides for Infrared Transmission. *Journal of Lightwave Technology,* Vol.LT-2, No.2, pp.116-126.

Miyagi, M., Hongo, A., Yoshizo, A. & Kawasami. S. (1983). Fabrication of germanium-coated nickel hollow waveguides for infrared transmission. *Appl. Phys. Letters,* Vol.43, No.5, pp.430-432.

Morrow C. and Gu G. (1994). *Soc. Photo – Opt. Instrum. Eng.*, v. 2131, p. 18.

Nishihara, H.; Inoue, T., Koyama, J. (1974). Low-loss parallel-plate waveguide at 10.6 μm. *Appl. Phys. Letters,* Vol.25, No.7, pp.391-393.

Nubling, R.K., Harrington, J.A. (1996). Hollow-waveguide delivery systems for high-power, industrial CO2 lasers. *Applied Optics,* Vol.34, No.3, pp. 372-380.

Pearce, G.J., Pottage, J.M., Bird, D.M., Roberts, P.J., Knight, J.C., Russell, P.St.J. (2005). Hollow-core PCF for guidance in the mid to far infra-red, *Optics Express,* Vol.13, No.18., pp. 6937-6946.

Pearce G. J.; Wiederhecker G. S.; Poulton C. G.; Burger S.; Russell P. St. J. (2007). Models for guidance in kagome structured hollow core photonic crystal fibres. *Optics Express,* Vol. 15, pp. 12680 - 12685.

Pottage, J.M., Bird, D.M., Hedley, T.D., Birks, T.A., Knight, J.C., Russell, P.St.J. (2003). Robust photonic band gaps for hollow core guidance in PCF made from high index glass, *Optics Express,* Vol.11, No.22, pp.2854-2861.

Pryamikov A. D.; Biriukov A. S.; Kosolapov A. F.; Plotnichenko V. G.; Semjonov S. L.; E. M. Dianov. (2011). Demonstration of a waveguide regime for a silica hollow core microstructured optical fiber with a negative curvature of the core boundary in the spectral region > 3.5 μm. *Optics Express,* Vol.19, pp. 1441 - 1448.

Pryamikov A.D., to be prepared.

Roberts P. J., Couny F., Sabert H., Mangan B. J., Williams D. P., Farr L., Mason M. W., Tomlinson A., Birks A., Knight J.C., Russell P. St. J. (2005). Ultimate low loss of hollow - core photonic crystal fibers. *Opt. Express,* Vol. 13, pp. 236 - 244.

Roberts, P.J., Williams, D.P., Mangan, B.J., Sabert, H., Couny, F., Wadsworth, W.J., Birks, T.A., Knight, J.C., Russell, P.St.J. (2005). Realizing low loss air core photonic crystal fibers by exploiting an antiresonant core surround, *Optics Express,* Vol.13, No.20, pp. 8277-8285.

Saitoh K.; Mortensen N. A.; Koshiba M. (2004). Air core photonic band gap fibers: the impact of surface modes. *Optics Express,* Vol. 12, pp. 394 - 400.

Shaw, L.B., Sanghera, J.S., Aggarwal, I.D., Kung, F.H. (2003). As-S and As-Se based photonic band gap fiber for IR laser transmission, *Optics Express,* Vol.11, No.25, pp.3455-3460.

Shephard, J.D., MacPherson, W.N., Maier, R.R.J., Jones, J.D.C., Hand, D.P., Mohebbi, M., George, A.K., Roberts, P.J., Knight, J.C. (2005). Single-mode mid-IR guidance in a hollow-core photonic crystal fiber, *Optics Express,* Vol.13, No.18. pp.7139-7144.

Smith C. M.; Venkataraman N.; Gallagher M. T.; Muller D.; West J. A.; Borrelli N. F.; Allan D. C.; Koch K. W. (2003). Low - loss hollow - core silica/air photonic bandgap fibre. *Nature,* Vol.424, pp. 657 - 659.

Skorobogatiy M.&Yang J. (2009). *Fundamentals of Photonic Crystal Guiding,* Cambridge University Press, ISBN - 10: 0521513286.

Snyder A. W. (1971). Continuous mode spectrum of a circular dielectric rod. *IEEE Transactions on Microwave Theory and Techniques,* Vol.MTT - 19, pp. 720 - 727.

Steinvurzel P.; de Sterke C. M.; Steel M. J.; Kuhlmey B. T.; Eggleton B. J. (2006). Single scatterer Fano resonances in solid core photonic band gap fibers. *Optics Express,* Vol.14, pp. 8797 - 8811.

Temelkuran B.; Hart S. D.; Benoit G.; Joannopoulos J. D.; Fink Y. (2002). Wavelength - scalable hollow optical fibres with large photonic bandgap for CO_2 laser transmission. *Nature,* Vol. 420, pp. 650 - 653.

Wait J. R. (1955). Scattering of a plane wave from a circular dielectric cylinder at oblique incidence. *Canadian Journal of Physics,* Vol. 33, pp. 189 - 195.

Wang Y.; Couny F.; Roberts P. J.; Benabid F. (2011). Low loss broadband transmission in hypocycloid - core kagome hollow core photonic crystal fiber. *Optics Letters,* Vol. 36, pp. 669 - 671.

West, J.A., Smith, C.M., Borrelli, N.F., Allen, D.C., Koch, K.W. (2004). Surface modes in air-core photonic band-gap fibers, *Optics Express,* Vol.12, No.8, pp.1485-1491.

White T. P.; Kuhlmey B. T.; McPhedran R. C.; Maystre D.; Renversez G.; de Sterke C. M.; Botten L. C. (2002). Multipole method for microstructured optical fibers. *Journal of the Optical Society of America B,* Vol.19, pp. 2322 - 2330.

Wood R. W. (1902). On remarkable case of uneven distribution of light in a diffraction grating spectrum, *Philosophical magazine Series 6,* Vol. 4, pp. 396 - 402.

Vincetti L.; Setti V. (2010). Waveguide mechanism in tube lattice fibers. *Optics Express*, Vol.18, pp. 23133 - 23146.

Yeh P.; Yariv A.; Marom E. J. (1978). Theory of Bragg fiber. *Journal of the Optical Society of America*, Vol.68, pp. 1196 - 1201.

8

Heterodyne Interferometer for Measurement of Electron Density in High-Pressure Plasmas

Keiichiro Urabe and Kunihide Tachibana
Kyoto University
Japan

1. Introduction

Conventional material processes using plasmas generated in low pressure gaseous media are recently being transposed to high-pressure plasma processes, because of the potential of high-pressure plasmas to reduce costs for vacuum systems in industrial applications. In many kinds of high-pressure plasma sources, small-scale atmospheric-pressure plasmas (APPs) having a property of thermal non-equilibrium, have been especially attracting much interest of researchers over the last 20 years (Becker, 2005). In such a high pressure gaseous medium, generating the small-scale plasmas in mm or μm order is effective to keep its ignition voltage low and discharge behavior stable, following a famous rule on discharge ignition called a "Paschen's law" (Paschen, 1889; von Engel, 1994). Using these small-scale high-pressure plasmas, localized maskless processes, for example etching (Ichiki et al., 2004) and deposition (Babayan et al., 1998), have been reported. Also, nanomaterial synthesis was realized by the APP utilizing their property of short residence time of source particles inside the plasma (Nozaki et al., 2007). In addition to the inorganic solid material processes, process objects of the APPs are spreading toward liquids (Bruggeman & Leys, 2009) and biocells (Kong et al., 2009) which cannot present in low-pressure conditions.

In characterization and comparison of plasma properties, electron density is one of the most important parameters. This is because that electrons play a major role for carrying external energy to heavy particles inside the plasma, and all other excited species can be calculated theoretically from the plasma parameters of electron density, electron energy distribution, and gas composition. Diagnostics of electron density have in the APPs have been reported by many researchers using Langmuir probe methods (Chang, 1973; Chang & Laframboise, 1976), Stark broadening measurement (Laux et al., 2003), and laser Thomson scattering measurement (Kono & Iwamoto, 2004). However, these methods have limitations for the APP measurements due to their finite sensitivities, expected perturbations or interferences, spatiotemporal resolutions, etc. For instance, the Langmuir probe is difficult be applied to the small-scale APPs since its theory in a collision dominant condition is not well developed and there are discharge perturbations by the metallic probe. The Stark broadening spectroscopic method enables us to derive electron density only for over 5×10^{13} cm^{-3} due to the large pressure broadening superposed over the Lorentzian shape (Laux et al., 2003). The laser Thomson scattering method is not applicable to molecular gases although its spatial resolution is high enough, because large Raman scattering components overlap with the Thomson scattering signal.

For the purpose of diagnostics in the small-scale high-pressure plasmas, refractive-index measurement using electromagnetic (EM) waves is appropriate since they can provide plasma property's information with negligible perturbation to the plasma. Also, in the refractive-index measurement, we can ignore excitation and scattering processes that are largely dependent on gas compositions and densities, which are indispensable for the above listed other measurement methods. For typical APPs whose electron densities range from 10^{12} to 10^{15} cm^{-3}, microwaves and millimeter-waves are suitable to detect absorption of the probing EM waves and derive the electron density from the absorption ratio (Tachibana et al., 2005(a), 2005(b); Sakai et al., 2005; Ito et al., 2010). However, diagnostics using those EM waves cannot have good spatial resolutions for the small-scale plasmas because of their diffraction limits. Meanwhile, a heterodyne interferometer of CO_2 laser beam, which is a theme in this chapter, can be in a category of the refractive-index measurement methods and have a good spatial resolution at the same time. This interferometer detects the laser beam's phase shift caused by the presence of tested plasma and provides line-integrated information of electron density with a spatial resolution in sub mm order (Leipold et al., 2000; Choi et al., 2009).

This chapter is focused on descriptions of the CO_2-laser heterodyne interferometer for the measurement in small-scale high-pressure plasma sources, because detailed descriptions of the interferometer for the low-pressure plasmas have been written in other chapters and textbooks (*for example* Hutchinson, 2002). In high-pressure plasmas, large contribution of gas-particle density (atoms and/or molecules in ground states) to the change of the refractive index is expected due to Joule heating in the discharge region, and this must be accurately separated from the signal in order to derive the absolute value of electron density. Therefore, we firstly explain how to divide the two components in the CO_2-laser beam's phase shift, which are the phase shifts due to electron generation and gas heating. Then, the fundamental properties of our CO_2-laser heterodyne interferometer, for example spatial resolution and lower limit of electron-density detection, are verified reviewing experimental measurements of the small-scale APPs driven by DC applied voltages (Choi et al., 2009). Finally, a combination measurement method composed of the CO_2-laser heterodyne interferometer and a millimeter-wave transmission method is introduced as a solution of spatiotemporally resolved electron-density measurement in small-scale APPs with high-speed temporal evolution of electron density (Urabe et al., 2011).

2. Fundamentals of heterodyne interferometer

This section includes brief introduction of a Mach-Zehnder principle and a heterodyne technique used in our interferometer and theoretical descriptions of the phase shift in the CO_2 laser beam induced by electrons in a tested plasma source, and explanations how to derive electron density in small-scale high-pressure plasmas eliminating influence of gas heating from total phase shift signals.

2.1 Mach-Zehnder heterodyne interferometer

Measurements of refractive index in tested materials are often done by some forms of interferometer. Most of interferometers are in Michelson, Fabry-Perot, and Mach-Zehnder configurations. Mach-Zehnder interferometer is a two-beam interferometer having two paths in which the laser beams travel in only one direction. The original laser beam is separated by a beam splitter, and the beam's phase change is caused by variations of the refractive index of the tested material placed only in one path. Difference of the refractive indexes between the two paths is derived from the merged beam by a second beam splitter.

A heterodyne technique is a reducing method of observing frequency from light frequency to external oscillator frequency in a instrumentally manageable range. When this method is applied to the Mach-Zehnder interferometer, the light frequency in one of two paths is modulated, and in the other path the frequency is not modulated but its phase is shifted by the tested material. After merging the two laser beams, a beat signal at the modulation frequency in the merged beam can be isolated from the original output signal of a solid-state detecting device, $I(t)$, which is expressed in a following equation.

$$I(t) = \left| U_1(t) + U_2(t) \right|^2_{\omega=0} = E_1^2 + E_2^2 + E_1 E_2 \cos(\Delta\omega t - \Delta\Phi), \tag{1}$$

where $U_1(t) = E_1\cos(\omega t + \Delta\omega t)$ is the electric field of the laser beam passing through the frequency modulating device with a modulation frequency at $\Delta\omega$, $U_2(t) = E_2\cos(\omega t + \Delta\Phi)$ is the electric field of the beam passing through the tested material, and $\Delta\Phi$ is the phase shift of the beam by the tested material. Components in a light frequency range are automatically eliminated from the signal because of the response-speed limitation of detectors. From the output signal, $I(t)$, we can derive the phase shift by the tested material $\Delta\Phi$ using a phase detecting device and a reference signal which is separated from the frequency-modulation signal.

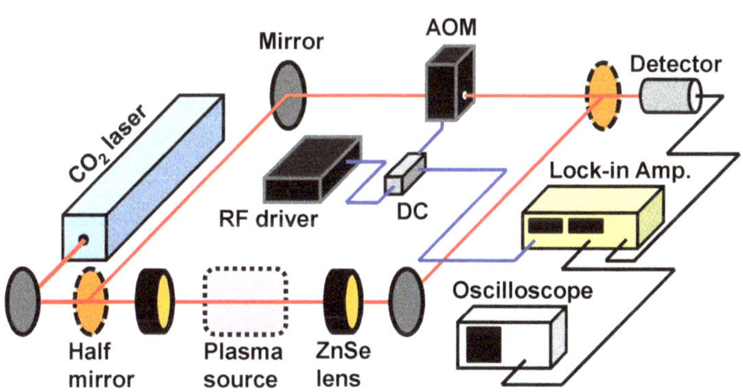

Fig. 1. Experimental setup of CO_2-laser heterodyne interferometer in Mach-Zhender principle for measurement of electron density inside high-pressure plasma source. AOM is acousto optical modulator shifting CO_2-laser frequency at frequency of RF driver's signal. DC is directional coupler taking small amplitude of RF signal driving AOM to input reference signal to lock-in amplifier.

A practical experimental setup of the CO_2-laser heterodyne interferometer used in our studies is shown in Fig. 1. The original CO_2 laser beam was split by a ZnSe half mirror. The beam frequency in one path was shifted an acousto-optical modulator (AOM) whose modulation frequency $\Delta\omega$ = 40 MHz. In the other path, the beam was focused on a tested plasma source by a pair of ZnSe lens, and the beam phase was shifted by the tested plasma. These two beams were superposed again at another ZnSe half mirror, and their beat signal was detected by a HgCdTe IR detector operated at room temperature. The output signal of the IR detector, $I(t)$, was put into a lock-in amplifier with the reference signal, $\cos(\Delta\omega t)$, which was divided from a RF signal driving the AOM by a directional coupler (DC). The lock-in amplifier outputs a signal proportional to the phase shift, $\Delta\Phi$, and its temporal evolution was recorded in an oscilloscope.

2.2 Phase shift of CO_2 laser beam by electrons inside plasmas

In refractive-index measurement inside a tested plasma source having no gas temperature fluctuation, for example plasmas generated in gaseous media at very lower pressure than atmosphere, the phase shift by tested plasma, $\Delta\Phi$, measured using the Mach-Zehnder heterodyne interferometer becomes

$$\Delta\Phi = \int (k_{\text{plasma}} - k_0)\,dl = \int (N_{\text{plasma}} - 1)\frac{2\pi}{\lambda}\,dl \,, \tag{2}$$

where k_{plasma} and k_0 are the wavenumbers of the plasma and vacuum, N_{plasma} is the refractive index of the plasma, and λ is the wavelength of the CO_2 laser beam. Absolute electron density inside the plasma, n_e (m), can be derived using a following relationship.

$$N_{\text{plasma}} = \sqrt{1 - \omega_{\text{plasma}}^2 / \omega^2} \approx 1 - \frac{1}{2}\left(\omega_{\text{plasma}}^2 / \omega^2 \right) = 1 - \frac{1}{2}\frac{n_e e^2}{m_e \varepsilon_0}\frac{\lambda^2}{4\pi^2 c^2} \,, \tag{3}$$

where ω_{plasma} and ω are the electron plasma frequency and the angular frequency of laser beam, m_e is the mass of electron, ε_0 is the permittivity of vacuum, c is the velocity of light (Hutchinson, 2002). We used an approximate expansion in this calculation because electron density is sufficiently smaller than gas-particle density in weakly ionized plasmas generated in laboratories. When we assume that electron density inside the tested plasma is spacially homogeneous and its length along the laser path is d (m), the relationship between the phase shift and the electron density becomes

$$\Delta\Phi = (N_{\text{plasma}} - 1)\frac{2\pi}{\lambda}d = -\frac{e^2\lambda}{4m_e\varepsilon_0\pi c^2}n_e d \approx -\left(3.0\times10^{-20}\right) n_e d \ \text{(rad)}, \tag{4}$$

in the CO_2-laser heterodyne interferometer at $\lambda = 10.6$ μm.

2.3 Electron-density measurement in high-pressure plasmas

In this subsection, influence of dense gas particles in high-pressure plasmas on the calculation of electron density from the phase shift of the CO_2 laser beam, which can be ignored in the measurement of low-pressure plasmas, is introduced. It should be noted that increase of gas temperature corresponding to decrease of gas-particle density is promoted in the high-pressure plasmas because of high collision frequencies between electrons and gas particles. This change of gas-particle density results in the change of the refractive index in gaseous medium. The refractive index of the gaseous medium, N_{gas}, is

$$N_{\text{gas}} = 1 + A\left(1 + \frac{B}{\lambda^2}\right)\frac{n_g}{n_{g0}} \,, \tag{5}$$

where A and B are the specific values for gas species, n_{g0} is the gas particle density under the standard temperature and pressure (STP) conditions, and n_g is the gas-particle density inside the tested plasma source (Allen, 1973). Considering the influence of gas-particle density on the refractive index, the phase shift of the CO_2 laser beam in the high-pressure plasmas is described by two components of the electron density, $\Delta\Phi_e$, and the gas-particle density, $\Delta\Phi_g$, as a following equation.

$$\Delta\Phi = \Delta\Phi_e + \Delta\Phi_g = -\frac{e^2\lambda}{4m_e\varepsilon_0\pi c^2}n_e d + \frac{2\pi A}{\lambda n_{g0}}\left(1+\frac{B}{\lambda^2}\right)n_g d. \tag{6}$$

Table 1 lists change directions of the phase shift in the CO_2-laser heterodyne interferometer. Whereas we used the gas-particle density n_g for the calculation in Eq. (6), $\Delta\Phi_g$ can be also expressed using gas temperature in the plasma. Each component of the phase shift in the high-pressure plasma, $\Delta\Phi_e$ and $\Delta\Phi_g$, changes to negative direction due to the electron production and the decrease of gas-particle density by the Joule heating. The separation of these two components becomes a crucial problem for accurate derivation of the electron density in practical measurements of the high-pressure plasmas, because these two components are only measured together in the heterodyne interferometer and start decreasing at the same timing.

	Increase	Decrease
Electron density	Negative	Positive
Gas-particle density	Positive	Negative
Gas temperature	Negative	Positive

Table 1. Change directions of CO_2-laser beam's phase shift by changes of electron density, gas-particle density, and gas temperature, when they increase or decrease.

For the separation of two components in the phase shift of the CO_2 laser beam, difference of time constants of the two phenomena is utilized, because the change of electron density is much faster than that of gas temperature in the high-pressure discharge (Leipold et al., 2000; Choi et al., 2009). Applicability of this technique can be confirmed by experimental evidence as explained below. Also, in general, the two terms of electron density, n_e, and gas-particle density, n_g, can be derived directly by solving a system of two equations using a two-wavelength heterodyne interferometer, because the phase-shift components due to the electron and gas-particle densities depend differently on the laser wavelength. Some groups have tried to measure both electron and gas-particle densities inside the plasma by an additional heterodyne interferometer using a He-Ne laser beam at 633 nm (Adler & Kindel, 2003; Acedo et al., 2004).

Figure 2 shows an example waveform of the output phase-shift signal of the lock-in amplifier obtained in a small-scale high-pressure plasma source operated at 70 Torr driven by a pulsed DC voltage at 300 V with a 300-μs pulse duration. The phase-shift signal recorded in the oscilloscope is proportional to the phase shift in a specification ratio of the lock-in amplifier. Around 50 μs after the discharge ignition, the phase shift decreased rapidly in 150 μs, and then a slower change followed until the discharge termination. The 50-μs delay of the phase-shift signal is due to calculating delay time of the phase shift in the lock-in amplifier. The initial faster falling part is attributed to the increase of electron density, and the second slower part is to the decrease of gas-particle density by the Joule heating. Faster and slower slopes of the phase shift similar to the ignition timing were also observed in the termination timing. Subtracting the phase-shift component of gas-particle density, $\Delta\Phi_g$, shown in a blue dashed-dotted curve in Fig. 2, the component of electron density, $\Delta\Phi_e$, was obtained as shown in a red dashed curve. The absolute value of electron density can be derived from the amplitude of electron-density component.

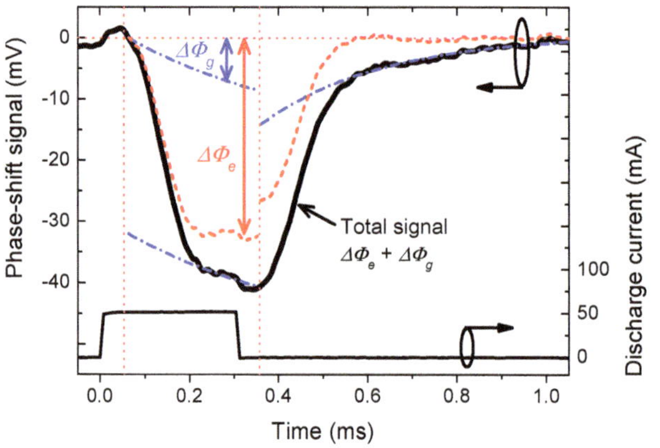

Fig. 2. Temporal evolutions of phase-shift signal from lock-in amplifier and discharge current. Tested plasma source is small-scale discharge driven by 300-V pulsed DC high voltage at 70 Torr of He gas (Choi et al., 2009).

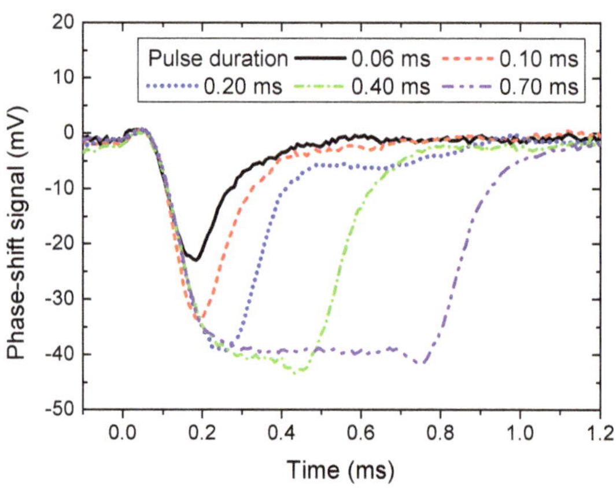

Fig. 3. Temporal evolutions of phase-shift signal measured changing voltage pulse duration from 60 to 700 μs (Choi et al., 2009).

As explained in the previous paragraph, because the temporal changes of electron and gas-particle density can be observed only in the discharge ignition or termination timings, there is a need to pulse modulation of continuous discharge for the separation of phase-shift components measured in the CO_2-laser heterodyne interferometer. In order to understand appropriate range of duration times for the pulse modulation, especially minimum duration time, it is important to measure the phase-shift signal decreasing the modulating pulse duration. Figure 3 shows waveforms of the phase-shift signal measured in our heterodyne interferometer using five pulse duration times ranging from 60 to 700 μs. From the measured waveforms, it could be confirmed that the pulse modulation with the duration time shorter than 200 μs does not give us accurate information of electron density in our setup, and we used the modulation pulse with duration times larger than 500 μs in our studies. This temporal behavior of the output signal of the heterodyne interferometer is of course different in each IR detector and phase measuring system.

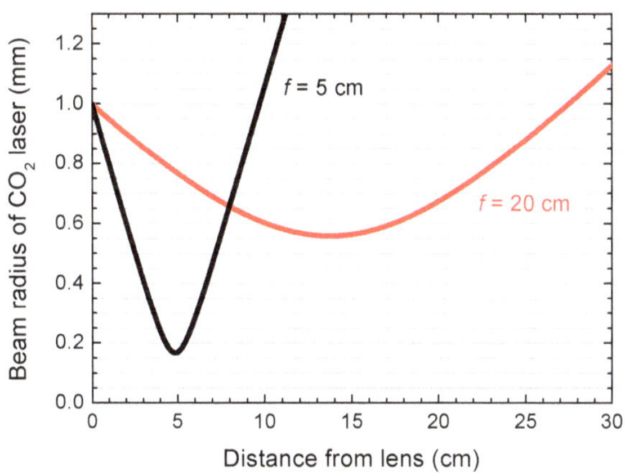

Fig. 4. Calculated beam spot radii of CO_2 laser beam at 10.6 μm as a function of distance from lens (Yariv, 1997). Initial spot radius at 1.0 mm and focal lengths of lens at 5 and 20 cm correspond to practical conditions used in our experiments.

For measuring the spatial distributions of electron density, evaluation of the spot radius of the CO_2 laser beam inside the tested plasma source is required. In order to measure the spatial distribution of electron density inside the high-pressure plasmas whose scale are usually below cm order, the CO_2-laser beam has to be focused in μm order using a pair of ZnSe lens as shown in Fig. 1. Assuming that the laser beam is a Gaussian beam, we can calculate a profile of the beam spot radius along the laser path from the beam parameter and focal length of the ZnSe lens using an ABCD matrix analysis (Yariv, 1997). Figrue 4 shows the beam-radius profiles after the two kinds of ZnSe lens used in our experimental study. The location and length of the tested plasma source and a scanning pitch of the laser beam should be determined by reference to these kinds of beam-profile evaluations.

3. Spatial distribution of electron density in high-pressure plasmas

Some of the specific measurement results of our CO_2-laser heterodyne interferometer are introduced in this section. Influence of gas heating on detected phase-shift signals, spatial resolutions of the heterodyne interferometer, and the combination measurement method for AC-voltage driven APPs, will be discussed in addition to the measured electron-density distributions in the tested plasma sources.

3.1 Small-scale DC discharge at high pressure

A short hollow cathode (HC) discharge tube shown in Fig. 5(a) was used for basic experiments of our CO_2-laser heterodyne interferometer system with variable pressure in pure He gas. The results can be good references to measurement of other plasmas operated at atmospheric pressure introduced in following subsections. Two electrodes for anode and cathode had a 2-mm bore and a 4-mm hole length, and these electrodes were separated by a 1.5-mm thick ceramic disk also having the 2-mm bore. Two of ZnSe windows were equipped to seal the discharge region and transmit the CO_2 laser beam. Gas pressure inside the small chamber was controlled in a range from several tens to hundreds Torr.

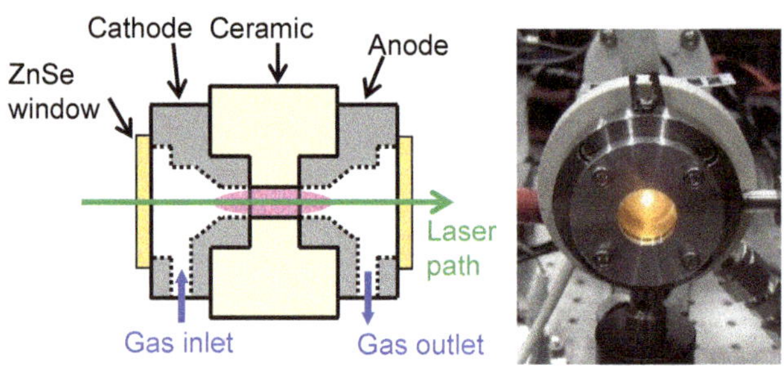

(a)

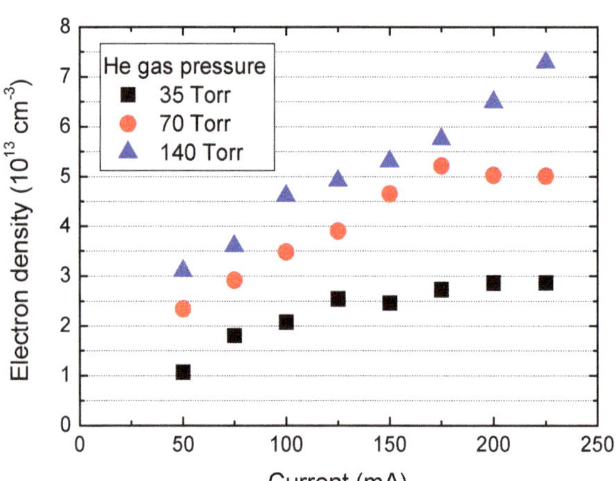

(b)

Fig. 5. (a) Cross-sectional diagram and photograph of short hollow cathode (HC) discharge. (b) Dependence of electron density on discharge current in the short HC discharge in He gas at three values of gas pressure.

Figure 5(b) shows dependences of electron density in the short HC discharge on the discharge current measured at several He gas pressures. In the calculation of electron density from the phase shift signal, we used two assumptions that the plasma was uniform along the CO_2 laser path and its effective length was 9.5 mm. The electron density in the discharge increased monotonically with the increase of the discharge current and the gas pressure. This result indicates that the CO_2-laser heterodyne interferometer is able to measure appropriate dependence of electron density on the gas pressure, because the electron density inside the high-pressure plasma must be increased at higher pressure by the slower drift velocity of electrons due to frequent collisions with gas particles when the discharge current is the same as that at lower pressure.

Here, the minimum sensitivity of electron density in our CO_2-laser heterodyne interferometer is introduced. The minimum electron density which we recorded in the short HC discharge was corresponds to a line integrated electron density $n_e d$ of 7×10^{12} cm^{-2} with a detected phase shift around 0.1 degree (Choi et al., 2009). However, the signal to noise ratio, for instance shown in Fig. 3, showed that the minimum sensitivity should be at least six times better than that. Therefore, it could be estimated that the minimum detectable phase shift in our system was about 0.02 degree, corresponding to $n_e d$ of 1×10^{12} cm^{-2}, which is much smaller than the minimum sensitivity of Stark broadening measurement (Laux et al, 2003).

3.2 Atmospheric-pressure DC discharge in open space

Figure 6(a) shows a schematic diagram of DC discharge in open space operated in a He gas flow ejected from a tubular cathode to a pin anode. The cylindrical tubular cathode, whose outer and inner diameters were 3.1 and 2.1 mm, and the pin anode with a 0.5-mm diameter were faced to the cathode with a 3-mm gap. The measurement point between the cathode and the anode was scanned by putting the whole discharge device on a three-dimensional mechanical movement stage. The influence of temporal evolutions of the gas-particle density on the measured phase shift can be seen in Fig. 6(b) showing the phase-shift signals observed in the atmospheric-pressure DC discharge using two different flow rates of He gas (1 and 2 L/min). This measurement result clearly suggests that the contribution of the Joule heating on temporal evolution of the phase shift becomes much greater in the discharge operated at atmospheric pressure and the gas-particle component in the phase shift, $\Delta\Phi_g$, has large dependence on the gas flow velocity. The fast time constant and large amplitude of $\Delta\Phi_g$ in both rising and falling periods of applied voltage in lower gas flow rate was due to less cooling effect by neutral gas particles.

In order to obtain radial distributions of electron density in the high-pressure plasmas, which cannot be evaluated by photographic observations, spatial distributions of line-integrated electron density and inverse Abel transformation of the distributions are required. This diagnostic method of radial distribution is also applied for excited species measurements in the high-pressure plasmas by a laser absorption spectroscopy (LAS) and a laser induced fluorescence (LIF) methods (Urabe et al., 2010). Figure 7(a) shows two spatial distributions of the line-integrated electron density, $\int n_e dl(x)$, measured near the tubular cathode and the pin anode in the DC discharge scanning the laser beam perpendicularly to the gas-flow axis. Using the CO_2-laser heterodyne interferometer with appropriate control of the beam radius around the tested plasma source, the spatial distribution of electron density with enough quality for the inverse Abel transformation can be measured.

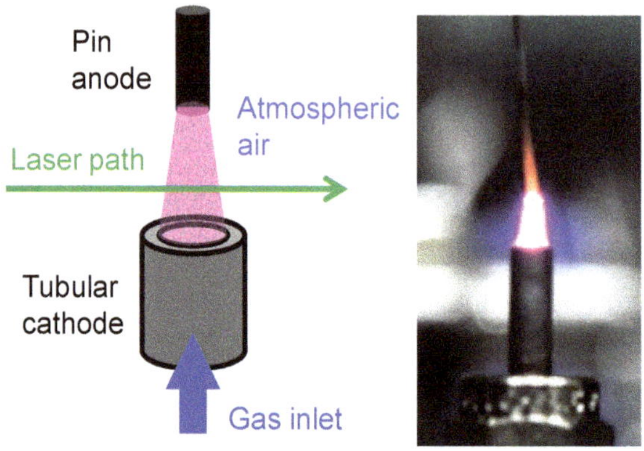

(a)

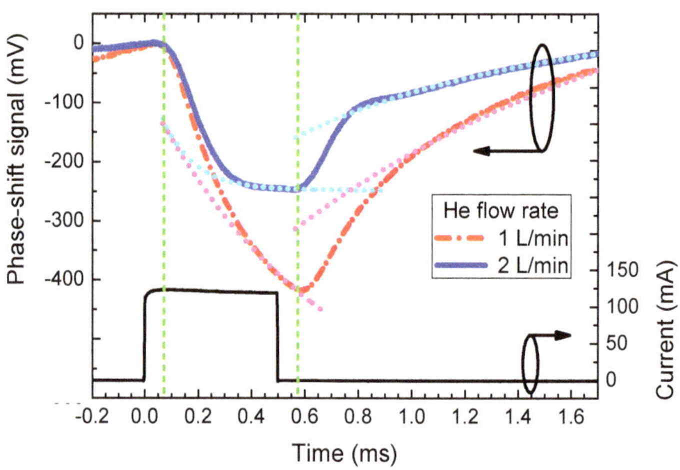

(b)

Fig. 6. (a) Schematic diagram and photograph of atmospheric-pressure DC discharge in open space. (b) Temporal evolutions of phase-shift signal in the DC discharge measured changing He gas flow rate.

Calculation results of the inverse Abel transformation corresponding to the radial distribution of electron density, $n_e(r)$, at the measurement points in the DC discharge are shown in Fig. 7(b). The calculated radial distributions showed two different structures which were a hollow shape near the cathode and a center-peaked shape near the anode, having a good agreement with the electrode structures. The inverse Abel transformation was performed using a following equation (Lochte-Holtgreven, 1968),

$$n_e(r) = -\frac{1}{\pi}\int_r^\infty \frac{d}{dx}\left(\int n_e\, dl(x)\right)\frac{dx}{\sqrt{x^2 - r^2}}. \tag{7}$$

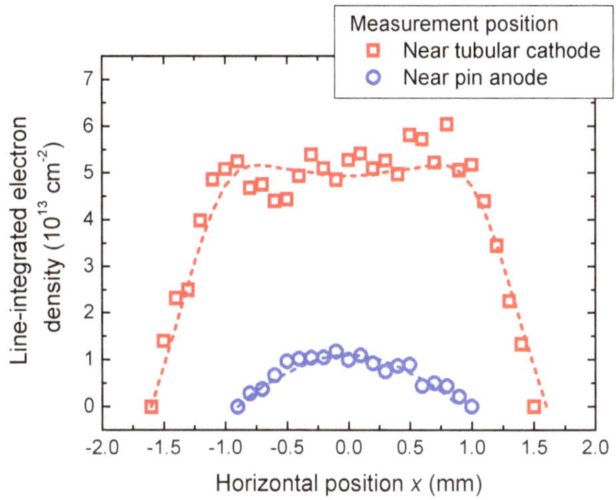

(a)

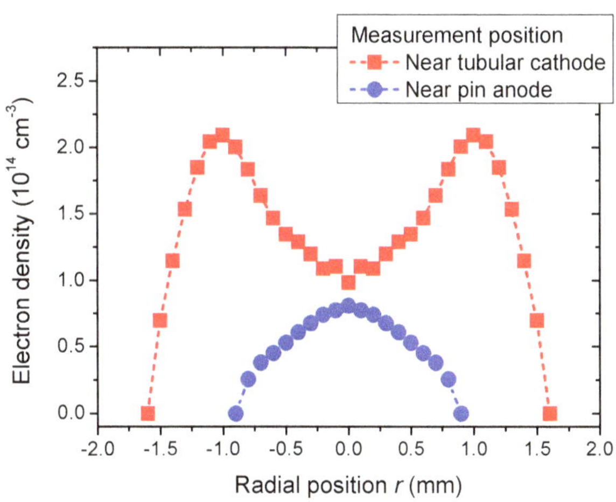

(b)

Fig. 7. (a) Spatial distributions of line-integrated electron density measured in DC plasma jet near tubular cathode and pin anode. (b) Radial distributions of electron density derived from line-integrated electron density using inverse Abel transformation (Choi et al., 2009).

The calculated results in Fig. 7(b) indicate that total amounts of the electrons near the cathode and the anode are significantly different in this DC discharge and there must be another kind of negatively charged particles delivering the discharge current and keeping its continuity. Considering that electron density near the central axis of the gas flow was not different at both measurement points, there were some losing mechanisms of the electrons outside of the He gas flow where ambient air contaminates into the flow. The electron loss was mainly due to an electron attachment process with O_2 molecules presenting in the ambient air in the DC discharge, because of µm-order short diffusion lengths of electrons in the atmospheric-pressure gas conditions which could be derived from calculations of electron diffusion coefficient (Hargelaar & Pitchford, 2005) and measurement results of electron lifetime (Moselhy et al., 2003). Whereas this negative-ion density, for example O_2^- ions, cannot be detected by the CO_2-laser heterodyne interferometer, this selectivity of negatively charged particles in the interferometer enables us to distinguish kinds and fractions of the negatively charged particle inside the high-pressure plasmas with the estimation of total amount of the charged particles from the discharge current amplitude.

3.3 Pulsed AC discharge at atmospheric pressure

For the measurement of electron density inside the high-pressure plasmas driven by kHz-order AC applied voltage, the temporal resolution of CO_2-laser heterodyne interferometer using the lock-in amplifier is often insufficient for direct measurement of temporal evolutions of electron density. In order to depict spatiotemporal structures of electron density inside such plasma sources, for example dielectric barrier discharges (DBDs) (Kogelschatz, 2003; Becker et al., 2005) and ns-order short pulsed discharges (Namihira et al., 2003; Walsh & Kong, 2007), it has been confirmed that an amplitude modulation of the kHz-order applied voltage at a frequency of a few hundreds Hz and additional measurement of a millimeter-wave (mm-wave) transmission method are effective.

Figure 8(a) shows a schematic diagram of atmospheric-pressure glow discharge (APGD) tested in our group. The APGD is a major kind of DBDs generating homogeneous plasmas with atmospheric-pressure He gas and kHz-order AC applied voltage (Kanazawa et al., 1988). Powered (upper side) and grounded (lower side) stainless-steel electrodes were round, and their diameters were 60 mm. Dielectric barriers of 1-mm thick alumina were placed on the electrodes' surface and a gap distance between the two barriers was set at 6.0 mm. Whole electrode setup was installed in a vacuum chamber with a pair of ZnSe windows to control the gas compositions and pressures. In the measurement of APGD, we used a pair of ZnSe lens with longer focal length (20 cm), in order to get constant beam shape in whole discharge region with 60-mm length along the CO_2 laser path. Therefore, the spatial resolution of the interferometer was worse than the measurement of small-scale DC plasmas explained in above subsections.

To divide the phase-shift signal into two components of the electron and gas-particle densities in a similar way to that used for the small-scale DC discharges, we used a square-pulse amplitude modulation at 125 Hz for the 30-kHz AC applied voltage, whose modulation-signal waveform is shown in Fig. 8(b). Using this amplitude modulation, the temporal evolution of phase shift has the fall and rise slopes in the both ends of modulation signal similar to that in the measurement of pulsed DC discharges. However, in the measurement of AC discharge, the phase-shift signal at the start-up timing of applied voltage (around −3.5 ms in the abscissa axis of Fig. 8(b)) is unsuitable for the derivation of the electron-density component in the phase shift. At this timing, the signal associated with the electron density was smoothed by the step-wise increase due to intermittent discharges in the APGD, and the influence of this smoothing effect on the phase-shift signal could not be resolved. Therefore, the measured phase-shift waveform should be divided into two components only at the cut-off timing of applied voltage (around 0 ms in Fig. 8(b)). Calculation procedure of the temporally-averaged electron density after dividing the two components is the same as the calculation in DC discharges.

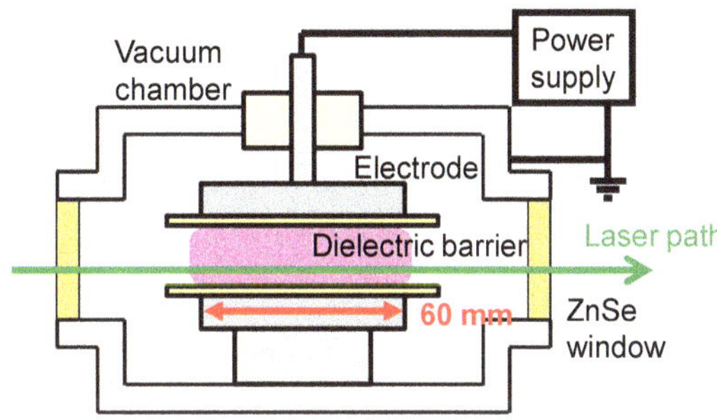

(a)

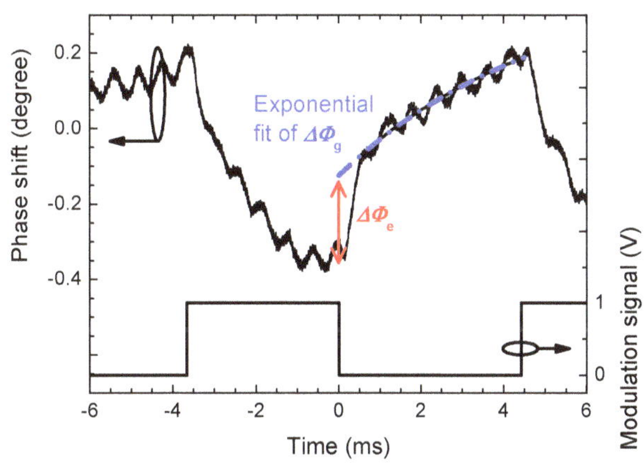

(b)

Fig. 8. (a) Cross-sectional diagram of parallel-plate dielectric barrier discharge (DBD). (b) Example temporal evolution of phase shift by the parallel-plate DBD together with amplitude modulation signal of 30-kHz AC applied voltage (Urabe et al., 2011).

The spatial distribution of the electron-density component in the phase shift and the calculated temporally-averaged electron density from the four signal waveforms measured under the same condition are shown in Fig. 9. The points and both ends of error bars indicate the averaged value and the maximum and minimum values in each measurement. The electron-density distribution inside the APGD was localized near the dielectric barriers, and this result had a good agreement with reported results of computational simulations in similar geometries (Massines et al., 1998, 2003; Martens et al., 2009). The electron density near the powered electrode was approximately two times larger than that near the grounded electrode in each gas composition. This asymmetric distribution was probably caused by diffusion of the discharge current flow into the chamber wall from the powered electrode not flowing into the grounded electrode.

To get temporally resolved information of the electron density inside the APGD, we inserted a mm-wave at 55 GHz through the vacuum chamber and measured temporal evolutions of the transmitted mm-wave intensity using a pair of horn antennas. In this mm-wave transmission method, spatial distributions of electron density cannot be measured because of the diffraction limit of the mm-wave. Details of the experimental setup and the calculation procedure of spatially-averaged electron density are introduced in our previous paper (Urabe et al., 2011). The temporal evolution of spatially-averaged electron density in the APGD derived from absorption ratio of the mm-wave in the plasma is shown in Fig. 10,

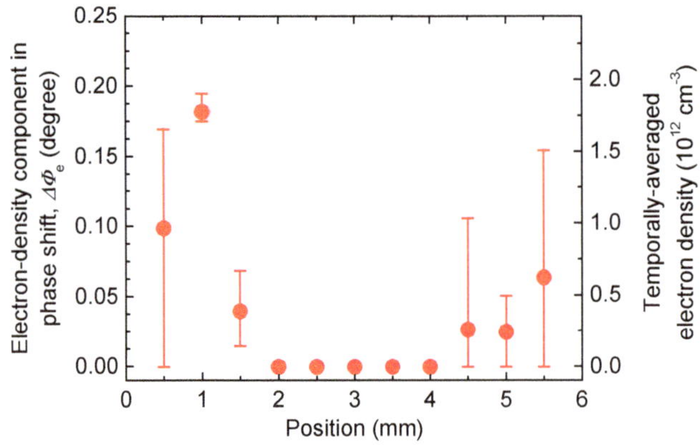

Fig. 9. Spatial distribution of electron-density component in phase shift and calculated temporally-averaged electron density in APGD (Urabe et al., 2011).

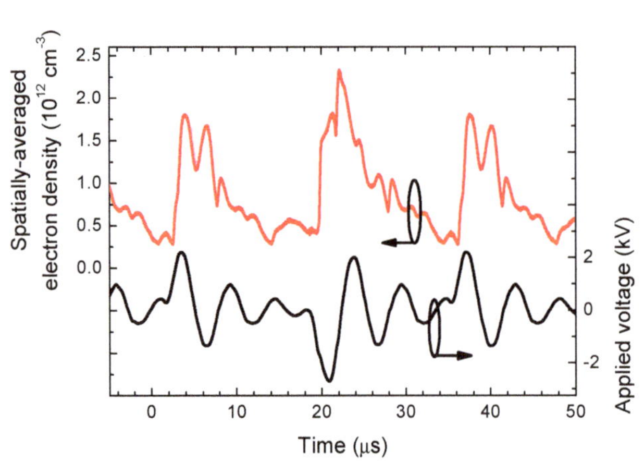

Fig. 10. Temporal evolution of spatially-averaged electron density in APGD measured by mm-wave transmission method under same conditions as interferometer measurement (Fig. 9), together with applied-voltage waveform (Urabe et al., 2011).

together with the waveform of applied voltage. The electron density increased after the positive- and negative-main pulses around 2.6 and 19.0 μs in the abscissa time axis, and there were small increases in electron density after the main pulses caused by weak discharges in the ringing part of applied voltage.

From the measurement results of the CO_2-laser heterodyne interferometer and the mm-wave transmission method applied to the same plasma source, spatial distribution of temporal-peak electron density can be calculated dividing the temporally-averaged electron densities by a duty ratio of plasma. The duty ratio of plasma is the ratio of temporally-averaged electron density to temporal-peak density in the result of the mm-wave transmission measurement, and it indicates the temporally averaging effects of the CO_2-laser heterodyne interferometer. In an example case of the APGD measurement shown in Figs. 9 and 10, the calculated duty ratio of plasma was 0.33. Then, the temporal-peak electron densities near the dielectric barriers were derived approximately 5×10^{12} cm^{-3} on the side of the powered electrode and 2×10^{12} cm^{-3} on the grounded electrode.

4. Concluding remarks

In this chapter, we reviewed experimental studies for electron-density measurement in small-scale plasmas operated at high and atmospheric pressures using the CO_2-laser heterodyne interferometer, from brief theoretical introduction of the interferometer to specific measurement results in the high-pressure plasma sources. It should be noted that separation of the CO_2 laser beam's phase shift into two components, which are due to changes of electron and gas-particle densities, is the most important procedure for the measurement in high-pressure plasmas, and pulse modulation of applied voltage is indispensable for the separation at the rise and fall timings of the modulation signal.

From the experimental results of the interferometer in high-pressure plasmas driven by pulsed DC voltage, fundamental properties of our interferometer including the minimum sensitivity of line-integrated electron density and the spatial resolution could be evaluated. Because these properties are changed due to specifications of the laser source and the phase detecting system, they must be confirmed in each setup of the interferometer before practical measurements. In addition to the interferometer, a mm-wave transmission method with a good temporal resolution was used to AC-voltage driven plasmas having ns-order fast temporal behavior. This novel combination method has potentials to be applied to the refractive-index measurements requiring both spatial and temporal high resolutions.

5. Acknowledgments

Our studies on the CO_2-laser heterodyne interferometer were partially supported by Grant-in-Aid for Scientific Research from the MEXT of Japan and Global Center of Excellence program on Photonics and Electronics Science and Engineering at Kyoto University. The authors would like to thank Prof. Osamu Sakai, Dr. Nobuhiko Takano, Dr. Joon-Young Choi, and Dr. Yosuke Ito at Kyoto University for their substantial supports. The author U.K. would like to acknowledge support of Research Fellowship from Japan Society for the Promotion of Science.

6. References

Acedo, P.; Lamela, H.; Sanchez, M.; Estrada, T. & Sanchez, J. (2004). CO_2 (λ_m=10.6μm) He-Ne (λ_c=633nm) two-color laser interferometry for low and medium electron density measurements in the TJ-II Stellarator. *Review of Scientific Instruments,* Vol. 75, pp. 4671-4677

Adler, F. & Kindel, E. (2003). Absolute Determination of Electron Densities in a Micro Hollow Cathode Discharge by Dual Wavelength Interferometry. *Proceedings of XXVIth International Conference on Phenomena in Ionized Gases,* Greifswald, Germany, July 15-20, 2003

Allen, C.W. (1973). *Astrophysical Quantities,* The Athlon Press, London, UK

Babayan, S.E.; Jeong, J.Y.; Tu, V.J.; Park, J.; Selwyn, G.S. & Hicks, R. F. (1998). Deposition of silicon dioxide films with an atmospheric-pressure plasma jet. *Plasma Sources Science and Technology,* Vol. 7, pp. 286-288

Becker, K.H.; Kogelschatz, U.; Schoenbach, K.H. & Barker, R.J. (Eds.). (2005). *Non-Equilibrium Air Plasmas at Atmospheric Pressure,* Institute of Physics, Bristol, UK

Bruggeman, P. & Leys, C. (2009). Non-thermal plasmas in and in contact with liquids. *Journal of Physics D: Applied Physics,* Vol. 42, pp. 053001-1-28

Chang, J.S. (1973). The inadequate reference electrode, a widespread source of error in plasma probe measurements. *Journal of Physics D: Applied Physics,* Vol. 6, pp. 1674-1683

Chang, J.S. & Laframboise, J.G. (1976). Probe theory for arbitrary shape in a large Debye length, stationary plasma. *Physics of Fluids,* Vol. 19, pp. 25-31

Choi, J.Y.; Takano, N.; Urabe, K. & Tachibana, K. (2009). Measurement of electron density in atmospheric pressure small-scale plasmas using CO_2-laser heterodyne interferometry. *Plasma Sources Science and Technology,* Vol. 18, pp. 035013-1-8

Hagelaar, G.J.M. & Pitchford, L.C. (2005). Solving the Boltzmann equation to obtain electron transport coefficients and rate coefficients for fluid models. *Plasma Sources Science and Technology,* Vol. 14, pp. 722-733

Hutchinson, I.H. (2002). *Principles of Plasma Diagnostics,* Cambridge University Press, Cambridge, UK

Ichiki, T.; Taura, R. & Horiike, Y. (2004). Localized and ultrahigh-rate etching of silicon wafers using atmospheric-pressure microplasma jets. *Journal of Applied Physics,* Vol. 95, pp. 35-39

Ito, Y.; Sakai, O. & Tachibana, K. (2010). Measurement of electron density in a microdischarge-integrated device operated in nitrogen at atmospheric pressure using a millimetre-wave transmission method. *Plasma Sources Science and Technology,* Vol. 19, pp. 025006-1-9

Kanazawa, S.; Kogoma, M.; Moriwaki, T.; & Okazaki, S. (1988). Stable glow plasma at atmospheric pressure. *Journal of Physics D: Applied Physics,* Vol. 21, pp. 838-840

Kogelschatz, U. (2003). Dielectric-barrier Discharges: Their History, Discharge Physics, and Industrial Applications. *Plasma Chemistry and Plasma Processing,* Vol. 23, pp. 1-46

Kong, M.G.; Kroesen, G.; Morfill, G.; Nosenko, T.; Shimizu, T.; van Dijk, J. & Zimmermann, J.L. (2009). Plasma medicine: an introductory review. *New Journal of Physics,* Vol. 11, pp. 115012-1-35

Kono, A. & Iwamoto, K. (2004). High-Spatial-Resolution Multichannel Thomson Scattering Measurements for Atmospheric Pressure Microdischarge. *Japanese Journal of Applied Physics,* Vol. 43, pp. L1010-L1013

Laux, C.O.; Spence, T.G.; Kruger, C.H. & Zare, R.N. (2003). Optical diagnostics of atmospheric pressure air plasmas. *Plasma Sources Science and Technology,* Vol. 12, pp. 125-138

Leipold, F.; Stark, R.H.; El-Habachi, A. & Schoenbach, K.H. (2000). Electron density measurements in an atmospheric pressure air plasma by means of infrared heterodyne interferometry. *Journal of Physics D: Applied Physics,* Vol. 33, pp. 2268-2273

Lochte-Holtgreven, W. (Ed.). (1968). *Plasma Diagnostics,* North-Holland Publishing, Amsterdam, Netherlands

Martens, T.; Brok, W.J.M.; van Dijk, J. & Bogaerts, A. (2009). On the regime transitions during the formation of an atmospheric pressure dielectric barrier glow discharge. *Journal of Physics D: Applied Physics,* Vol. 42, pp. 122002-1-5

Massines, F.; Rabehi, A.; Decomps, P.; Gadri, R.B.; Segur, P. & Mayoux, C. (1998). Experimental and theoretical study of a glow discharge at atmospheric pressure controlled by dielectric barrier. *Journal of Applied Physics,* Vol. 83, pp. 2950-2957

Massines, F.; Segur, P.; Gherardi, N.; Khamphan, C. & Ricard, A. (2003). Physics and chemistry in a glow dielectric barrier discharge at atmospheric pressure: diagnostics and modeling. *Surface and Coatings Technology,* Vol. 174–175, pp. 8-14

Moselhy, M.; Petzenhauser, I.; Frank, K. & Schoenbach, K.H. (2003). Excimer emission from microhollow cathode argon discharges. *Journal of Physics D: Applied Physics,* Vol. 36, pp. 2922-2927

Namihira, T.; Wang, D.; Katsuki, S.; Hackam, R. & Akiyama, H. (2003). Propagation Velocity of Pulsed Streamer Discharges in Atmospheric Air. *IEEE Transactions on Plasma Science,* Vol. 31, pp. 1091-1094

Nozaki, T.; Sasaki, K.; Ogino, T.; Asahi, D. & Okazaki, K. (2007). Microplasma synthesis of tunable photoluminescent silicon nanocrystals. *Nanotechnology,* Vol. 18, pp. 235603-1-6

Paschen, F. (1889). Ueber die zum Funkenübergang in Luft, Wasserstoff und Kohlensäure bei verschiedenen Drucken erforderliche Potentialdifferenz. *Annalen der Physik,* Vol. 273, pp. 69–96

Sakai, O.; Sakaguchi, T.; Ito, Y. & Tachibana, K. (2005). Interaction and control of millimetre-waves with microplasma arrays. *Plasma Physics and Controlled Fusion,* Vol. 47, pp. B617-B627

Tachibana, K.; Kishimoto, Y. & Sakai, O. (2005(a)). Measurement of metastable He*(23S1) density in dielectric barrier discharges with two different configurations operating at around atmospheric pressure. *Journal of Applied Physics,* Vol. 97, pp. 123301-1-7

Tachibana, K.; Kishimoto, Y.; Kawai, S.; Sakaguchi, T. & Sakai, O. (2005(b)). Diagnostics of microdischarge-integrated plasma sources for display and materials processing. *Plasma Physics and Controlled Fusion,* Vol. 47, pp. A167-A177

Urabe, K.; Morita, T.; Tachibana, K. & Ganguly, B.N. (2010). Investigation of discharge mechanisms in helium plasma jet at atmospheric pressure by laser spectroscopic measurements. *Journal of Physics D: Applied Physics,* Vol. 43, pp. 095201-1-13

Urabe, K.; Sakai, O. & Tachibana, K. (2011). Combined spectroscopic methods for electron-density diagnostics inside atmospheric-pressure glow discharge using He/N2 gas mixture. *Journal of Physics D: Applied Physics,* Vol. 44, pp. 115203-1-11

von Engel, A. (1994). *Ionized Gases,* American Institute of Physics, New York, USA

Walsh, J.L. & Kong, M.G. (2007). 10 ns pulsed atmospheric air plasma for uniform treatment of polymeric surfaces. *Applied Physics Letters,* Vol. 91, pp. 241504-1-3

Yariv, A. (1997). *Optical Electronics in Modern Communications,* Oxford University Press, New York, USA

Part 3

Material Processing

9

Application of Laser-Burnishing Treatment for Improvement of Surface Layer Properties

Joanna Radziejewska
Institute of Fundamental Technological Research, Polish Academy of Sciences, Warsaw Poland

1. Introduction

Laser Beam Machining (LBM) has been successfully applied for improvement of surface layer properties of machine elements. Some of laser treatments are based on melting of surface with a laser beam. Among them is laser hardening, as well as cladding and alloying. The results of broad scope research have shown that surface roughness of elements which underwent the laser melting is too high to apply the process without an additional abrasive machining, even at the optimum parameters of the laser treatment. In most cases after the surface melting with laser beam the tension stresses are observed. That is demonstrated by the cracks in the surface layer and deterioration of its properties (Dietrich Lepski et al., 2009).

A new hybrid treatment was elaborated for laser treated materials. The treatment, combining the laser melting with the burnishing process was performed simultaneously at the laser stage. The aim of the hybrid treatment was to reduce surface roughness formed in the laser process and induce compressive stresses. A surface smoothing effect was the result of plastic deformation of the surface layer in high temperature, while a reduction of the tensile stresses within the surface layer was due to cold work (Radziejewska&Skrzypek 2009).

For many years the surface burnishing has been used as smoothing and strain hardening finishing. The strain hardening, favourable compressive stress and smooth surface is obtained as a result of plastic deformation of the surface layer of elements made of homogenous material, as well as material with surface layer formed in order to obtain operating properties of superior requirements (Shiou & Hsu, 2008; Milad, 2008). The main limitation of the use of burnishing is high hardness and low plasticity of material after alloying and high roughness of surface. Przybylski, 1987, Shiou & Chen, 2003 showed that as a result of burnishing such materials undergo slight deformation due to the process. Strain hardening degree and thickness of plastic zone are small; cracks often form, whereas expected roughness could not be obtained. For this reason, the process of burnishing is not applied in industry as finishing of layers produced by laser beam. The study (Abbas & West, 1991; Arutunian et al., 1989; Meijer, 2004) demonstrated that modification of surface layers of metals by laser beam - such as hardening, alloying or cladding - provide very hard, resistant to abrasive wear, erosion and corrosion layers.

After laser remelting the stresses in the surface layer are generated in accordance with a hot stress model and they are mostly large tensile stresses, leading to the formation of micro-cracks in extreme cases (Grum & Sturm, 2004; Robinson, 1996) Anthony and Cline 1977 proved theoretically that the surface topography is characterized by relatively high asperities and study (Radziejewska, 2006) where waviness and roughness after laser alloying was examined confirmed this. Such a state of surface implies a need for an additional machining in order to improve surface smoothness.

The classical burnishing process applied after laser treatment was proposed in the works (De Hossonand & Noordhuis 1989, Ignatiev et al. 1993). The reduction of surface roughness and tensile stresses was obtained in the case of thin layer produced by laser alloying of titanium. Ignatiev et al. proposed another solution - the application of classical shot-peening process after laser hardening. As a result of shot-peening the change of stresses, from tensile stresses to compressive, in surface layer 70 μm thick was obtained.

The laser heating process is successfully applied to support the mechanical and plastic working of materials which are difficult for machining. Such a hybrid method was applied for cutting and turning of hard ceramic (Tsai & Ou, 2004). The research on local heating with laser beam during turning, milling and grinding of titanium alloys, cast iron and special steel was conducted. The hybrid treatment - laser-assisted burnishing (LAB) - was elaborated by Tian and Shin 2007. The laser heating process was applied for the burnishing of steel. It provided the reduction of the burnishing force, as well as the tool wear. It was shown that LAB can form better surface roughness and higher hardness than conventional burnishing.

In the work (Radziejewska 2007) the new method to modify surface layer, combing the laser melting with the slide burnishing, was proposed. The smoothing of surface was carried out by plastic deformation of surface layer at high temperature, whereas transformation process of stresses, from compressive to tensile stresses, was performed by plastic deformation at low temperatures. All machining operations - LBM, high and low temperature burnishing - are performed simultaneously on the laser station, in one pass. Temperature changes while the cooling of material that undergoes the laser beam treatment, are used. It does not extend duration of treatment. It was stated that multiple alloying combined with slide burnishing generated compressive stresses of about – 600 MPa at the surface. Because of the adopted type of burnishing – the slide burnishing and high hardness of material, the relatively small thickness of textured zone, about 30 μm, was obtained. In the case of thick layers it can be insufficient. According to (Przybylski, 1987) high degree of strain hardening of surface is possible to provide using dynamic burnishing.

The current work presents the analysis of the plastic deformation of surface layer when the new type of the laser-burnishing process was applied. The surface layer was generated by laser alloying and dynamic burnishing. Design of new head allows for simultaneous surface burnishing in two different temperatures on the laser station providing high-intensity of the burnishing process. That allowed treatment hard and low plasticity materials. Thus such a technical solution should result in high degree of surface deformation as well as its large depth. The proposed solution is designed for thick layers, above 1mm, mostly produced by LBM.

The aim of this study was to evaluate the influence of the hybrid processing parameters on plastic deformations of surface layer and the analysis of the correlations between treatment parameters and surface layer state. Establishing relationships between process parameters and the state of the surface layer of tested material will determine the application area of the proposed hybrid treatment for other materials and coatings. Obtaining improved features of surface layer, such as material microstructure, hardening and the compressive stresses, while reducing the amount of surface roughness, will eliminate additional finishing after the laser improving. New features of surface layer affected the basic functional properties of machine parts. Hybrid treatment improved such properties as surface roughness, contact stiffness and erosive wear mainly required in operating condition. It will extend application of the hybrid treatment on cases in which the surface layers have to meet high durability requirements.

2. The experiment description

2.1 Laser alloying

The study was performed using the CO2 laser with the maximum power 2.5 kW. The axially-symmetric beam, of the mode close to TEM10, was focused with use of the ZnSe lens. The focal length of it was 5". The set of treatment parameters, such as laser power, feed rate, and diameter of the beam focused on the metal surface, type of shielding gas and speed of the air flow to ensure a sufficient power density to obtain the remelting and optimal results, was selected. Based on previous experience (Radziejewska&Skrzypek, 2009) the ranges of parameters of laser processing were identified in the first stage of the study. The optimization criteria, such as penetration depth, surface roughness and hardness of the resulting surface layer, were taken into consideration.

The alloying process was carried out on steel 304. Prior to the alloying a layer of Stellite 6, with a thickness of about 200 μm, was formed on the surface with plasma spraying method. Preliminary studies have shown that in this case, the alloying process takes place preferably at the following parameters: laser power 2 kW, laser beam diameter of 3 mm, feed rate of sample against the laser beam from 150-900 mm/min.

2.2 Laser-mechanical treatment

On the laser station the burnishing process was carried out simultaneously with the alloying process. The dynamic burnishing process with use of micro-hammers was applied. The technology of micro-hammering was based on a dynamic centrifugal burnishing. For micro-hammering a high rotational head was developed, providing the possibility of working directly on the laser processing. Processing concept and principle of operation of the head is described in work (Radziejewska et al., 2005). In this study the modified version of the head was applied. Two rows of 8 micro-hammers allow providing greater intensity of the process and the simultaneous treatment at two different temperatures. In order to obtain more uniform deformation of surface material, the oscillation motion of the sample in a direction perpendicular to the direction of feed was introduced. The motion was generated using an oscillating table. The oscillations eliminated the problem of the formation of unfavourable geometric surface structure - the grooves occurring in earlier solution. A constant velocity, of 15 oscillations per second and the amplitude of 2 mm, was used. The small radii of micro-hammers allow to obtained high surface plastic deformation at low forces. The scheme of the head is shown in Figure 1a while the laser-mechanical treatment presented in Fig. 1b. Figure 1c shows the temperature distribution on the surface along x axis with the selected range of temperatures in which the burnishing process was conducted.

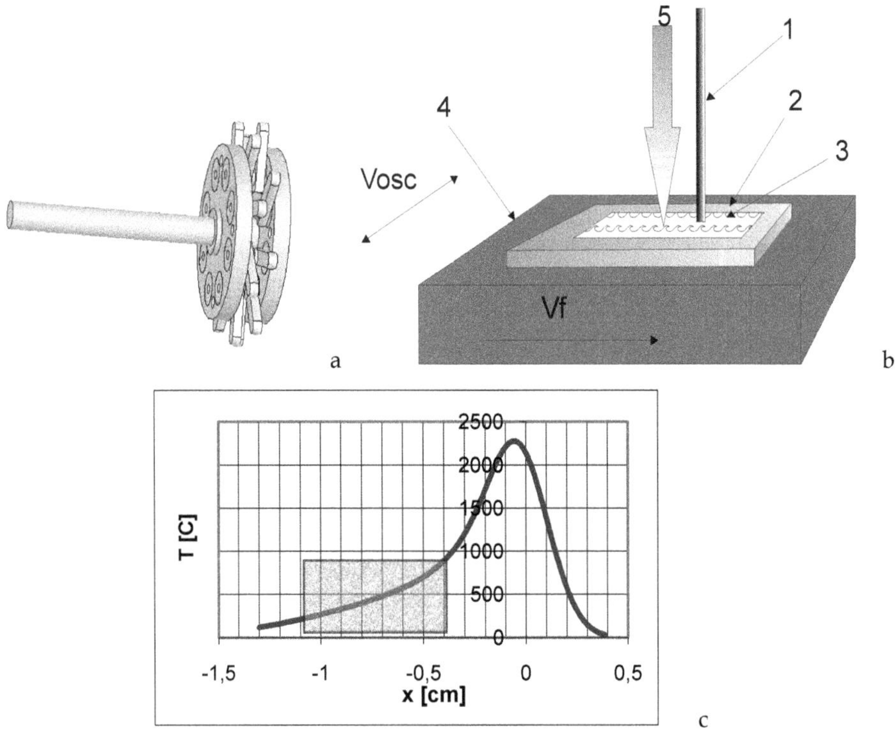

Fig. 1. A - dynamic burnishing head, b - scheme of the station for laser-mechanic treatment: 1-laser beam, 2-sample, 3-laser path, 4-oscillation table, 5-dynamic burnishing head; c - temperature distribution on the surface, along x axis, with burnishing area.

The treatment with the head is based on cyclical impacts of the burnishing elements onto the machined surface. The micro-hammers are made of bearing steel, and their working part has radius of 1.5 mm. They are placed evenly between body shields of the head and rotary mounted on the axes, providing the swinging motion of the hammers in relation to the head as well as the rotary motion with head. The compact head enables the processing of flat surfaces and curved of small sizes. The head is designed as the smoothing and strengthening treatment of laser modified parts. The head was mounted in a grip of portable grinding tool, which is mounted on the laser treatment station together with the system of the head adjustment. The station enables controlling wide range parameters of the process:

- impact forces on the tool surface by controlling the rational speed of the head - V_{rev} and its distance from the surface undergoing treatment,
- temperature of the process zone due to changes of the distance between the impact of micro-hammers and the axis of the laser beam - X,
- intensity of the surface hardening by adjusting the feed rate of the sample - V_f, and the rotational speed of micro-hammers - V_{rev}.

Before basic studies of the preliminary tests were carried out in order determine the optimal position of the tool in relation to the machined surface. The distance between the burnishing element and the laser beam axis was determined initially on the basis of calculations of temperature distribution on material surface, when a moving source of heat, which has a Gaussian distribution of energy density, was considered. The distribution in a half-space has an approximate character because of numerous simplified assumptions (Anthony&Cline, 1977). The distribution in a half-space has an approximate character because of following simplified assumptions:

- material properties, such as heat conduction, density, specific heat and coefficient of reflection are constant and temperature independent,
- there is no loss of energy,
- heat of phase transformation is not taken into consideration,
- convectional heat transfer in liquid metal is not considered.

The calculations were done for parameters of laser treatment, the same as those were applied in preliminary investigations. The results of calculations have been verified by microstructure changes of material. The tools interacted with material within temperature range about 200–850^{0}C.

2.3 Testing methods

Study of surface layer state is time-consuming. Therefore determination the effect of treatment conditions on the state of surface layer was based on the theory of planned experiment, which minimizes the number of studies (Filipowski 1996; Montgomery, 1997). An experiment based on a static, determined, multi-factorial, rotatable planned program with repetition PS/DS- λ was carried out. The aim was to find functional relations between the parameters of the hybrid treatment and the state of the surface layer.

The selection of input variables and ranges of their variation was based on preliminary results of the laser- mechanical process taking their suitability to control the hybrid treatment as an additional criterion. As the measurable, controllable input variables the following quantities were considered:

- distance between the head and the laser beam axis (X) that determines the temperature of the burnishing process,
- rotational speed of the burnishing head (V_{rev}); the parameter affects the impact force of hammers on the surface,
- feed rate of the sample (V_f), which contributes both to the intensity of the burnishing process and temperature in the region of treatment.

As the output factors characterizing the state of surface layer and the effects of hybrid treatment were: change in microhardness compared to the microhardness after laser alloying, thickness of the plastic deformation zones and the ratio of thickness of the plastic deformation zones to thickness of alloyed zone. On the basis of preliminary studies the areas of treatment parameters variability and the intervals of variation of input data were determined. Indications and values of variation ranges of input data are contained in Table 1.

The statistical analysis of experimental results included a selection of the regression function, a statistical verification of the approximating function adequacy and verification of significance of the approximating function coefficients. The attempts of approximation using the power function and the first-degree polynomial have been made. The correlation and significance evaluation has been determined according to criteria based on I. P. Guilford theory. The confidence level $\alpha = 0.1$ was adopted. Credibility of the equations was assessed based on the following criteria:

- critical value of the *F*-statistic, $F_{kr} = 2.71$,
- critical value of the *T*-Student coefficients, $T_{kr} = 1.41$ for $a = 0.1$.

Parameter	Input values					Unit
V_{rev}	3500	4200	5000	5950	7100	rev/min
V_f	150	230	360	570	900	mm/min
X	5	6	7	8,5	10	mm

Table 1. The input values of the hybrid treatment experiment for steel 304 alloyed with Stellite 6.

2.3.1 Microstructure and macro-stress analysis

Microstructure analysis and measurements of a size of the melted zone was performed on an optical microscope at magnifications from 50 to 1000 X, and for selected samples on a scanning microscope. For this purpose, after the laser alloying and the laser-mechanical treatment all samples were cut perpendicular to the treated surface and metallographic micro-sections were made in a direction perpendicular to the direction of feed. For selected samples the micro-sections in a direction parallel to the feed samples were made additionally. Surface analysis of chemical composition was also carried out for selected samples after the alloying and the hybrid treatment.

A study of internal macro-stresses and phase composition was conducted on Bruker's D-8 Advance diffractometer, with the Mo anode lamp. The measurements of internal macro-stresses were carried out on surface at a distance of 0.3 and 0.5 mm from the surface. Calculations were performed for elasticity indexes E=210 GPa and ν=0.28. An evaluation of the degree of plastic deformation, caused by surface burnishing process, was carried out on the basis of changes in microhardness of the material. Measurements were made at load of 0.2 N, in the central zone of the melting for both samples - alloyed with the oscillations as well as laser-mechanically treated with oscillations. The microhardness result is an average of 5 measurements.

2.3.2 Surface roughness measurements

The surface topography was examined. The measurements were conducted on a scanning profilometer Form Talysurf after laser alloying and hybrid process performed at different parameters. Surface roughness measurements were performed for each track of the laser alloying, and the laser alloying with oscillations combined with micro-hammering. The 3D roughness measurements were conducted in central area of the melting path. The values of surface topography parameters were determined for scanned area 1.4×4 mm. The measurements were conducted at steps dx = 0.5 μm, dy = 5 μm, with the stylus radius of 2 μm. Profile measurements were carried out in the middle of the zone of the melting on measuring section equal to 4 mm, parallel to direction of the feed rate. Roughness parameters are the average values of 16 measured profiles.

2.3.3 Contact stiffness

The process of contact deformation was carried on the stand using a measuring method of the surface approach proposed by Demkin 1956. The device enables measurement the approach a with accuracy 1 μm as a function of applied nominal pressure q. Contact is realized between flat and rough surface of the sample, and the smooth and rigid surface of the counter-sample, made of sintered carbide. The counter-samples have three punches, 5mm diameter and nominal area of 58.875 mm^2 each. The sample undergoes deformation under the punches. Both samples are placed in a specially constructed device, which is mounted inside the laboratory precision hydraulic press that allows applying the normal pressure up to 1000 MPa. The applied pressure are measured using the compression proving ring, while the approach of samples is measured using the inductive sensor. The results of measurements were recorded in the form of the approach value a [μm] for the given loading F [N].

The study was carried out to nominal pressure 270 MPa, then the unloading to 0 N. The testing enabled to determine the values of total deformation, denoted as a_1 and a_2, and plastic deformation, a_{pl}, as well as the elastic one, a_e. It also allowed determining the curves of approach - nominal pressure relation. The values of deformation are the averaged values from three areas of measurement.

Due to the large differences in the surface geometric structure, which has a significant influence on the contact stiffness, the surfaces of the samples were subjected to grinding in order to form similar surface topography. Application of the grinding allowed to eliminate the influence of differences in surface roughness on contact stiffness due to the various processes applied previously. It allowed for the analysis of properties of the surface layer, in this case the burnished layer, on the process of contact deformation. After grinding the surface roughness measurement was repeated (Radziejewska, 2011).

2.3.4 Wear test

The slurry erosive wear test was conducted on test stand equipped with the chamber, the sprinkler head with nozzle of 10 mm diameter, the mixer and a sample holder. The pressure in the power system filled with compressed air was 5 Barr. The erosive tests were performed using 15% aqueous suspension of SiC particles. The size of SiC particles was 42.5 - 46.5 μm. The samples were placed 100 mm from the nozzle. The suspension angle between the sample and surface was 90^0. Before the test the samples were polished to reach the same roughness of surface. A special protection was applied to provide the erosive testing only within chosen limited area. The preliminary test allowed selecting erosive test duration. It also gave information that depths of eroded material were less than thickness of tested layers. The duration of each test was 90 s. All erosive tests were performed in the same conditions.

The erosion was determined using scanning profilometer. The eroded areas, and non-eroded area that was treated as a reference surface, were 3D measured. The depth and volume of eroded region was calculated based on TalyMap Platinium 5.0 program. For each examined surface layers the testing was performed on three areas. Final erosive losses were determined as an average value for medium depths and volumes.

3. Plastic deformations due to laser-burnishing

3.1 Microstructure and size of plastic deformation zone

Studies of the microstructure showed that both after the alloying and the hybrid treatment the surface layer is homogeneous, free of pores and micro-cracks. A very fine dendrites structure, oriented in the direction of heat dissipation, is formed. The oscillations caused an increase in the width of the melted zone in relation to the alloyed samples without oscillations by the value of oscillation amplitude. Additionally the application of oscillations resulted in more even thickness of surface layer. Figure 2a shows the shape of the melted zone, while Figure 2b presents the melted zone at same processing parameters using oscillations.

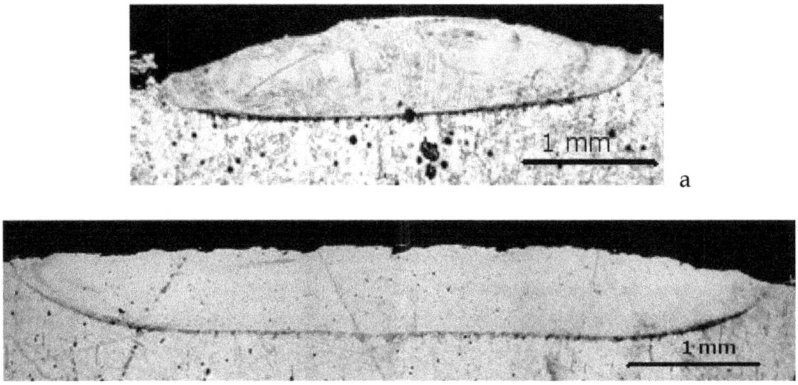

a

b

Fig. 2. Shape of melted zone after: a - laser alloying of steel 0H18N9 with Stellite 6 at treatment parameters: P = 2 kW, d = 3 mm, V_f = 900 mm/min; b - alloying using the oscillations in perpendicular direction.

The study proved constant thickness of the melting zone. When the treatment is conducted at the lowest speed, 150 mm/min, minor changes in the thickness of the melting, related to waves of the bottom of the zone melting, can be observed (Fig. 3).

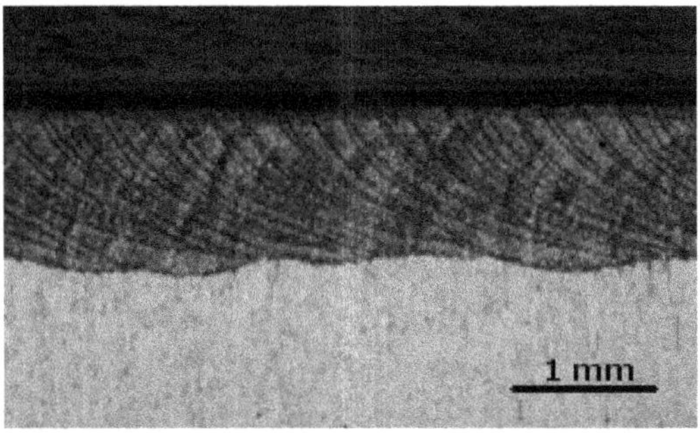

Fig. 3. The surface layer after laser alloying of steel 0H18 N9 with Stellite 6 at treatment parameters: P = 2kW, d = 3 mm, V_f = 150 mm/min using the oscillations. The cross-section parallel to the direction of the sample motion.

Surface plastic deformation caused a complete reconstruction of material microstructure near the surface. The microstructure, after laser alloying, is shown in Figure 4. There are equiaxed grains of a size of several micrometers. The intensive flattening of grains in the direction perpendicular to the surface is presented, Fig.5. There are no cracks or chipping. The coating particles, which were not melted by laser, can be observed on the surface in some places. They were probably stuck in the surface of already melted and solidified material by the micro-hammers. The largest deformation of grains is in the zone about 100-150 µm from the surface. As it was expected the varied plastic deformation thickness of the zone was determined depending on the impact of the micro-hammers on the surface and the temperature at which the process was carried out.

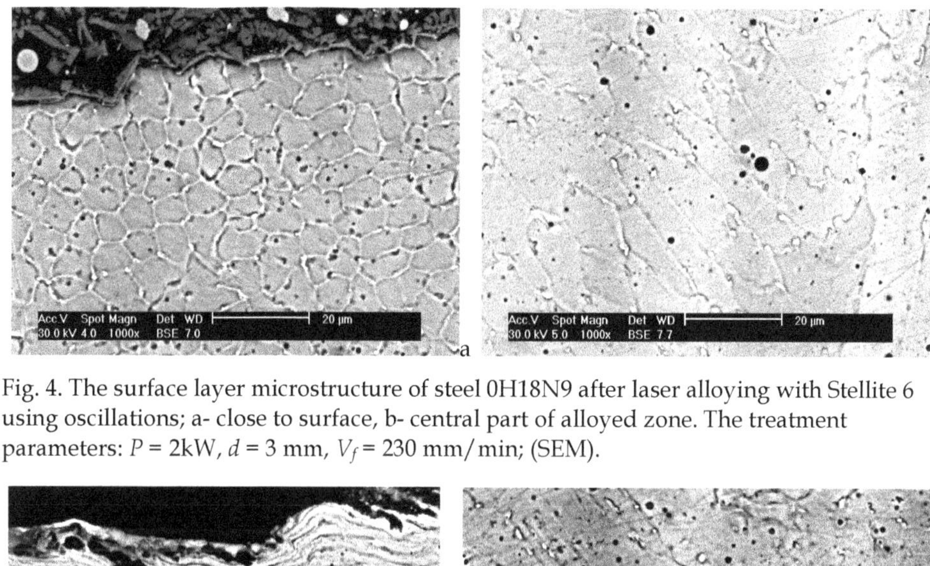

Fig. 4. The surface layer microstructure of steel 0H18N9 after laser alloying with Stellite 6 using oscillations; a- close to surface, b- central part of alloyed zone. The treatment parameters: P = 2kW, d = 3 mm, V_f = 230 mm/min; (SEM).

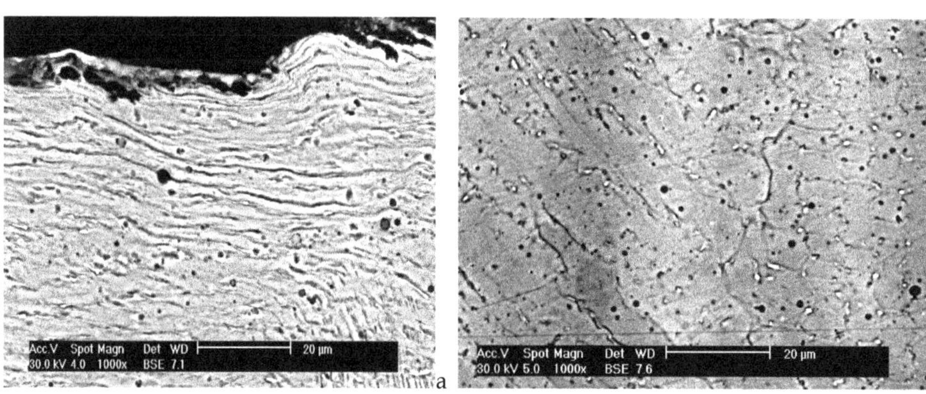

Fig. 5. The surface layer microstructure after hybrid treatment at the parameters: P = 2kW, d = 3 mm, V_f = 230 mm/min, V_{rev}= 5950 rev/min, X = 6 mm, using oscillations; a- close to surface, b- central part of alloyed zone; (SEM).

The X -ray diffraction XRD showed the presence of cobalt austenite, tungsten carbides and chromium as well as chromium oxides, cobalt and cobalt ferrite. In almost all samples the dominant phase was cobalt austenite. The analysis of chemical composition in the melting zone showed the increase in concentration of Co, W, Ni, and C in relation to the core. A homogeneous distribution of elements was found in the melted zone for both groups of samples - laser alloyed and hybrid treated. In Fig.6 the exemplary surface distributions of concentration of iron and cobalt for the hybrid treated sample, at the edge of the hybrid laser path, are shown.

a

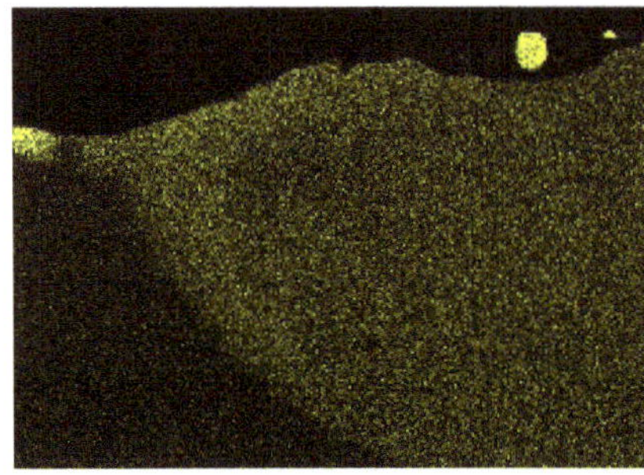
b

Fig. 6. Surface distributions of elements in the melted zone after laser-mechanical treatment: a - Fe, b – Co; (EDS-SEM).

Measurements of dimensions of the alloyed zone showed that thickness of the melted zone ranges from 0.43 mm to 0.87 mm, while its width is from 3.68 mm to 4.6 mm. The size of the zone depends on the sample feed rates. The largest size of melted zone was found for the lowest feed rate of 150 mm/min. The oscillation motion caused an increase in the width of the melting zone by the value of the oscillation amplitude, i.e. 2 mm, compared to the laser alloying without oscillations. Shape changes of the melting also resulted from the oscillations, shown in Fig. 3.

3.2 Microhardness of strain hardening zone

Table 2 presents the results of microhardness study. Microhardness values refer to five distances from the surface: 0.02, 0.2, 0.3, 0.4 and 0.6 mm. After laser alloying the surface layer material had microhardness 300 - 420 HV. In most cases the increase in microhardness was recognized at the surface compared to the bottom of the melting. This is due to the presence of fine grains in the subsurface zone. Differences in microhardness observed for the same depth of the melted zone are due to different thickness and porosity of Stellite layer deposited before the process of laser-mechanical stated. These differences determine the chemical and phase composition of the alloyed layer. In order to eliminate these distortions all calculations, related to the assessment of surface plastic deformation and thickness of the hardening zone, were carried out for one laser "path" where the zone only laser-alloyed and alloyed with micro-hammering existed.

Microhardness of layer generated by laser alloying combined with micro-hammering is 530-670 HV at the material surface and about 400 HV at the melting of the bottom. For all tested samples the increase in microhardness at the surface can be observed. It is related to the process of burnishing. The thickness of the strain hardening zone varies depending on the burnishing force as well as the intensity and temperature of the process. In all cases it is thicker than a textured zone observed on the metallographic cross-sections.

Sample	Treatment parameter			Microhardness after hybrid treatment at selected levels from surface				
	V_{rev} [rev/min]	V_f [mm/min]	X [mm]	HV 0.02	HV 0.2	HV 0.3	HV 0.4	HV 0.6
1	4200	230	6	527	480	430	410	350
2	5950	230	6	530	520	460	430	400
3	4200	570	6	540	430	380	360	290
4	5950	570	6	610	550	480	450	370
5	4200	230	8,5	640	520	440	420	290
6	5950	230	8,5	700	570	490	440	360
7	4200	570	8,5	580	550	510	440	350
8	5950	570	8,5	630	600	540	440	400
9	3500	360	7	630	500	560	520	410
10	7100	360	7	680	570	460	410	310
11	5000	150	7	660	480	450	410	400
12	5000	900	7	570	430	410	370	270
13	5000	360	5	540	460	420	380	290
14	5000	360	10	500	450	420	410	280
15	5000	360	7	610	480	430	410	280
16	5000	360	7	670	570	380	410	280
17	5000	360	7	630	490	490	420	380
18	5000	360	7	640	480	430	410	290
19	5000	360	7	620	530	450	400	350
20	5000	360	7	610	510	440	420	290

Table 2. The microhardness as function of depth for samples after hybrid treatment for variable process parameters.

Figure 7 shows the exemplary microhardness distributions after alloying and the hybrid treatment for the samples 1 and 2. The treatment process was carried out at the same feed rate, V_f = 230 mm/s, and at the same temperature. Both samples exhibit a substantial increase in microhardness of the hybrid treated material in relation to only alloyed. Additionally, the diagram Fig.7 shows the differences in microhardness of alloyed layers generated in the same conditions.

On the basis of the results the relative percentage increase in microhardness, caused by surface plastic deformation, was determined for each sample, and the thickness of the zones of plastic deformation was estimated under the assumption that minimal increase in the layer is 10% (Tab.3). The micro-hammering caused the relative increase in microhardness of SL of about 32-100% at the surface compared to microhardness of SL due to laser alloying. This effect is due to surface strain hardening. The smallest increase in microhardness, 32%, at the surface was found for the sample that was burnished at the lowest temperature when distance between the hammers and the beam axis was 10 mm. In this case, the smallest depth of the plastic deformation zone was also recognized. The increase in microhardness, over 60%, was found for most of the samples (2, 4, 5, 6, 10, 11, 15-20) burnished with large impact forces of micro-hammers on surface, rotational speeds of the head above 5000 rev/min and temperature of treatment from the middle range. For the samples (1, 13) burnished at high temperature the degree of strain hardening is the order of 40-50%, which is probably related to the partial recovery of the material at high temperature.

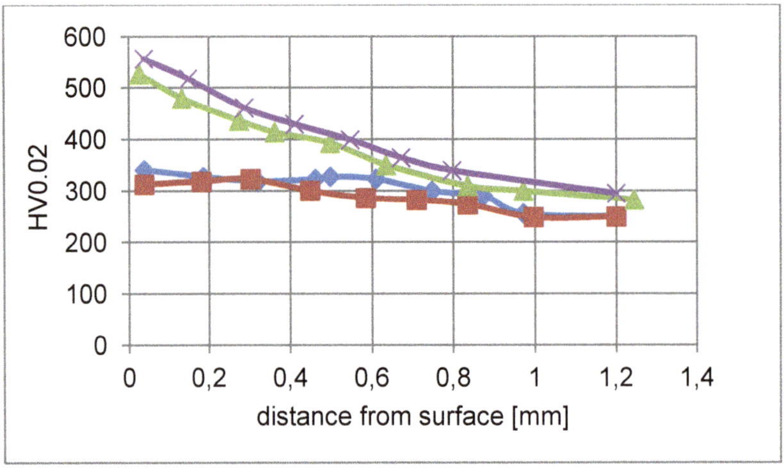

Fig. 7. Microhardness after: ♦ ■ laser alloying of steel 0H18 N9 with Stellite 6 treatment parameters: $P = 2$ kW, $d = 3$ mm, $V_f = 230$ mm/min and alloying combined with micro-hammering ▲ $X = 6$ mm, $V_{rev} = 4200$ rev/min, × $V_{rev} = 5950$ rev/min,

The thickness of the plastic deformation zone is from 0.25 to 1.2 mm, depending on the parameters of laser-mechanical treatment. It grows with increase of the impact forces of burnishing elements on the surface and the rise of temperature in which the process of burnishing was undergoing. This effect is due to the increase in material plasticity with temperature growth.

Plastic deformation of the surface layer caused changes in residual stresses. In order to assess the state of stresses in the surface layer the preliminary studies of internal macro-stresses were performed. The selected for study samples were laser alloyed at feed rate of 230 mm/min and dynamically burnished at temperature from the upper range, $V_{rev} = 5950$ rev/min, $X = 6$ mm. The measurements were carried out on the surface and at the depth of 0.3 and 0.5 mm beneath the surface. The results confirmed the stresses change, from tensile +398 MPa after the alloying process to compressive stresses -800 MPa caused by the burnishing process. After the dynamic burnishing the compressive stresses across the whole depth of the melted zone were found.

Sample	Treatment parameter			Relative change in microhardness [%]					Thickness of plastic deformation zone [mm]	G_{pl}/G_{la}
	V_{rev} [rev/min]	V_f [mm/min]	X [mm]	HV 0.02	HV 0.2	HV 0.3	HV 0.4	HV 0.6		
1	4200	230	6	46	45	34	28	9	0.55	0.75
2	5950	230	6	71	63	44	43	38	1.2	1.62
3	4200	570	6	59	13	0	0	0	0.25	0.54
4	5950	570	6	74	67	60	50	42	1.2	2.67
5	4200	230	8.5	78	30	13	14	-	0.5	0.77
6	5950	230	8.5	89	36	32	16	38	0.7	1.08
7	4200	570	8.5	32	45	34	10	-	0.4	0.75
8	5950	570	8.5	50	43	32	13	14	0.65	1.30
9	3500	360	7	62	39	56	33	8	0.5	0.94
10	7100	360	7	100	68	53	37	3	0.55	1.00
11	5000	150	7	89	55	50	37	33	1.1	1.26
12	5000	900	7	68	39	24	12	4	0.45	1.05
13	5000	360	5	54	44	35	27	-	0.5	0.77
14	5000	360	10	32	18	2	8	4	0.25	0.43
15	5000	360	7	74	41	23	28	-	0.5	0.88
16	5000	360	7	91	78	9	17	-	0.4	0.67
17	5000	360	7	70	29	32	17	12	0.55	0.80
18	5000	360	7	83	41	23	28	-	0.5	0.82
19	5000	360	7	77	66	29	14	3	0.5	0.85
20	5000	360	7	65	34	19	17	-	0.5	0.74

Table 3. The percentage change in the microhardness value at different depths, the thickness of the plastic deformation zone, G_{pl}, and the ratio of the plastic deformation thickness to the thickness of the alloyed zone, G_{pl}/G_{la}.

3.3 Statistical analysis of results

The influence of hybrid treatment parameters on features characterizing plastic deformation zone, e.g. thickness, change in microhardness compared to the microhardness after laser alloying and ratio of thickness of the plastic deformation zone to thickness of alloyed zone, was presented in form of the regression function.

$$G_{pl} = 0.0026\, V_{rev}^{1.01}\, V_f^{-0.19}\, X^{-1.16} \tag{1}$$

$$\Delta HV = 0.53\, V_{rev}^{0.66}\, V_f^{-0.082}\, X^{-0.61} \tag{2}$$

$$G_{pl}/G_{la} = 0.00035\, V_{rev}^{1.02}\, V_f^{0.2}\, X^{-1.09} \quad (3)$$

Table 4 contains the value of multiple correlation coefficients *R*, the value of the Fisher's number *F* and *T*-Student coefficient describing the significance of subsequent independent variables *T1, T2, T3,* and *T4.*

Relation	R	F	T1	T2	T3
1	0.78	8.4	3.15	1.5	3.59
2	0.68	4.6	2.65	0.85	2.45
3	0.74	6.3	2.81	1.46	2.97

Table 4. The regression function for the thickness of the plastic deformation zone, G_{pl}, change in microhardness ,ΔHV, and ratio of thickness of the plastic deformation zones to thickness of alloyed zone G_{pl}/G_{la} of hybrid treated steel 304.

Multiple correlation coefficients of the equations are high and the relation between the studied properties is significant. For all equations, the condition $F > F_{kr}$ is fulfilled. For the first and third equations all the factors are significant $t > t_{kr}$ at confidence level $\alpha = 0.1$. Only in the case of the function 2 which shows the relation of the microhardness increase the factor *T2* describing influence of feed rate is insignificant for the assumed level of confidence.

Figure 8 shows the graphical interpretation of the thickness of strain hardening depending on the rotational speed of the burnishing head and the distance between the tool from the axis of the laser beam for fixed feed rate, $V_f = 360$ mm/s, according to the relation 1, Table 4. The strain hardening thickness almost linearly increases with speed growing and decreases with increasing distance from the axis. With increasing of the head rotational speed the intensity of the burnishing process and the impact forces of micro-hammers on machined surface increases. It induced an enlargement of the plastic deformation depth of the material in the entire range of temperatures applied in the burnishing. The effect of temperature on the depth of the plastic deformation zone is stronger for raised values of the impact forces and higher intensity of the burnishing process.

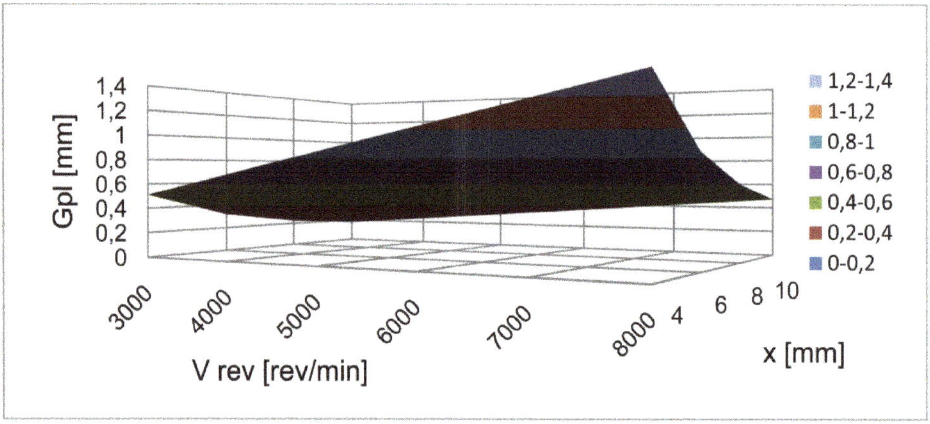

Fig. 8. Effect of rotational speed and the distance between laser beam and the head on the thickness plastic deformation zone at fixed feed rate, $V_f = 360$ mm/min.

Figure 9 shows the relation between the increase in material microhardness at the depth of 0.2 mm and the rotational speed of the head, and the distance from the tool axis to the laser beam at constant feed rate 360 mm/min. The microhardness growth, characterizing the degree of strain hardening at the surface, depends mainly on the rotational speed of the head and the temperature of the process zone, which is a function of the distance between the burnishing tool and the laser beam.

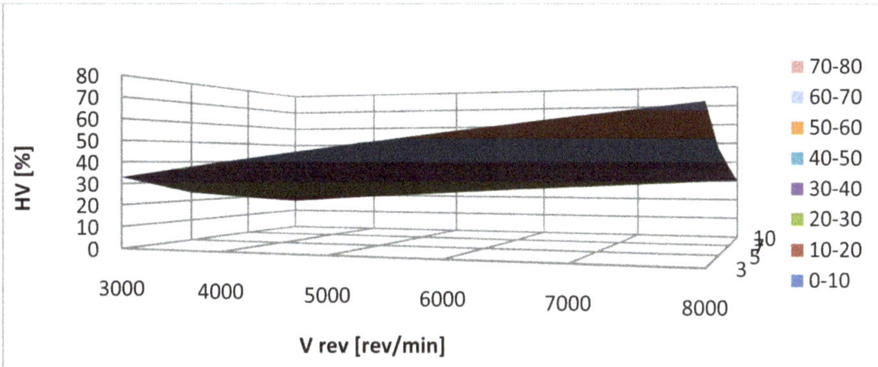

Fig. 9. Dependence of microhardness increase at a depth of 0.2 mm on rotational speed of the head and distance between the tool axis *X* and the laser beam, for fixed feed rate 360 mm/min.

Figure 10 shows the influence of the rotational speed of the head and the distance of the tool from surface, at a fixed feed rate V_f = 360 mm/min, on the G_{pl}/G_{la} parameter. It is evident that at the highest speed and the distances of less than 9 mm the ratio is bigger than 1. At speeds below 4000 rev/min and the distance of less than 8 mm the thickness of the strain-hardening zone in relation to the thickness of the alloyed layer is less than 0.5, although it does not ensure the presence of compressive stresses in the entire alloyed zone.

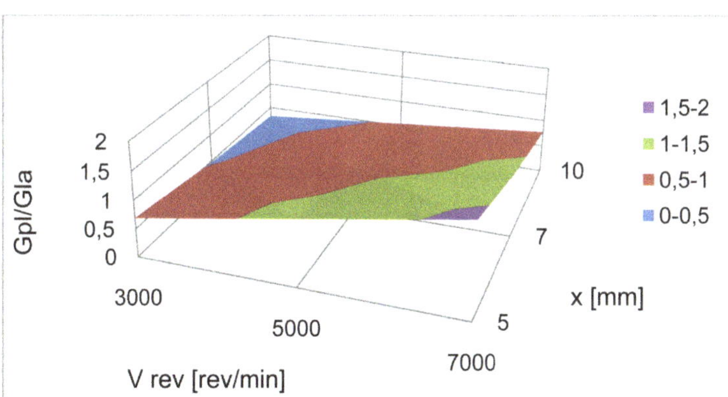

Fig. 10. Influence of rotational speed and the distance between the beam axis and the tools on the ratio of the thickness of strain hardening zone to the thickness of alloyed zone G_{pl}/G_{la} at a fixed feed rate V_f = 360 mm/min.

4. Surface roughness after hybrid treatment

4.1 Surface topography

The study showed that of the laser alloying combined with the burnishing can provide reduction of surface roughness in relation to the heights obtained in the laser melting process. The shape of the melted zone has improved as well. In the case of profile parameters cut-off 0.8 was used for all surfaces. In Figure 11a the examples of the surface topography views of the laser path and section of the central part of the path are shown. Below the profiles perpendicular and parallel to the direction of movement are visible. Figure 11b presents geometrical features of hybrid treated surface. Table 5 shows the 3D parameters *Sa, Sz* , their changes compared to the parameters after laser alloying $Sa_{(la)}/Sa_{(h)}$, $Sz_{(la)}/Sz_{(h)}$ and the 2D roughness parameters *Ra, RSm* of the surface which underwent hybrid treatment.

A change in shape of the melted zone due to the burnishing is noticeable. The heights of asperities at the alloying zone boundary decreased significantly, while the shape of asperities in the central area of the melting underwent "flattening" in comparison with only alloyed material.

No	Process parameters			3D topography parameters				2D roughness parameters	
	V_{rev} [rev/min]	V_f [mm/min]	X [mm]	Sa [µm]	Sz [µm]	$Sa_{(la)}/Sa_{(h)}$	$Sz_{(la)}/Sz_{(h)}$	Ra [µm]	RSm [mm]
1	4200	230	6	8.16	60.6	2,70	1,85	2.68	0.195
2	5950	230	6	11.1	87.6	2,00	1,28	4.83	0.201
3	4200	570	6	8.64	81.2	3,00	1,60	4.36	0.203
4	5950	570	6	10.1	77.6	2,56	1,68	3.95	0.19
5	4200	230	8.5	8.67	61.6	2,54	1,82	1.62	0.251
6	5950	230	8.5	11.4	73	1,95	1,53	1.79	0.242
7	4200	570	8.5	7.61	57.6	3,40	2,26	1.67	0.226
8	5950	570	8.5	8.4	52.8	3,08	2,46	1.87	0.282
9	3500	360	7	6.16	44.6	3,90	2,80	1.71	0.231
10	7100	360	7	9.01	66.4	2,66	1,88	2.9	0.25
11	5000	150	7	13.3	82.4	2,49	1,61	3.58	0.238
12	5000	900	7	9.93	80.2	3,55	1,90	4.44	0.207
13	5000	360	5	10.6	86.4	2,26	1,45	4.3	0.213
14	5000	360	10	7.96	51.2	3,02	2,44	0.87	0.28
15	5000	360	7	9.44	67.8	2,54	1,84	2.22	0.221
16	5000	360	7	12.1	68.6	1,98	1,82	2.44	0.219
17	5000	360	7	14.8	95	1,62	1,32	2.93	0.235
18	5000	360	7	5.97	45.6	4,02	2,74	2.03	0.234
19	5000	360	7	5.81	49.6	4,13	2,52	1.69	0.212
20	5000	360	7	7.84	57.6	3,06	2,17	2.24	0.245

Table 5. Surface topography and roughness parameters after the hybrid treatment.

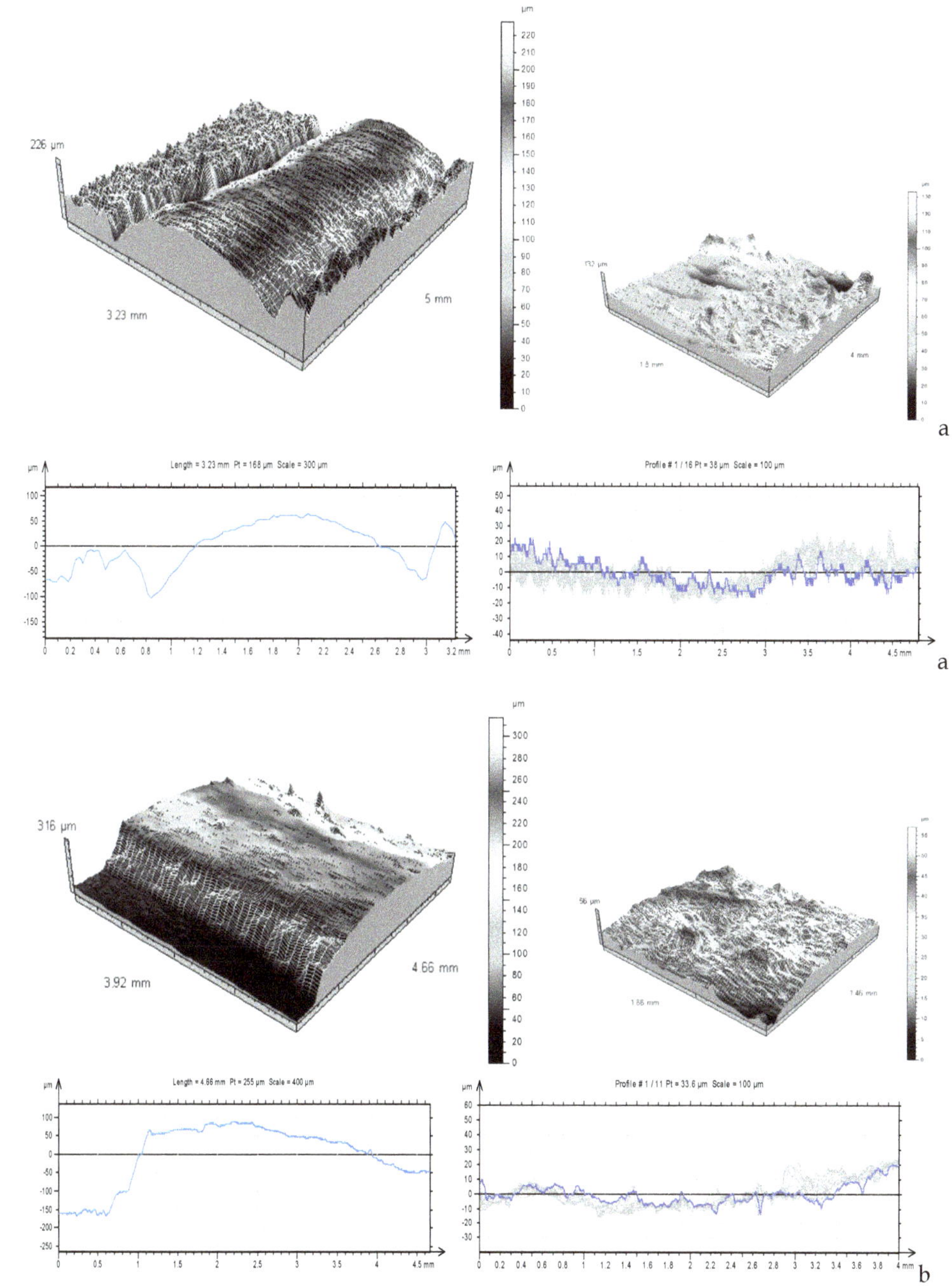

Fig. 11. The surface topography views of the laser path and section of the central part of path, and perpendicular and parallel profiles to the laser path and: a - laser alloyed surface, b - after hybrid treatment.

4.2 Correlations between roughness and treatment parameters

The analysis of the influence of process parameters (X, V_f, V_{rev}) on roughness parameters *Ra*, *RSm* and changes on topography parameters *Sz/Szh* was done, a good correlation was also stated.

The following relation between *Ra* and the treatment parameters was found:

$$Ra = 6+0.00037V_{rev} + 0.0017V_f - 0.76X \quad (4)$$

Multiple correlation coefficients, describing the relation between *Ra* and the parameters of hybrid treatment are high $R=0.83$. The dependence between studied properties is significant: $F=12.1$, $F > F_{kr}$. All coefficients are significant at the accepted level of confidence. Figure 3 shows a graphical interpretation of polynomial function that describes the relation between R_a and the rotational speed of the burnishing head, V_{rev}, and the distance between the tool axle and the axis of the laser beam, X, for fixed feed rate $V_f = 360$ mm/s. The parameter *Ra* grows with increasing, V_{rev}, V_f, and it lowers with the X distance increasing. With rise of the rotational speed of head, V_{rev}, the intensity of the burnishing process, and the forces of impact of micro-hammers on machined surface, grow. The increase in feed rate reduces temperature and intensity of the process of burnishing that means the number of micro-hammer strokes per unit area, leading to roughness asperity enlargement. The increasing of the X distance causes temperature reduces in the zone of mechanical treatment. This is also associated with the lowering of plastic properties of material, smaller plastic deformations of surface asperities and increase of the height parameter, *Ra*. high temperature of treatment occur.

The influence of the rotational speed of the head, as well as of the distance between the tool axle and the laser beam axis, on the selected surface topography parameters, *Ra* for constant value of feed speed V_f=360 mm/min, are shown in Figure 12.

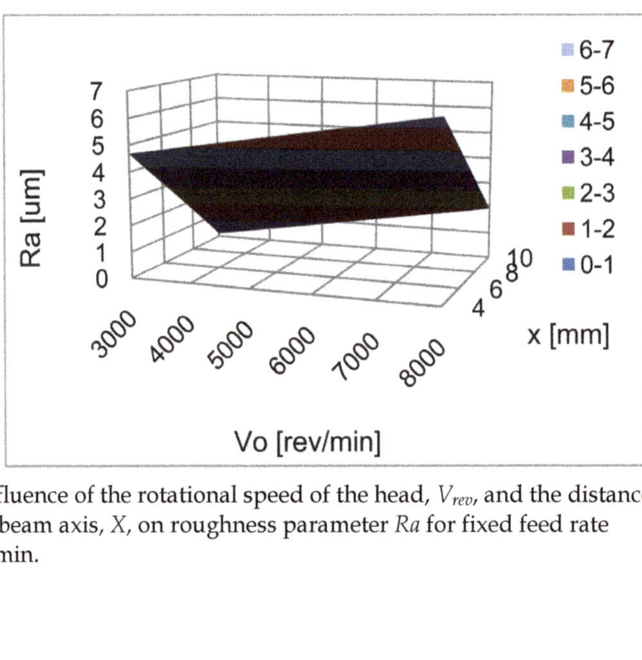

Fig. 12. The influence of the rotational speed of the head, V_{rev}, and the distance between tool axle and laser beam axis, X, on roughness parameter *Ra* for fixed feed rate $V_f = 360$ mm/min.

When analyzing the influence of process parameters on roughness sampling, *RSm,* a good correlation was also stated; $R=0.82$, $F=11.3$.Polynomial function, for which all the coefficients of the equation are significant at the confidence level of 0.1, shows good fit.

$$RSm = 0.085-0.000004\,V_{rev}+ 0.00002\,V_f + 0.017\,X \quad (5)$$

With an increase in the rotational speed of the head the roughness spacing decreases, while the increasing distance between the head and the laser beam and in the feed rate, causes lowering of RS_m. Analysis of the equation shows that the rise of temperature of the burnishing process affects the spatial property of surface geometrical structure increasing the distance between micro-asperities.

The analysis of the parameter $Sz_{(la)}/Sz_{(h)}$ allows the assessment of the degree of plastic deformation of surface asperities due to the micro-hammering.

$$Sz_{(la)}/Sz_{(h)} = 3.26+0.00039V_{rev}-0.00026\,V_f-0.29\,X \quad (6)$$

The biggest change in the asperity magnitude is obtained using the highest rotational speeds of the burnishing head, small feed rates and small *X* distances. At these parameters the large forces of micro-hammer impact on surface and high temperature of treatment occur.

The influence of the rotational speed of the head, as well as of the distance between the tool axle and the laser beam axis, on the parameters $Sz_{(la)}/Sz_{(h)}$, for constant value of feed speed V_f=360 mm/min, is shown in Figure 13.

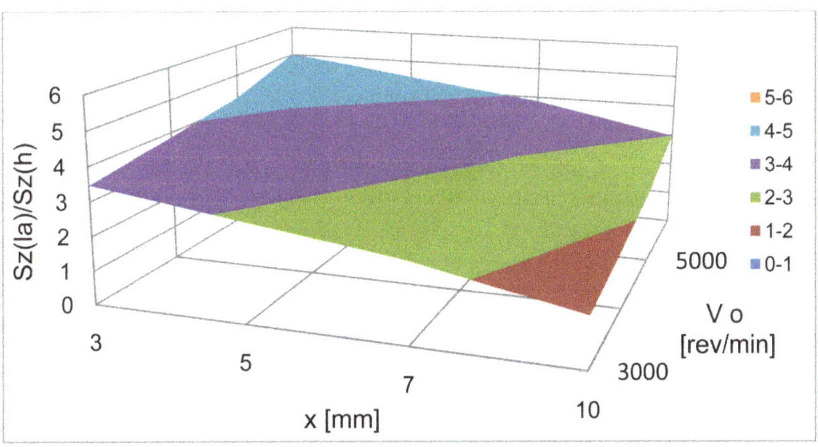

Fig. 13. The influence of rotational speed of the head, V_{rev}, and the *X* distance on changes of the surface topography parameter $Sz_{(la)}/Sz_{(h)}$ for fixed feed rate V_f =360 mm/min.

5. Surface layer properties

5.1 Wear resistance

The erosive tests were performed for sample 15 after laser-mechanical treatment carried at parameters presented in Table 4. It was stated that the increase of hardness compared to only laser alloyed Stellite layer was 70% while depth of plastic deformation was about 0.5

mm. The tests were done also for the plasma sprayed Stellite 6 layer and the layer formed by laser alloying at feed rate 360 mm/min.

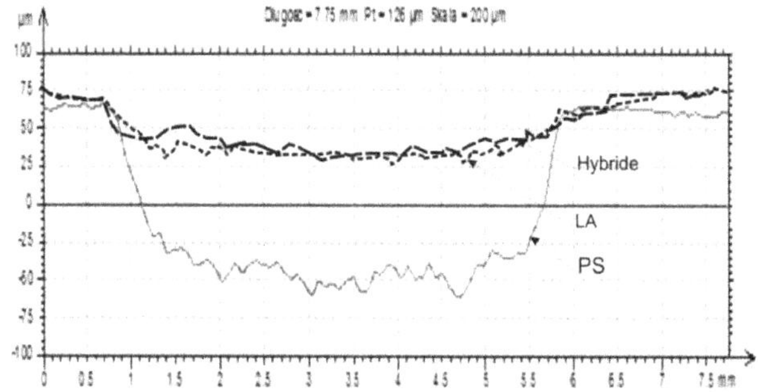

Fig. 14. The profiles after erosive wear test: PS-plasma sprayed Stellite 6 layer, LA- laser alloyed layer (P=2 kW, d=3 mm V_f =360 mm/min), Hybrid - sample 13 after laser-mechanical treatment (P=2 kW, d=2.5 mm V_f =360 mm/min, V_{rev} = 5000 rev/min, X = 5 mm).

Figure 5 shows exemplary surface profiles after erosive wear after laser-mechanical treatment, laser alloying and plasma sprayed Stellite 6 layer. The eroded surface is visible in the middle. The non-eroded side material parts were used as the reference surface. The profiles of examined surfaces show significant differences in erosive depth. The maximal depth was smaller than the thickness of tested layers. The measurement of surface topography was performed on three selected areas and the medium depth and volume per 1 mm^2 was chosen for quantitative analysis. This allowed compare erosive wear resistance of different surface layers of Stellite 6. Table 6 contains the average value of depth and volume of eroded surfaces whereas Figure 5 presents average eroded materials volumes loss.

Among Stellite 6 layers produced by plasma spraying, laser alloying and laser alloying combined with burnishing; the surfaces after hybrid treatment had the highest resistance to slurry erosive wear. The examined surfaces after laser-mechanical treatment showed the average depth of erosive loss 16.2µm. Surfaces after laser alloying showed little less of erosive resistance, 13-30 %, than hybrid treated surfaces. Their erosive depth was 22 µm.

The largest erosive loss was stated for plasma sprayed Stellite 6 layer, for which the average depth was 82 µm. Its erosive resistance is five times worse than the one of hybrid treated layer.

Treatment	The depth of erosive loss [µm]	The volume of loss [mm^3]
Laser alloying	22.2	0.0992
Hybrid treatment	16.2	0.0699
Plasma spraying	81.9	0.3621

Table 6. The average depth and volume of erosive losses of Stellite 6 layer after slurry erosive test

5.2 Contact stiffness

The results of contact strain and contact stiffness are shown in Table. The analysis of the total value of elastic and plastic strain shows that the surfaces after the burnishing undergo significantly lower elastic and plastic deformation in contact with counter sample. For samples without grinding this effect is associated primarily with surface topography. In the process of burnishing the rebuilding of surface topography occurs: surface irregularities have smaller heights, while their shapes become preferable. Therefore, the real contact area is increased and the unit pressure on individual asperities is reduced. At nominal pressure of 270 MPa the plastic deformation is 88 μm for samples that were only alloyed, whereas the burnished surface is deformed by 52 μm in height. It is the effect of more profitable geometric structure and the strengthening of surface layer due to the burnishing.

Grinding process has provided similar geometric structure: the surface heights and the shape of asperities are similar for all samples. In this case, examining the process of deformation of the surface under the influence of contact pressure, the differences in elastic and plastic properties of the surface layer material can be established.

Treatment	Contact strain at nominal pressure 270 MPa			Contact stiffness
	a_2 [μm]	a_{2e} [μm]	a_{2pl} [μm]	q=270 [MPa]
Laser alloying	138	90	88	3.38
Hybrid	80	28	52	1.96
Laser alloying & grinding	40	28	12	6.75
Hybrid & grinding	24	14	10	11.25
Plasma sprayed	57	36	27	4.74
Plasma sprayed & grinding	40	28	12	7.11

Table 7. The contact strain and the contact stiffness for nominal pressures 270 MPa. The hybrid treatment parameters: P=2 kW, d=3mm V_f =360 mm/min, V_{rev}=5000 rev/min, X=7 mm.

6. Discussions

The proposed new type of hybrid treatment, using the head for dynamic burnishing and applying the oscillatory motion, has lead to large plastic deformation of surface layer of steel 304 alloyed with Stellite 6. Application of the new head with two rows of hammers highly intensified the process and enabled the burnishing at various temperatures in one operation. The forces of the micro-hammer impact were changing in a wide range. The introduction of the oscillations increased the width of the melted zone by the value of the oscillation amplitude, provided more uniform surface plastic deformation, and also allowed for obtained favourable surface topography. The design of the head allowed the treatment in high as well as low temperature in the single pass. The temperature of the burnishing was around 850-200 K, depending on treatment parameters. The studies of residual stresses have shown the temperature in the zone of plastic deformation is sufficient for transforming the tensile stresses into compressive once at the material surface even at the highest temperature of the burnishing.

The results indicate that due to the burnishing at high temperature large plastic deformation of surface layer is possible to be obtain, without cracks and other defects of loosen structure that are characteristic for the classical burnishing of hard and brittle materials.

Good correlation between the process parameters and the features of plastic deformation zones and surface roughness was found. It enables controlling of the hybrid treatment. The variable degree of plastic deformation, strain hardening and thickness of plastic deformation zone can be governed by the controlling the impact force of micro-hammers on the surface and the temperature of metal in the treated zone. The application of high temperatures lowers the hardness while the plasticity of the material undergoes increasing, which in turn provides greater degree of plastic deformation and strain hardening of material. The hardness, depth of strain hardening zone increased, and residual stresses changed. Further increase of plastic deformation is possible to be obtained by the use of higher forces of micro-hammer impact on the surface. For the tested material the strain hardening, 32-100%, at the surface was obtained. Despite high degree of deformation there were no cracks no spallings. Fine grain material, homogeneous chemical and phase composition, was found.

The thickness of the strain hardened zone varying from 0.25 to 1.2 mm was obtained. This range is similar to the thickness of the typical mostly produced by LBM alloyed and cladding layers. It was found that with proper selection of process parameters it is possible to obtain the depth of strain hardening zone greater than the depth of the melting. This ensures the presence of compressive stress across the alloyed layer and provides greater durability, especially of parts subjected to fatigue during operation life.

The results of measurement of surface topography, contact stiffness and slurry erosive wear showed that laser-mechanical treatment allows the attainment better properties than the plasma sprayed and laser alloyed Stellite 6 layers have, and it can be used in specific industrial applications.

On the basis of regression equations and well-known effect of temperature on plastic properties of the material, it is possible to select parameters of the hybrid treatment in order to obtain the expected degree and thickness of the plastic deformation zone for other materials and layers. The hybrid treatment enables to combine in a single operation the advantages of laser treatment in the form of preferred microstructure, good adhesion, beneficial chemical composition of the surface layer and burnishing treatment. It ensures the increase of material hardness, improved surface topography and favourable compressive stresses of formed layer.

7. Conclusion

1. The hybrid treatment with the new dynamic burnishing head provides the extended range of plastic deformation of surface layer of steel 304, alloyed with Stellite 6. The increase in microhardness, caused by surface strain hardening, was 32-100% depending on the impact forces of micro-hammers on the surface and the temperature of the metal in the treated zone.
2. Due to the treatment at high temperature, despite the high degree of plastic deformation, no cracking and spallings or other phenomena proving loosen microstructure of the material were recognized.

3. The decisive influence on the thickness of the plastic deformation zone has temperature of the material in the region of burnishing. Thickness from 0.25 to 1.2 mm of the strain hardened zone was determined depending on the applied parameters of hybrid treatment. It enables the use of the hybrid treatment for the majority of layers produced by LBM.
4. The residual stress measurements showed the change in stresses within the melting zone from tensile stresses after the laser alloying to compressive ones after the hybrid treatment.
5. The hybrid treatment causes an increase in surface smoothness compared to the laser alloying. More than a threefold decrease in the average height of roughness, *Sa*, due to the burnishing process, was observed.
6. The increase of the contact stiffness in relation to the laser alloying and the Stellite 6 layer, formed by plasma spraying after hybrid treatment was stated.
7. The better slurry erosive wear resistance than for the plasma sprayed layer and the lower erosive rate compared to the laser alloyed layer were recognised.
8. The study of the correlation of treatment parameters with the state of surface layer showed that the dependences between the investigated properties are significant. Therefore, the controlling the hybrid treatment is possible and its industrial applications are recommended.

8. References

Abbas, G., West, D.R. (1991). Laser Surface Cladding of Stellite and Stellite-SiC Composite Deposits for Enhanced Hardness and Wear. *Wear*, Vol. 143, pp. 87-95, ISSN 0043-1648

Anthony, T.R., Cline, H.E. (1977). Surface Rippling Induced by Surface-Tension Gradients During Laser Surface Melting and Alloying. *J Appl. Phys.*, Vol. 48, pp. 1265-1272, ISSN 1089-7550

Arutunjan R,W., Baranow W.Yu., Bolszow L.A. Majuta D.D., Sebrant A.Yu.,(1989). *Laser beam effects on materials,* Nauka, ISBN -5-02-000747-X, Moscow, Russia

De Hosson, J. Th. M., Noordhuis J. (1989). Surface Modification by Means of Laser Melting Combined with Shot Peening. *Material Science and Engineering,* Vol. A121, pp. 1211-1220, ISSN 0267-0836

Demkin, M.B. (1959). A device for measuring the deformation at the point contact of two surfaces under compression. *Bulletyn Izobretanii,* Vol. 19, pp. 15-19

Filipowski, R. (1996). Application of matrix calculus for determining the coefficients of the linear regression for varying degrees of a matrix describing the set of normal equations. *The Archive of Mechanical Engineering,* Vol. 43, pp. 5-17, ISSN 0137-4478

Grum, J. Sturm, R. (2004). A new experimental technique for measuring strin and residual stresses during a laser remelting process. *J of Materials Processing Technology,* Vol. 147, pp. 351-358, ISSN 0924-0136

Ignatiev, M., Kovalev, E., Melekhin, I., Sumurov, I., Surlese, S. (1993). Investigation of the hardening of titanium alloy by laser nitriding. *Wear,* Vol. 166, pp. 233-236, ISSN 0043-1648

Meijer, J. (2004). Laser Beam Machining (LBM), state of the art and new opportunities. *J of Materials Processing Technology,* Vol. 149, pp. 2-17, ISSN 0924-0136

Milad, M., Zreiba, N., Elhalolouani, F., Baradai, C. (2008). Effect of cold work on structure and properties of AISI stainless steel. *J of Material Processing Technology*, Vol. 303, pp. 80-85, ISSN 0924-0136

Montgomery D.C. (1991). *Design and analyses of experiments*, ISBN 0471520004, 3th ed. N.Y., Wien,

Nowicki, B. (2007). *The dynamic micro-hammering head for surface metals treatment*, Patent No P 377600, Poland

Przybylski, W. (1986). *Burnishing technology*, ISBN 83-204-0742-7, WNT, Warsaw, Poland

Radziejewska J., Kalita W., Bartoszewicz A. Modification of surface layer properties by laser alloying combined with burnishing. *Proceedings of Laser Technologies in Welding and Materials Processing*, pp. 162-164, ISBN 966-8872-01-0 Katsiveli Crimea, Ukraine, May 2005

Radziejewska, J. (2006). Surface layer morphology due to laser alloying process. *J of Engineering Manufacture Part B, Proc. IMechE.*, Vol. 220, pp. 447-454, ISSN 0954-4054

Radziejewska J., Skrzypek, S. (2009). *J of Materials Processing Technology*, Vol. 209, pp. 2047–2056

Radziejewska J., Skrzypek, S. (2009). Microstructure and residual stresses in surface layer of simultaneously laser alloyed and burnished steel. *J of Materials Processing Technology*, Vol. 209, pp. 2047–2056, ISSN 0924-0136

Radziejewska J. (2011). Influence of laser-mechanical treatment on surface topography, erosive wear and contact stiffness. *Materials and Design*, Vol.32 pp. 5073-5081, ISSN 0261-3069

Robinson, J.R., Van Brussel, A.B., De Hosson, J. Th.M., Reed, R.C. (1996). X ray measurement of residual stresses in laser melted Ti-6Al-V alloy, *Material Science and Engineering*, Vol. A208, pp. 143-147, ISSN 0921-5093

Sai W. Bouzid, J.L. Lebbrun, J. (2002). *J of Materials Engineering and Performance*, Vol. 12, pp. 37-40

Shiou F-J., Chen Ch-H., (2003). Freeform surface finish of plastic injection mold by using ball-burnishing process. *J of Materials Processing Technology*, Vol. 140, pp. 248–254, ISSN 0924-0136

Shiou F-J., Hsu Ch-C., (2008). Surface finishing of hardened and tempered stainless tool steel using ball grinding, ball burnishing and polishing process on machine centre. *J of Material Processing Technology*, Vol. 205, pp. 249-58, ISSN 0924-0136

Tsai, C.-H. , Ou, C.-H. (2004). Machining a smooth surface ceramic material by laser fracture machining technique. *Materials Processing Technology*, Vol. 155–156, pp. 1797–1804, ISSN 0924-0136

Tian, Y., Shin, Y.C. (2007). Laser-assisted burnishing of metals. *Int. J of Machine Tools and Manufacture*, Vol. 47(1), pp. 14-22, ISSN 0890-6955

10

Covering with Carbon Black and Thermal Treatment by CO_2 Laser Surfaces of AISI 4340 Steel

G. Vasconcelos[1], D. C. Chagas[1] and A. N. Dias[2]
[1]Institute of Advanced Studies - IEAv, EFO-L, S. J. dos Campos, SP,
[2]University of Vale do Paraíba - UNIVAP, S. J. dos Campos, SP
Brazil

1. Introduction

The application of photo-absorbing coatings is a common practice, especially when lasers of low density power are used. These materials normally MoS_2, graphite and carbon black, futher the coupling of incident radiation, reducing the losses by reflection, common to the process, when CO_2 lasers are used as radiation source.

In a previous work, using graphite coatings, it was observed that part of the coating, after irradiation, remained on the metal surface. In pin on disc tests, it was observed a reduction in the coefficient of friction surface with this coating. REIS, J. L., (2009) improvement on the surface hardness, even using laser low of energy density power. This hardenning process was attributed to better coupling in the region of beam interaction with the metal surface. The laser hardening consists in heating and rapid cooling the steel surface. If the power density is enough, a layer on the steel surface will reach the austenitizing temperature (during heating) and then with rapid cooling, place the formation of martensites (Ganeev, R. A., 2002). The depth of the surface treated is determined by the law of thermal conductivity, where the propagation of heat occurs in a region of higher temperature to a region of lower temperature (Benedeck, J.; Shachrai, A.; Levin, L., 1980). The laser hardening allows the hardening of specific areas with controlled depth and with minimal surface deformation when compared to other methods. It also promotes, improves the mechanical properties and fatigue resistance, attraction, wear (reducing the friction factor) and increased resistance to corrosion (Dohotre, N. B., 1998; Machado, I. F., 2006). This work will evaluate the use of carbon black to replace the graphite used in the work of REIS, J. L. 2009, to eliminate the stage of solution preparation, grinding mills at high energy.

2. Methodology

The steel used in this work is AISI 4340. Its chemical composition was assessed by the optical spectrometer Thermo Scientific, Model ARL 3460 OES Metals Analyzer, presented in Table 1.

	Fe	*C*	*Mn*	*Si*	*Cr*	*Ni*	*Mo*	*P*	*S*
Steel 4340	95.79	0.361	0.638	0.261	0.794	1.702	0.221	0.024	0.008

Table 1. Chemical composition of steel AISI 4340 -% mass

Carbon black is formed by fine particles obtained by the process of pyrolysis or partial combustion of hydrocarbon gases or liquids. These nano-particulate structure, favors the coating with thin layers (Sector Report N 09, 1998). The shape of the particles was observed by scanning electron microscopy (SEM), Zeiss / EVO MA10, as shown in Figure 1.

Fig. 1. SEM of particles of carbon black (1320X)

The particle size of the lubricant can influence the thickness of the coating deposited and after irradiation with the laser beam. In order to determine the size distribution of particles, carbon black was subjected to particle size analysis through testing by laser diffraction (CILAS 1064L, range from 0.04 to 500µm). The results of this analysis are presented in Figure 2.

Samples of AISI 4340 steel with a thickness of 3mm and 20mm diameter, previously sanded (SiC paper 600), were coated with a solution prepared with 10g of carbon black and 0.1g of carboxilmetilcelulose in 100ml of ethanol.

This solution was mechanically mixed for 20 minutes in a plastic container with metal balls to the homogenization of the solution. Subsequently, the solution was sprayed with a pneumatic pistol on the surface of steel samples previously heated to 60°C. Then the samples are irradiated with a beam of CO_2 laser (50W) and beam diameter of 300µm. In the region of action beam on the sample surfaces, we used a flow of nitrogen to prevent oxidation. Figure 3 shows the laser used and Figure 4, the diagram of the experimental set up of the treatment process.

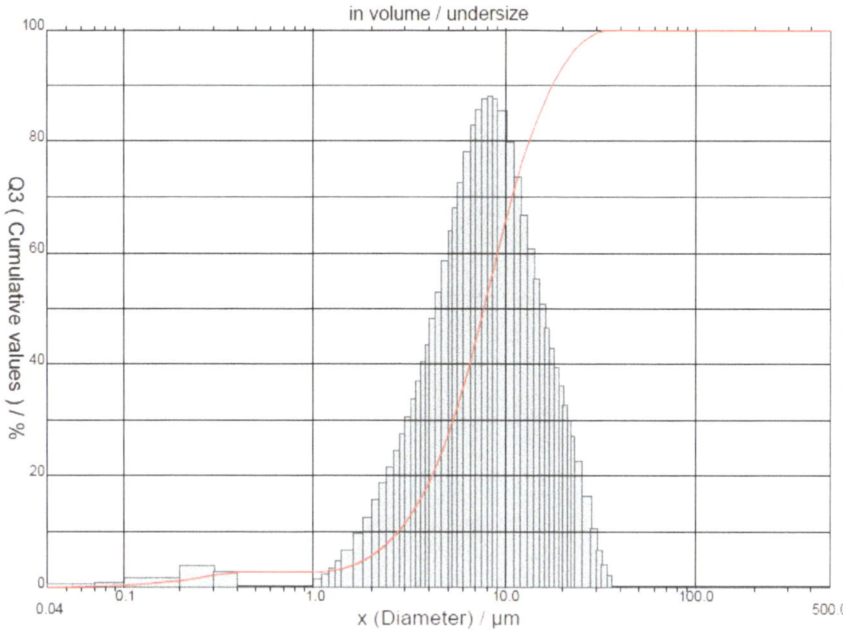

Fig. 2. Histogram of particle size distribution of carbon black.

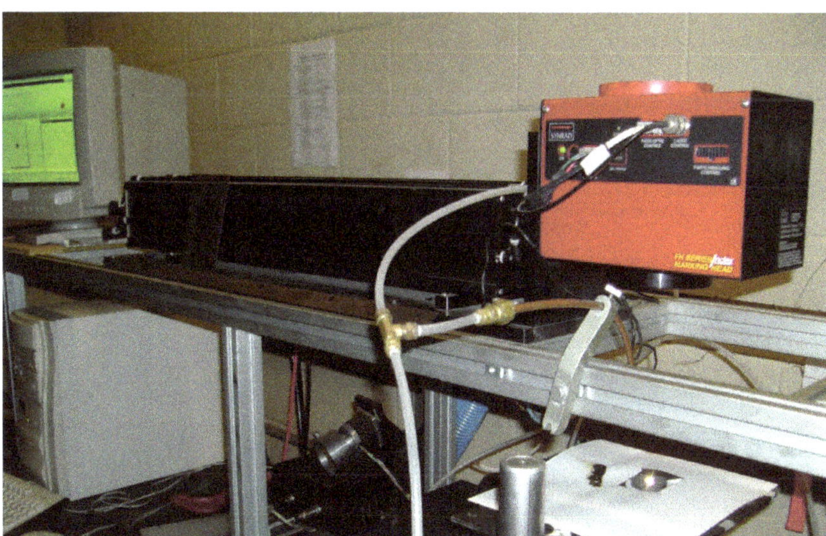

Fig. 3. Experimental set-up. CO_2 laser. Highlighted in red rectangular box, located galvanometric mirrors.

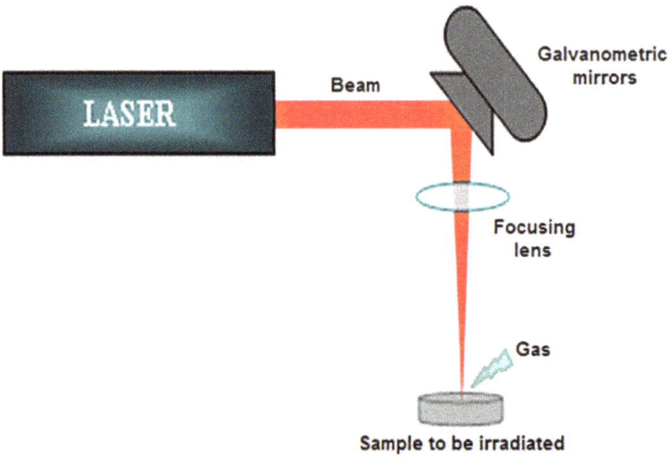

Fig. 4. Fitting of the experimental process. The laser beam is guided by a set of mirrors galvanometric controlled by software

The speed of scanning laser beam (mm/s), the resolution in pulses per inch (ppp) and number of heating cycles (NC) to be used in this experiment were selected from tests previously conducted (Chagas, D. C.; et al 2010). The Table 2 shows the parameters of the laser beam used in the treatment of the samples.

Samples	Speed (mm/s)	Resolution (ppp)	Number of cicles
P1	40	300	5
P2	60	300	5
P3	80	300	5

Table 2. Parameters of laser used for surface hardening of AISI 4340.

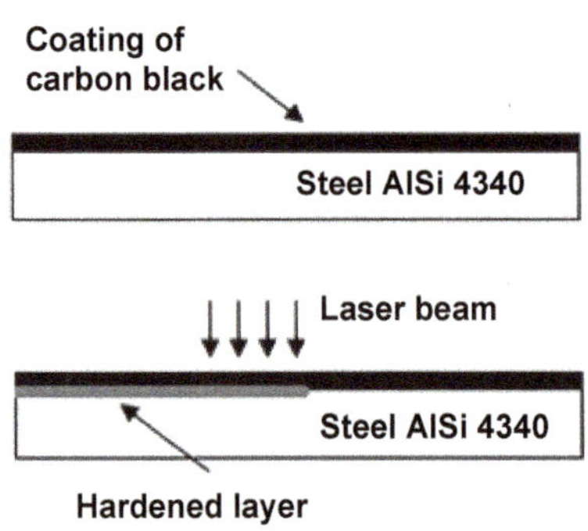

Fig. 5. Illustrates the layout of the treatment process, with laser beam, the samples previously coated with carbon black.

3. Results and discussions

The irradiated samples according to Table 2 were selected one that showed lower surface ablation and greater extension of the layer treated. These parameters were evaluated by measurements of roughness and optical microscopy (OM), respectively. Figure 6 shows a cross section of the treated region of the sample P2.

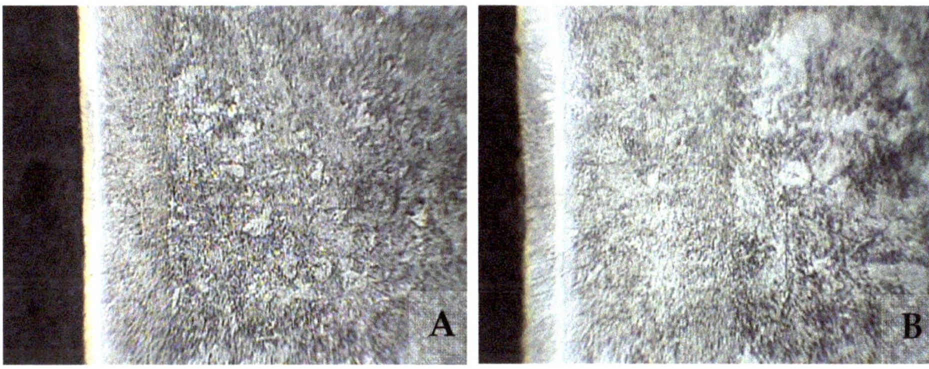

Fig. 6. Cross section of the irradiated surface of the sample P2. Optical microscopy a) OM-200X, b) OM-500X.

The average microhardness of AISI 4340 steel without heat treatment is 286 $HV_{0,05}$. After the thermal treatment with CO_2 laser, we can observe that the hardness of the material increased significantly, reaching an average of about 760 $HV_{0,05}$. Figure 7 shows the microhardness profile of the cross section of the treated region. These results were obtained by means of microhardness Future-Tech / FM-700.

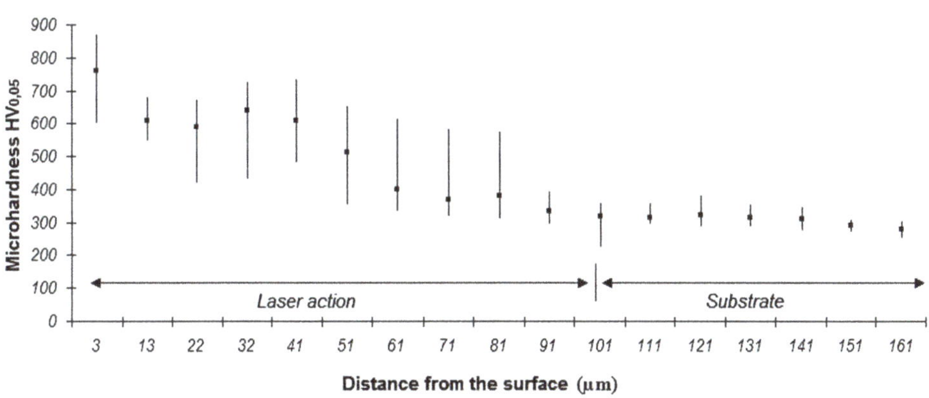

Fig. 7. Microhardness profile of cross section of the treated region after treatment via laser-coated carbon black.

According to Figure 7, there is increased hardness, occurred due to surface hardening process, resulting in heating and cooling of the sample.

The sample was subjected to tribological tests to evaluate the friction coefficient. The parameters used in the test were: linear velocity of 10cm/s, the track radius of 5mm, 52100 steel balls with 6mm diameter, number of rounds equal to 2000 and load of 5N. In this test, the sample is supported on a support rotation and pressed with a steel ball with known load, as shows the scheme of Figure 8.

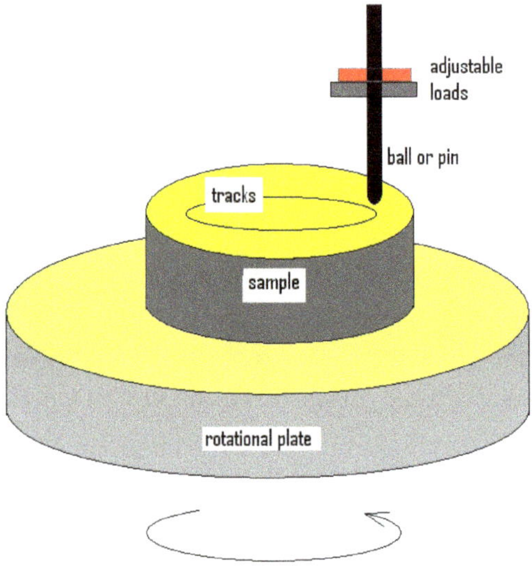

Fig. 8. Schematic drawing of the tribological tests

Then, the sample is rotated to evaluate the friction and the results obtained from the tribological tests, are presented in Figure 9.

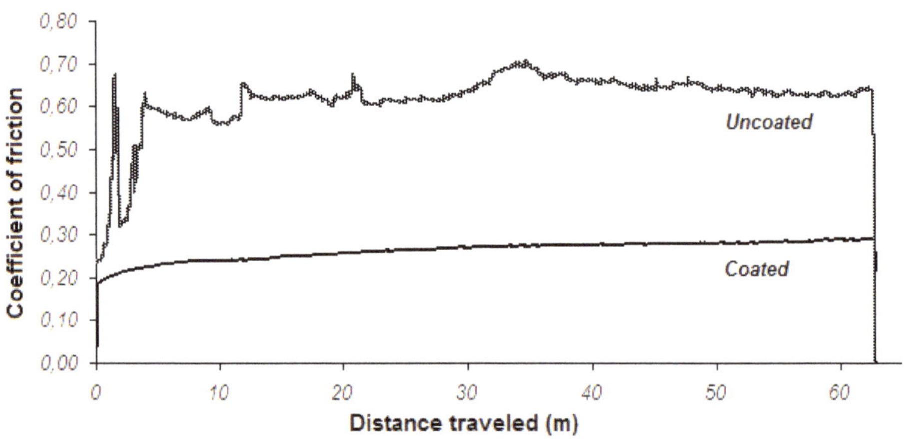

Fig. 9. Results of the friction coefficients of steel ball (52100) in 4340 and uncoated.

The tests of atomic force microscopy (AFM) were performed to evaluate the morphology and surface topography. The results obtained by AFM indicate a possible crystallization of carbon black, coalescence of grains and the appearance of new phases from the process of heating by laser, as shown in Figure 10.

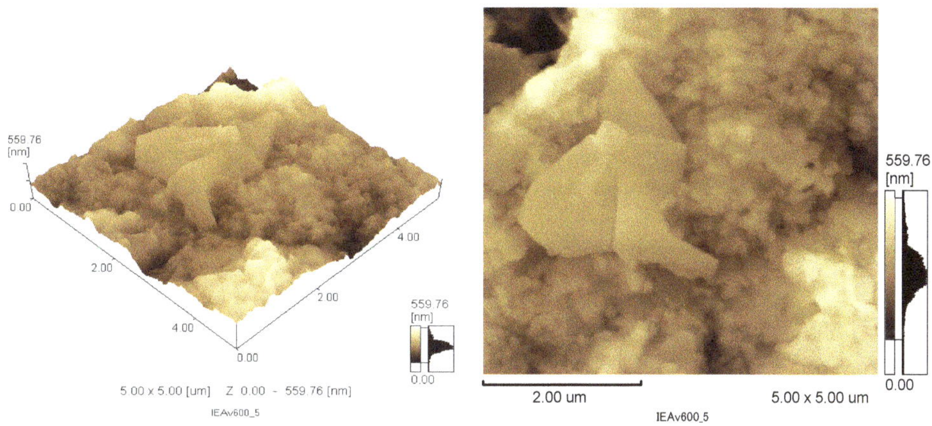

Fig. 10. AFM image obtained by the carbon black surface after irradiation with lasers. The bright spots in the figure correspond to higher regions and the darker the lower regions, according to the scale shown on the right side of the figure.

In roughness tests obtained by AFM, there is heterogeneous across regions with different surface roughness, where in the region obtained a RA= 87.999nm and region B was obtained RA= 124.115nm.

4. Conclusion

The experiments conducted indicate that the use of carbon black is feasible, presenting results of microhardness, friction coefficient, tempera and extent of the treated layer, similar to results reported by REIS (2009), and also eliminates the grinding step, which is required when using graphite.

The use of nanoparticles of carbon black aids in the absorption of radiation incident on steel, influences the alteration of microstructure and promotes surface temperature of the surface and still attached to the steel surface acting as a lubricant.

With the different laser parameters presented, it appears that high rates of speed in the beam of laser irradiation, the steel AISI 4340, showed no significant change in hardness, where the increase in hardness was 268 $HV_{0.05}$ to 405 $HV_{0.05}$. Unlike the lowest parameters that showed a homogeneous microstructure and with greater depths of layers treated, increasing by up to three times the surface hardness of steel, where the increase in hardness was 268 $HV_{0.05}$ to 760 $HV_{0.05}$. The extent of the treated layer, the homogeneity of the steel and its microstructure can be controlled by varying the parameters of the laser.

In tribological test, it was observed that the uncoated sample has a higher coefficient of friction of 0.70 and the samples coated with carbon black and further treated with lasers, have coefficient of friction of the order of 0.20, favoring better properties mechanical equipment and increase the service life.

In the trial by AFM showed that the coating presents heterogeneity throughout the area, with variations in surface roughness in different regions, possible crystallization of carbon black, coalescence of grains and the appearance of new phases resulting from via laser heating process.

5. Acknowledgement

Thanks to CNPq by financial support, the Group DEDALO-IEAv, Dr. J. R. Martinelli of the IPEN-USP and to Mr. A. Zanatta of the CCM-ITA.

6. References

Benedeck, J.; Shachrai, A.; Levin, L., 1980, Case hardening of steel by a CO_2 laser beam, Optics and Laser Technology. October.

Chagas, D. C.; Dias, A. N.; Vasconcelos, G.; Antunes, E. F., 2010, Surface treatment and covering by carbon black AISI 4340 by CO_2 laser action, ISSN 1983-1544. Ativ.P&D IEAv, v.3, p.84.

Dohotre, N. B., 1998, Lasers in Surface Engineering: Surface Engineering Series, Volume 1, ASM International - The Materials Information Society, Chapter 1 and 3.

Ganeev, R. A., 2002, Low-power laser hardening of steels, Journal of Materials Processing Tech., 121, 414-419.

Machado, I. F., 2006, Technological advances in steels heat treatment, Journal of Materials Processing Technology, 172, 160-173.

Reis, J. L., 2009, Thermal treatment of AISI M2 Steel by CO_2 Laser, 104f. Master Dissertation in Physics and Chemistry, Aeronautics Institute of Technology, São José dos Campos - SP.

Sector Report N 09, 1998, Chemical Complex, Carbon Black, BNDES.

11

Welding of Thin Light Alloys Sheets by CO_2 Laser Beam: Magnesium Alloys

Afia Kouadri-David
PSM Team, European University of Brittany, France, INSA of Rennes, LGCGM
France

1. Introduction

Laser welding is an important joining technique for magnesium alloys with their increasing applications in aerospace, aircraft, automotive, electronics and other industries. In this document the research and progress in laser welding of magnesium alloys are critically reviewed from different perspectives. Some important laser processing parameters and their effects on weld quality are discussed. The microstructure, metallurgical defects and mechanical properties encountered in laser welding of magnesium alloys, such as porosity, grains size, crystallographic texture and loss of alloying elements are described. Mechanical properties of welds such as hardness, residual stresses and other important structural properties are discussed. The aim of the chapter is to review the recent progress in laser welding of magnesium alloys and to provide a basis for follow-on research.

2. General principle of laser beam welding

Laser Beam Welding (LBW) consists in the laser beam focalisation on the workpiece surface. The high power density then created, induces metal ionisation and then plasma is formed. The vaporisation of the surface progressively forms a depression in the workpiece and then a keyhole, which allows the laser energy in-depth absorption. The melted metal will progressively fill the keyhole during the laser displacement, to form the weld. The two laser sources available are CO_2 and Nd: YAG. Laser CO_2 consists in a mixture of CO_2, N_2 and noble gases. The nitrogen discharges in CO_2 molecules activate the laser emission. The Nd: YAG (neodymium-doped yttrium aluminium garnet) consists in Nd^{3+} ions inserted in YAG crystal, the excitation is supplied by laser diodes. Nd: YAG laser light (λ = 1.06 μm) has a much higher absorption degree than CO_2 laser light (λ=10.6 μm). Both CO_2 and Nd:YAG lasers operate in the infrared region of the electromagnetic radiation spectrum, invisible to the human eye. The Nd:YAG provides its primary light output in the near-infrared, at a wavelength of 1.06 microns. This wavelength is absorbed quite well by conductive materials, with a typical reflectance of about 20 to 30 percent for most metals. The near-infrared radiation permits the use of standard optics to achieve focused spot sizes as small as .001" in diameter. On the other hand, the far infrared (10.6 micron) output wavelength of the CO_2 laser has an initial reflectance of about 80 percent to 90 percent for most metals and requires special optics to focus the beam to a minimum spot size of .003" to .004" diam. However, whereas Nd:YAG lasers might produce power outputs up to 500 watts, CO_2

systems can easily supply 10,000 watts and greater. The two laser processes are differentiated by the fabrication and the shape of the beam. It is generally accepted that the heat input parameter, defined as the ratio of beam power to beam travel speed, is well suited for describing LBW process. However, our results and those of the literature show that this parameter was not convenient, and that, the effect of the laser power and the welding speed parameters have to be differentiated, in particular for the light alloys such as magnesium or aluminium alloys. As a result of these broad differences, the two laser types are usually employed for different applications. The powerful CO_2 lasers overcome the high reflectance by keyholing, wherein the absorption approaches blackbody. The reflectivity of the metal is only important until the keyhole weld begins. Once the material's surface at the point of focus approaches its melting point, the reflectivity drops within microseconds.

3. Optimisation and influence of CO_2 laser beam parameters on thin sheets

The present investigation is concerned with laser power, welding speed, defocusing distance and type of shielding gas and their effects on the fusion zone shape and final solidification structure of magnesium alloys.

3.1 Laser power, P (kW)

The laser power is a critical parameter to obtain a full penetration depth and to control the weld bead profile. High power density at the workpiece is crucial to achieve keyhole welding and to control the formation of welds. Studies realized in this domain showed this effect of laser power on the penetration depth and weld width. The increasing beam power led to deeper and wider beads. Figure 1 shows the effect of laser power on the penetration depth (Fig. 1A) and weld width (Fig. 1B) for WE43 alloy welded at a speed of 33 mm/s and a focused diameter of 0.25mm (Dharhi et al., 2001a, 2001b as cited in Cao et al., 2006).

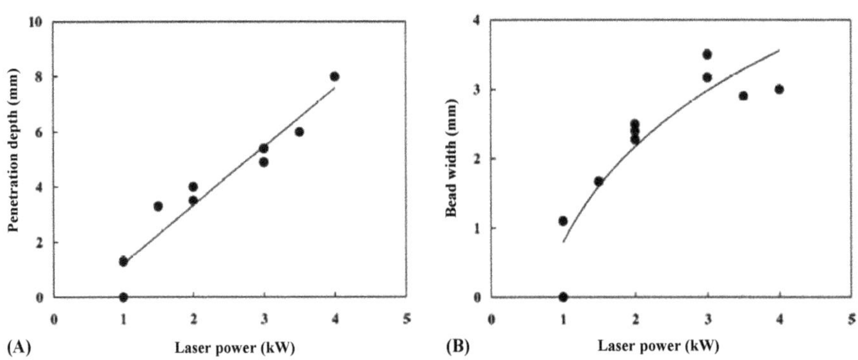

Fig. 1. Influence of laser power on the penetration depth (A) and on the weld width (B) for WE43 alloy (Dharhi et al., 2001a, 2001b)

The penetration depth and weld width increased with increasing laser power due to higher power density. Our experiences showed too the same evolution: the weld width becomes larger with increasing laser power. For example the threshold power to achieve full penetration is 2,5 kW (i.e. a power density of 2 MW/cm^2) for 3 mm AZ91 plates welded at a speed of 4 m/min and a focused diameter of 0.25 mm.

Then, when laser power was too low, lack of penetration was observed whereas high laser power produced laser cutting. These observations are consistent with the LBW literature. From macrostructure point of view, many authors observed the same evolutions where at low beam powers, some chevronlike pattern which is also called "ripples" (Marya et al., 2001). The mechanism of ripples formation is related to the effects of surface tension on the weld pool during solidification (D'Annessa, 1970 as cited in Cao et al., 2006). They also observed that the thickness in the weld area was slightly higher, a phenomenon which is called "crowning" or "humping". High beam powers led to deep and wide beads, and reduce both ripples and crowning (Marya et al., 2001). However, others authors showed at high beam powers, spatters and the evaporative losses would be produced. Authors (Weisheit et al., 1997) investigated the laser parameters for several magnesium alloys. They reported that for thin AZ31 plates (1, 8 mm), a 1.5 kW beam power was sufficient for achieving full penetration. In our experiences, for achieving full penetration, 3 mm AZ91 plates welds were produced at 4 kW (Kouadri & Barrallier, 2006).

3.2 LBW welding speed, V (mm/min)

The penetration depth and weld width both increase linearly with decreasing welding speed and decrease with increasing welding speed. The obtained results in the literature and reported on the figure 2A and 2B confirm the effects of welding speed on penetration depth and weld width at different levels of power for CO_2 lasers: The penetration depth and weld width both decrease linearly with increasing welding speed.

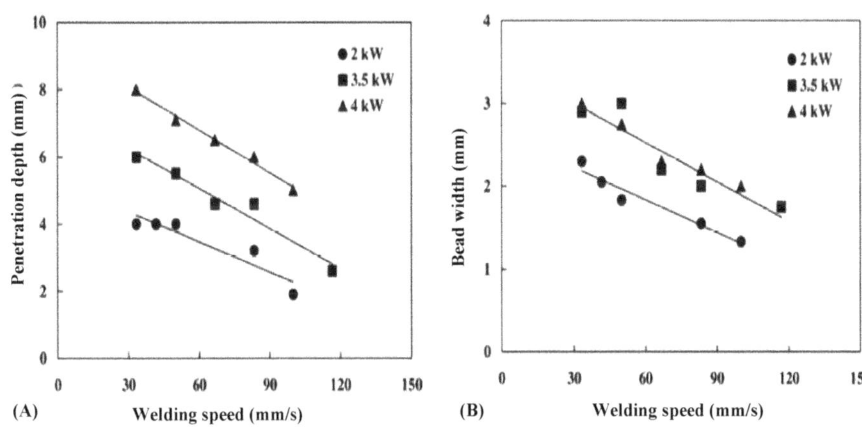

Fig. 2. Influence of welding speed on the penetration depth (A) and on the weld width (B) for WE43 magnesium alloy (Dharhi et al., 2001a, 2001b)

Moreover, it was reported that the speed lead too either to improve or to decrease the weld quality, in particular by the formation of the defects such as cracking or the pores formation.

Indeed, at low speeds, the interaction time between molten metal and surrounding air is large enough to allow pores to nucleate in large quantity, grow and escape from the molten pool as a result of buoyancy and convection flow. Moreover, when the welding speed is too slow, the bead produced by the superfluous heat exchange will extend to the side, and the heat influenced area will become too heat and extended, the seam metallographic structure crystal becomes thick, sometimes the cracking will appear, which will seriously influence the welding quality. When the welding speed achieves the lower limitation, the superfluous power absorption also will induce local evaporation loss and hollows. Moreover, using lower welding speeds induced no real change in the penetration depth but wider the weld width and especially the heat affected zone (HAZ) (Marya & Edwards, 2000).

From pores formation point of view, at high speeds, the pores do not have enough time to nucleate. The influence of the welding speed on pore formation was studied (Marya & Edwards, 2001). They found that the pore fraction goes to a maximum with increasing welding speed. The obtained results showed that using higher welding speeds reduced ripples but greatly increased crowning phenomena (Marya & Edwards, 2001), and the fusion zone appeared to be far more brittle (Watkins, 2003 as cited in Cao et al., 2006). Moreover, they showed a dependency between crowning and pores content, so that crowning is actually a relevant parameter to assess the weld quality. When welding speed was too high, lack of penetration was observed, whereas low welding speed produced laser cutting. This one is explained by the fact the power density increases with decreasing welding speed (Dharhi et al., 2000 as cited in Cao et al., 2006).

However these observations must be readjusted because the results depend on the nature of used magnesium alloys. Indeed, though similar welding parameters are used, various magnesium alloys exhibit different welding performance due to their different metallurgical and thermophysical properties. For example, die cast AZ91D has a lower thermal conductivity of 51 W/m K as compared with 139 W/m K for wrought AZ21A alloy. Thus, for similar welding parameters, the AZ91D alloy has a higher weld depth and weld volume compared with AZ21A alloy. It was also reported that greater penetration depth could be reached in AM50 alloy compared with AZ91 alloy welded under similar conditions using a 6kW CO_2 laser (Marya & Edwards, 2001). These observations explain why it is needed to systematically investigate the laser-welding characteristics of different magnesium alloys because of the difference in their thermal properties. In the same way, it is needed to take account of the geometry and thickness of the plates for readjusting the speed of welding. Many authors reported that a welding speed of 2.5-3 m/min was suitable for thin plates, when using 1.5 kW laser beam. Therefore, welding speed above 3 m/min should be achievable during CO_2 laser welding of 2 mm thick plates (Weisheit et al., 1997).

3.3 LBW density (W/cm^2)

It's the one more important parameter: the power density is one of most pivotal parameters in laser weld. When the laser power density is lower than 10^6 W/cm^2, the laser weld belongs to category of heat exchange weld. When the laser power density achieves only 10^6 W/cm^2, the deep penetration weld can be formed and "keyhole effects" appears.

The "keyhole effects" is closely correlative to the laser power density which is more low, the "keyhole effects" is more unstable even can not be formed, and the melting pool is also small. The melting depth of laser weld is directly correlative to the laser output power density and which is the function of incidence beam power and beam diameter. Therefore, to enhance the power density, we can enhance laser power or decrease the laser speed. A good balance had to be found to avoid laser cutting when the power density was too high and lack of penetration when the power density was too low. Concerning LBW, increasing the laser power (P) and decreasing the welding speed (V) result in an increase of the power density. This tendency is consistent with all the previous studies on laser welding.

3.4 Beam diameter

This is also a very important technical parameter, because in a certain output power, it will decide the density of beam power which is the key factor for laser weld. But for the laser beam with high power, it is difficult to measure, what is produced by the nature of the beam diameter. For laser weld, the condition of high effective deep penetration weld is that the power density on the laser focus must exceed 10^6 W/cm^2. We can adopt two methods to enhance the power density, one is to enhance the laser power, and the other one is to reduce the diameter of the beam. The power density has linear relation with the laser power, and has inverse-square ratio relation with beam diameter, so the effect of reducing beam diameter is better. In ours experiments, to realize full deep penetration weld of 2mm thick plates, we choose beam diameter of 1 mm with 4 kW CO_2 at 3 m/min.

3.5 LBW focal distance, f (mm)

For the sake of simplicity, the focal distance is defined as the distance between the focal point and the top surface of the sample. The position of focal points has an important influence on welding process and quality. The focal plane should be set where the maximum penetration depths or best process tolerances are produced. The laser welding usually needs some focus distance, because too high power density of the beam center at the laser focus is easy to vaporize and become bores. When the focus distance reduces to a certain value, the melting depth will suddenly change, which will establish necessary conditions for producing penetration pores. These most results in this domain showed that the focus distance influences not only the laser beam on the weld piece surface, but also the incidence direction of beam, so it has important influences to the melting depth and seam shape.

In our experiences, the most acceptable weld profile was obtained at defocusing distance of - 0.2 mm for 3 mm thickness where weld bead depth / width ratio is maximum and fusion zone size is minimum. In order to obtain the optimum value, complete penetration butt welds were made using previously obtained optimum laser power (4 kW) and optimum welding speed (3 m/min) (Kouadri & Barrallier, 2006). Sound welds were achieved with a focal point on the surface, which is consistent with what we found for thin plates. The weld width increases with moving the focal point away from the surface (i.e. increasing focal distance) which was also observed by other authors. These results indicated that the most effective range of defocusing distance to get maximum penetration with acceptable weld profile lies between zero and - 1 mm.

However, this distance has to be adjusted to obtain the best quality of welding. For example, the optimum defocusing distance to attain acceptable weld profile for 5 mm thickness was 0.4 mm under the surface of the workpiece (Cao et al., 2006). These results were consistent with the literature study (Dharhi et al., 2000, 2001, 2002 as cited in Cao et al., 2006). They studied 1-5kW CO_2 laser welding of 2mm AZ91 and 4mm WE43-T6 alloys. Their results showed that an adequate weld could be obtained for a focal position on or 1mm under the surface of the workpiece. Focal position on the workpiece surface had the smallest weld width while the weld width became larger when the focal position deviated above or below the surface.

Then, the optimal focal point position to weld thin plates lies on the top surface of the workpiece. Indeed, Weisheit et al., (1997, 1998) investigated 2.5kW CO_2 laser welding of some magnesium alloys. For thin plates (2.5 and 3 mm), the best welds according to penetration depth, aspect ratio and sag were achieved when the focal point was adjusted on the surface of workpiece, whereas for thick plates (5 and 8 mm) a position of 2mm below the surface of workpiece proved to be the best. Thus, the focal position should be moved deeper into the material for thicker work pieces and the following used process. Lehner et al., 1999 further researched the tolerance of focal position. For 3mm AZ91 and AM50 die castings welded using a 3 kW Nd:YAG laser, the best focal position is approximately 0.8mm below the workpiece surface, with a tolerance of ±0.5 mm. For 5mm material, the focal position has to shift to about 1.2±0.2mm below the surface.

3.6 LBW shielding gas flow, V (l/min)

Shielding gas selection produces a best weld quality. With the welding laser, the welding gas is flushed onto the workpiece through a nozzle system in order to protect molten and heated metal from the atmosphere. Gases have different chemical reactions and physical properties, which affect their suitability as assist gases for different welding tasks. At least three important points must be considered: tendency to form plasma, influence on mechanical properties and shielding effect.

Three main types of shielding gases are used: helium, argon and nitrogen. Helium is a gas characterized by minimum molecular weight, maximum thermal conductivity, and maximum ionization energy, thereby making it the most suitable gas for suppressing plasma formation. Argon, on the other hand, becomes ionized relatively easily and is therefore more prone to forming excessive amounts of plasma, in particular at CO_2 laser power over 3 kW. Carbon dioxide and nitrogen, on the other hand, are reactive gases, which may react with the weld metal to form oxides, carbides, or nitrides and get trapped in pores. This can result in welds with deficient mechanical properties. As a result, pure carbon dioxide or nitrogen are unsuitable as welding gases in certain applications in particular for the aluminium or magnesium alloys due to the oxidation.

To reduce the plasma effect, in these cases, it is advantageous to use inert gases such as helium or argon as welding gases, because there is no reaction on the weld metal and do not affect weld metallurgy. Indeed, in general, when the laser beam interacts with the workpiece, a hole is drilled through the thickness of the material. This hole or cavity is filled with plasma and surrounded by molten metal, thus, the high energy density of the focused beam could be lost easily. Weisheit et al., 1997 investigated the effectiveness of these three shielding gases and reported that a helium gas flow was the best choice. This plasma effect was reduced as a result of the higher ionization potential of helium, and then the weld profile was improved.

These welding gases have other functions, too. It protects the focusing optics against fumes and spatters and, in the case of CO_2 lasers, also controls plasma cloud formation. Leong et al., 1998 when welding 1.8 mm thick AZ31B-H4 used helium top shielding gas to protect from oxidation, combined with helium back shielding and nitrogen shielding to protect the optics. Wang et al., 2006 studied the influence of gas flow rate on weld width and reported that increasing gas flow up to 20 l/min is needed to affect the susceptibility to oxidation.

As magnesium is highly susceptible to oxidation, a protective atmosphere is required during welding. Surface cracking leading to laser welding was observed without gas protection. This is due to the oxide formation during welding in the magnesium alloys. To increase the magnesium alloy weldability, argon or helium are the most common choices. Argon is heavier than air so it provides a better shield than helium, but it ionizes easily and has much lower thermal conductivity than helium. This causes a problem with high power CO_2 welding: The metal vapour emerging from the keyhole is partially ionized, with charged atoms and free electrons. The free electrons absorb some of the laser light, reducing the power available for welding. As the vapor absorbs energy, it heats up, increasing the number of free elections and further increasing absorption. Helium shield gas is more effective than argon in suppressing this effect because it cools the vapor plume and does not contribute many electrons itself. This welding gas often plays an active role in the welding process, such as increasing the welding speed and improving the mechanical properties of the joint. Weisheit et al., 1997 investigated the effectiveness of these three shielding gases and confirmed that a helium gas flow was the best choice.

In addition, often to increase the weld quality, helium/argon mixtures combining the benefits of both gases, i.e. the higher density of argon and the higher ionization potential of helium, may be used to obtain better protection of the weld zone in CO_2 laser welding. Hiraga et al., 2001 studied 1.7 mm thick AZ31B-H24 butt joints and get some improvements using argon back shielding in addition to the helium centre shielding. With these two gases, weld profile is remarkably improved where fusion zone interfaces are almost parallel to each other. The melting depth increases with the increase of gas flux, but too much gas flux will induce the surface hollow even penetration of the melting pool. Indeed, higher porosity content was observed for He gas flow higher than 50 l/min. Using Ar back shielding gas allowed us to produce sound welds at lower welding speed, reducing sag of the weld pool. Our study led to the same conclusions and sound welds were produced (Kouadri & Barrallier, 2006). Then, the optimum shielding system consists in a top helium flow superior to 20 l/min and Ar back shielding. By adding single-sided access, laser welding is even more strategically advantageous.

4. Application of laser beam CO_2 on thin sheets of magnesium alloy

The presented material is a cast magnesium alloy (AZ91D) welded by laser CO_2 processing. The alloy used for the study of the laser welding is a ternary magnesium - aluminium - zinc of designation AZ91, according to standard ASTM. Laser welding of magnesium alloys appears to be a challenge itself. Indeed, the ability to produce laser welds depends on the properties of the material to be welded. Then, magnesium being characterised by quite unfavourable properties (i.e. low absorptivity of laser beams, strong tendency to oxidize, high thermal conductivity, high coefficient of thermal expansion, low melting temperature, wide solidification temperature range, high solidification shrinkage, a tendency to form low melting-point constituents, low viscosity, low surface tensions, high solubility for hydrogen in the liquid state), processing is expected to be an issue.

The used magnesium alloy in this study were obtained by high pressure die casting under neutral gas and did not undergo heat treatment to match the conditions generally encountered in automobile applications. The provided plates were sheared to recover 3-mm-thick samples. Their edges were machined by milling. The plates were welded together side by side using a laser beam CO_2, which penetrated through the thickness of the plates. The welding was performed using a 4000 W CO_2 laser in an inert helium atmosphere. The speed of welding was optimised in the range of 1.0 - 4.25 m/min and the power in the range of 1 - 4 kW.

The objective of this part is to show the evolutions of the metallurgical and mechanical properties generated by the laser CO_2 in thin AZ91 magnesium alloy sheets. The presented results were obtained with optimized parameters of CO_2 laser beam welding. This part shows the microstructure modifications (characterization of the grain size, chemical properties and the crystallographic texture) occurring during laser welding in every zone of the welded sheet. From mechanical properties point of view, we present the evolution of the hardness and the residual stresses. These one have been performed by taking into account the crystallographic texture. The strain measurements and the characterization of the crystallographic texture have been performed using X-ray diffraction techniques. The set of results demonstrated that laser welding induces the presence of several distinct zones which have distinct microstructural and mechanical properties.

4.1 Study of metallurgical properties

4.1.1 Macrostructure analysis

From macrostructure point of view, a narrow weld joint is an important characteristic of high power density welding. The 4 kW CO_2 laser welding in an inert helium atmosphere (2 bars) with a speed of welding of 2 m/min of 3mm AZ91D plates showed that the fusion zones have widths of approximately 0.8 1.6 mm (Kouadri & Barrallier, 2006). The region with a width of about 200 - 500 μm between the base metal (BM) and the fusion zone (FZ) can be recognized as the heat affected zone (HAZ). However, the width of the HAZ is defined according to the variations of laser beam parameters. For example, in the literature, the 6 kW CO_2 laser welding with a speed of welding of 3.5 m/min of wrought AZ31B alloy indicated that the width of the HAZ was 50–60 μm, but can be doubled at substantially slower speeds (Leong et al., 1998; Sanders et al., 1999). These results showed that the width of the HAZ is tightly connected to laser process parameters.

4.1.2 Microstructure analysis

From microstructure point of view, the microstructure of the laser welds is characteristic of a high-speed process in which heat is rapidly extracted from the molten fusion zone by surrounding base material. In our study, the mean grain sizes of the base metal (BM) range from 50 to 200 μm (figure 3a).

The BM is heterogeneous and characterised by a mixture of a large primary α-Mg phase and of a (α-Mg + β-$Mg_{17}Al_{12}$) eutectic phases. This later constituent is a so-called abnormal eutectic (Kouadri & Barrallier, 2006, 2011; Dubé et al., 2001; Luo, 1996) because of its lamellar shape. The base metal exhibits small precipitates dispersed in the matrix but mainly located at the grain boundaries. These precipitates are β-$Mg_{17}Al_{12}$ and to a lesser degree

Al_8Mn_3. The typical overall microstructure of the HAZ is shown in figure 3b. The microstructure of HAZ has coarse grain polygonal Mg as the base metal. Nevertheless, eutectic grains disappeared whereas a continuous β-$Al_{12}Mg_{17}$ phase was created at grain boundary. At the fusion boundary, where a relatively large thermal gradient and small growth rate are established, the microstructure is predominantly cellular (Marya & Edwards, 2000). Grains usually grow epitaxially from the Fusion Zone (FZ) –Heat Affected Zone (HAZ) interface.

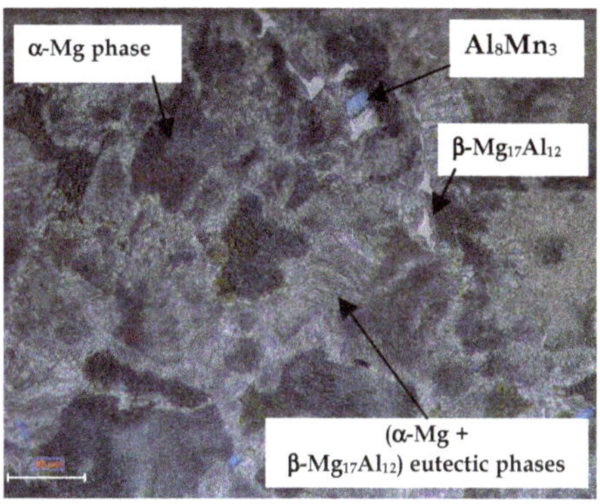

a

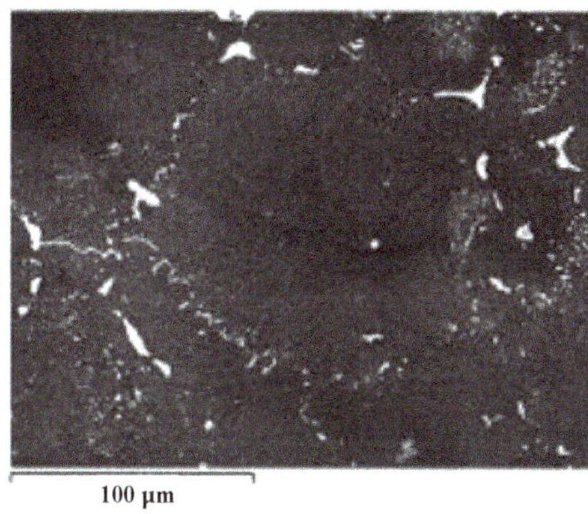

b

Fig. 3. a) Microstructure of the base metal, solid solution α-Mg and (α-Mg + β-$Mg_{17}Al_{12}$) eutectic phase and precipitates (OM), b) Microstructure of the Heat Affected Zone (HAZ) by SEM (Kouadri & Barrallier, 2006, 2011)

In the fusion zone (FZ), the microstructure is very different from the base metal (Figure 4a, 4b).

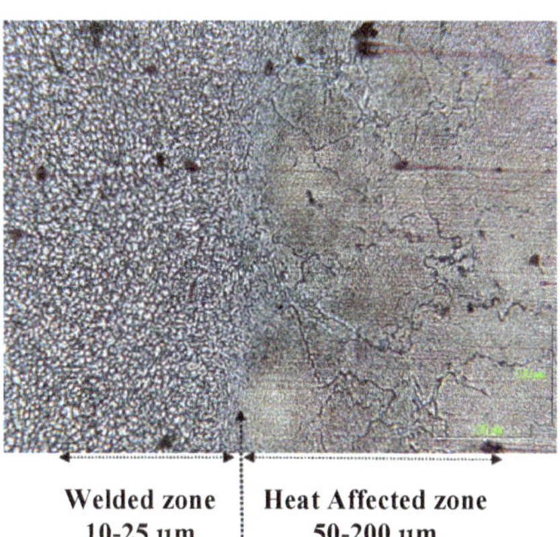

a

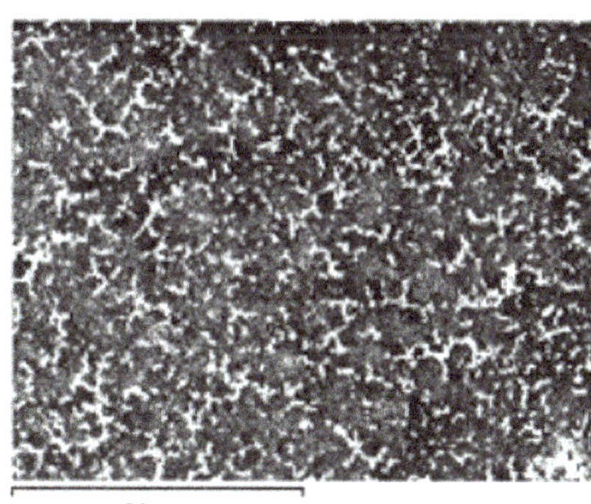

b

Fig. 4. a) Microstructure in the welding zone and in the heat affected zone by (OM), b) Microstructure of the fusion zone (SEM) (Kouadri & Barrallier, 2006, 2011)

The α-Mg microstructure is much finer ranging between 10μm and 25μm. There is β-$Mg_{17}Al_{12}$ precipitates too, clearly present and located in grain boundaries. So, the microstructure appears to be more homogenous at scale lengths of a few micrometers. This fine equiaxed grains in the fusion zones formed by cellular growth were also observed by the others authors (Cao et al., 2006) in Zr-containing ZE41A alloy. Weisheit et al., 1997, 1998 have also observed a cellular morphology in all joints except for the WE54 alloy which showed a more globular grain shape. It was further observed the equiaxed morphology in AM60B alloy occurring at low welding speeds. At higher welding speeds, however, the morphology changes from equiaxed to dendritic forms (Pastor et al., 2000). In the same way, our observations showed that the rapid cooling experienced during laser welding leads to a significant grain refinement with cellular growth in the fusion zone. In brief, the laser welding leads to a grain refinement some is the initial structure (Haferkamp et al., 1996, 1998). Only the grain morphology changes following the used laser parameters. Indeed, it was reported that the original microstructure has little influence on the fusion zone structure though magnesium alloys can be welded in different conditions (Weisheit et al., 1998).

4.1.3 Distribution of grain size and volumetric fractions of phases by image analysis

An example of the statistical distribution of the different grain sizes obtained by grain count is presented on the figure 5a.

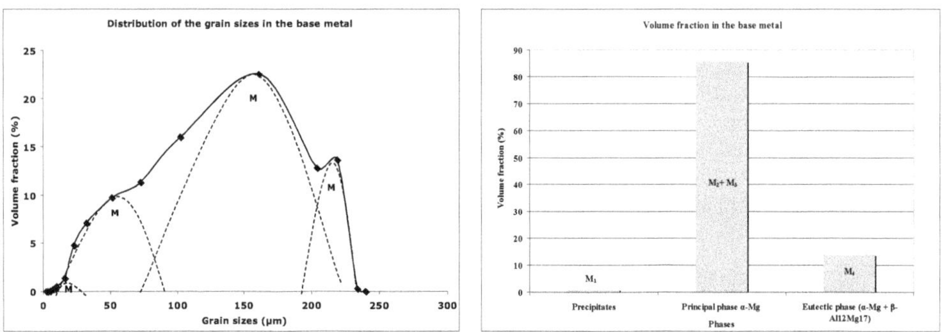

Fig. 5. (a) Statistical distribution of the grain sizes of AZ91 alloys in the base metal. (b) Volume fraction of the α-Mg grains, of the eutectic (α-Mg grains β-$Mg_{17}Al_{12}$) phase and of the precipitates in the base metal (Kouadri & Barrallier, 2011).

This distribution shows the presence of four modes; M_1, M_2, M_3 and M_4: The precipitates (M_1), whose main mode is about 10 μm and the principal α-Mg (M_2 + M_3) phase, whose main modes are 50 and 160 μm. Finally the principal mode of the eutectic phase (α-Mg + β-$Al_{12}Mg_{17}$) is about 220μm. The volumetric fractions calculation (figure 5b) demonstrates that the volumetric fraction of the base metal for the principal α-Mg phase is estimated at 85.4%, that of the eutectic phases at 13.8% and those of the precipitates at 0.8%. These results are in line with those in the literature and the diagram of the alloy phase AZ91 (StJohn et al., 2003). The large reduction of grains in the fusion zone is confirmed by statistical distribution of the grains (Figure 6a) where the principal mode is 16μm. The volumetric fraction of the principal phase represented by modes 2 and 3 constitutes 96% of the matrix (Figure 6b). The eutectic phase has almost disappeared in the welded zone.

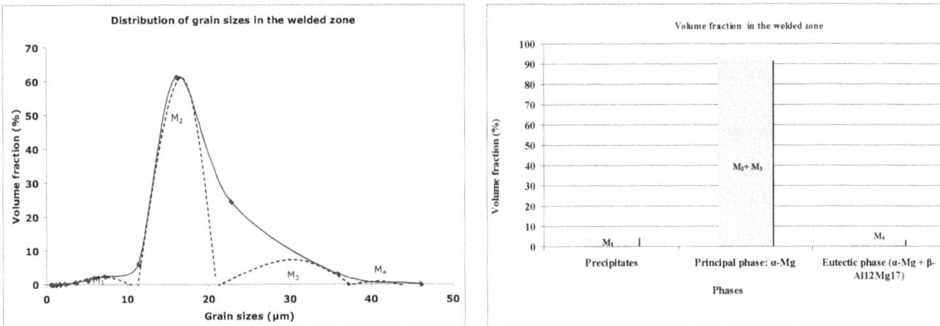

Fig. 6. a) Statistical distribution of the grain sizes of AZ91 alloys in welded zone, b) Volume fraction of the α-Mg grains and of the β-$Mg_{17}Al_{12}$ precipitates in the welded zone (Kouadri & Barrallier, 2010).

4.1.4 Chemical analysis

Generally, the laser welding leads to the redistribution of chemical composition or evaporative losses in the fusion zone. In the case of magnesium alloys, the temperatures reached within keyholes are far greater than the boiling temperatures of magnesium, aluminium or zinc (Liu et al., 2000). Thus, during welding, the preferential evaporative losses consist primarily of zinc and magnesium and then aluminum. This evaporation causes a variation of chemical composition in the fusion zone, especially at high laser power density which leads to a new chemical redistribution.

In our studies, the welding of AZ91 D plates, the chemical composition (% weight.) of each individual phase in the different phases was studied using EDS analysis. The base metal constitutes a reference state for the comparison with the welded area. In the table 1 energy dispersive spectroscopy (EDS) analysis indicated that the α-Mg grain contains up to 8.1 and 1.15 wt. % of Al and Zn, respectively. The α-phase contains ≈ 32 wt. % Al (Kouadri & Barrallier, 2006). All these results were in complete agreement with the literature on AZ91 alloy microstructure, which has been widely studied.

Chemical Composition (weight %)		Al	Mg	Zn	Mn	Fe	Si
Base metal	Matrix ***α, (α + β)***	8,1	90,6	1,15	0,15	-	-
	Precipitates β	32,5	67,5	-	-	-	-
	Precipitates Al_8Mn_3	32,8	14,5	-	52,7	-	-

Table 1. Chemical composition of AZ91 alloy in base metal, BM, (EDS analysis) (Kouadri & Barrallier, 2006).

The tables 2 and 3 show the evolution of the chemical composition in the HAZ and in the welded zone.

In the HAZ, higher Al and Zn concentrations in the α-phase were measured as compared to the base metal and to the welded zone (Kouadri & Barrallier, 2006). So, this confirms that β-$Mg_{17}Al_{12}$ and Al_8Mn_3 precipitates were diluted in a matrix made of over-saturated α-phase.

In the welded zone, the chemical composition of phases was much the same as in base metal (Kouadri & Barrallier, 2006) except in a thin superficial layer close to the surface. The Al-content in the β-phase decreases from 30 % (weight) in the BM down to 17% (weight) in the fusion zone. Likewise, we could see a strong decrease of the Al-content in every crystalline phase. We did'nt observe evaporative losses due to a good optimisation of the laser parameters. Indeed, it is known that higher energy densities lead to greater evaporative losses, increased spatter, and uneven weld beads. Thus, minimizing the irradiance incident upon the workpiece would reduce the loss of high vapor pressure elements. For example, larger reductions of both Mg and Zn were also reported at slower travel speeds (Leong et al., 1998; Sanders et al., 1999). We can conclude in the case of the magnesium alloys that there aren't evaporative losses of Mg and Zn if the laser parameters are optimized. There is only a chemical redistribution of the overall Al quantity for example in our case, due to the solidification conditions (Kouadri & Barrallier, 2006, 2011; Dubé et al., 2001; Luo, 1996). This redistribution should then be carefully controlled and optimized by manipulation of welding parameters.

Chemical Composition (weight %)		Al	Mg	Zn	Mn	Fe	Si
HAZ	Matrix α	8,51	90,31	1,15	0,04		-
	Precipitates β	26,83	69,63	3,50	0,04		-

Table 2. Chemical composition of AZ91 alloy in the HAZ, (EDS analysis) (Kouadri & Barrallier, 2006).

Chemical Composition (weight %)		Al	Mg	Zn	Mn	Fe	Si
welded zone	Matrix α	8,3	90,3	1,1	0,2	-	0,1
	Precipitates β	29,9	69,1	-	-	-	-

Table 3. Chemical composition of AZ91 alloy in welded metal, (EDS analysis) (Kouadri & Barrallier, 2006).

4.1.5 Texture characterisation

In general, texture develops in a metal as a result of processes such as crystallisation, plastic deformation.... The practical importance of preferred crystallographic orientation results from the dependence of many mechanical and physical properties on crystal direction. Thus, a textured material will have, in general, anisotropic values for a number of parameters including the yield strength, Young's modulus and Poisson's ratio. In order to understand how preferred crystallographic orientations might occur in laser welding, it is necessary to consider the formation and structure of the fusion zone in detail.

Initially, in our study, the base metal is characterized by a random orientation. Likewise, there is no more texture in the HAZ with 90% of the grains being randomly oriented. This is consistent with Coelho et al., 2008, study. In the fusion zone, the microstructure consists of fine and randomly oriented equiaxed dendrites nucleated. The texture analysis showed clearly that there is no texture. Indeed, ODF calculation indicates that more than 99% of the grains are randomly oriented (Kouadri & Barrallier, 2006, 2011).

However, close to the surface, the AZ91 alloy exhibits two preferential orientations concerning 77% of the grains (Figure 7a).

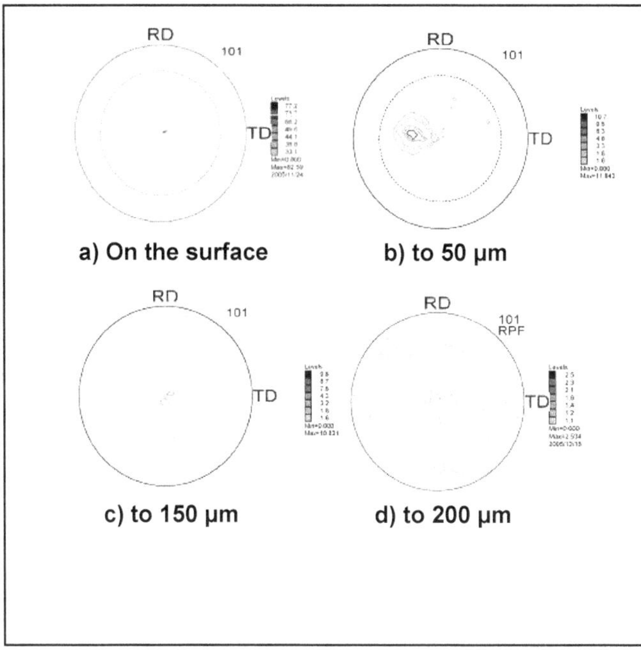

Fig. 7. Evolution of the crystallographic texture (pole figure 101) from the surface to a depth of 200 μm: a) Texture to the surface, b) Texture to 50 μm, c) Texture to 150 μm, d) Texture to 200 μm (Kouadri & Barrallier, 2006, 2011).

The $\{10\bar{1}1\}$ pole figure showed that a large fraction of grains are oriented with pyramidal $\{10\bar{1}1\}$ planes parallel to the surface of the sheet. The intensity of the center of this pole figure is a tenfold improvement in intensity compared with other peaks. A close look at the position reveals that the normal axis is tilted at an angle of ± 4° from the normal sheet direction to the transverse direction and around the welding direction. The ODF calculation allows us to say that 71% of the grains have an orientation $\{10\bar{1}1\}\langle\bar{3}4\bar{1}3\rangle$ corresponding to the 3 Euler angles (φ_1 = 65,6°, ϕ = 60,8°, φ_2 = 54,4°). Likewise, the $\{10\bar{1}0\}$ pole figure shows the presence of a smaller proportion of grains with $\{10\bar{1}0\}$ planes parallel to the surface. Such grains show two other poles $P_i^{\{10\bar{1}0\}}$ and $P_j^{\{10\bar{1}0\}}$, on both sides of the centre, due to the multiplicity of the hexagonal symmetry, which is equal to three. These poles are tilted at 60° around the centre. The ODF calculation allows us to say that 6% about of the grains have an orientation $\{10\bar{1}0\}\langle 0\bar{3}34\rangle$ corresponding to the 3 Euler angles (φ_1 = 65°, ϕ = 90°, φ_2 = 60°). The most likely explanation for the variation of texture between the surface and the depth of the fusion zone is that the differences of the thermophysical and thermomechanical properties of the investigated location affect the process of solidification and plastic deformation, leading to different final out-comes. In our study, the nature of this texture has been explained by the thermodynamic conditions of minimisation of surface energies

(Kalinyuk et al., 2003; Matysina, 1999) which results in the presence of columnar grain growth at the surface of the welded zone (Kurtz et al., 2001). Between the surface and the 200 µm depth (Figure 7b, 7c, 7d) the texture decreases to disappear completely from 200 µm. This change underlines the presence of a transition from columnar growth to equiaxial grain growth. These results showed that the laser welding led to a complex microstructure and induced high temperature and deformation gradients which may cause changes in crystalline orientations. The study of the texture evolution is then required to understand the anisotropic characteristic of the welds and its influence on mechanical properties. Compared with the literature, little study takes into account the texture due to laser welding. So our results showed that the laser welding can form a crystallographic texture and that it is necessary to study it thoroughly to apprehend the mechanical properties as well as possible.

4.2 Experimental results of the mechanical properties

4.2.1 Hardness characterization

From laser parameters point of view, hardness in the fusion zone was found to increase almost linearly with welding speed because higher welding speeds lead to a more significant refinement of the microstructure and more alloying elements into the matrix, even though hard intermetallics are reduced and more finely distributed at high cooling rates. The average hardness of CO_2 laser welded joints decreases with slower welding speeds. Indeed, at low welding speeds the weld structure and hardness were nearly the same or sometimes lower as those in the base die-cast material. The decrease in the hardness of the HAZ was due to grain growth. However, these results depend too on the laser power and the nature of used alloy. In the literature, it was also reported that there was a gradual decrease in hardness of 6 kW CO_2 laser welded joints from the BM to the HAZ to the FZ of AZ31BH24 alloy, with a minimum value in the FZ (Leong et al., 1998; Sanders et al., 1999). Dhahri et al., 2001 investigated WE43-T6 alloy using 5 kW CO_2 laser. The hardness at the top and bottom of the welds was similar but the hardness in the middle of the bead was lower. In our studies, the 4 kW CO_2 laser welding of die cast AZ91D alloy showed that there is an increase in hardness of the fusion zone but little variation in hardness occurs in the HAZ according to the localization of the measurements. Figure 8 shows an example of the hardness results, measured close to the surface on both sides of the linear weld in a profile including the base metal passing through the heat affected zone and the welded zone. The same measurement has been realized along the same profile and at a depth around 200 µm.

Close to the surface, the hardness varies from around 90 HV in the base metal and around 95 HV in the heat affected zone to 110 HV in the welded zone. The hardness in the HAZ is higher than in the BM, even though the size of the grains is identical. This augmentation of microhardness has in part been explained by the contribution of added elements and particularly the increase in the level of aluminium in this zone (10%). Other studies have also demonstrated the presence of precipitates which are formed in this zone considered to be a zone of diffusion which contributes towards augmenting the hardness (Shaw et al., 1997). In the fusion zone, compared to the base metal, the increase in hardness was probably due to its finer microstructure and higher volume fraction of intermetallics such as $Mg_{17}Al_{12}$ (Watkins, 2003).

Of the same form, through the thickness at a depth around 200 μm, the results show a different evolution than the surface concerning the BM and the HAZ, the hardness is lower. In the depth, the evolution of the microhardness shows no statistical variation of the microhardness between the base metal, the heat affected zone and the welded zone contrary to the surface. We can see that the hardness between the base metal, the heat affected zone and the core of the welded zone at a depth around 200 μm stays stable with a value of about 85 HV. Though significant grain coarsening occurred in the HAZ of AZ91 alloy, the hardness in the HAZ was still almost the same as that in base metal. These results join those of literature. Weisheit et al., 1997, studied 2.5kW CO_2 laser welding of cast magnesium alloys such as AZ91 AM60, ZC63, ZE41, QE22 and WE54 and wrought alloys (AZ31, AZ61, ZW3 and ZC71). For as-cast alloys, there is an increase in hardness of the FZ but little variation in hardness occurs in the HAZ. These observations demonstrate that laser welding induces particular profiles in the zones studied. These differences in hardness distribution over laser weld joints indicate the inhomogeneity of the joints following used parameters. Hardness is influenced by the laser welding parameters but too the initial chemical composition and also depends on the manufacturing process of the magnesium alloy. The choice of the laser process CO_2 or Nd:YAG influences too the hardness. This last point has been demonstrated by Hiraga et al., 2001, studied 2 kW CO_2 and Nd:YAG laser welding of wrought AZ31B-H24 butt joints of 1.7mm thickness. Their results showed that the Nd:YAG laser welded fusion zone is slightly harder than the FZ produced by CO2 laser.

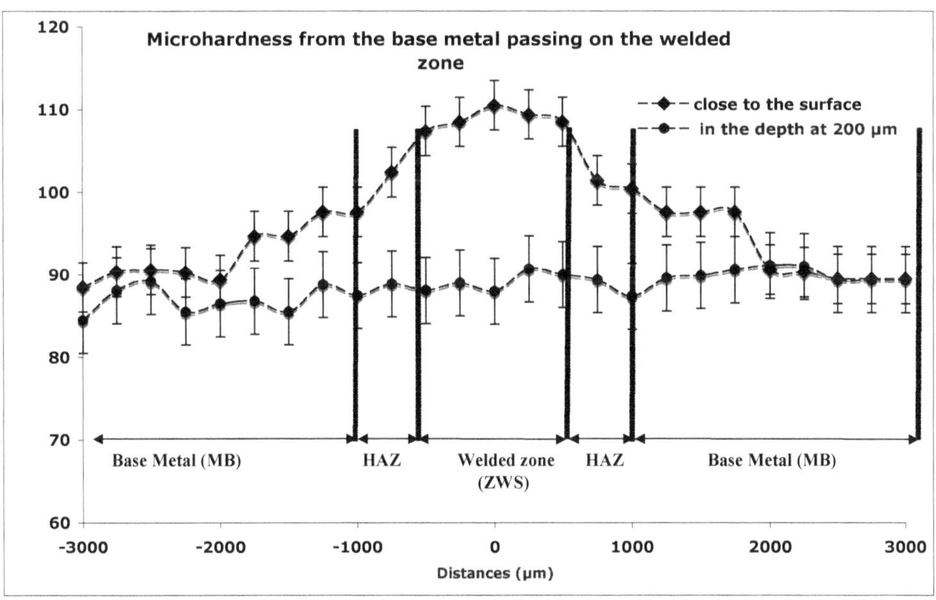

Fig. 8. Hardness from the base metal passing on the heat affected zone and the welded zone close to the surface and in the thickness at a depth about 200 μm (Kouadri & Barrallier, 2011).

4.2.2 Residual stresses results

The origin of residual stresses and their evolution within a welded joint is difficult to evaluate because they are the result of a number of competing mechanisms: shrinkage changes in phase and microstructure (Dai & Shaw, 2003). In our case, the magnesium alloy does not undergo a phase transformation, as is the case for aluminium alloys. Numerous studies have shown that when this is the case residual stresses are primarily a consequence of an inhibited shrinkage in the weld line and of the modified microstructure which is linked to strong temperature gradients, and to their distribution within the material (Wagner, 1999; Cho et al., 2003; Mao et al., 2006).

4.2.2.1 Distribution of the residual stresses at the surface of the assembled sheets

The measurements were undertaken in the welded zone, perpendicular to the weld line towards the base metal. Figure 9 shows an example of the obtained results.

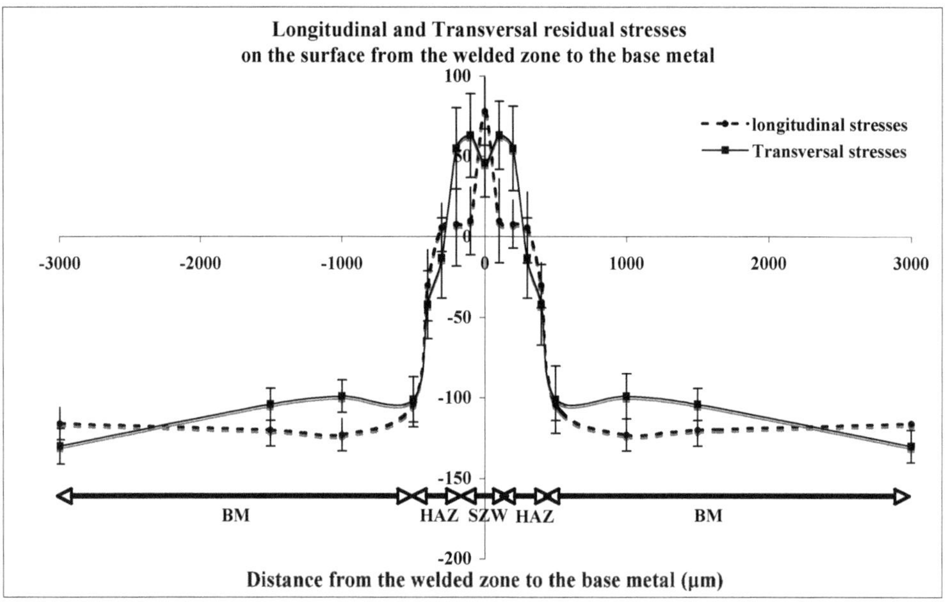

Fig. 9. Longitudinal and transverse residual stresses close to the surface from the base metal until the welded zone (Kouadri & Barrallier, 2011)

At the surface, the results demonstrated that the base metal presents a state of compression, whereas the weld line is submitted to residual traction stresses. This state of compression has been attributed to the nature of the cooling, linked to the moulding process. With laser welding, cooling occurs by the diffusion of heat through the outer surfaces of the plates which are in contact with the mould walls. Furthermore the machining by milling of the surface of the base metal before welding accentuates the state of compression and explains the raised values (- 120 MPa) observed at the surface of the base metal. However, in the weld line the heat is evacuated by the plates and not the free surfaces. This leads to traction stresses. This evolution is in line with mechanical equilibrium. The welded zone exhibits residual traction stresses which are counter balanced by compression stresses in the base metal (Pryds & Huang, 2000).

Furthermore, study of the state of surface stresses demonstrates some anisotropy: the residual stresses are not equibiaxial. We observe that the longitudinal component decreases from the center of the weld zone towards the base metal, whereas the transverse component remains high before a sudden reduction. These changes occur in the thermally affected zone, and are associated with numerous factors and by many authors with a zone of relaxation. The evolution of the longitudinal stress can be connected with the heat flow resulting from the mobile heat source that follows the welding direction. We can see two explanations of these evolutions. On the one hand, the effect of temperature and cooling speed gradients arise from the anisotropic heat flow (Teng et al., 2002), on the other, the anisotropy can be the result of a shrinking structure. The negligible thermal dilation of both plates prevents the free shrinkage of the weld line along the direction of the weld line. The same applies to the transverse component. The restricted shrinkage in the transverse plane arises from the clamping of the plates during welding. Even if the influence of the clamping is hard to evaluate experimentally, digital studies have shown that the field of residual stresses is strongly influenced by the geometry of the assembly (Jensen et al., 2002; Dai & Shaw, 2003).

4.2.2.2 Distribution of the residual stresses though the thickness of the welded zone

The profiles of the average stresses through the thickness are plotted in the figure 10 for the longitudinal and transversal residual stresses.

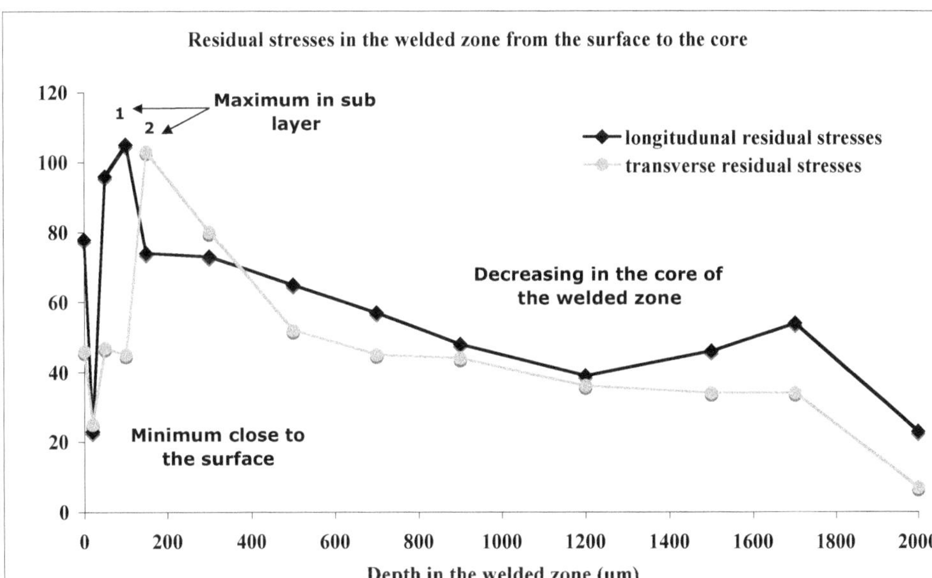

Fig. 10. Longitudinal and transverse residual stresses from the welded zone close to the surface to the core of the welded zone (Kouadri & Barrallier, 2011)

The study through the thickness of the welded zone shows that in general the profiles of the stresses reproduce the asymmetry of the welding process. Their behaviour in tension and their variation have in part been explained by the influence of the thermal cycle on the origin of residual stresses and their evolution within the material. The residual stresses on the face exposed to the laser beam are elevated (up to 80 MPa) whereas the opposite face creates stresses of only 23 and 7 MPa respectively for the longitudinal and transverse stresses. This effect can also be explained by the fact that using inert gas ensures very rapid cooling of the superior face whereas the inferior face cools more slowly (Dong et al., 2004).

Finally we have noticed that over the first 200 micrometers the residual stresses present a particular evolution. The intensity of the stresses is not maximum at the surface of the welded zone (78 and 45 MPa respectively for the longitudinal and transverse constraints) as expected, but at a depth of 200µm the stress is 100 MPa. The presence of this maximum stress demonstrates that there is a stress gradient between the textured layer and the heart of the isotropic welded zone. This specific evolution has partly been explained by the plastic deformation of the superficial layer. By presenting a strong texture and an important loss of aluminum, the superficial layer is more sensitive to plastic deformation in the plane compared to the heart of the weld line which is isotropic (Hsiao et al., 2000). These results can be compared to a thin coat deposit because these thin coatings are textured and the maximum stresses are found at the interface (Pina et al., 1997; Cevat Sarioglu 2006).

In our study there is a transition zone with a continual evolution in properties, in particular an evolution of the texture between the outer surface and the depth at about 200 µm. It appears that the development of this texture affects the distribution of stresses with a relaxation of the stresses at the surface and a maximum in the under layer. We explain these modifications by the fact that the level of plastic flow, related to local stresses, is dependent on grain orientation (Su et al., 2002; Agnew & Duygulu, 2005; Wu et al., 2007).

In conclusion, these results showed that the laser welding processes influence the residual stress distribution. Whereas compressive stresses are obtained in the base metal, tensile stresses are obtained in the LBWelds due to thermal gradients and high residual stresses are observed in the LBW fusion zone. These results showed too that it is important to take into account the crystallographic texture to evaluate the residual stresses.

5. Conclusions

The influence of various welding parameters during continuous wave CO_2 laser beam welding of thin plates of magnesium alloys was investigated in this chapter. It is known that the weldability of such materials is usually not excellent and lasers can be utilized to achieve good quality welds. The obtained results and the realized synthesis from the literature showed that the CO_2 laser welding possesses comprehensive performances such as good technology and the technology of laser welding magnesium alloys plates is well appropriated. The keyhole welding mode is likely to be encountered in the laser welding of thin sheet magnesium. The results of a detailed investigation showed the influence of different parameters of the laser which have to be tightly combined to obtain a weld quality.

However, during laser welding of magnesium alloys, therefore, some processing problems and weld defects can be encountered such as an unstable weld pool, substantial spatter a strong tendency to drop-through for large weld pools (Leong et al., 1998; Haferkamp et al., 1998; Sanders et al., 1999), sag of the weld pool (especially for thick workpiece), undercut (Dubé et al., 2001), porous oxide inclusions, loss of alloying elements (Leong et al., 1998; Sanders et al., 1999), excessive pore formation (particularly for die castings) (Pastor et al., 2000; Zhao & DebRoy, 2001) and solidification cracking (Marya & Edwards, 2000). These defects are generally decreased by a good optimization of the laser parameters. In view of the results achieved in this study, the use of high-power intensity focused CO_2 laser beam with optimized parameters and careful material preparation prior to welding can produce welds with high quality for the most magnesium alloys, in particular for AZ91D of our study (Kouadri & Barrallier, 2006, 2010). Welding speed of 2 m/min and laser power of 4kW let to a full penetration of 3mm thickness welded joint. Optimum weld profile was obtained when focal point was placed on the top surface. In comparison with the literature, all the investigated magnesium alloys showed tendencies for porosity and solidification cracking particularly, at high welding speed (≥4m/min). Porosity was prevented by accurate cleaning of the base metal before welding and optimizing the flow rate of argon shielding gas. In order to maintain the mechanical properties when welding magnesium alloys, the heat input and time of exposure to very high temperatures must be minimized. For LBW, the laser power (P) and weld speed (V) directly influence the heat input. This relationship is often used to determine the heat input.

Beyond of the optimization of laser parameters, it is believed that the efficiency of CO_2 laser beam welding of magnesium alloys could be improved by cleaning the workpiece surface prior to welding. This is due to increasing surface roughness that means decreasing surface reflectivity and enhancing the laser energy coupling during welding. Recent efforts on $C0_2$ laser beam welding have resolved several of the initial problems associated with the welding of magnesium alloys. Consistent and repeatable welds can now be obtained without resorting to meticulous edge preparation. Moreover, elimination or reduction of the plasma is recommended for optimal welding of magnesium. This effect of plasma formation which affects the weld quality and the optics during welding has been clarified: trouble-free operation of the optics has been achieved with the use of inert gas shielding such as helium.

However, several results showed that the weldability of thin magnesium plates was significantly better with the Nd:YAG laser. These observations were attributed to the higher absorption of the Nd:YAG beam, which in turn reduced the threshold irradiance required for welding and produced a more stable weldpool. Indeed, an advantage of Nd:YAG laser processing is its shorter wavelength; consequently, because of the dependency of the material's emissivity on the wavelength, energy is absorbed by the material more readily than for the CO_2 laser and a lower energy can be used for welding, allowing greater control of the heat input. This is particularly useful when working with thin materials. Recently, tremendous efforts have been made to clarify the fundamental laser weldability of different types of magnesium alloys using both Nd:YAG and CO_2 lasers. It is pointed out that improvements in the laser weldability of a range of magnesium alloys are possible by increasing the power density of the focused spot, and this can be achieved through higher average powers, improved beam focusing system, and decreasing beam reflectivity on workpiece surface. In conclusion, the weldability problems of magnesium alloys are much more easily overcome when using Nd:YAG than CO_2 laser. However, our studies showed that CO_2 laser is more appropriate to weld cast magnesium alloy than wrought magnesium alloy.

From technology point of view, in comparison with the traditional welding method such as arc welding processes, the laser welding has high efficiency, small welding distortion, low labor costs and convenient construction, is easy to realize automatization, and can be the effective measure to enhance the weld quality. Laser welding processes offer great benefit over other welding processes, e.g., arc welding, resistance welding, etc., since less heat is coupled into the workpiece. The low-heat input will tend to keep a very narrow HAZ then, retaining some to the strength of the material. The benefits of low-heat input and extremely rapid cooling rate, all of which help to minimize the metallurgical problems in the fusion zone. For example, high cooling rate will tend to slow down the development of blisters because of the short time in which the diffusion of hydrogen can take place. In comparison with electron beam welding, even though this process offers the advantages of a high energy density welding process, a vacuum chamber is required, which is not always practical.

All advantages explain their integration in the advanced technology industries such as in aerospace, aircraft, automotive, electronics and other industries. Indeed, today, laser beam welding is being used in an increasingly wider range of industries, from the production of medical devices and microelectronics to shipbuilding. The automotive industry, in particular, takes advantage of this technology's benefits: low heat input, small heat-affected zone (HAZ), low distortion rate, good repeatability, reduced need for post processing and high welding speed. This last point is becoming critical for a successful application in the automotive industry because the increase in welding speed provided by laser welding has resulted in the need for an automated system. Another application of the laser process is the aircraft where weight and cost reduction in civil aircrafts by replacing rivets by advanced welding techniques has now been realized for skin-stringer joints. In this context the laser beam welding technology has proved to be very suitable for a number of reasons, for example, low distortion while processing at high speeds. These benefits have made laser welding the process of choice for many applications that previously used resistance welding. Compared for example with TIG welding, the welding speed with laser is generally three times higher.

However, although laser materials processing has gained widespread acceptability, the mechanisms and main factors controlling the process remain controversial and need further theoretical and experimental studies. Further work is needed to develop the weld process parameters necessary to achieve the materials characteristics required for the use of magnesium alloys in industrial applications. Improved gas shielding requirements are expected to be critical to obtaining welds with the required materials properties. Further studies are needed to determine the parameters controlling weld quality. Indeed, laser beam welding involves many variables: laser power, welding speed, defocusing distance and type of shielding gas, any of which may have an important effect on heat flow and fluid flow in the weld pool. This in turn will affect penetration depth, shape and final solidification structure of the fusion zone. These final states affect the mechanical behaviour. It is for that this technique stays still in the developing stage and many mechanisms need to be studied because of many parameters which govern this process. However, with a good optimization, laser welding for the magnesium alloys seems to be the most appropriated joining technique and can promote their wider uses in aerospace, aircraft, automotive, electronics and other industries.

6. References

Agnew S.R., Duygulu Ö. (2005). Plastic anisotropy and the role of non-basal slip in magnesium alloy AZ31B. *International Journal of Plasticity*, Vol. 21, Issue 6, pp. 1161-1193.

Cao, X., Jahazi, M., Immarigeon, J.P., Wallace, W. (2006). A review of laser welding techniques for magnesium alloys. *Journal of Materials Processing Technology*, Vol. 171, pp. 188–204.

Cevat Sarioglu C. (2006). The effect of anisotropy on residual stress values and modification of Serruys approach to residual stress calculations for coatings such as TiN, ZrN and HfN. *Surface and coatings technology*, Vol. 201, Issue 3- 4, pp.707-717.

Cho, J.R., Conlon, K.T., Reed, R.C. (2003). Residual Stresses in an Electron Beam Weld of Ti-834: Characterization and Numerical Modelling. *Metallurgical and Materials Transactions A*, vol. 37, Issue 2, pp. 2935-2946.

Coelho, R.S., Kostka, A., Pinto, H., Riekehr, S., Koçak, M., Pyzalla, A.R. (2008). Microstructure and mechanical properties of magnesium alloy AZ31B laser beam welds. *Materials Science and Engineering A*, Vol. 485, Issue 1-2, pp.20–30.

D'Annessa, A. T. (1970). Sources and effects of growth rate fluctuation during weld metal solidification. *Welding Journal*, Vol. 49, Issue 2, pp 41–45.

Dai, K., Shaw L. (2003). Finite-Element Analysis of Effects of the Laser-Processed Bimaterial Component Size on Stresses and Distortion. *Metallurgical and Materials Transactions A*, vol. 37A, Issue 2, pp. 1133-1145.

Dhahri, M., Masse, J.E, Mathieu, JF, Barreau, G, Autric, M. (2000). CO_2 laser welding of magnesium alloys. *In Proceedings of the SPIE: High power lasers in manufacturing*, pp. 725–732.

Dhahri, M., Masse, J.E, Mathieu, JF, Barreau, G, Autric, M. (2000). CO_2 laser welding of magnesium alloys. *In Proceedings of the SPIE: High power lasers in manufacturing*, pp. 725–732.

Dhahri, M., Masse, J.E, Mathieu, JF, Barreau, G, Autric, M. (2001). Laser weldability of WE43 magnesium alloys for aeronautic industry. *In third LANE 2001: Laser Assisted Net Shape Engineering*, pp. 297–310,

Dhahri, M., Masse, J.E, Mathieu, JF, Barreau, G, Autric, M. (2001). Laser welding of AZ91 and WE43 magnesium alloys for automotive and aerospace industries. *Advanced engineering materials*, Vol. 3, Issue 7, pp. 504–507, 2001.

Dhahri, M., Masse, J.E, Mathieu, JF, Barreau, G, Autric, M. (2002). Laser welding of magnesium alloys for automotive and aerospace applications. *http://webdb.dgm.de/dgm lit/*, October 2002.

Dong, W., Kokawa, H., Tsukamoto, S., Sato Y. S., Ogawa, M. (2004). Mechanism governing nitrogen absorption by steel weld metal during laser welding. *Metallurgical and Materials Transactions B*, pp. 331-338.

Dubé, D., Fiset, M., Couture, A., Nakatsugawa, I. (2001). Characterization and performance of laser melted AZ91D and AM60B. *Materials science and Engineering A*, vol. 299, pp. 38-45.

Haferkamp, H., Bach, Fr.-W., Burmester, I., Kreutzburg, K., Niemeyer, M. (1996). Nd:YAG laser beam welding of magnesium constructions. *In Proceedings of the Third International Magnesium Conference, UMIST,* Manchester, UK, 10–12 April 1996, pp. 89–98.

Haferkamp, H., Dilthey, U., Trager, G., Burmester, I., Niemeyer, M. (1998). Beam welding of magnesium alloys. *In Proceedings Conference: Magnesium Alloys and Their Applications,* Wolfsburg, Germany, 28–30 April 1998, pp. 595–600.

Hiraga, H., Inoue, T., Kamado, S. Kojima, Y. (2001). Effect of the shielding gas and laser wavelenght in laser welding magnesium alloy sheet. *Quart. J. Weld. Soc.*, Vol. 19, Issue 4, pp. 591–599.

Hsiao, I.C., Su, S.W., Huang, J.C. (2000). Evolution of texture and grain misorientation in an Al-Mg alloy exhibiting low-temperature superplasticity. *Metallurgical and Materials Transactions A*, Vol. 33, Issue 5, pp. 2169-2180.

Jensen, M.V.R. S., Dye, D., James, K. E., Korsunsky, A. M., Roberts, S. M., Reed R. C. (2002). Residual stresses in a welded superalloy disc : characterization using synchrotron diffraction and numerical process modelling. *Metallurgical and Materials Transactions A*, Vol. 37, Issue 2, pp. 2921-2931.

Kalinyuk, A.N., Trigub, N.P, Zamkov, V.N., Ivasishin, O.M., Markovsky, P.E, Teliovich, R.V., Semiatin, S.L. (2003). Microstructure, texture, and mechanical properties of electron-beam melted Ti-6Al-4V. *Materials Science and Engineering A*, Vol. 346, pp. 178-188.

Kouadri A., Barrallier L. (2011). Study of mechanical properties of AZ91 magnesium alloy welded by laser process taking into account the anisotropy: micro-hardness and residual stresses by X-ray diffraction., *Metallurgical and Materials Transactions A*, Vol. 42, Issue 7, pp. 1815-1826.

Kouadri, A., Barrallier, L. (2006). Texture characterization of hexagonal metals: magnesium AZ91 alloy, welded by laser processing. *Materials science and Engineering A*, Vol. 429, Issues 1-2, pp. 11- 17.

Kurtz, W., Bezençon, C., Gäumann, M. (2001). Columnar to equiaxed transition in solidification processing. *Science and Technology of Advanced Materials*, Vol. 2, pp. 185-191.

Lehner, C., Reinhart, G., Schaller, L. (1999). Welding of die cast magnesium alloys for production. *J. Laser Appl.* Vol. 11, Issue 5, pp. 206–210.

Leong, K.H., Sanders, P.G., Keske, J.S., Kornecki, G. (1998). Laser beam welding of AZ31BH24 alloy. *ICALEO 98: Laser Material processing conference*, Orlando, FL, 16-19 November 1998, pp. 28–36.

Liu, S., Edwards, G.R., Olson, D., Marya, M. (2000). Laser processing research at the Colorado School of Mines. *In Asian Pacific Welding Conference: Proceedings NZIW 2000 Annual Conference and WTIA 48th Annual Conference*, Melbourne, 29 October to 2 November 2000, pp. 1–18.

Luo, A. (1996). Heterogeneous nucleation and grain refinement in cast Mg(AZ91)/SiCp metal matrix composites. *Canadian Metallurgical Quarterly*, Vol. 35, pp. 375-383.

Mao, W.G., Zhou, Y.C., Yang L., Yu, X.H. (2006). Modeling of residual stresses variation with thermal cycling in thermal barrier coatings. *Mechanics of materials*, Vol.38, Issue 12, pp. 1118-1127.

Marya, M., Edwards, G.R. (2001). Factors controlling the magnesium weld morphology in deep penetration welding by a CO_2 laser. *Journal of Materials Engineering and Performance*, Vol. 10, Issue 4, pp.435–443,

Marya, M., Edwards, G.R. (2000). The laser welding of magnesium alloy AZ91. *Weld. world*, Vol. 44, Issue 2, pp.31–37.

Matysina, Z.A. (1999). The relative surface energy of hexagonal close-packed crystals. *Materials Chemistry and Physics*, Vol. 60, pp. 70-78.

Pastor, M., Zhao, H., DebRoy, T. (2000). Continuous wave- Nd:yttrium–aluminium–garnet laser welding of AM60B magnesium alloys, *J. Laser Appl.*, Vol. 12, Issue 3, pp. 91–100.

Pina, J. Dias, A., François, M., Lebrun, J.L. (1997). Residual stresses and crystallographic texture in hard-chronium electroplated coatings. *Surface and coatings technology*, Vol. 96, pp.148-162.

Pryds N.H., Huang, X. (2000). The effect of cooling rate on the microstructures formed during solidification of ferritic steel. *Metallurgical and Materials Transactions A*, Vol. 33, Issue 5, pp. 3155-3166.

Sanders, P.G., Keske, J.S., Leong, K.H., Kornecki, G. (1999). High power Nd:YAG and CO2 laser welding of magnesium. *J. Laser Appl.*, Vol. 11, Issue 2, pp. 96–103.

Shaw, C., Jones, H. (1997). The contributions of different alloying additions to hardening in rapidly solidified magnesium alloys. *Materials Sciences and Engineering A*, Vol. 226-228, pp. 856-860.

StJohn, D.H., Dahle, A.K., Abbott, T., Nave M.D., Qian, M. (2003).Solidification of cast magnesium alloys. *The minerals, Metals and Materials Society*, pp. 95-100.

Su, S.F., Huang, J.C., Lin, H.K., Ho N.J. (2002). Electron-Beam Welding Behavior in Mg-Al-Based Alloys. *Metallurgical and Materials Transactions A*, Vol. 33, Issue 5, pp. 1461-1473.

Teng, T.L, Fung C.H., Chang, P.H. (2002). Effect of weld geometry and residual stresses on fatigue in butt-welded joints. *International Journal of Pressure Vessels and Piping*, Vol. 79, Issue 7, pp. 467-482.

Wagner, L. (1999). Mechanical surface treatments on titanium, aluminum and magnesium alloys. *Materials Science and Engineering A*, Vol. 263, Issue 2, pp. 210-216.

Wang H.Y., Li, Z.J. (2006). Investigation of laser beam welding process of AZ61 magnesium based alloy. *Acta metallurgica sinica (english letters)*, Vol. 19, Issue 4, pp. 287–294.

Watkins, K.G. (2003). Laser welding of magnesium alloys. *In H.I. Kaplan, editor, Magnesium Technology, TMS Annual Meeting and Exhibition*, pp. 153–156, San Diego.

Weisheit, A., Galun, R., Mordike, B.L. (1997). Weldability of various magnesium alloys using a CO2 laser, *IIW seminar, Trends Weld, Lightweight automobile railroad vehicles*, pp. 28–41.

Weisheit, Galun, R., Mordike, B.L. (1998). CO2 laser beam welding of magnesium-based alloys. *Weld. Res. Suppl.*, Vol. 74, Issue 4, pp. 149–154.

Wu, X., Kalidindi, S. R., Necker, C., Salem, A.A. (2007). Prediction of crystallographic texture evolution and anisotropic stress–strain curves during large plastic strains in high purity titanium using a Taylor-type crystal plasticity. *Acta Materialia*, Vol.55, pp. 423-432.

Zhao, H., DebRoy, T. (2001). Pore formation during laser beam welding of die cast magnesium alloy AM60B - mechanism and remedy. *Weld. J.*, Vol. 80, Issue 8, pp. 204S–210S.

12

Infrared Lasers in Nanoscale Science

Rui F. M. Lobo[1,2]
[1]*Group for Nanoscale Science and Nanotechnology (GNCN), Physics Department, Faculdade de Ciencias e Tecnologia - New University of Lisbon, Caparica,*
[2]*Institute for Science and Technology of Materials and Surfaces (ICEMS)*
Portugal

1. Introduction

In the nearly half a century scientists have already realized that, just as Feynman predicted, there is plenty of research room at the bottom of the matter world in a tiny universe so small that new methods for viewing it are still being discovered. Actually, nanoscience and nanotechnology have evolved into a revolutionary area of technology-based research, opening the door to precise engineering on the atomic scale and affecting everything from healthcare to the environment. Nanoscience research and education lead to nanotechnology, the manipulation of nanometer-length atoms, molecules, and supramolecular structures in order to generate larger structures with superior features. Because all natural materials and systems exist at a nanoscale level, nanotechnology impacts a variety of scientific fundamental and applied disciplines, from physics to medicine and engineering. Nanomaterials consisting of nano-sized building blocks exhibit unique and often superior properties relatively to their bulk counterpart. Due to the fact that most of the novel properties of nanomaterials are size-dependent, synthesis methods leading to better control of size, distribution and chemical content of the nanoparticles are imperative in modern nanotechnologies.

On its turn, the laser has been one of the top applied physics inventions that played a significant role in many fields of science and technology. It has been used in tackling and solving many scientific and technological problems, including interesting applications in the field of nanotechnology, biotechnology/medicine, environment, material characterization, and energy.

There are several gaseous molecules which serve as good laser media and the majority of them are simple molecules which provide emission in the ultraviolet. Infrared molecular gas lasers fall into two general categories, namely the middle- and far-infrared lasers, which occur on rotational-vibrational transitions or on pure rotational transitions.

The N_2 laser is known as a pulse ultraviolet laser and in addition it covers some lines in the infrared up to 8,2 μm. Normally, the pulse width is a few nanoseconds and a high-voltage power supply of 30-40 kV is necessary to excite it. The HF is a high power chemical laser media with an emission wavelength of about 2,7 μm, a laser pulse of the order of μs in duration and the output energy ranges from 1 J to more than 1 kJ per pulse. The DF and HBr chemical lasers emit larger wavelengths than the HF laser, and their output power is lower [1,2].

The CO_2 laser is a gas laser electrically pumped, that emits in the mid-infrared. It gives a cw output at 10 μm in the infrared with a high efficiency and it is the most practical molecular laser. There are a large number of CO_2 lasers, varying in structure, method of excitation and capacity, which can provide hundreds of laser lines, the main ones being between 9 and 11 μm. The output power of even a small CO_2 laser is about 1 kW and large ones give over 10 kW. The usual way of obtaining single-line oscillation is to use a diffraction grating in conjunction with a laser resonator. If only mirrors are used, simultaneous oscillation on several lines in the neighborhood of 10,6 μm is commonly obtained [1,2]. Transverse excited atmosphere (TEA) CO_2 lasers have a very high (about atmospheric) gas pressure. As the voltage required for a longitudinal discharge would be too high, transverse excitation is done with a series of electrodes along the tube. TEA lasers are operated in pulsed mode only, as the gas discharge would not be stable at high pressures, and are suitable for average powers of tens of kilowatts [1,2].

Although N_2O and CO laser have a lower output power than the CO_2 laser, they have about one hundred laser lines each in the ranges 10-11 μm and 5-6,5 μm, respectivelly (considering the main isotopic species). The molecules NH_3, OCS, CS also have quite a few laser lines in the infrared. With the SO_2, HCN, H_2O, many laser lines are obtained in the infrared from 30 μm up to submilimeter wavelengths [1,2].

Dye lasers are convenient tunable lasers in the visible but not so far in the infrared, mainly due to the lack of appropriate dyes, and in addition, since the dye laser medium is liquid, it is very inconvenient to handle.

In face of the real advantage of the laser as a very intense heating source that can be applied to a very small area, the most significant areas in which the CO_2 laser has shown remarkable applications are in the general fields of materials processing and medical applications. This includes cutting, cauterizing, drilling, material removal, melting, welding, alloying, hardening, surgery, cancer treatment and so forth.

The carbon dioxide laser, invented by Patel [3], operates on rotational-vibrational transitions and is still one of the most useful among all the infrared molecular lasers. In general, it is one of the most powerful lasers currently available. It operates in the middle infrared wavelength region with the principal wavelength bands centering around 9.4 and 10.6 micrometers. It is also quite efficient, with a ratio of output power to pump power as large as 20%. It can operate at very high pressures because the energies of the upper laser levels are much closer to the ground state of the CO_2 molecule than are energies of the upper laser levels of atomic lasers and so the electron temperature can be much lower, thereby allowing higher operating gas pressures. Higher operating gas pressures means a much greater population in the upper laser level per unit volume of the laser discharge and therefore much higher power output per unit volume of laser gain media. These lasers have produced cw powers of greater than 100 kW and pulsed energies of as much as 10 kJ. The gain occurs on a range of rotational-vibrational transitions that are dominated by either Doppler broadening or pressure broadening, depending upon the gas pressure.

Although laser radiation is obtainable with pure CO_2 gas, the usual CO_2 laser uses a mixture of He, N_2 and CO_2. The population inversion in the laser is achieved by a sequence of fundamental processes starting by vibrational excitation of nitrogen molecules in the electric discharge. The nitrogen molecules are left in a lower excited state and their transition to ground state takes place by collision with cold helium atoms. The resulting hot helium atoms must be cooled in order to sustain the ability to produce a population inversion in the carbon dioxide molecules. In sealed lasers, this takes place as the helium atoms strike the walls of the container. In flow-through lasers, a continuous stream of CO_2 and nitrogen is excited by the plasma discharge and the hot gas mixture is exhausted from the resonator by pumps.

The CO_2 laser transitions are the 961 cm^{-1} transition of the 10,4 μm band and the 1064 cm^{-1} transition of the 9,4 μm band. Owing to the symmetry of the CO_2 molecule, laser transitions occur to lower energy levels whose rotational quantum numbers are even, resulting in more than 30 laser lines in each of the two branches P and R [1,2].

In general, there are two common different types of CO_2 laser configurations. In one of them (longitudinally excited laser), the CO_2 laser is excited by direct current and when the pressure raised from 10^3 Pa to 10^4 Pa, a peak power is obtainable by using a pulsed discharge. This is an arc maintained by an anode and a cathode at the ends of a long discharge tube. Another possibility is the transversely excited atmospheric pressure laser (TEA), excited by an arc discharge at roughly atmospheric pressure [4]. The current in the arc flows at right angles to the axis of the laser [5]. A TEA laser is always pulsed and many CO_2 lasers are TEA lasers.

Some CO_2 TEA lasers have been developed with additional techniques enabling us to achieve tuneable wavelengths, and in particular may reach oscillation threshold for several atomic or molecular transitions. The laser can then simultaneously oscillate on these transitions. In order to reach single mode operation, one has to first select a single transition.

Because the laser transitions are actually on vibration-rotation bands of a linear triatomic molecule CO_2, the rotational structure of the P and R bands can be selected by a tuning element in the laser cavity. Because transmissive materials in the infrared are rather lossy, the frequency tuning element is almost always a diffraction grating. By rotating the diffraction grating, a particular rotational line of the vibrational transition can be selected. The finest frequency selection may also be obtained through the use of an etalon.

A procedure towards optimization performance of a CO_2 pulsed tuneable laser was developed which allows the power and the energy to be optimized [5]. The MTL3-GT is a very compact grating tuneable TEA laser version (Figure 1), and represents a significant improvement in performance and portability [6]. Combining a pulse mode with a grating tuning facility, it enables us to scan the working wavelength between 9.2 and 10.8 μm (operating on more than 60 lines), with repetition rates ranging from single-shot to 200 Hz. The maximum energy for this version is 50 mJ/pulse on the strongest lines. The MTL3-GT CO_2 infrared laser works with a a gas mixture (40% He : 30% CO_2 : 30% N_2) and a chiller for high repetition rates. Actually, above 20 Hz, the number of HV discharges increases and the laser needs to be cooled down in order to lower the temperature in the optical cavity.

Following an adequate procedure, the energy values could be optimized in intensity and stability, and therefore indirectly laser power. In addition, the same procedure allows to check the wavelengths of the laser emission lines in the absence of a spectrometer, using a previously established conversion table of the grating position versus line designation. With such method, many experiments can be performed in real time with simultaneous control of power/energy and wavelength, and taking advantage of the full laser power for each selected wavelength.

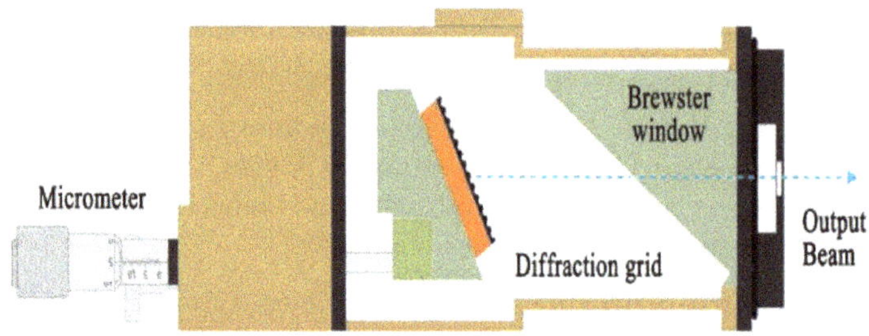

Fig. 1. Schematics of the MTL3-GT TEA laser from Edinburgh Instruments [6]

One could observe, after improving the procedure, that energy values are more stable in all four emission bands (9P, 9R, 10P and 10R). This behaviour was also observed regardless of the repetition rate, even for higher ones around 100 Hz. Besides energy, power was also measured and improved following the same procedure. This procedure can also be used on other infrared lasers with some minor adaptations regarding the software and energy detectors used.

In order to overcome several error sources which are the causes of non-reproducibility in these type of lasers, the procedure used a continuous measurement of the energy line, making use of an infrared detector and power meter acquisition software. Such a display method reflects the inherent error associated with the grating tuning motion and therefore the micrometer hysteresis. The method is suitable to obtain the energy and power values for each emission laser TEA CO_2 line optimized. The experimental set-up consists of the tuneable TEA CO2 laser, a pyroelectric energy detector connected to a handheld power/energy meter and a computer for acquisition purposes [5].

Since the laser is tuneable by wavelength, some specific emission lines of the CO_2 molecule can be selected, making use of a micrometer. The correspondence between such emission lines and the micrometer driving position can be previously verified with an infrared spectrometer, in order to check those mentioned in the user's manual. Changing the position of the micrometer, one varies the angular position of the diffraction grid. This allows to scan among several emission lines, and so to choose the working wavelength.

Using a graphite target block, the pulse shape can be observed while the micrometer is moving. When the correct position is achieved, the focus should be round and symmetric ($\approx$5 mm in diameter), displaying a strong luminosity and without sudden changes for consecutive shots. However, this method proved to be somewhat inaccurate and not very user friendly. To overcome these drawbacks, one must look at the real-time graphic line display of energy on the computer and follow its behaviour during the micrometer rotation, as well. The higher value of the energy line display corresponds, for each wavelength, to the desired position of the micrometer. This can be confirmed at any time by crossing the laser beam with the graphite target [5]. However, the micrometer hysteresis makes the procedure unreliable, tedious and very sensitive to the rotation speed of the micrometer and stability of the energy signal. Thus, two acquisitions were made for each repetition rate, one by turning the micrometer clockwise and another counter clockwise. An average of them was calculated and definitive energy values were registered. This procedure was repeated in the opposite direction, in order to obtain an average, and also to confirm the reproducibility of the result. It was actually confirmed for every emission line and several values of repetition rates [5].

Acquisitions recorded without concerns about the external factors and in different days revealed instability in energy values for each repetition rate measured (singleshot, 5 Hz, 10 Hz or 20 Hz), as displayed in Figure 2 (A). For the other three emission bands available (10R, 9P and 9R), the problem is also present. A variation in the power measured was observed in all possible cases, emission bands and repetition rates available. The values could vary from 1 mJ up to 5 mJ for energy and 10 mW to 50 mW for power (repetition rate was 10 Hz in this measurement). The final acquisitions were recorded taking into account the improved procedure regarding the verification of a correspondence between the micrometer drive readings and wavelengths. It can be observed in Figure 2 (B) that the energy values for each repetition rates available were smoother and without significant deviations [5].

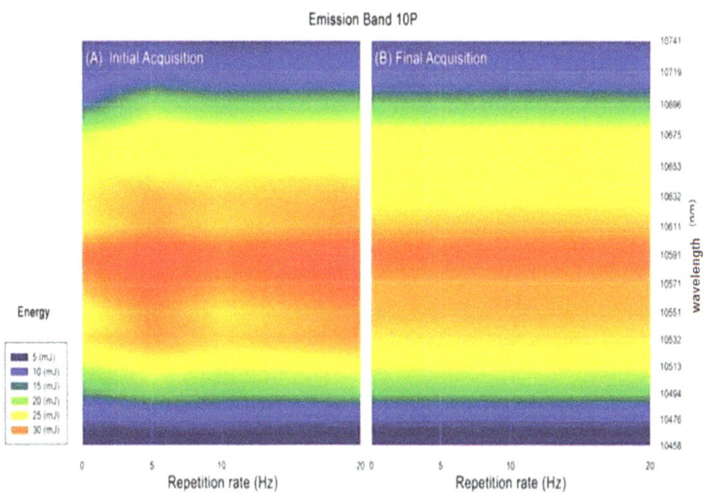

Fig. 2. Relationship between emission line, energy and repetition rates
(10P emission band-repetition rates single-shot, 5 Hz, 10 Hz and 20 Hz)

This has also been verified for the power values measured for the same repetition rates. The values still vary with the new procedure but in a much lower interval, between 0.5 mJ and 1 mJ for energy, and 5 mW and 10 mW for power values. The same improvement was verified for higher repetition rates up to 100 Hz. The confirmation could be observed not only for the 10P emission band (Figure 3), but also for the other three emission bands available (10R, 9P and 9R) [5].

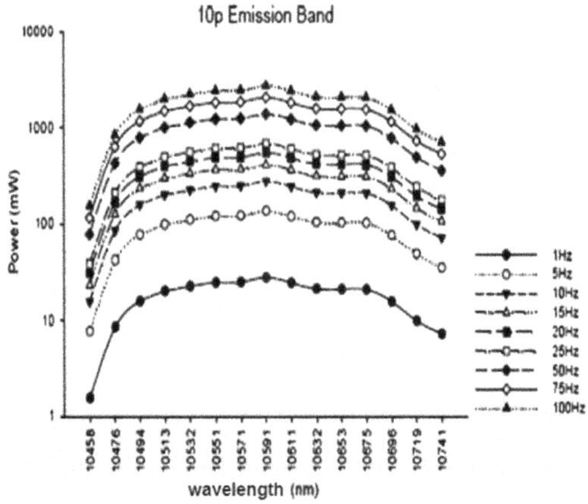

Fig. 3. Power versus repetition rate for 10P emission band.

Along this book chapter, several examples of CO_2 lasers applications to nanoscale science and nanotechnology, are explored and generally explained. These include examples in different topics, namely molecular photodynamics, tailored-size nanoparticles production, optical spectroscopy of nanopowders, infrared irradiation of nanostructures, desorption kinetics, photodynamic therapy, among others.

2. Laser spectroscopy and photodynamics

The combination of pulsed lasers, pulsed molecular beams and time-of-flight mass spectrometry represents a powerful technique for studying excitation, ionization and fragmentation of wanted molecules out of a large variety of different species present in a molecular beam [7]. The success of these two combined techniques is mainly due to the increase in the spectral resolution of absorption and fluorescence spectra by using collimated molecular beams with reduced transverse velocity components, and also to the fact that internal cooling of molecules during adiabatic expansion of supersonic beams compresses their population distribution into the lowest vibrational-rotational levels. This particular aspect greatly reduces the number of absorbing levels and results in a huge simplification of the absorption spectrum [7].

In addition, the low translational temperature achieved in supersonic beams allows the generation and observation of loosely bound van der Waals complexes and clusters. The collision-free conditions in molecular beams after their expansion into a vacuum chamber facilitates saturation of absorbing levels, since no collisions refill a level depleted by optical pumping. This makes Doppler-free saturation spectroscopy feasible even at low cw laser intensities [1].

The structure of molecular complexes in their electronic ground state can be obtained from direct infrared laser absorption spectroscopy in pulsed supersonic-slit jet expansions. This allows one to follow the formation rate of clusters and complexes during the adiabatic expansion. Selective photodissociation of van der Waals clusters by infrared lasers could be used for isotope separation [1].

A typical example of a beam-gas collision is the process

$$Ca^*(^3P_J) + HCl \rightarrow CaCl(X;v'',J'') + H$$

The reaction with the ground state $Ca(^1S_0)$ is endothermic and this is why excited Ca atoms are required. When interrogating the centre of the reaction cell with a tuneable cw laser, Laser Induced Fluorescence (LIF) emission is observed on transitions in the CaCl(A-X) band system [2]. An example of a fraction of the related LIF excitation spectrum is shown in Figure 4.

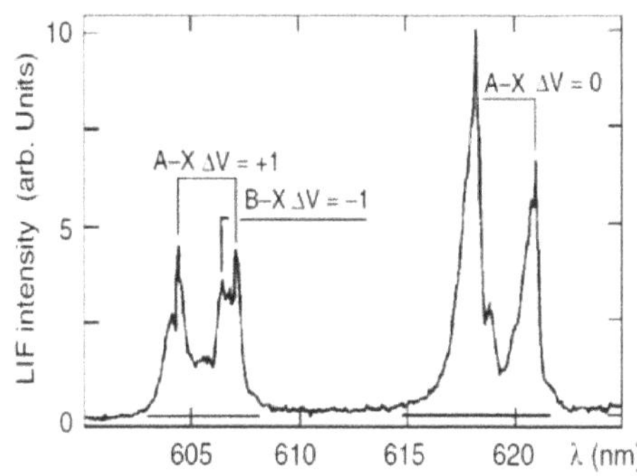

Fig. 4. LIF spectroscopy of the beam-gas reaction, revealing part of the rotational level population of the reaction product.

Important analytical applications are represented by measurements of the internal-state distribution of reaction products with LIF and spectroscopic investigations of collision-induced energy-transfer processes. The high output power of pulsed CO_2 lasers allows excitation of high vibrational levels by multiphoton absorption, which eventually may lead to the dissociation of the excited molecule. In some favorable cases the excited molecules or the dissociation fragments can even selectively react with other added components. Such selectively initiated chemical reactions can be induced by CO_2 lasers which are particularly advantageous due to their large electrical efficiency.

As an example, let us consider the synthesis of SF_5NF_2 by multiphoton absorption of CO_2 photons in a mixture of S_2F_{10} and N_2F_4, which proceeds according to the following scheme:

$$S_2F_{10} + nh\nu \rightarrow 2SF_5$$

$$N_2F_4 + nh\nu \rightarrow 2NF_2$$

$$SF_5 + NF_2 \rightarrow SF_5NF_2$$

This laser-driven reaction proceeds much more quickly than the conventional high-temperature synthesis without laser, even at the lower temperature of 350 K.

Another example of CO_2 laser-initiated reactions is the gas-phase telomerization of methyl-iodide CF_3I with C_2F_4, which represents an exothermic radical chain reaction

$$(CF_3I)^* + nC_2F_4 \rightarrow CF_3(C_2F_4)_nI \qquad (n = 1, 2, 3)$$

producing $CF_3(C_2F_4)_nI$ with low values of n. The CO_2 laser is in near resonance with the $\nu_2+\nu_3$ band of CF_3I. The quantum yield for this reaction increases with increasing pressure in the irradiated cell [2].

The infrared lasers have the advantage that the contribution of scattering losses to the total beam attenuation is much smaller than in the visible range. For measurements of very low concentrations, on the other hand, visible dye lasers may be more advantageous because of the larger absorption cross sections for electronic transitions and the higher detector sensitivity.

The applications of lasers to chemical reactions in gas phase are usually classified in two categories: laser induced chemical reactions and laser catalyzed chemical reactions. In the first ones, the laser supplies all the energy thermodinamically needed for the occurence of the reaction and they correspond typically to unimolecular processes (dissociation by multiphotonic absorption); in the second ones (typically bimolecular reactions) only a partial energy amount is supplied and then reaction proceeds by itself. The dissociation by multiphotonic absorption has seen a huge growth in the last decades [8] due to the availability of high power infrared lasers and important technological applications, like isotopic separation. As an example, since in SF_6 a mixture exists of $^{32}SF_6$ and $^{34}SF_6$, an infrared CO_2 laser with $\lambda = 10.61$ μm only gives rise to the excitation of vibrational states of $^{32}SF_6$ but not those of $^{34}SF_6$; thus, when the continuum of $^{32}SF_6$ vibrational states is reached after the absorption of 25 photons, only dissociation into $^{32}SF_5$ and F is produced. This dissociation is fast and corresponds to a statistical mechanism. On its turn, when the wavelength is tuned to 10.82 μm, the dissociation takes place in the $^{34}SF_6$ molecules.

The observation of dissociation phenomena in molecular beam apparatuses proves unequivocally that it is unimolecular and non-collisional, as it was shown through energetic and angular distributions of the SF_5 fragment formed in the dissociation of a SF_6 molecular beam, by a CO_2 laser pulse of 5 J/cm^3 [2]. Actually the results are consistent with RRKM unimolecular theory predictions. In this theory, it is assumed that energy is statistically distributed among the several available modes before dissociation takes place. This means that excitation energy is not localized in just one or a few modes, because in such cases, it will be not possible to reproduce with RRKM theory the above mentioned experimental results; in addition, the mean lifetime would be not about 10^{-8}s (as predicted by RRKM) but much smaller [2].

The dissociation of a polyatomic molecule by multiphotonic absorption is in fact a statistical process (i.e, non-selective), and allows to consider distinct types of selectivity with lasers, based on the existing relation between the different relaxation times which are involved in a vibrationally excited molecule:

$$\tau_{intra\ v\text{-}v'} << \tau_{v\text{-}v'} << \tau_{v\text{-}T}$$

where:

$\tau_{intra\ v\text{-}v'}$ - relaxation time for intramolecular vibrational energy transfer
$\tau_{v\text{-}v'}$ - relaxation time for vibrational energy transfer among distinct molecules
$\tau_{v\text{-}T}$ - relaxation time for vibration-translation transfer, i.e, the time needed to reach the complete thermal equilibrium

The first process can occur without collisions, but the other two are necesserelly collisional and therefore are pressure dependent. Considering v_{exc} as the vibrational excitation velocity of a molecule by multiphoton ionization (i.e, $1/v_{exc}$ will be the excitation time, which depends on radiation intensity and on the vibrational transition cross-section), a comparison of v_{exc} with the several excitation velocities, gives rise to four different situations:

- $v_{exc} >> 1/\tau_{intra\ v\text{-}v'}$; the excitation by absorption is faster than intramolecular relaxation, and so selective excitations are making possible; thus, a certain polyatomic molecule vibrational mode can reach a vibrational temperature higher than the other remaining modes.
- $1/\tau_{intra\ v\text{-}v'} >> v_{exc} >> 1/\tau_{v\text{-}v'}$; this is the more common molecular selectivity situation, where despite there is no vibrational equilibrium among the several molecules, intramolecular equilibrium exists for those molecules which interact with the infrared electromagnetic field.
- $1/\tau_{v\text{-}v'} >> v_{exc} >> 1/\tau_{v\text{-}T}$; vibrational equilibrium among all the molecules is guaranteed, but there is no relaxation for the translational modes which produce heating; in these conditions, photochemical vibrational experiments can be performed whenever the reactions have a characteristic time not longer than $\tau_{v\text{-}T}$ and a small energy barrier.
- $1/\tau_{v\text{-}T} >> v_{exc}$; the vibrational and translational temperatures of all molecules are equal due to the vibrational excitation relaxation in the molecular collisions, which heats the system up.

The excitation spectrum obtained by LIF of the CN fragment (produced by multiphoton absorption in the infrared with a high power CO_2 laser) of the gas $H_2C{=}CHCN$, shows clearly the rotational fine structure of the (0,0) band of the CN violet emission, which allows to conclude that the rotational distribution is statistical and characterized by a certain Boltzmann temperature, confirming that the excitation energy is statistically redistributed in the dissociation by multiphoton absorption [9].

The ability to energize a specific molecular bond and thereby promote a certain desired reaction pathway, has been a widely pursued goal, called mode-selective control in molecular physics. Actually, tunable infrared lasers are very convenient tools to divert a reaction from its dominant thermal pathway toward an envisaged possible product. However, the surplus of vibrational energy tends to be redistributed rapidly within a molecule. An initially excited, high-frequency localized mode can quickly de-excite by transferring its energy into combinations of lower frequency modes. In large molecules, in condensed phases, and at surfaces, huge numbers of low-frequency modes can accept energy, and energy randomization is very rapid (generally on the picosecond time scale or faster). This way, energy does not remain localized in a bond for a sufficiently long time to influence a chemical reaction. Therefore, the resulting chemistry is thermal rather than selective, which leads to the breaking of the weakest bond or to the reaction of the most reactive site. However, in small molecules with sparse vibrational modes, only a few or even zero combinations of low-frequency modes can accept the energy, and so the lifetime of the initially excited mode may be sufficiently long to allow mode-selective chemistry. This was already demonstrated for the outcome of the gas-phase reaction of H atoms with singly deuterated water (HOD) that can be controlled through laser excitation of specific HOD vibrational modes [10] and it is also illustrated in Figure 5 for a molecular case where one can take profit of the C - Cl bond being stronger than C - Br one.

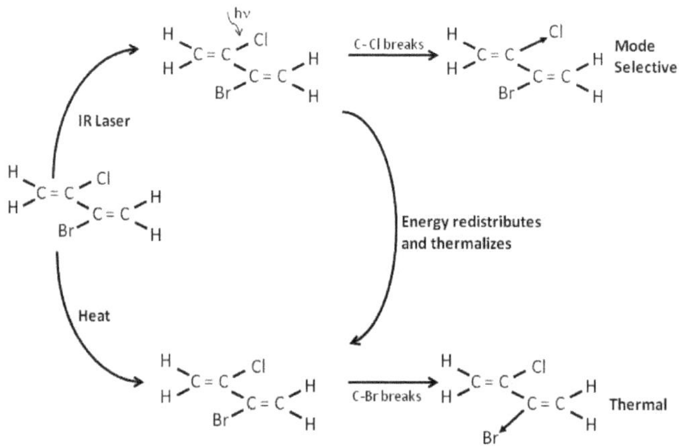

Fig. 5. LIF spectroscopy of the beam-gas reaction, revealing part of the rotational level population of the reaction product.

From the few attempts at IR mode-selective desorption of molecules from surfaces, that have been reported, perhaps the most successfull was the experimental findings of Liu et al [11], where the authors first created an absorbed layer of about 15% H atoms and 85% D atoms on an Si(111) surface. They then irradiated the surface with a free-electron laser tuned to the 4,8 μm Si-H stretching mode. They found that almost all desorbing atoms were in the form of H_2 and less than 5% of desorbing molecules were HD or D_2. This result rules out any local heating mechanism, which would produce a statistical mixture (2% H_2, 26% HD and 72% D_2). Largely as a result of the layer-molecular beam experiments of Y. T. Lee et al [12] the interpretation of the IR-MPD is now almost entirely clarified. The key to the understanding was the determination, under collision-free conditions, of the translational energy distribution of the photofragments as a function of laser intensity and, separately, laser fluence. It is found, indeed, that a suitable tailored statistical (RRKM) theory of unimolecular dissociation can explain most of the observations. All the evidence from these experiments suggests that mode selective molecular dissociation is only possible to achieve using a fast (ps) intense laser pulse.

Thermodynamic laws can be used to determine the equilibrium state of chemical reaction systems like the H_2/O_2 combustion ($2H_2 + O_2 = 2H_2O$). If chemical reactions are fast compared to all other physical processes (molecular diffusion, heat conduction and flow) thermodynamics alone allows the local description of even complex systems [13]. However, in most cases, chemistry occurs on time scales which are comparable with those of molecular transport. Thus, chemical kinetics information is required, e.g rate coefficients k(T) of the individual elementary reactions which occur on the molecular level have to be known. For example, the above mentioned H_2/O_2 combustion reaction consists of 38 elementary reactions involving a variety of reactive intermediates like H, O atoms and OH radicals [14]. Laser pump-and-probe techniques, which combine pulsed laser photolysis for reactive species generation with time-resolved laser-induced fluorescence (LIF) detection for reaction products, have paved the way for detailed studies of the molecular dynamics of the elementary reactions [1,2].

The experimental possibilities for studying processes in technical combustion devices have expanded a lot in recent years as a result of the development of various pulsed high-power laser sources which provide high temporal, spectral and spatial resolution. Laser spectroscopic methods are important for non-intensive measurements in systems where complex chemical kinetics are coupled with transport processes. One of the key factors for improving the performance of many technical combustion devices is an optimum control of the ignition process. Optimized reproducible ignition ensures an efficient and safe operation. Actually, experimental studies on CO_2 laser-induced thermal ignition of CH_3OH/O_2 mixtures have been performed. In a quartz cell equipped with SrF_2 windows, CH_3OH/O_2 mixtures are ignited using a cw CO_2 laser in the pulsed mode. The coincidence of the 9P(12) CO_2 laser line in the (001)-(020) band with the R(12) CO stretch fundamental band of the methanol molecule at 9.6 μm allows controlled heating and ignition of the mixture. OH radicals formed during flame propagation were excited in the (v' = 3, v* = 0) vibrational band of the OH ($A^2\Sigma^+$ - $X^2\Pi$) transition around 248 nm using two tunable KrF excimer lasers (laser wavelength tunable in the range 247.9 - 248.9 nm with a bandwidth of typically 0.5 cm^{-1}). The time delay between the two excimer laser pulses is 100 ns, in order to separate the signals induced by them. The fluorescence is collected using achromatic UV lens. Reflection filters are used to spectrally isolate the (v' = 3, v'' = 2) fluorescence band of the OH radical for detection. Fluorescence is detected by gated image-intensified CCD cameras. Excitation of two different optical transitions starting from the N'' = 8 and N'' = 11 rotational levels of the OH ($X^2\Pi$ - v'' = 0) vibrational state, allowed the measurement of spatially corresponding LIF image pairs. Assuming a Boltzmann distribution for the population of the OH ($X^2\Pi$ - v'' = 0) rotational states, the ratio of the two OH fluorescence images can be converted into a OH temperature field [1,2].

Interest in infrared laser-induced chemical reactions centered on the quest for mode selectivity was sparked by work on laser isotope separation in the 1970s. In much of this work it is common to find a strongly increasing yield with fluence. However, it is much less common to find an increase with pressure. The most dramatic example of an increasing yield with pressure is the IR laser-induced reaction of isomerization of methyl isocyanid to acetonitrile. [15]. It was then demonstrated that absorption and dissociation can be significantly enhanced through collisions and such reaction exhibits a sharp threshold pressure above which nearly complete isomerization occurs in a single pulse. Thus, the laser-initiated isomerization of methyl isocyanide is an ideal reaction for examination of collision-induced energy transfer phenomena. For the fluence dependence experiments, the laser was operated on the P(20) 944.19 cm^{-1} transition of the 10.6-μm band. CaF_2 flats and KCl windows were used to attenuate the beam for the lower fluence data. For the wavelength data, the laser was operated on the P(6) through P(34) lines of the 10.6-pm band (corresponding to the maximum of the R through the maximum of the P branch of the ν_4(C-

N) stretch of methyl isocyanide) with an average energy per pulse of 1.05 J at the sample. Analysis for reaction was performed by monitoring the 2165 cm^{-1} ν_2(C=N) stretch with a grating spectrophotometer. To determine the relationship between fluence and threshold pressure, the fluence was kept constant while the pressure was varied so as to bracket the threshold. Then the process was repeated with a new fluence. The results indicate a linear variation of the fluence with the inverse of the pressure. Analysis of the fluence dependence of the threshold pressure indicates that the inverse of the threshold pressure is directly proportional to the average number of photons absorbed per molecule. This balance between incident fluence (or average number of photons absorbed) and threshold pressure can be understood in terms of the usual model of increasing yield with fluence. Hence, if the yield is increased owing to an increase in fluence, the threshold pressure should exhibit a concomitant decrease, as is observed. The multiphoton absorption spectrum as well as the wavelength dependence of the threshold pressure reflects the structure of the linear absorption spectrum [1].

Interest in using infrared laser radiation for studying charge transfer processes at surfaces relies on the possibility of exciting vibrational modes of adsorbate molecules. It is well-known that vibrational excitation is very important in promoting endoergic gas-phase chemical reactions, as well as in controlling chemical processes occurring in adsorbed layers. The use of IR lasers for initiating gas-surface reactions when the gas consists of polyatomic molecules allows us to put, at the gas-surface interface, a large amount of energy due to IR multiphoton absorption. As is known from gas-phase experiments, at rather moderate for IR $C0_2$ laser energy fluences of about 1 J/cm^2, it is possible to excite to high vibrationally excited states (up to energy levels E > 1 eV) practically all irradiated molecule. Therefore, in spite of a rather small value of $C0_2$ laser quantum (0.1-0.12 eV) and longer pulse duration (>100 ns), one can induce effective charge transfer.

The use of a pulsed TEA C02 laser allowed us to apply the time-of-flight (TOF) technique for the detection of ion signal and, as a result, to distinguish between molecular negative ions and electron emission. A TEA $C0_2$ laser line tunable in the range 9-11 μm was used for the excitation of molecules at the Ba surface. The laser beam was directed to the Ba surface perpendicular to the SF_6 beam direction via a ZnSe window in the HVC and a KBr window in the UHVC [16]. The negative molecular ion signal was shown to be very sensitive to the SF_6 molecular absorption (to the exciting $C0_2$ laser frequency). Thus, enhancement factors of 10 or 4 were found for 10P(20) versus 10R(20) and 10P(16) versus 10P(22) lines, respectively. This support the vibrational selectivity (vibrational enhancement) of the SF_6 + Ba gas-surface IR laser-photoinduced ionization process [16].

Regarding applications in cluster spectroscopy, some experiments have been performed, in particular infrared photodissociation by crossing a continuous supersonic molecular beam of small methanol clusters with the radiation of a pulsed CO_2 laser [17] . Subsequent scattering by a secondary He beam disperses the cluster beam and allows the off-axis detection of selected cluster species, undisturbed by ionizer fragmentation artifacts. In the region of the ν_8 C-O stretching vibration, the dependence of IR photon absorption on laser frequency and fluence is investigated as a function of cluster size [17]. The predissociation spectrum of the dimer shows two distinct peaks at 1026.5 and 1051.6 cm^{-1} which correspond to the excitation of the two non-equivalent monomers in the dimer. The trimer spectrum features one single peak centered at 1042.2 cm^{-1}. This is consistent with a cyclic structure in which all three methanol molecules are equivalent. Higher cluster spectra are characterized by single peaks gradually blue-shifted with respect to the trimer line [17].

LIF detection is also used for various ultrasensitive techniques by probing reagents that are either autofluorescing or tagged with a fluorescent dye molecule. Applying different microscopic techniques with tight spatial and spectral filtering, various groups have directly visualized a variety of single fluorescent dye molecules (rhodamines and coumarins), dissolved in liquids by using coherent one- and two-photon excitation. Photophysical parameters and photobleaching play a crucial role for the accuracy of single-molecule detection by LIF and for the high sensitivity of fluorescence spectroscopy. Key properties for a fluorescent dye are its absorption coefficient, fluorescence- and photobleaching quantum yield. Photobleaching is a dynamic irreversible process in which fluorescent molecules undergo photoinduced chemical destruction upon absorption of light, thus losing their ability to fluoresce. Thus, for every absorption process there is a certain fixed probability ϕ_b of the molecule to be bleached. The probability P to survive n absorption cycles and become bleached in the $(n+1)^{th}$ cycle is given by:

$$P_{\text{(survival of n absorptions)}} = (1-\phi_b)^n\,\phi_b$$

This corresponds to the geometric distribution, which is the discrete counterpart of the exponential distribution. Due to this exponential nature of the photodestruction process with the standard deviation $(1-\phi_b)/\phi_b$, the relative fluctuation of the number of detected photons due to a single molecule transit can be as high as 100% [18]. The mean number of survived absorption cycles μ is equal to the standard deviation:

$$\mu = (1-\phi_b)/\phi_b = 1/\phi_b$$

Photobleaching is the ultimate limit of fluorescence-based single-molecule spectroscopy, and the quantum yield of photobleaching is defined by the ratio:

ϕ_b = number of photobleached molecules/total number of absorbed molecules

Unfortunatelly, the total number of absorbed molecules cannot be measured directly and precisely under single-molecule-detection, making it impossible to determine ϕ_b. However, the above definition can be expressed in kinetic terms. If a dye solution is illuminated, it is possible to measure the number of irreversibly photobleached molecules as a decrease in the dye concentration c(t) with time t. Under appropriate conditions the rate of this decrease is proportional to c(t), and so photobleaching reaction can be treated as a quasi-unimolecular reaction:

$$dc(t)/dt = -k_b\,c(t) \quad \rightarrow \quad c(t) = c(0)\,e^{-k_b t}$$

The rate constant k_b is dependent on cw laser irradiance, and the study of this dependence can lead to the evaluation of ϕ_b. The photostability of many organic dyes in organic solvents is higher than in water.

Pursuing the goal of single-molecule spectroscopy where ultra-low concentrations of the fluorophore in water (< 10 pM) are used, we should focus on photoreactions occuring under these conditions. In single-molecule spectroscopy, photostability and fluorescence saturation of the fluorophores impose limitations on the achievable fluorescence flow and the resulting signal-to-background ratio.

The rate constants for excitation from a state *i* to a state *f* are proportional to the irradiance *I* [W/cm^2] and to the absorption cross section $Q_{if}(\lambda)$ [cm^2] at a wavelength λ:

$$k_{Tif}(\lambda) = I\, Q_{if}(\lambda)\, \gamma$$

where = $\gamma/(hc)$, being *c* the velocity of light in vacuum and *h* the Planck constant.

Fluorescence saturation follows from the fact that a molecule cannot be in an electronically excited state and in the ground state at the same time; i.e. a single molecule can emit only a limited number of fluorescence photons in a certain time interval. Thus, the saturation characteristics of the fluorescent flows are determined by ground state depletion due to the finite excited state lifetimes of the S_1 and T_1 states (Figure 6).

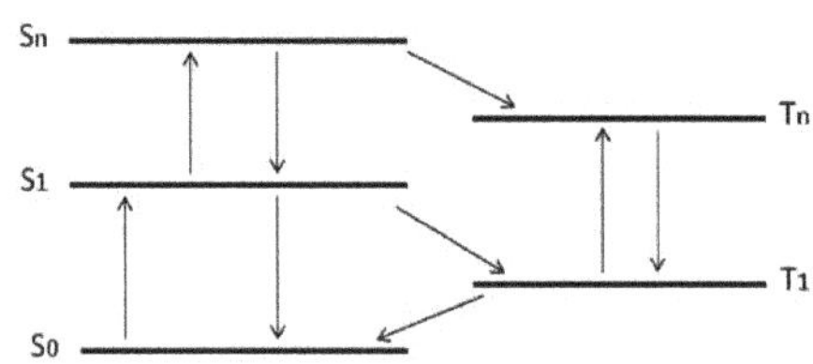

Fig. 6. Electronic Energy Diagram of a dye molecule with 5 electronic levels: S - singlet states; T - triplet states

The main reasons for using in life sciences near-infrared and infrared dyes as infrared fluorophores are threefold (although they photobleach more readily than dyes emitting in the visible):

- near-infrared dyes require excitation near IR which produces almost no auto-fluorescence from any endogenous components; hence, the sensitivity of detection, often limited by the auto-fluorescence background, is significantly improved.
- the laser wavelength also produce reduced scattering in the biological tissue, and thus increase both the penetration depth and the efficiency of collection of emission.
- available low-cost and compact NIR and IR diode lasers (e.g. 650 nm, 800 nm, 970 nm, etc...) can be used as excitation sources for these dyes.

Most of these dyes are cyanines, and they display a common problem which is their stability in biological fluids (composed mainly of water). They tend to aggregate, and so contribute to the quenching of emission. One method of preventing such aggregation is to isolate the dyes by encapsulation (in a nanobubble or a liposome) or to use chemical functionalization.

3. Nanoparticles and carbon nanotechnology

Given the importance of nanoscale particles in present technology, size distribution is a fundamental aspect as a quality control parameter. The production of nanoparticles using laser-induced gas-phase reactions techniques assures in general narrow size distributions contrarily to chemical techniques (such as precipitation or sol-gel processing) or to usual vapor-phase methods (furnace-heated vapor or arc-plasma), while the low reaction volume and ability to maintain steep temperature gradients allows for precise control of the nucleation and growth rates favoring the formation of very fine and uniform powders. When reaction occurs in the gas phase, far from polluting walls very pure nano-scale materials may be prepared, and so conditions are created permitting the homogeneous nucleation of particles by condensation from a supersaturated vapor phase. In addition, the laser processing is cleaner.

Pulsed laser ablation of materials in aqueous solutions of surfactants can also lead in some cases to the formation of ultrafine particles as in the case of TiO_2 crystalline anatase 3 nm nanoparticles, using the third harmonic of a Nd:YAG laser (355 nm) operating at 10 Hz [19].

Laser-induced production of silicon nanoparticles has been mostly based on the global reaction

$$SiH_4\,(g) + nh\nu \rightarrow Si(s) + 2H_2\,(g)$$

Silane strongly absorbs in the ν_4 band at 10.5 μm resonant with the P(20) line of the $00^{0}1$-$10^{0}0$ transition at 944.19 cm^{-1} of the CO_2 laser [20]. Experiments have been performed in high vacuum with a pulsed TEA CO_2 laser at fluences that varied between 0.5 J/cm^2 and 150 J/cm^2 (using different converging focal distance infrared lenses). The shape and size of the nanoparticles is then examined by electron microscopy (SEM and TEM) and Atomic Force Microscopy (AFM), and their structure by X-ray diffraction [21,22]. Figure 7 schematically describes the experimental set-up for laser pyrolysis (chemical decomposition by heat in the absence of oxygen). Silane gas is laminarilly flowing through the center of the laser pyrolysis reactor, surrounded by another laminar flow of helium. The focalized laser pulse (Δt = 100 ns, I = 30-40 mJ/pulse) decompose silane into silicon and hydrogen atoms which then recombine to form molecular hydrogen. Since a nozzle is placed close to the reaction zone, the silicon atoms and small silicon clusters are extracted in the majority helium supersonic expansion (like an ultra-fast cooler at a rate of 10^9 Ks^{-1}), giving rise by condensation to larger clusters and nanoparticles. Their formation is processing in bursts of some nanoseconds, according to the laser pulses characteristics. The nanoparticles size selection is assured by a synchronized chopper with the laser pulses, and the size distribution is measured in-situ by time-of-flight mass spectrometry (TOFMS). It is also possible to deposit these silicon nanocrystals on a removable target mica substrate [21,22].

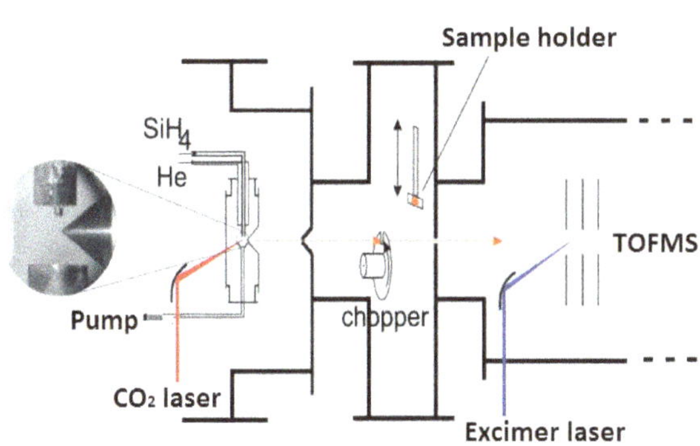

Fig. 7. Laser pirolysis set-up for size-selected nanoparticles production [21,22].

The kinetic energy of the clusters is small (less than 0.4 eV per atom for 4 nm size particles, which corresponds to 10% of the bonding energy) and so a Low Energy Cluster Beam Deposition (LECBD) is taking place, without changing the cluster properties in gas phase. Due to the geometry of the system, a distribution of nanoparticle sizes also appears on the substrate, and it can be guaranteed that all the sizes correspond to the same air exposition history. By varying some experimental parameters (pressure, flow, laser power, delay between laser pulse and chopper slit) it is possible to control the size distribution of the silicon nanoparticles deposited onto the substrate, for ultramicroscopic analysis (Figure 8).

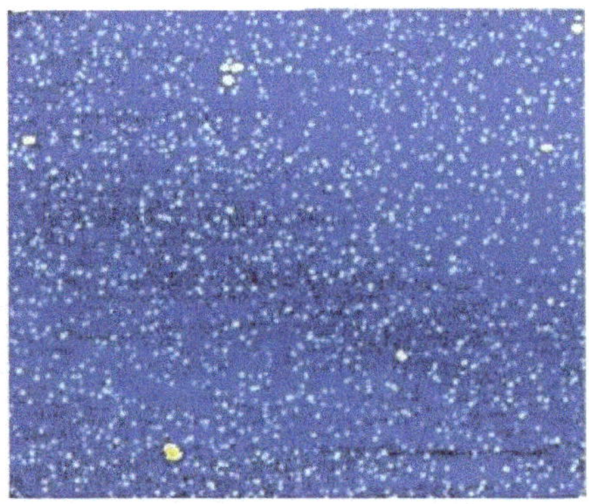

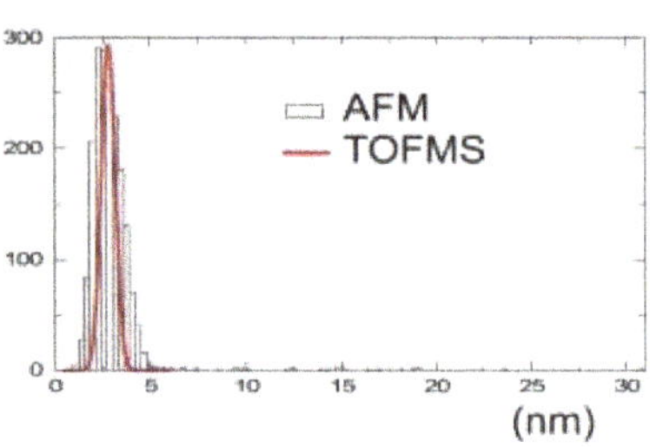

Fig. 8. AFM image of size-selected silicon nanoparticles together with their size distributions measured by TOFMS and AFM[21,22].

The advantage of using a TEA CO_2 laser in the pyrolysis becomes clear for the fine tuning wavelength adjust, in order to optimize the production process of several other types of nanoparticles [22].

Several synthesis reactors geometries based on the vaporization of a target (graphite/metal catalyst pellet) inside a oven at a fixed temperature (above 1000 K) by continuous CO_2 laser beam ($\lambda = 10.6\ \mu m$) have been developed to produce several types of carbon nanotubes. The laser power can be varied from 100 W to 1600 W and the temperature of the target is measured with an optical pyrometer. In general a inert gas flow carries away the solid particles formed in the laser ablation process which are then collected on a filter [23].

A typical mid-IR to near-IR absorbance spectrum taken on uniformly dispersed, purified CNTs (grown by CVD-mostly MWNT) at room temperature is displayed in Figure 9 [24]. In the measured IR absorbance spectrum, a prime intensity peak is seen at 1584 cm^{-1}. This is an

IR-active, graphite-like E_{1u} mode (also known as the *G* band) originating from the sp_2-hybridized carbon. The absorbance peak at 1200 cm^{-1}, is a disorder-induced one phonon absorbance band (D band-from the sp_3-hybridized carbon), which has been also observed in neutron irradiated diamonds [24]. This lattice mode arises from the disruption of the translational symmetry of the diamond lattice. In general, the presence of CH_x groups (evidenced in the IR active bands in the range of 3000 cm^{-1}) and non-conjugated carboxylic carbonyl groups (peak around 1725 cm^{-1}) can benefit a number of bio-sensing applications by offering a simple route to nanotube functionalization [25].

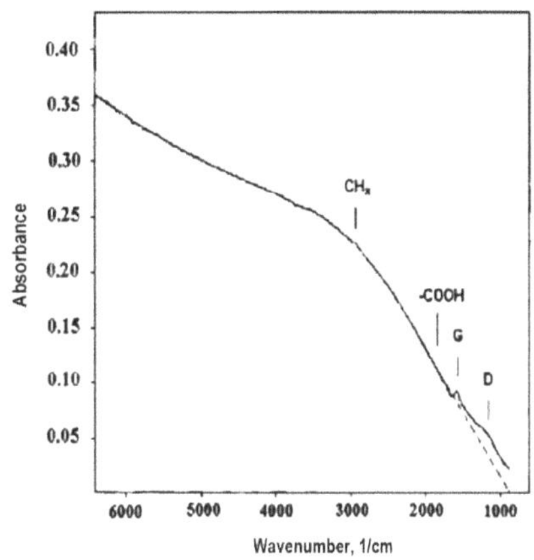

Fig. 9. Typical infrared absorbance spectrum obtained on 60 nm multiwall carbon nanotubes [25]

Production of single-wall carbon nanotubes (SWNT) by the laser-ablation technique using graphite, pitch and coke as carbonaceous feedstock materials has been reported [26]. This has been done with a 250W continuous-wave CO_2-laser at a wavelength of 10.6 μm and varying the nature and the concentration of the metal catalyst, the type and pressure of the buffer-gas as well as the laser conditions. The amount of SWNT material obtained is much higher when using graphite as a precursor than in the case of coke and carbonaceous feedstock [26].

Laser-assisted chemical vapour deposition (LCVD) for the formation and growth of carbon nanotubes has also been performed using a medium-power continuous-wave CO_2 laser to irradiate a sensitized mixture of $Fe(CO)_5$ vapour and

acetylene and to simultaneously heat a silicon substrate on which the carbon nanotubes were grown [27]. Electron microscopy (TEM and HRTEM) as well as atomic force microscopy (AFM) were used to analyze the as-grown films and samples specially prepared on TEM grids and AFM substrates. Carbon nanotubes with different structures (straight, curved and even branched), including single- and multi-walled nanotubes were observed. Some nanotubes were found to be partially filled with a solid material (probably metallic iron) that seems to catalyze the nanotube growth [27].

Using a CO_2 laser perpendicularly directed onto a silicon substrate, sensitized mixtures of iron pentacarbonyl vapour and acetylene were pyrolyzed in a flow reactor. The method involves the heating of both the gas phase and the substrate by IR radiation. The carbon nanotubes were formed via the catalyzing action of the fine iron particles produced in the same experiment by the decomposition of the organometallic precursor molecules [27]. The reactant gas, a mixture of iron pentacarbonyl vapour ($Fe(CO)_5$), ethylene (C_2H_4) and acetylene (C_2H_2), is admitted to the reaction cell through a rectangular nozzle, creating a gas flow close and parallel to the Si substrate and being pumped from the opposite side. The flow of ethylene is directed through a bubbler containing liquid iron pentacarbonyl at room temperature (and 27 Torr vapour pressure). Thus, the ethylene serves as carrier gas for the iron pentacarbonyl. The third gas, acetylene, is supplied by an extra line. Before entering the flow reactor, the gases are mixed in a small mixing vessel. The iron nanoparticles, needed to catalyze the formation of carbon nanotubes from carbon-containing precursors, are obtained by decomposing $Fe(CO)_5$ during the laser-induced reaction. Ethylene gas, introduced into the gaseous atmosphere, serves also as a sensitizer activating the laser reaction and speeding up the $Fe(CO)_5$ dissociation. C_2H_4 has a resonant absorption at the CO_2 laser emission wavelength (10.6 μm) and is characterized by a rather high dissociation energy. Under the present conditions, C_2H_4 is only expected to collisionally exchange its internal energy with the other precursor molecules that do not absorb the CO_2 laser radiation, thus heating the entire gas mixture [27].

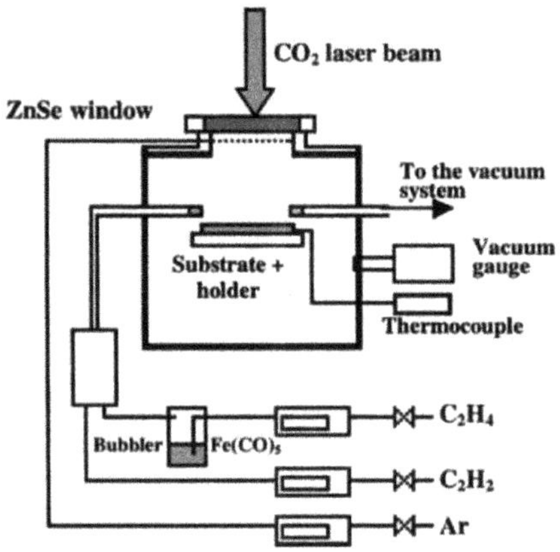

Fig. 10. Experimental set-up for the deposition of carbon nanotubes by LCVD
Adapted from [27]

The pressure inside the reaction chamber was keptat a constant value of 150 Torr. A flow of argon (500 sccm) was used to avoid contamination of the ZnSe entrance window during irradiation. At the heated surface and interface, iron pentacarbonyl is the first molecule to undergo dissociation, which can proceed until bare iron is obtained. Carbon nanotubes with straight, curved or even branched structures have been identified by ultramicroscopy. While some nanotubes were hollow, many of them were found to be partially or totally filled with nanoparticles most probably being metallic iron [27].

Films of vertically aligned MWCNTs of extremely high packing density were produced by this technique under very clean hydrocarbon supply conditions. Using an open-air pyrolitic LCVD system in which the role of gas-phase reactions are minimized, the growth of highly oriented and aligned single- and multiwall carbon nanotubes have been reported [28]. The

LCVD technique in general has several prominent advantages including high deposition rates that are favorable for scale-up production of CNTs. In contrast to the standard LCVD system, in which the in situ thermal decarbonylation of $Fe(CO)_5$ is used for obtaining Fe nanoparticles C-LCVD employs the catalytic activity of pre-deposited metal-based nanoparticles. For a better control of CNT growth conditions, it proved to be advantageous to carry out separately the catalyst deposition and carbon nanotube growth. Thus, as a main advantage, C-LCVD allows *ex-situ* prepared metal-based particles with the desired properties and dispersion degree to provide the nucleation conditions for the growth of CNTs [28]. The temperature is measured with a thermocouple positioned behind the substrate. Since the temperature of the area which is irradiated by the laser is expected to be considerably higher than the average temperature of the substrate holder, the temperature inside the laser spot is measured optically with a pyrometer. Under a total gas pressure of about 80 mbar, the temperature in the laser spot, measured with the pyrometer could reach values between 800 and 900 C, while the thermocouple indicates values that are about 100 C lower. Mainly depending on the ethylene concentration, nanotube mean diameters between 10 and 60 nm were found. By increasing ethylene precursor flow rate, not only larger mean diameters of the CNTs were found but also the distribution of the CNT diameters became broader [28].

Cementite (Fe_3C) is of great technological importance for the mechanical properties of steels and iron alloys and for its role as catalyst to produce various hydrocarbons (including olefins, from CO_2 and H_2) and preferred catalysts for carbon fibers, nanotubes and nanoparticles due to their low nanometric mean sizes and narrow size distributions. Pure Fe_3C nanomaterials have also been obtained by the pyrolysis of methyl methacrylate monomer, ethylene (as sensitizer), and iron pentacarbonyl (vapors) in a suitable range of laser intensities (by irradiating the same reactive mixture with a lower intensity radiation, the chemical content of the produced nanoparticles shifts towards mixtures of iron and iron oxides). Such nanopowders exhibited core (Fe_3C)-shell polymer-based morphologies and their magnetic properties are likely to display high values for the saturation magnetization.

Semiconductor photocatalysts have been used in different applications, and the combination of high photocatalytic activity, high stability and the benefit of environmental friendliness makes titanium dioxide the material of choice for such applications. In order to enhance the performance of this material for industrial purposes, coupling of TiO_2 with other semiconductors and immobilization of TiO_2 on porous materials have been studied as means for improving its photocatalytic activity. Nanocomposites of TiO_2 and multi-walled carbon nanotubes were prepared and deposited by sol-gel spin coating on borosilicate substrates and sintered in air at 300 °C. Further irradiation of the films with different CO_2 laser intensities was carried out in order to crystallize TiO_2 in the anatase form while preserving the MWNT's structure. The laser irradiation changed the crystal structure of the coatings and also affected the wettability and photocatalytic activity of the films. The anatase phase was only observed when a minimum laser intensity of 12.5 W/m^2 was used [29]. The contact angle decreased with the enhancement of the laser intensity. The photocatalytic activity of the films was determined from the degradation of a stearic acid layer deposited on the films. It was observed that the addition of carbon nanotubes themselves increases the photocatalytic activity of TiO_2 films. This efficiency is even improved when high CO_2 laser intensities are used during the sintering of the coatings [29].

The high aspect ratio combined with high mechanical and chemical stability of carbon nanotubes could help in enhancing the photocatalytic activity of TiO_2. Nanocomposites of TiO_2/CNTs were prepared by the addition of $-NH_2$ functionalized, 10 nm outer diameter MWNTs to the TiO_2-based sol, in a concentration of 3 mg/ml. After a fine dispersion of the nanotubes was achieved using a high shear processor, the sol was deposited on borosilicate glass substrates by a spin coating technique (2000 rpm, 10 s) [29].

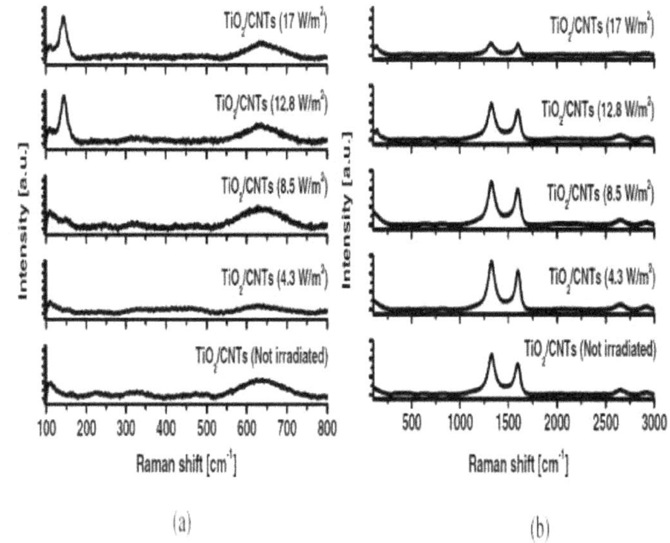

Fig. 11. Raman spectra in the ranges (a) 100-800 and (b) 300-3000 cm^{-1} of TiO_2/MWNT coatings not irradiated and irradiated with different CO_2 laser intensities [29]

Raman spectra of TiO_2/MWNT coatings not irradiated and irradiated with different CO_2 laser intensities are shown in Figure 11. Non-irradiated films sintered in air at 300°C have shown no Raman bands, as is clear from Figure 11 (a). Higher sintering temperatures were applied to the coatings inorder to obtain anatase TiO_2 phase structure, but oxidation of the MWNTs took place, limiting the study of the composites. Therefore, CO_2 laser irradiation was applied to the coatings with different density outputs as an alternative sintering process. This process is very fast, and because it is pulsed (highly spatially limited heated zone), it has not damaged the CNT's structure. Films irradiated with lasers of intensity lower than 12.8 Wm^{-2} have shown similar results to non-irradiated coatings in the range 100–1000 cm^{-1} [29]. However, a sharp and intense Raman band at 144 cm^{-1} corresponding to anatase TiO_2 phase was observed when such minimum laser intensity was applied during the sintering of the coatings, suggesting that the local temperature obtained using 12.8 Wm^{-2} is sufficient to crystallize the coating. Figure 11 (b) shows D and G bands corresponding to the presence of MWNTs at 1319–1328 and 1592–1601 cm^{-1}, respectively. The D band corresponds to defects present in carbonaceous materials. The addition of CNTs to the TiO_2

matrix improved the photocatalytic activity of TiO_2 coatings, as has been demonstrated by the degradation of a stearic acid layer deposited on the films [29]. In addition, higher CO_2 laser intensities during the sintering implies enhanced photocatalytic activity of the nanocomposites.

Laser Induced Breakdown Spectroscopy (LIBS) and Laser Induced Incandescence (LII) are two suitable techniques for analyzing small particles. In LII a pulsed laser rapidly heats the particles and by monitoring the rate of decay of the resulting incandescent radiation, one can extract particle size information, as the rate is related to the size of the particle. In order to get information on the particles chemical composition LIBS must be used instead. With it, the pulsed laser is tightly focused on the sample to induce a breakdown (microspark) of the material, and so by monitoring the emission of light from this plasma one can gather information about chemical compositions. Sensor technology is now able to capture light between 200 and 940 nm, a region where all elements emit. Since LIBS data is generated in real-time (response time of 1 second or less), one keep track of rapid changes in the composition of the particles during the actual production run. LIBS is very sensitive, having a resolution in the femtogram region and capable of detecting as few as 100 particles/cm^3.

Irradiating the surface of a solid with a laser, material can be ablated in a controlled way by optimizing intensity and pulse duration of the laser (laser ablation). Depending on the laser wavelength, the ablation is dominated by thermal evaporation (using a CO_2 laser) or photochemical processes (using an excimer laser). Laser-spectroscopic diagnostics can distinguish between the two processes. Excitation spectroscopy or resonant two-photon ionization of the sputtered atoms, molecules, clusters, nanoparticles or even microparticles allows their identification (Figure 12).

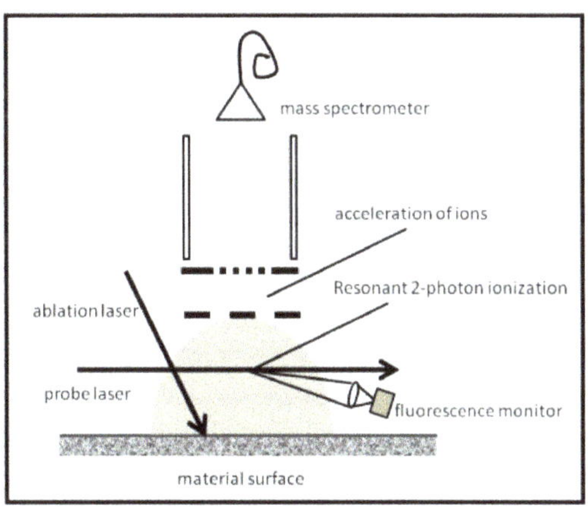

Fig. 12. Laser ablation from a surface

In addition, the velocity distribution of particles emitted from the surface can be obtained from the Doppler shifts and broadning of the absorption lines, and their internal energy distribution from the intensity ratios of different vibrational-rotational transitions. With a pulsed ablation laser, the measured time delay between ablation pulses and probe laser pulses allows the determination of the velocity distribution. Resonant two-photon ionization in combination with a TOFMS gives the mass spectrum. It is common to observe a broad mass range of clusters. The question is whether these clusters were emitted from the solid or whether they were formed by collisions in the evaporated cloud just after emission. Measurements of the vibrational energy distributions can give an answer. If the mean vibrational energy is much higher than the temperature of the solid, the molecules were formed in the gas phase, where an insufficient number of collisions cannot fully transfer the internal energy of molecules formed by recombination of sputtered atoms into kinetic energy.

Whereas laser ablation of graphite yields thermalized C_2 molecules with a rotational-vibrational energy distribution following a Boltzmann distribution at the temperature T of the solid, ablation of electrical insulators, such as AlO, produces AlO molecules with a large kinetic energy (≈1 eV), but a "rotational temperature" of only 500 K.

4. Technological applications

Photodynamic therapy (PDT) is used clinically to treat a wide range of medical conditions, including malignant cancers, and is recognised as a treatment strategy which is both minimally invasive and minimally toxic. Photosensitization is a process of transferring the energy of absorbed light. After absorption, the energy is transferred to the (chosen) reactants. This is part of the work of photochemistry in general. In particular this process is commonly employed where reactions require light sources of certain wavelengths that are not readily available. For example, mercury absorbs radiation at 1849 and 2537 angstroms, and the source is often high-intensity mercury lamps. It is a commonly used sensitizer. When mercury vapor is mixed with ethylene, and the compound is irradiated with a mercury lamp, this results in the photodecomposition of ethylene to acetylene. This occurs on absorption of light to yield excited state mercury atoms, which are able to transfer this energy to the ethylene molecules, and are in turn deactivated to their initial energy state.

In order to achieve the selective destruction of the target biological area using PDT while leaving normal tissues untouched, the photosensitizer can be applied locally to the target area. For instance, in the treatment of skin conditions, including acne, psoriasis, and also skin cancers, the photosensitizer can be locally excited by a light source. In the local treatment of internal tissues and cancers, after photosensitizers have been administered intravenously, light can be delivered to the target area using endoscopes and fiber optic catheters. Compared to normal tissues, most types of cancers are especially active in both the uptake and accumulation of photosensitizers agents, which makes cancers especially vulnerable to PDT. Since photosensitizers can also have a high affinity for vascular endothelial cells, PDT can be targetted to the blood carrying vasculature that supplies nutrients to tumours, increasing further the destruction of tumours.

The optimal spectral window for biological tissue penetration of irradiation is around 800 nm to 1 μm. The use of CO_2 laser relies on water (the largest constituent of most biological tissues) absorbing strongly at 10.6 μm; this strong absorption leads to a shorter optical penetration depth (≈13 μm), limiting its use to extremely thin tissues. A patient would be given a photo sensitive drug (photofrin) containing cancer killing substances which are absorbed by cancer cells. During the surgery, the light beam is positioned at the tumor site, which then activates the drug that kills the cancer cells, thus photodynamic therapy.

A sensitizer in chemoluminescence is a chemical compound, capable of light emission after it has received energy from a molecule, which became excited previously in the chemical reaction. A good example is when an alkaline solution of sodium hypochlorite and a concentrated solution of hydrogen peroxide are mixed, a reaction occurs:

$$ClO^-(aq) + H_2O_2(aq) \rightarrow O_2^*(g) + H^+(aq) + Cl^-(aq) + OH^-(aq)$$

$O_2^*$is excited oxygen - meaning, one or more electrons in the O_2 molecule have been promoted to higher-energy molecular orbitals. Hence, oxygen produced by this chemical reaction somehow 'absorbed' the energy released by the reaction and became excited. This energy state is unstable, therefore it will return to the ground state by lowering its energy. It can do that in more than one way:

- it can react further, without any light emission
- it can lose energy without emission, for example, giving off heat to the surroundings or transferring energy to another molecule
- it can emit light

New types of photosensitizers used in photodynamic therapy, which are based on photon upconverting nanoparticles, have been developed. Such photosensitizers are excitable with infrared irradiation, which has several times larger tissue penetration depth than the currently available ones. Photon upconverting materials convert lower-energy light to higher-energy light through excitation with multiple photons. For instance, such materials would adsorb infrared irradiation and emit visible light to further excite the photosensitizing molecules. Photon upconverting nanoparticles (PUNPs) play here a crucial role. They can be first coated with a porous, thin layer of silica through sol-gel reaction. During the coating process, photosensitizing molecules with high absorbance in the spectral window matching the emission of the PUNPs are doped, so that the resulting silica layer contains a certain amount of these photosensitizing molecules. Finally, an antibody, specific to antigens expressed on the target cell surface, is covalently attached to the silica-coated nanoparticles. When the thus-prepared nanoparticles are irradiated by infrared light, emission from the PUNPs will be absorbed by the photosensitizing molecules coated on their surfaces. Subsequently, excited photosensitizing molecules will interact with surrounding groundstate molecular oxygen, generating singlet oxygen, leading to oxidative damage of the neighboring cells to which the nanoparticles are attached via specific antigen-antibody binding.

PUNPs made from $NaYF_4:Yb^{3+},Er^{3+}$ have been recognized as one of the most efficient photon upconverting phosphors [30].

When excited by an infrared (974 nm) source, strong visible bands appear around 537 nm and 635 nm. Merocyanine 540 (M-540) is used as photosensitizing molecule, and doped into the silica layer during the coating process. M-540 is a molecule that can produce singlet oxygen and other reactive oxygen species, and has been used before in photodynamic therapy as a photosensitizer with a visible light source. Both the emission spectrum of $NaYF_4:Yb^{3+},Er^{3+}$ nanoparticles and the absorption spectrum of M-540 show a good overlap between the nanoparticles' emission and M-540's absorption [30]. The photon upconverting property of the nanoparticles was not affected by the silica coating, as confirmed by their photoluminescence spectrum. The presence of M-540 in the silica coating could be readily confirmed by the change in color of the nanoparticles, to slightly yellowish.

Photosensitizers drugs, should ideally be specific to the target, highly effective in producing reactive oxygen species (ROS) when exposed to appropriate illumination, and excitable by a wavelength close to the near-infrared region (800 nm to 1 μm),where tissue penetration of the illumination is at a maximum.

Regarding the last desired feature, single photons with infrared wavelengths are usually too weak energetically to generate reactive oxygen species ($^1O_2$). Thus, multiphoton excitation would be needed for infrared light to be used as illumination source. It has been reported the synthesis and characterization of a type of nanomaterial capable of generating $^1O_2$ under continuous wave infrared excitation, based on photon upconverting nanoparticles (PUNPs). The results demonstrate that such nanoparticles have great potential for becoming a new type of versatile PDT drugs for photodynamic therapy [31]. Although photon upconverting materials do not directly produce ROS, one utilizes the fact that they adsorb infrared photons and emit visible ones to further excite the photosensitizing molecules, thus indirectly causing the photosensitizing molecules to generate $^1O_2$ under infrared excitation. The design and synthesis of the PUNP-based photosensitizers follow a schema depicted in Figure 13.

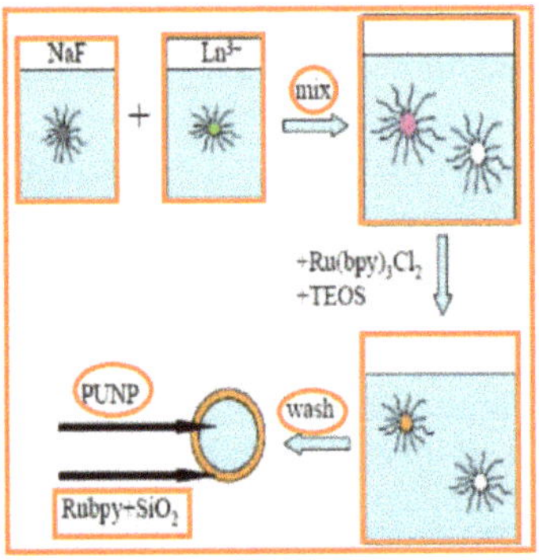

Fig. 13. PUNP-based photosensitizer preparation.

The core is a $NaYF_4:Yb^{3+},Tm^{3+}$ nanoparticle, a photon upconverting material capable of emitting blue light (≈ 477 nm) upon excitation by an infrared light source (≈ 975 nm). The nanoparticle was then coated by a thin layer of tris(bipyridine)ruthenium(II)-doped silica which generates $^1O_2$. The $NaYF_4:Yb^{3+},Tm^{3+}$ nanoparticles were synthesized by a microemulsion method [31].

The photoluminescence spectra of the $NaYF_4:Yb^{3+},Tm^{3+}$ nanoparticles before and after beingcoated with Ru(bpy)3-doped silica, under 975 nm excitation, are shown in Figure 14.

One of the most sensitive methods to detect volatile compounds released by the plants is laser photoacoustic spectroscopy (LPAS), which allows the identification of many molecules signaling plant defence mechanisms [32]. The technique is based on the photoacoustic effect, i.e. the generation of acoustic waves as a consequence of light absorption. The absorption of photons of a suitable wavelength and energy by the gas molecules excites them to a higher ro-vibrational state. The absorbed energy is subsequently transferred by intermolecular collisions to translational energy, and thereby to heat. When a gas sample is collected in a closed cell, the heating of the gas molecules will increase the cell pressure. Hence, by modulating the light intensity pressure variations are produced that generate a sound wave, which can be detected with a sensitive microphone. A schematic view of a typical LPAS experimental system for the detection of volatile molecules is shown in Figure 15.

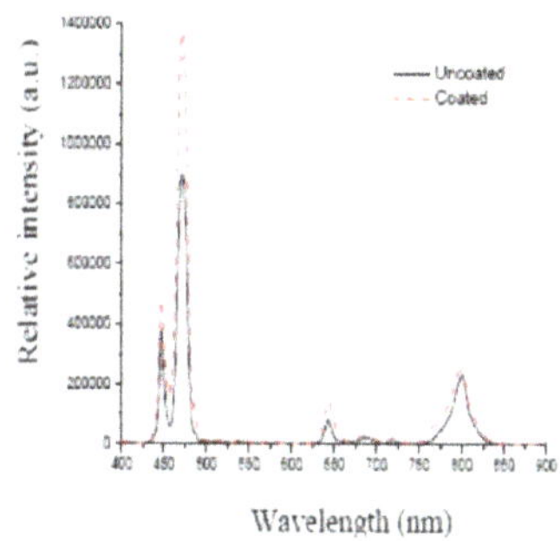

Fig. 14. Photoluminescence spectra of $NaYF_4$:Yb^{3+},Tm^{3+}nanoparticles before and after being coated with $Ru(bpy)_3$-doped silica [31]

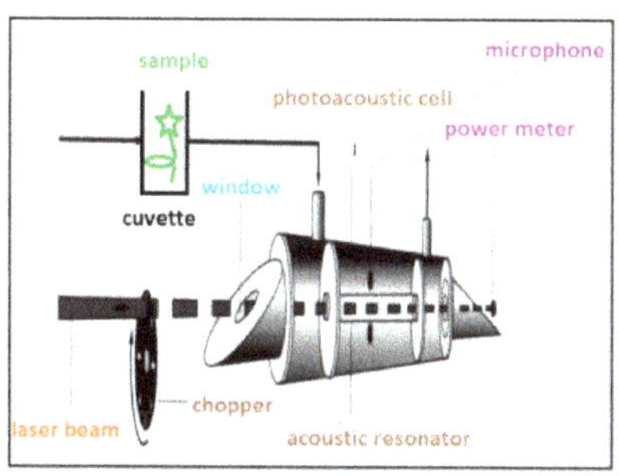

Fig. 15. Schematic view of a LPAS set-up for the detection of volatile molecule emission from plants.

The photoacoustic signal depends on the number of absorbing molecules present in the gas, the absorption strength of the molecules at a specific light frequency, and the intensity of the light. Then, for trace gas detection, the light source should have a narrow bandwidth and be tuneable (in order to match the specific molecular absorption feature), and it should have a high intensity to ensure a good signal-to-noise ratio. Since the absorption processes of interest involve ro-vibrational transitions, it is normally necessary to work in the IR region. In this spectral range each molecule has its own *fingerprint* absorption spectrum,whose strength can vary rapidly over a short wavelength interval. Specifically, the preferred range for spectroscopic applications lies in the range 3–20 μm. Specifically, CO_2 and CO lasers serve as the most frequently used light sources for photoacoustic detection of gases because they provide relatively high CW powers, typically up to 100 W and 20 W respectively, over this wavelength region. LPAS shows a large versatility of applications not only in plant science, but also in other fields, e.g. in environmental chemistry [33]. It has been shown to be a reliable method for the detection of ethylene in several plant physiological processes at parts per trillion concentration levels (e.g. from a cherry tomato under different conditions) [34].

One of the major analytical problems with fruit and vegetable samples is the detection and identification of non-volatile organic compounds present in low concentration levels, as happens for most of the phytoalexins produced by plants. Mass spectrometry is widely used in the analysis of such compounds, providing exact mass identification. However, the difficulty with their unequivocal identification and quantitative detection lies in their volatilization into the gas phase prior to injection into the analyser. This constitutes particular problems for thermally labile samples, as they rapidly decompose upon heating. To circumvent this difficulty a wide range of techniques have been applied for non-volatile compound analysis, including LD (Laser Desorption). Recently, LD methods have been developed in which the volatilization and ionization steps are separated, providing higher sample sensitivity. In particular, REMPI-TOFMS is considered to be one of the most powerful methods for trace component analysis in complex matrices [2]. The high selectivity of REMPI-TOFMS stems from the combination of the mass-selective detection with the resonant ionization process, i.e. the ionization is achieved by absorption of two or more laser photons through a resonant, intermediate state. This condition provides a second selectivity to the technique, namely laser wavelength-selective ionization. In addition, it shows an easy control of the molecular fragmentation by the laser intensity and the possibility of simultaneous analysis of different components present in a matrix.

As an example, it is possible to perform fast and direct analysis of non-volatile compounds in fruit and vegetables, particularly *trans*-resveratrol in grapes and vine leaves. The method is based on the combination of LD followed by REMPI and TOFMS detection, often identified by its sum of acronyms, i.e. LD-REMPI-TOFMS [35]. *Trans*-Resveratrol is an antioxidant compound naturally produced in a huge number of plants, including grapes. Analysis of *trans*-resveratrol is generally carried out by high-performance liquid chromatography. Its analysis in grapes and wines requires the use of pre-concentration prior to analysis and/or multi-solvent extraction techniques, due to the complexity of the matrices and to the low concentration of the analyte. The extraction methods generally employed are liquid extraction with organic solvents or solid-phase extraction. It is generally accepted that the sample preparation is the limiting step in *trans*-resveratrol analysis, not only because of the need for costly and time-consuming operations, but also because of the error sources introduced during this operation. These error sources can largely be overcome when applying the method of LD-REMPI-TOFMS. The experimental set-up used in this analysis method basically consists of two independent high vacuum chambers; the first chamber is used for both laser desorption and laser post-ionization of the samples, and the second chamber for TOFMS [2].

Some other relevant technological applications of infrared lasers will be generally described here in the following paragraphs.

Katzir was the first researcher to apply the carbon dioxide laser, coupled to optical fibers made from silver halide, for wound closure under a tight temperature control. The fibers deliver the laser's energy to heat the bonded cut and are used for controlling the temperature. They also make it possible to bond tissues inside the body. Sutures or stitches are not water tight, and blood or urine can pass through cuts, causing severe infection. Laser-bonded tissues heal faster, with less scarring. Even using today's microsurgery techniques, the treated wounds are open to infection, and the patient is inevitably left with permanent and unsightly scars. The near-infrared light is just the right wavelength to excite vibrations in chemical bonds in the water molecules (via first-overtone excitation in the OH-stretch manifold); the vibrations quickly turn into heat.

Keeping the heat from the laser at exactly the right temperature for optimal wound healing, allows surgeons to seal cuts both on our skin and inside our bodies with less scarring, and less exposure to infection. When the laser begins to overheat and risks burning the tissue, the device reduces laser power, and if the temperature is too low to complete a closure, laser power is increased appropriately.

There is also an enormous potential of a CO_2-laser system for rapidly producing polymer microfluidic structures. The common polymer poly (methyl methacrylate) (PMMA) absorbs IR light in the 2.8–25 μm wavelength band, so CO_2 lasers have been used in recent years for fabricating microfluidic devices from it, with channel widths of a few hundred micrometers. The narrowest produced channel was 85 μm wide. A solvent-assisted thermal bonding method proved to be the most time-efficient one. These systems provide a cost effective alternative to UV-laser systems and they are especially useful in microfluidic prototyping due to the very short cycle time of production.

Furthermore, surface heat treatment in glasses and ceramics, using CO_2 lasers, has drawn the attention to several technological applications, such as lab-on-a-chip devices, diffraction gratings and microlenses. Microlens fabrication on a glass surface has been studied mainly due to its importance in optical devices, as fiber coupling and CCD signal enhancement. Using microlens arrays, recorded on the glass surface, can enable the bidimensional codification for product identification. This would allow the production of codes without any residues (like the fine powder generated by laser ablation) and resistance to an aggressive environment, such as sterilization processes. Microlens arrays can be fabricated using a continuous wave CO_2 laser, focused on the surface of flat commercial soda-lime silicate glass substrates.

Silicon micromachining is a very important technology in microfabrication and microelectromechanical system (MEMS) industry. Nd:YAG laser has a wavelength of 1.06 μm, which is adsorbed by silicon, and is easily used for direct silicon machining. But the cost is very high. Although CO_2 laser is cheap, its wavelength of 10.64 μm is not absorbed by silicon. However, a silicon sample put on the top of a glass, instead of pure silicon, is used for CO_2 laser micromachining. The silicon on the top of a glass may absorb the CO_2 laser and become able to be etched, even through the wafer. Commercial available air-cooled CO_2 laser equipment can be used with a maximum laser power of 30 W. A glass below the silicon changes the absorption of silicon to CO_2 laser during machining. The silicon on the top of a glass may be etched by CO_2 laser even through the wafer due to the absorption variation. The etching depth increases with the pass number at constant laser power and scanning speed.

Several techniques are available that allow the size distribution of an aerosol to be determined in real time, but the determination of chemical composition, which has traditionally been done by impaction methods, is slow and yields only an average composition of the ensemble of particles in a given size range. It is essential, therefore, that new techniques be developed to allow the characterization of both the physical properties and chemical composition of aerosols, and that these operate on a time-scale that allows changes in the aerosol composition to be determined in real time. Several variants of the aerosol TOFMS (ATOFMS) instrument have been described in the literature [2]. The principle of the most sophisticated instrument reported to date employs two laser systems; first, a tuneable IR laser (OPO) is used to desorb material selectively from the particle, and then a second (VUV) laser is used to ionize the molecules that are produced [2]. With this approach, greater control over the particle ablation and ionization steps is possible, and by using low IR laser energy for the first evaporation step it is possible to depth profile heterogeneously mixed aerosol particles. Molecular information can be obtained by tuning the laser energy to just above the threshold required for desorption.

5. Conclusions

Several applications have been demonstrated for CO_2 lasers, despite it is impossible to use photoelectric emission to detect this radiation (photon energy of about 0.1 eV is only about five times room temperature, and cryogenically cooled photoconductors are necessary to achieve fast low-level detection) and little engineering has been done in the mid-infrared region.

All applications mentioned require a stable, single-frequency source of radiation. A kilowatt of radiation at 10 microns, focused down to its diffraction limit, is a power density of 1 gigawatt per square cm. Because most materials absorb at 10 microns, considerable interest has been shown in CO_2 lasers in many applications, and any problem which requires controlled surface heating or burning might find a potential solution with the CO_2 laser.

The MTL3-GT CO_2 laser ability to combine the laser pulsed mode with tune-ability introduces new perspectives to perform different experiments, which require a suitable, reliable and user-friendly procedure. With the method described in this work, many experiments can be performed in real time with simultaneous control of power/energy and wavelength, and taking advantage of the full laser power for each selected wavelength. One could observe, after improving the procedure, that energy values are more stable in all four emission bands (9P, 9R, 10P and 10R). This behavior was also observed regardless of the repetition rate, even for higher ones around 100 Hz. Besides energy, power was also measured and improved following the same procedure. This procedure can also be used on other infrared lasers with some minor adaptations regarding the software and energy detectors used. The observation of the energy line variation in real time is very important for understanding the behavior of the laser under certain external conditions and also to make sure that internal mechanisms are also error free.

It was demonstrated that CO_2 laser applications in the fields of nanoscience and nanotechnology are very promising, with particular relevance on spectroscopy, photodynamic kinetics and photodynamic therapy, ultra-pure and size-selected nanoparticle production, surgery and biomedicine, nanoanalysis, nanomaterials processing and composite nano-engineered catalysts.

6. References

[1] W. Demtröder, *Laser Spectroscopy*, 3rd Ed. Springer, 2003.
[2] H. H. Telle, A. Gonzalez Ureña, R. J. Donovan, *Laser chemistry,* Wiley, (2007).
[3] C. N. Patel, *Phys. Rev. Lett.* 12 (1964) 588.
[4] A. J. Beaulieu *Appl. Phys. Lett.* 16 (1970) 504.
[5] J. F. Ribeiro and R. F. M. Lobo, *Eur. J. Phys.* 30 (2009) 911.
[6] Edinburgh Instruments Ltd, *TEA CO2 MTL3-GT Laser User's Manual*, Issue G September, 2003.
[7] G. Scoles, *Atomic and Molecular Beam Methods* , Oxford University Press, New York, 1992.
[8] J. Steinfeld, *Laser-Induced Chemical Processes,* Plenum Press, New York, 1981.
[9] C. Miller, R. Zare, *Chem. Phys. Lett.*, 71 (1980) 376.
[10] M. Bronikowski, W. Simpson, R. Zare, *J. Phys. Chem.* 97 (1992) 2194.
[11] Z. Liu, *Science* 312 (2006) 1024.
[12] Y. T. Lee, Y. R. Shen, *Phys. Today* 33 (1980) 11.
[13] W. Gardiner, *Combustion Chemistry*, Springer, New York, 1994.
[14] J. Warnatz, U. Maas, R. Dibble, *Combustion*, Springer, Heidelberg, 1996.
[15] M. J. Shultz, E. J. Rock, R. E. Tricca, L. M. J. Yam, *Phys. Chem.* 88 (1984) 5157.
[20] I. Herman, *Chem. Rev.* 89 (1989) 1323.
[16] J. Castano, V. Zapata, G. Makarov and A. Gonzalez Urena, *J. Phys. Chem.* 99 (1995) 13659.
[17] Friedrich Huisken, Martin Stemmler, *Chem. Phys.Lett.* 144 (1988) 391.
[18] M. Köllner, *Appl. Opt.*, 32 (1993) 806.
[19] C. Liang, Y. Shimizu, T. Sasaki, N. Koshisaki, *J. Mater. Res*, 19 (2004) 1551.
[20] I. Herman, *Chem. Rev.* 89 (1989) 1323.
[21] M. Ehbrecht and F. Huisken, *Phys. Rev. B*, 59 (1999) 2975.
[22] G. Ledoux, R. Lobo, F. Huisken, O. Guillois, C. Reynaud, *Photoluminescence of Silicon Nanocrystals Synthetised by Laser Pyrolysis*, in Trends in Nanotechnology Research, ed. E. Dirote, Nova Siience Publish, New York, 2004.
[23] B. Bhushan, *Handbook of Nanotechnology*, Springer, New York, 2004.
[24] N. Kouklin, M. Tzolov, D. Straus, A. Yin, and J. M. Xu, *Appl. Phys. Lett.* 85 (2004) 4463.
[25] L. Zhang, V. U. Kiny, H. Peng, J. Zhu, R.F.M. Lobo, J. L. Margrave, V. N. Khabashesku, *Chem. Mat.*, 16 (2004) 2055.
[26] W. Maser, A M Benito, E Munoz, G M de Val, M T Martınez, A Larrea and G de la Fuente, *Nanotechnology* 12 (2001) 147.
[27] R. Alexandrescu, A. Crunteanu, R.-E. Morjan, I. Morjan, F. Rohmund, L. Falk , G. Ledoux, F. Huisken, *Infrared Physics & Technology* 44 (2003) 43.
[28] I. Morjan, I. Soare, R. Alexandrescu, L. Gavrila-Florescu, R Morjan, G. Prodan, C. Fleaca, I. Sandu, I. Voicu , F. Dumitrache, E. Popovici, *Infrared Physics & Technology* 51 (2008) 186.
[29] M. S. Castro, E. D. Sam, M. Veith and P. W. Oliveira, *Nanotechnology* 19 (2008) 105704.
[30] S.I. Klink, H. Keizer and V. Veggel, *Angewandte Chemie International Edition* 39 (2000) 4319.
[31] Y. Guo, M. Kumar and P. Zhang, *Chem. Mater. 19* (2007) 6071.

[32] F. J. M. Harren and J. Reuss, *Photoacoustic Spectroscopy* in Encyclopedia of Applied Physics, vol 19, G L Trigg (ed.) , VCH, Weinheim,1997.
[33] M. W. Sigrist, A. Bohren, T. Lerber, M. Nagel, M. Romann, *Anal. Sci.,* 17 (2001) S511.
[34] De Vries, F J M Harren and J Reuss, *Biol. Technol.*, 6 (1995) 275.
[35] J. M. Orea, C. Montero, J. B. Jiménez, A. G. Ureña, *Anal Chem.*, 73 (2001) 5921.
[36] C. Montero, J. M. Orea, M. Soledad Muñoz, R. F. M. Lobo, A. González Ureña, *Applied Phys. B,* 71 (2000) 601.

13

CO2 Laser and Micro-Fluidics

Mohammadreza Riahi
Shahid Beheshti University/Laser and Plasma Research Institute
Iran

1. Introduction

Microfluidic chips have attracted significant attention over the past decade due to their wide range of potential applications in the biomedical and chemical analysis field such as drug delivery, Point of care diagnostics (Jakeway et al, 2004), flow cytometry (Fu LM et al 2004; Chen & Wang, 2009; Lin et al, 2009), polymerize chain reaction (Suna et al, 2007; Sun & Kwok, 2006; Hsieh et al, 2009), electrophoresis (Fu et al, 2007, 2009) and many other applications.

Traditionally, silicon and glass are the predominant materials employed in the design of microfluidic systems. This was primarily due to their excellent chemical, physical, electrical and optical properties. But fabrication of a microfluidic device on these materials needs standard photolithography equipments such as Reactive Ion Etching (RIE) system which are very expensive and increases the production costs specially in single-use applications which are desired in order to avoid contamination.

In recent years application of polymeric materials for microfluidic device fabrication is becoming more and more important. Different methods for microfluidic fabrication on polymers such as hot embossing (Gerlach et al, 2002), injection molding (Rotting et al, 2002), soft lithography (Xia et al, 1998) and laser micromachining can be applied.

Different kind of lasers such as UV (Ball et al, 2000) and Infrared lasers is used for laser micromachining of polymers. In infrared regime, CO2 laser has a predominant application due to it's excellent absorption in polymers.

In this chapter, we will deal with application of a CO2 laser in microfluidic device fabrication. The application of CO2 laser for fabrication of a optofluidic device and application of a optofluidic device for CO2 laser characterization is also presented.

2. Interaction of a CO2 laser with polymers

Application of the CO2 laser for microfluidic device fabrication was first proposed in 2002 by H. Klank et al (Klank et al, 2002).

CO2 laser emits radiation with the wavelength of 10.6 micrometer. A CO2 laser mostly interacts with a polymer, photo-thermally. When a CO2 laser is irradiated on a polymer surface, it is strongly absorbed and raises the temperature of the polymer. The polymer is then melted, decomposed and leaving a void in a workpiece.

Different kind of polymers can be used for microfluidic applications by taking a choice care that the fluid in the device do not interact chemically with the device. However just some of the polymers can be machined with CO2 laser. Most of the polymers leave contamination and soot when exposed to CO2 laser irradiation. For example, polycarbonate (PC) leaves a brownish residue after exposing to the CO2 laser.

Among different kind of polymers, poly methyl methacrylate (PMMA) is the most suitable polymer for CO2 laser machining. When PMMA heats up by the CO2 laser, after passing the glass temperature, the material turns into a rubbery material and by increasing the temperature, the chains are broken by depropagation process (Arisawa & Brill, 1997; Ferriol et al, 2003), and decompose with a non-charring process to it's MMA monomer which is volatile

$$Solids \rightarrow Volatiles \tag{1}$$

Fig. 1 shows the decomposition process of PMMA polymer.

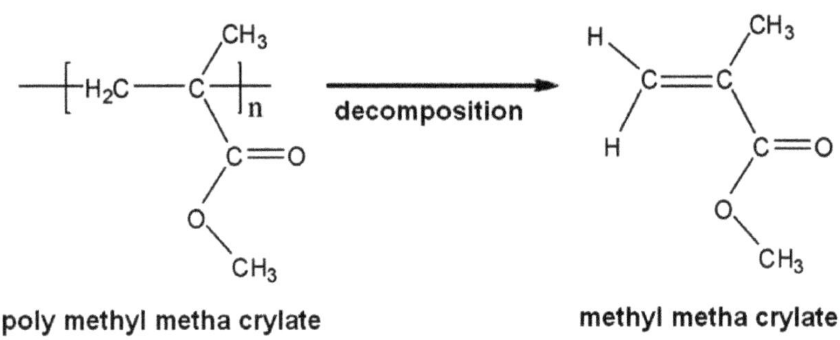

Fig. 1. Decomposition process of PMMA.

Decomposition of PMMA into volatile MMA monomers makes a hole in the workpiece. The shape and size of the hole, highly depends on the thermal properties of PMMA, focusing parameters, laser beam profile, exposure time and even exposure strategy. At the beginning of the exposure, the shape of the channel is very similar to the laser beam profile but as time goes up, the shape of the hole becomes more conical.

3. Fabrication of a channel on PMMA utilizing a CO2 laser

Fabrication of a channel on the surface of PMMA can be performed by scanning a CO2 laser over the surface of the workpiece. Commercial CO2 engraving systems with laser powers about a few watts to a few tens of watts and scanning speeds from a few tens of mm/sec up to a few hundreds of mm/sec, are good choices for micro channel fabrication.

By scanning the PMMA surface with CO2 laser, different channels and cavities can be fabricated. However, the ablated structures may be very rugged such that those can not be used for microfluidic structures with optical detection. Martin et al. reported that the roughness of the machined surfaces depends on the grade of the PMMA sheets. He reported roughness of 1.54 microns and 0.42 microns for two different grades of PMMA (Martin et al, 2003). Presence of the different roughness should probably be sought in the chemical additives of the different types of PMMA.

Cheng et. al. reported that the roughness of the machined channels can be treated by thermal annealing of the samples (Cheng et al 2004). Fig. 2. shows the surface of their work piece before and after annealing.

Hong et. al. also reported that the roughness of the microfluidic structures can be drastically reduced by out of focus machining of PMMA (Hong, et al, 2010).

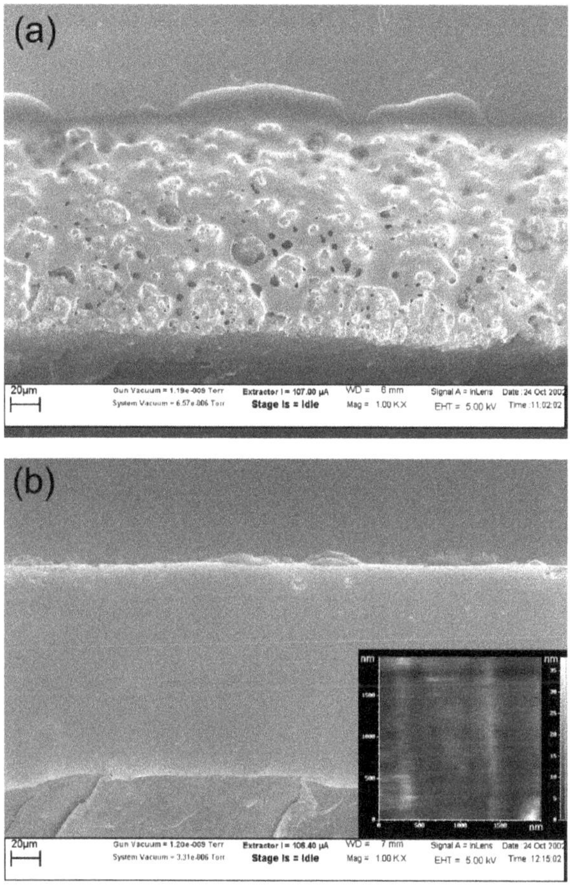

Fig. 2. The SEM pictures showing the rugged interior surface of the trench after laser machining (a) and smooth surface after thermal annealing (b). The AFM topography of the annealed surface is shown in the inset with full scale of 38 nm in the Z-axes. The viewing angle is perpendicular to the plane of the side wall (Cheng et al 2004) - Reproduced by permission of Elsevier under the license no. 283342005275.

PMMA micro fluidic structures can then be top covered by other polymers like PMMA or poly carbonate (PC) utilizing thermal-bonding process. Thermal bonding is a process of joining two materials by the mechanism of diffusion; and unity of the materials. This process is accomplished through the application of pressure at temperature higher that the glass temperature of the polymers.

In addition to fabrication of the holes, channels and cavities, CO2 laser machining can be used to make some complicated structures, like bending holes. These structures can also be molded with other materials such as PDMS to get the negative of the PMMA structures. In the next section fabrication technique of the other complicated structure with 3D structure is presented.

4. Fabrication of a 3D Mixer with CO2 laser machining of PMMA and PDMS molding

In this section we present application of a CO2 laser for fabrication of a 3D mixer with bending cones (Riahi, 2012). Mixers are the elements in microfluidic and micro total analysis systems which are used for mixing two or more liquids in biological and chemical analyses. Mixers can be divided into the two categories, active and passive. In active mixers, an external actuation mechanism is used to mix liquids in a microfluidic chamber. In passive mixers, there is no energy consumption and the structure of these devices is simpler than that of active devices. Different schemes such as a Tesla structure (Hong et al, 2004), a T mixer (Hoe et al, 2004), a 3D serpentine (Liu et al, 2000) and twisted shapes (Bertsch et al, 2001) are also used in passive micro mixers.

The technique which is presented here is based on the application of the CO2 laser for fabrication of some bending and straight cones on PMMA followed by PDMS molding. The designed mixer is shown in Fig. 3.

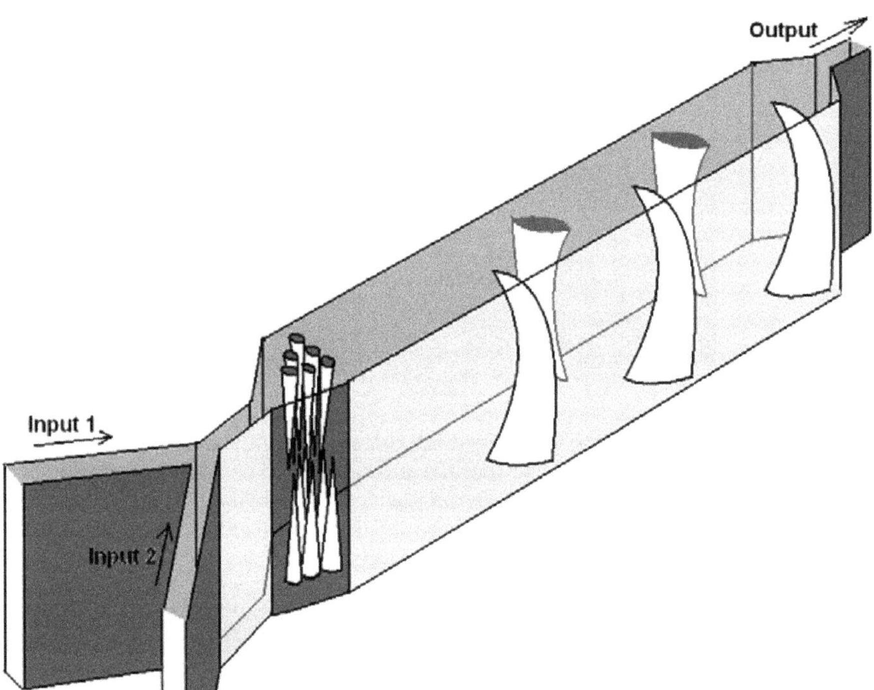

Fig. 3. Schematic of the designed 3D mixer (Riahi, 2012).

4.1 Fabrication of the bending cones

When a shape, is engraved in a raster scan mode by a CO2 laser engraving system on PMMA, the system scans a shape, row by row which each row has a certain overlap with a previous row. During the first row scan, a symmetrical V-shape channel is ablated on the PMMA surface. When the laser scans the subsequent rows, a small portion of the laser beam reflects from the wall of the channel produced by the previous scan to the bottom of the hole in the opposite side of the scanning direction as shown in Fig. 4a. After several scans, the reflected beam can ablate a considerable amount of PMMA material at the bottom of the hole at the opposite side of the scanning direction which can causes a bending shape in the structure (Fig. 4b).

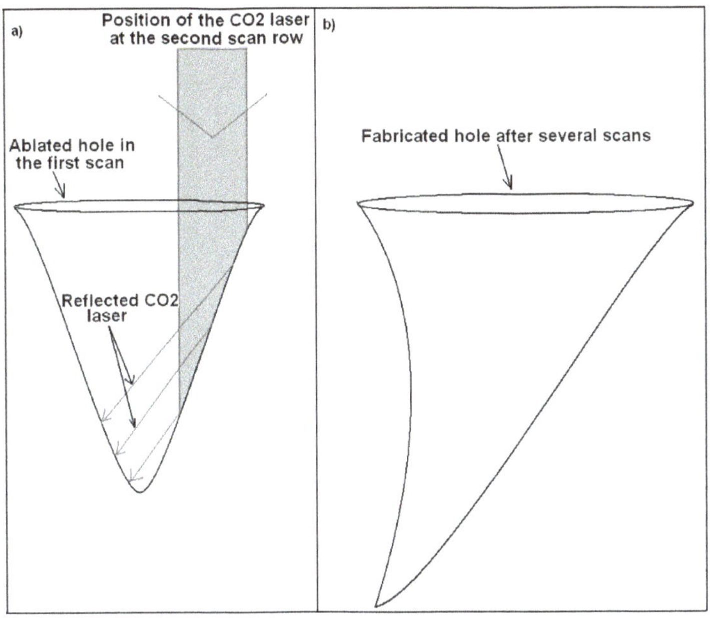

Fig. 4. Ablation of a PMMA hole with CO2 laser. a) Reflection of the laser from the walls of an ablated hole. b) The shape of the hole after several scans (Riahi, 2012).

It is found that the shape of the holes can be controlled by adjusting the scanning parameters such as resolution, power and scan speed. Some of the fabricated holes have very bent shapes and some are straight. Fig. 5 shows the ablated bending holes for different scan parameters.

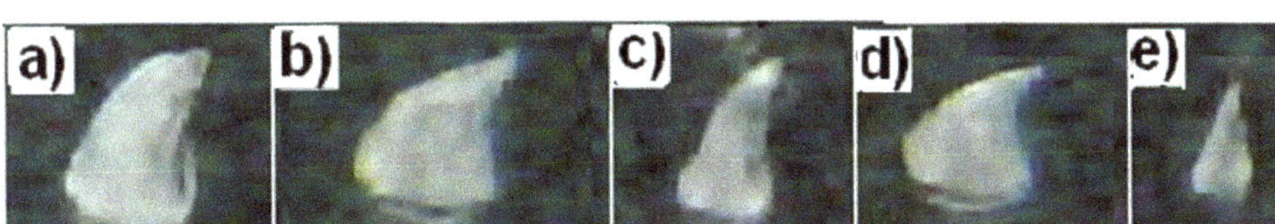

Fig. 5. The ablated bending holes for different scan parameters (Riahi, 2012).

4.2 Ablation of the mixer structure

To fabricate the mixer, a few straight cones and bending cones are ablated with CO2 laser on two different PMMA sheets. One of the PMMA sheets is CO2 laser cut to form a channel with two inputs and one output. The structures are then molded with PDMS and one is placed upside down on top of the other. Fig. 6. shows the fabricated channels and holes on the PMMA sheet and the molded PDMS structure.

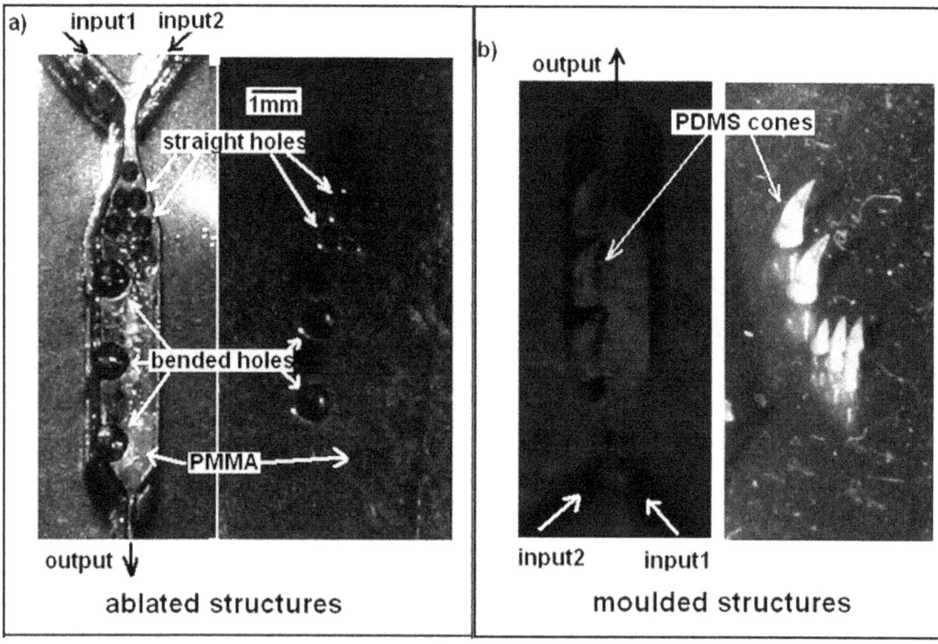

Fig. 6. a) Fabricated structures on PMMA sheets. b) The PDMS molds of structures shown in part a. The straight and bending cones are clear (Riahi, 2012).

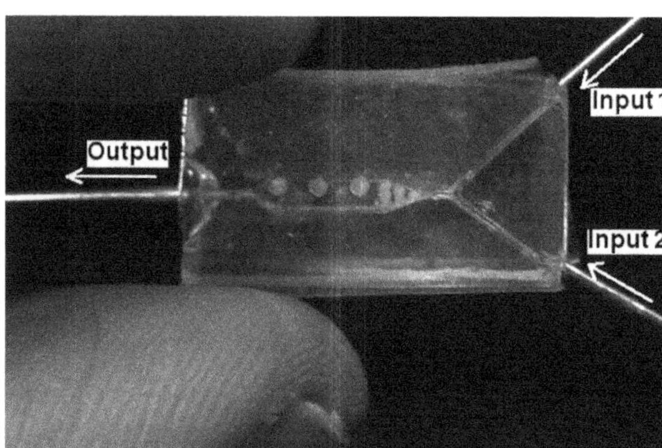

Fig. 7. The fabricated mixer (Riahi, 2012).

The molded PDMS structures are then stacked to each other and three steel tubes are inserted into the input and output channels and the voids are filled with PDMS to form the final mixer structure. The fabricated mixer is shown in Fig. 7.

5. Fabrication of the structures for optofluidics applications

Optofluidics refers to a science that uses the optical property of fluids for adjusting, measuring the properties of a device. Some examples of such devices are, liquid mirrors (Wood, 1909), liquid-crystal displays (Haas, 1983) and liquid lenses (Kuiper & Hendriks, 2004).

Several techniques are used to fabricate a tunable lens array (Dong et al, 2006; Jeong et al, 2004; Xu et al 2009)

In this section we show how a CO2 laser can be used for fabrication of an optofluidic device, liquid micro lens array (Riahi, 2011).

The liquid microlens array is an array of tunable liquid lenses which can be used for Medical stereoendoscopy, Telecommunication, Optical data storage, Photonic imaging, etc.

Fig. 8. shows the basic structure of the liquid lens array which is presented here. An array of the hexagonal holes with about 2mm width each, are first fabricated on a 1mm thick PMMA sheet. A thin layer of PDMS with the thickness of about 50 microns is fabricated and placed on the array of holes. A 1mm depth reservoir with an inlet and outlet for the fluid is also fabricated. The whole of the collection are placed on top of each other.

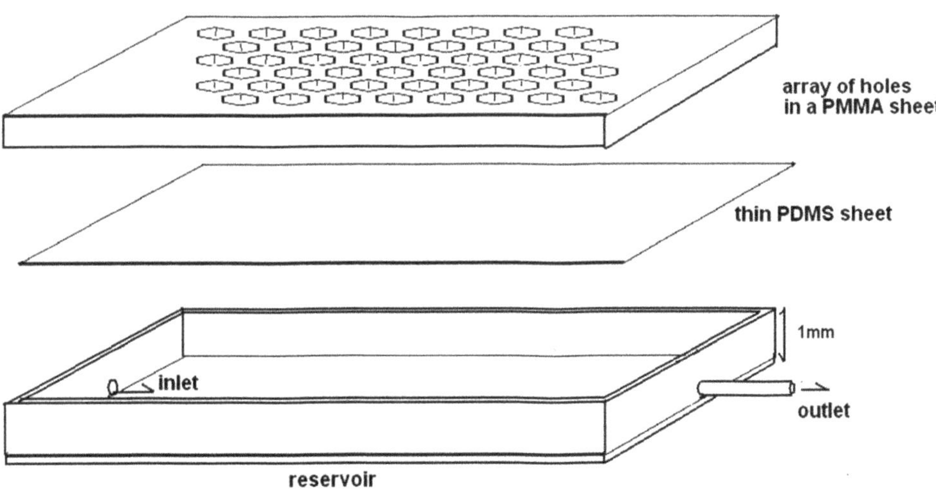

Fig. 8. The structure of a tunable liquid lens array.

By introducing a liquid into the reservoir and changing the pressure inside, the curvature of the PDMS sheet in place of the holes changes and produces convex lenses as shown in Fig. 9.

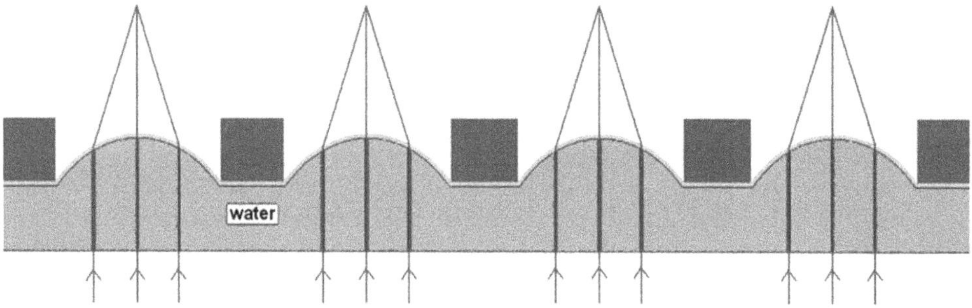

Fig. 9. Mechanism of convex micro lens creation by applying pressure in a water reservoir limited by PDMS and PMMA walls.

As shown in Fig. 10, a commercial CO2 laser engraving system is used for producing the patterns on PMMA sheets. This engraving machine is also used for fabrication of the reservoir on PMMA sheets.

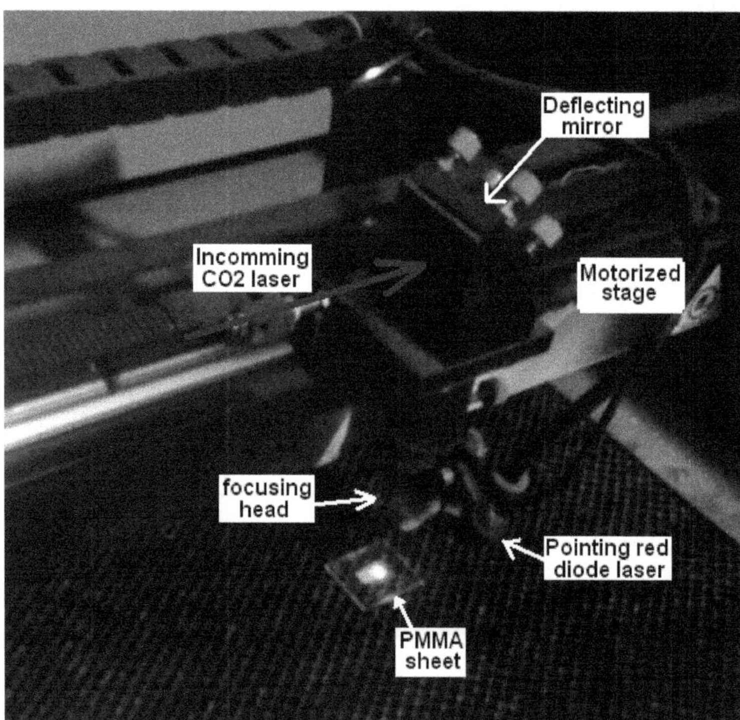

Fig. 10. The commercial CO2 laser engraving system in production process of an array of hexagonal holes on a PMMA sheet.

Fig. 11 shows the fabricated tunable microlens array with this technique. Fig. 12 also shows this microlens array in imaging from a "B" letter.

Fig. 11. The fabricated tunable liquid lens array.

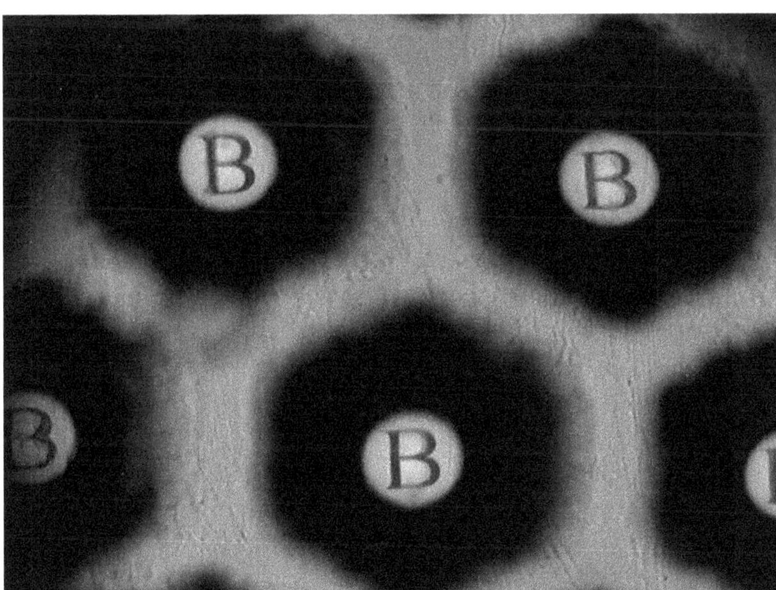

Fig. 12. Imaging from the letter "B" with the fabricated liquid lens array.

6. Fabrication of a beam profiler using the optical properties of liquids

In the previous sections we focused on the application of the CO2 laser for fabrication of the devices used in microfluidics and optofluidics. In this section we look at the application of a fluid device which is used for CO2 laser characterization. We present a device called thermally tunable grating (TTG), which can be used as a CO2 laser beam profiler.

Thermally tunable grating is a family of the gratings which some of their specifications can be adjusted by the user. The tuning ability of a diffractive grating can be divided into two categories: first, gratings in which the diffractive angle can be tuned, and second, gratings in which the intensity of diffraction orders can be modulated which are called grating light valves (GLV). Electrostatic actuation is one of the methods used in MEMS based grating light valves system (Trisnadi et al, 2004). In this grating light valve system, tiny suspended ribbons are put together to form a specular surface. Electrostatic actuation lowers some of the ribbons, and a diffractive grating is formed. Electric field actuation has also been used to actuate an electro-optically controlled liquid crystal based GLV (Chen et al, 1995).

But in TTG device, thermal method is used for actuation of a grating which contains a liquid in it's grooves. Increasing temperature, changes the refractive index of the liquid and consequently the diffraction efficiency of the grating (Riahi et al, 2008).

6.1 Principle of the method

As shown in Fig. 13a, we suppose that the grooves of a transparent binary grating with refractive index n1 are filled with another transparent material with refractive index n2. Assume that a laser beam with wavelength λ is incident on this grating. If the period of the grating is large enough compared to the wavelength of light, the rays that pass through the n1 and n2 materials will have phases φ1 and φ2, respectively and the phase difference Δφ= φ1- φ2 as shown in Fig. 13b.

By changing Δφ, the intensity of the diffraction orders is changed as shown in Fig. 13c,d,e. It can be shown that the intensity of the first order of diffraction can be calculated as follow (Riahi et al, 2009):

$$I = I_{\max} Sin^2(\frac{\Delta\varphi}{2}) \tag{2}$$

It is clear now that if n1 or n2 are changed, the intensity of the first order of diffraction changes sinusoidally (Fig. 13f).

6.2 Fabrication method

Standard lithography technique is used for fabrication of the binary grating on a glass substrate (n=1.52). As shown in Fig 14, the grooves are then filled with nitrobenzene and a thin glass sheet with 250 microns thickness is placed on it. The high boiling point (T= 210:8 °C), low specific heat capacity (1:51 J/gK), and high dn/dT (−4:6 × 10−4 K−1 at 626.58 at T= 288 K) [36] make nitrobenzene suitable for this work. The refractive index of nitrobenzene is 1.546 at 656:28nm at 293:15 K.

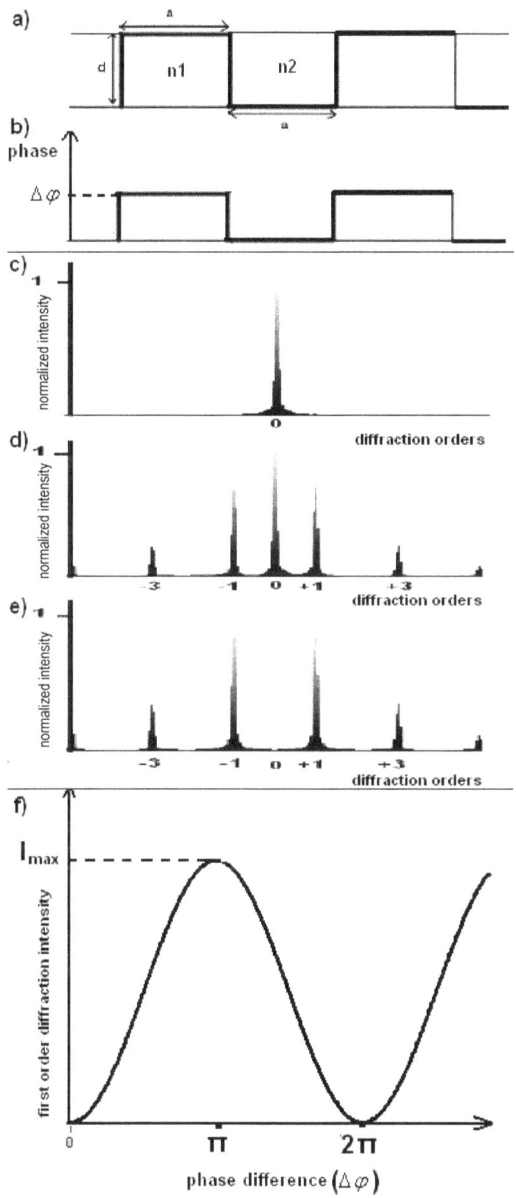

Fig. 13. (a) Square-well grating with n1 and n2 for the refractive indices of the land and the groove. (b)Wavefront of an incoming ray immediately after passing through the grating. (c), (d), (e) Simulation results of diffraction from the grating shown in (a) for $\gamma = 0$, $\gamma = \pi=2$, and $\gamma = \pi$, respectively. On the vertical axes, the maximum intensity has been normalized to unity. (f) Results of simulation of the intensity of the 1st order of diffraction versus phase difference (Riahi et al, 2008).

The diffraction pattern and intensity of the first order of diffraction versus temperature has been presented in Fig. 15.

6.3 Measurement of the beam profile of a CO2 laser

By changing the temperature, the intensity of the 1st order of diffraction is changed. The temperature of the TTG changes upon radiation by a CO2 laser beam. Radiation of a CO2 laser beam on a substrate warms it up and produces a temperature profile on the surface of the substrate. The temperature profile depends on the intensity profile of the laser beam. For example, if the laser profile is circular Gaussian, the temperature profile on the surface will be circular Gaussian in ideal case. Now if another visible laser is expanded and diffracted from the surface of the grating, the laser will be diffracted in different amounts from different parts of the grating, containing information on the temperature profile on the grating.

The setup shown in Fig. 16 is used to measure the beam profile of a CO2 laser. In this setup, a CO2 laser and a 658nm diode laser are made collinear with each other using a ZnSe window, and finally both lasers are irradiated on a 4mm × 4mm TTG device. The diode laser is expanded to about 3 cm diameter to cover the 4mm × 4mm TTG device with uniform intensity. The CO2 laser is passed through a shutter so that the irradiation time can be controlled. Immediately after the CO2 laser pulse, the CCD camera takes a picture from diffracted diode laser by a 4f imaging system using the 1st order of diffraction. It takes about 1 min for the device to get cool enough to repeat the experiment. The heat gun shown in Fig. 16. is used to keep the working area between point A and B as specified in Fig. 15d.

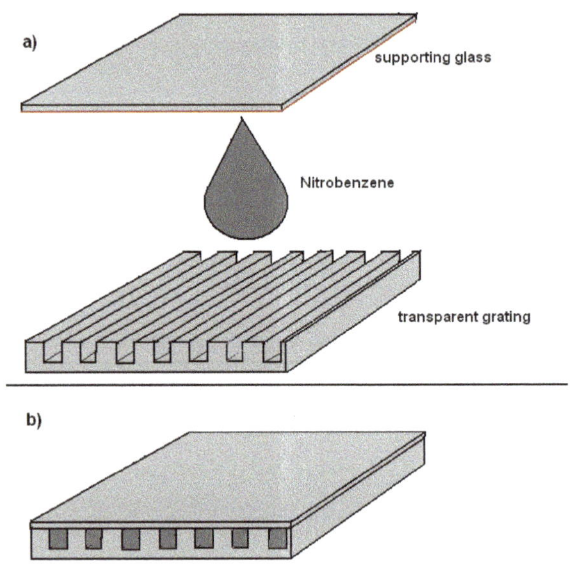

Fig. 14. Fabrication of the TTG device: (a) the grooves of the grating are filled with nitrobenzene and (b) a supporting glass is placed on the device (Riahi et al, 2008).

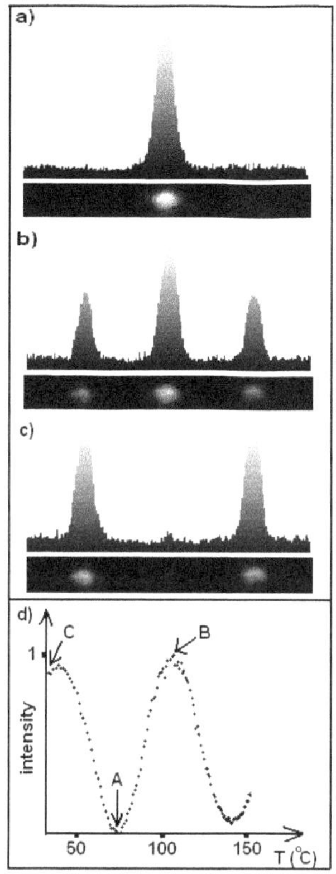

Fig. 15. Diffraction order intensities at different temperatures: (a) T = 77 °C, (b) T = 108 °C, and (c) T = 140 °C. The maximum intensity is normalized to unity. (d) Experimental result of the intensity of the 1st order of diffraction versus temperature. The maximum intensity is normalized to unity (Riahi et al, 2008).

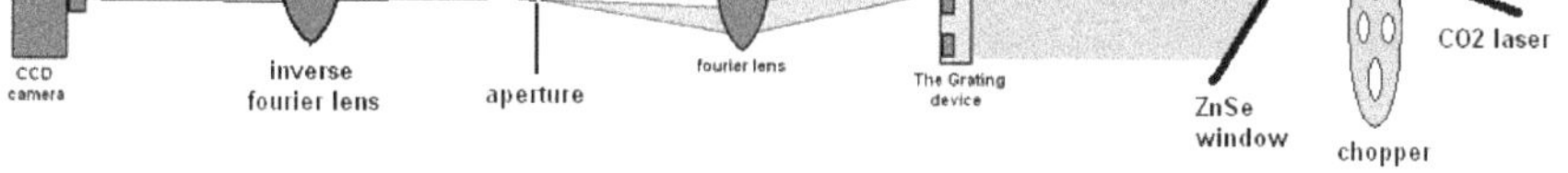

Fig. 16. Setup used for measurement of the temperature profile of the CO2 laser (Riahi et al, 2008).

The Image produced on the CCD camera and measured beam profile of the CO2 laser is shown in Fig. 17.

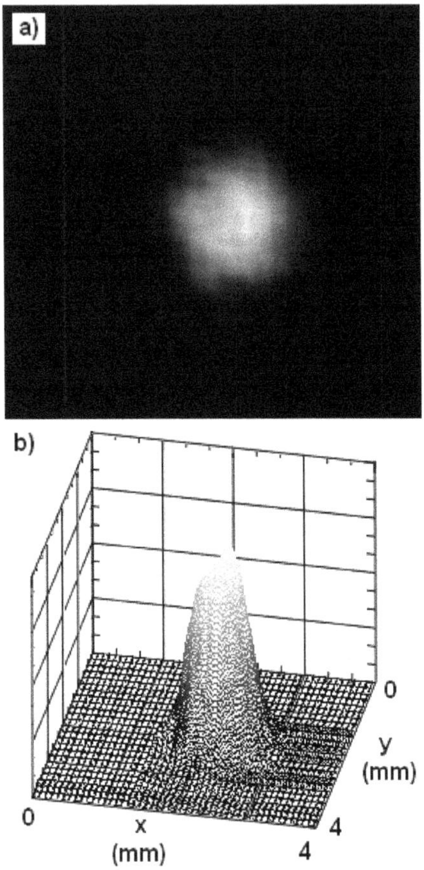

Fig. 17. (a) Image produced on the CCD camera. (b) 3D intensity profile of "a" will be the same as the beam profile of the CO2 laser (Riahi et al, 2008).

The followings are some errors presented in this experiment.

- additional temperature Gradient on the Thermally Tunable Grating because of the environmental errors
- Thickness of the Supporting Glass
- Aberrations of the Grating
- Expansion of the Grating
- Beam Profile of the Diode Laser

Some of these errors are so small to affect the beam profile, but some of them might be important and have to be corrected.

7. Real time measurement of the CO2 laser beam profile utilizing TTG

In method we presented in the previous section, after each measurement, time had to be taken for the grating to cool down and get ready for another measurement. This can be a big problem for real time measurement. In this part, a thermally tunable grating with fast response time is presented, which makes the real time measurements feasible (Riahi & Latifi, 2011).

The principle of this method is the same as what was mentioned in the previous section except that the device becomes a reflective instead of transitive and a thin supporting glass in the device is replaced by a double side polished silicon wafer. The silicon wafer plays the role of a reflector at 532 nm (40% of reflection) and also as an optical window for the CO2 laser. But the most important characteristic of the silicon is it's high thermal diffusivity. The thermal diffusivity of silicon is $\alpha_{Si} = 0.95$ (cm^/sec). It has to be mentioned that the thermal diffusivity of copper which is used as a very good heat sink is $\alpha_{cu} = 1.1$ (cm^2/sec) which is just a bit more than for silicon.

However, silicon can plays a role of a heat sink during the measurements and maked the real time measurements feasible.

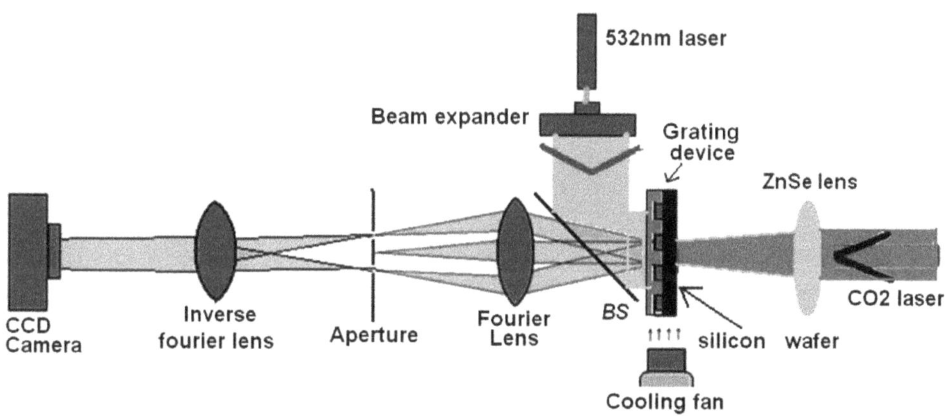

Fig. 18. Schematic setup for real time measurement of the CO2 laser beam profile (Riahi & Latifi, 2011)

To measure the beam profile of a CO2 laser, a setup as shown in Fig. 18 was used. In this setup, a CO2 laser beam incident on the grating device from the silicon side is absorbed in the grating structure and warms it up. A 532 nm laser is expanded and irradiates the grating, from the glass side. After passing through the grating, the visible light reflects back from the silicon slab and is directed to a 4*f* imaging system. A high pass spatial filter is used to keep the first order of diffraction for imaging.

The response time of this system can be measured. For this reason the same setup as in Fig. 18 is used, except that a chopper is placed in front of the CO2 laser and a fast photo-detector is used instead of the CCD camera. By chopping the CO2 laser beam, the signal of the photo-detector was monitored by an oscilloscope. As seen in Fig. 19, the response time of this device is about 10 milliseconds.

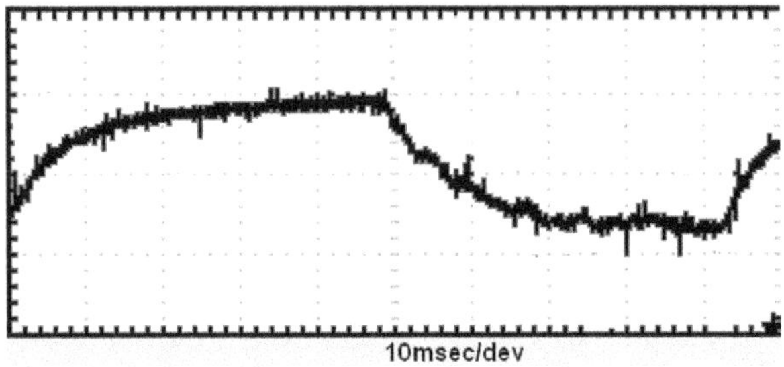

Fig. 19. Detected signals from photo-diode when CO2 laser is chopped off and on

8. Conclusion

In this chapter, the relation between CO2 laser and fluid applications was presented. First, application of a CO2 laser for fabrication of microfluidic and optofluidic structures on PMMA polymer was presented. Then application of a fluidic device for measurement of a characteristic of a CO2 laser was discussed.

Application of the CO2 laser for microfluidic fabrication is a simple and low cost method which can be performed by a commercial CO2 laser engraving system. This method makes the final products very cheap which are suitable for single use applications.

Also it seems that the application of the CO2 laser in microfluidics shows a good potential for fabrication of some complicated structures even 3D structures for future works.

9. References

Arisawa H and Brill T B (1997) Kinetics and mechanisms of flash pyrolysis of poly (methyl methacrylate) (PMMA) *Combust. Flame* Vol. 109, pp. 415–26, ISSN 0010-2180

Ball J. C., Scott D. L., Lumpp J. K., Daunert S., Wang J. and Bachas L. G., (2000), Electrochemistry in Nanovials Fabricated by Screen Printing and Laser Micromachining *Anal. Chem* , Vol. 72, pp. 497–501, ISSN 0003-2700.

Bertsch A. , Heimgartner S., Cousseau P., Renaud P. (2001) 3D micromixers— downscaling large scale industrial static mixers, *Proc. IEEE MEMS Workshop*, Interlaken, Switzerland.

Chen HT, Wang YN (2009) Optical microflow cytometer for particle counting, sizing and fluorescence detection. *Microfluid Nanofluid* Vol. 6, pp. 529–537, ISSN (printed) 1613-4982. ISSN (electronic) 1613-4990

Chen J., Bos P. J., Vithana H., and Johnson D. L. (1995), An electrooptically controlled liquid crystal diffraction grating, *Appl. Phys. Lett.* Vol. 67, pp. 2588–2590, ISSN (print) 0003-6951 ISSN (online) 1077-3118.

Cheng Ji-yen, Cheng-Wey Wei, Kai-Hsiung Hsua (2004), Direct-write laser micromachining and universal surface modification of PMMA for device development, *Sensors and Actuators B* , Vol. 99, pp. 186–196, ISSN 0925-4005.

Dong L., Abhishek K. Agarwal1, David J. Beebe2 & Hongrui Jiang, (2006), Adaptive liquid microlenses activated by stimuliresponsive hydrogels, *Nature*, Vol 442, pp. 551-554, ISSN 0028-0836.

Ferriol M, Gentilhomme A, Cochez M, Oget N and Mieloszynski J L (2003), Thermal degradation of poly (methyl methacrylate) (PMMA): modelling of DTG and TG curves *Polym. Degrad. Stab.* Vol. 79, pp. 271–81, ISSN 0141-3910.

Fu LM, Yang RJ, Lin CH, Lee GB, Pan YJ. (2004), Electrokinetically driven micro flow cytometers with integrated fiber optics for online cell/particle detection. *Anal Chim Acta,* Vol. 507, pp. 163–169, ISSN 0003-2670.

Fu LM, Leong JC, Lin CF, Tai CH, Tsai CH (2007), High performance microfluidic capillary electrophoresis devices, *Biomed Microdevices,* Vol. 9, pp. 405–412, ISSN 1387-2176 (Print) ; 1572-8781 (Electronic) ; 1387-2176 (Linking).

Fu LM, Wang JH, Luo WB, Lin CH (2009), Experimental and numerical investigation into the joule heating effect for electrokinetically driven microfluidic chips utilizing total internal reflection fluorescence microscopy, *Microfluid Nanofluid,* Vol. 6, pp. 499–507, ISSN (printed) 1613-4982. ISSN (electronic) 1613-4990.

Gerlach A, Knebel G., Guber A. E., Heckele M., Herrmann D., Muslija A. and Schaller Th. (2002), Microfabrication of single-use plastic microfluidic devices for high-throughput screening and DNA analysis, *Microsyst. Technol.* Vol. 7, pp.265–268, ISSN: 0946-7076 (print version) ISSN: 1432-1858 (electronic version)

Haas, W. E. (1983), Liquid-crystal display research — the first 15 years. *Mol. Cryst. Liq. Cryst.* Vol. 94, pp. 1–31, Print ISSN: 1542-1406 Online ISSN: 1563-5287.

Hoe Wong Seck, Michael C.L. Ward, Christopher W. Wharton (2004), Micro T-mixer as a rapid mixing micromixer, *Sens. Actuators B,* Vol. 100, pp. 359–379, ISSN 0925-4005.

Hong Chien-Chong, Jin-Woo Choi, Chong H. Ahn (2004), A novel in-plane passive microfluidic mixer with modified Tesla structures, *Lab Chip,* Vol. 4, pp. 109–113, ISSN (printed) 1473-0197. ISSN (electronic) 1473-0189.

Hong Ting-Fu, Ju Wei-Jhong, Wu Ming-Chang, Tai Chang-Hsien, Tsai Chien-Hsiung, Fu Lung-Ming (2010), Rapid prototyping of PMMA microfluidic chips utilizing a CO2 laser, *Microfluid Nanofluid* Vol.: 9, pp. 1125-1133, ISSN (printed) 1613-4982. ISSN (electronic) 1613-4990.

Hsieh TM, Luo CH, Wang JH, Lin JL, Lien KY, Lee GB (2009), A two-dimensional, self-compensated, microthermal cycler for one-step reverse transcription polymerase chain reaction applications, *Microfluid Nanofluid,* Vol. 2, pp.357–360, ISSN (printed) 1613-4982. ISSN (electronic) 1613-4990.

Jakeway s. C., A. J. de Mello and E. L. Russel, Fresenius' J. (2000), Miniaturized total analysis systems for biological analysis. *Anal. Chem.* Vol. 366, pp. 525–539, ISSN 0003-2700.

Jeong Ki-Hun, Gang L. Liu, Nikolas Chronis and Luke P. Lee (2005), Tunable microdoublet lens array, *Opt. Express.,* Vol. 12, pp. 2494-2500, ISSN 1094-4087.

Klank H., Jörg P. Kutter and Oliver Geschke (2002), CO2-laser micromachining and back-end processing for rapid production of PMMA-based microfluidic systems, *Lab Chip,* Vol. 2, pp. 242–246, ISSN (printed) 1473-0197. ISSN (electronic) 1473-0189.

Kuiper, S. & Hendriks, B. H. W. (2004), Variable-focus liquid lens for miniature cameras. *Appl. Phys. Lett.* Vol. 85, pp. 1128–1130, ISSN (print) 0003-6951 ISSN (online) 1077-3118.

Lin CH, Lee CY, Tsai CH, Fu LM (2009), Novel continuous particle sorting in microfluidic chip utilizing cascaded squeeze effect, *Microfluid Nanofluid,* Vol. 7, pp. 499–508, ISSN (printed) 1613-4982. ISSN (electronic) 1613-4990.

Liu Robin H., Mark Stremler A., Kendra V. Sharp, Michael G. Olsen, Juan G. Santiago, Ronald J. Adrian, Hassan Aref, David J. Beebe (2000), Passive mixing in a threedimensional serpentine microchannel, *J. Microelectromech. Syst.* Vol. 9, pp.190–197, ISSN: 1057-7157.

Martin F. Jensen,ab Mikkel Noerholm,bc Leif Højslet Christensena and Oliver Geschkeb (2003), Microstructure fabrication with a CO2 laser system: characterization and fabrication of cavities produced by raster scanning of the laser beam, *Lab Chip*, Vol. 3, pp. 302–307, ISSN (printed) 1473-0197. ISSN (electronic) 1473-0189.

Riahi M. (2012), Fabrication of a passive 3D mixer using CO2 laser ablation of PMMA and PDMS moldings *Microchemical Journal,* Vol. 100, pp. 14–20, ISSN 0026-265X.

Riahi M. (2011), Fabrication and characterization of a tunable liquid lens array in water-PDMS sheet interface by applying pressure , 1st EOS conference on optofluidics. (EOSOF2011) 23-25 may, Munich, Germany

Riahi M., Latifi H., and Moghimislam G. (2008), Fabrication of a thermally actuated tunable grating and its application as a CO_2 laser beam profile analyzer, *Appl. Opt.* Vol. 47, pp. 5175-5181, ISSN: 1559-128X (print), ISSN: 2155-3165 (online)

Riahi M., Latifi H., Madani A., Moazzenzadeh A. (2009), Design and fabrication of a spatial light modulator using thermally tunable grating and thin film heater, *Appl. Opt.* Vol. 48, pp. 5647-5654, ISSN: 1559-128X (print), ISSN: 2155-3165 (online)

Riahi M., Latifi M. (2011), Fabrication of a 2D thermally tunable reflective grating for measuring a CO2 laser beam profile, *Optica Applicata* , Vol. 41. pp. 735-742, ISSN: 0078-5466.

Rotting O, Ropke W, Becker H and Gartner C. (2002), Polymer microfabrication technologies, *Microsyst. Technol.* Vol. 8, pp. 32–36, ISSN: 0946-7076 (print version) ISSN: 1432-1858 (electronic version)

Sun Y.,Y.C.Kwok (2006), Polymeric microfluidic system for DNA analysis, *Anal. Chim. Acta* Vol. 556, pp. 80–96, ISSN 0003-2670.

Suna Yi, Satyanarayan M.V.D., Nguyenb Nam Trung, Kwok Yien Chian (2008), Continuous flow polymerase chain reaction using a hybrid PMMA-PC microchip with improved heat tolerance, *Sensors and Actuators B*, Vol. 130, pp. 836–841, ISSN 0925-4005.

Trisnadi J. I., Carlisle C. B., and Monteverde R. (2004), Overview and applications of Grating Light Valve based optical write engines for high-speed digital imaging, presented at Photonics West2004—Micromachining and Microfabrication Symposium, 26 January, San Jose, California, USA.

Wood, R. W. (1909), The mercury paraboloid as a reflecting telescope. *Astrophys. J.*, Vol. 29, pp. 164–176.

Xia Y. and Whitesides G. M. (1998), soft lithography, *Annu. Rev. Mater. Sci.*, Vol. 28, pp. 153–184, ISSN: 0084-6600

Xu S., Lin Yeong-Jyh, and Wu Shin-Tson (2009), Dielectric liquid microlens with well-shaped electrode, *Opt. Express,* Vol. 17, pp. 10499-10505, ISSN 1094-4087.

Part 4

Medical Applications

14

CO2 Laser: Evidence Based Applications in Dentistry

Pinalben Viraparia[1,2], Joel M. White[1] and Ram M. Vaderhobli[1,3]
[1]University Of California San Francisco,
[2] Lutheran Medical Center-Advanced Education in General Dentistry,
[3]Lutheran Medical Center/UCSF,
Dept. Preventive and Restorative Dental Sciences, UCSF
USA

1. Introduction

Ever since Kumar Patel introduced lasers in 1960s', researchers have been looking into its possible applications in the field of dentistry. Researchers have investigated the effects of laser radiation on teeth, bone, pulp and oral mucosal tissues (Taylor, Shklar, & Roeber, 1965). CO2 lasers have been used extensively in medical field and the first laser to be approved by FDA for dental application was Nd:YAG (Neodymium-Yttrium-Aluminum-Garnet) in 1990s. Since then many types of lasers including CO2, Er:YAG (Erbium-Yttrium-Aluminum-Garnet), Diode, Er Cr:YSGG (Erbium-Chromium-Yttrium-Scallium-Gallium-Garnet) have been approved for dental use. FDA approved Er:YAG for dental hard tissue in 1997 and has approved other types of lasers for soft and hard tissue procedures in many area of dentistry.

Many authors have reported the use of Carbon Dioxide (CO_2) lasers for soft tissue applications in dentistry (Pick & Pecaro, 1987a; White et al., 1998). The Food and Drug Administration (FDA) granted clearance for marketing CO_2 lasers for soft tissue procedures such as frenectomy, gingivectomy, biopsies, and removal of benign and malignant lesions because CO2 laser energy is well absorbed by water. Specific indications for use in dentistry include apthous ulcer treatment, coagulation of extraction sites, sulcular debridement and intraoral soft tissue surgeries such as ablating, incising, and excising (U.S. FDA 510(k) marketing clearance) as shown in Table 1.

In this chapter, we will discuss basic design, tissue interactions, evidence based clinical applications, and future of dental applications of CO2 lasers.

2. Basic equipment design & tissue interactions of CO2 laser

The growth of CO2 laser applications in Dentistry has grown substantially with its wavelength bands ranging from 9.4 and 10.6 micrometers. The laser medium consists of water or air cooled gas discharge (Carbon dioxide, nitrogen, hydrogen, xenon, helium) that helps in producing a beam of infrared light by activating the footswitch. The original CO_2 lasers were continuous wave or interrupted pulse durations of about 0.5 sec to 50 msec with non contact delivery and large beam diameters up to 1mm and larger. Because, the delivery mode is non-contact this results in lack of tactile sensation to the operator. Previous studies with these continuous wave CO_2 lasers showed a variety of structural and ultrasonic changes of the hard tooth structure. These included cracking, flaking, crater formation, charring, melting, and recrystallizaton due to the highly efficient absorption of CO_2 wavelengths by the apatite mineral of hard tissues (Boehm, Rich, Webster, & Janke, 1977; J. D. B. Featherstone & D. G. A. Nelson, 1987; McCormack, 1995; Stern & Sognnaes, 1964; Stern, Vahl, & Sognnaes, 1972). All dental tissues have different absorption coefficient for various wavelength depending on water, blood, pigment, and mineral content. For example, Nd:YAG and Diode lasers are absorbed by dark pigments making them ideal for soft tissue procedures. Tissue component that maximally absorbs CO2 wavelength is water followed by apatite (Gouw-Soares et al., 2004). Because of this CO2 lasers have been proven to be the gold standard for intra-oral soft tissue applications for decades. Thermal effects and various parameters settings of CO2 lasers have also been studied extensively (Leighty, Pogrel, Goodis, & White, 1991; Malmström, McCormack, Fried, & Featherstone, 2001). These studies indicated that application of CO_2 laser created unacceptable thermal damage to adjacent tissue. Because of these reasons early CO_2 laser system had been limited by their continuous wave operations and delivery system constraints. Lasers parameters such as power, repetition rate, average power and highest peak power play a role in surgical and collateral effects. Studies have concluded that high repetition rate, high peak power and lower average power yield favourable clinical results (Wilder-Smith, Dang, & Kurosaki, 1997).

Device Trade Name	Wavelength (micro mm)	Most Absorption	Recommended Use	FDA Approval
Opus 20 Dental laser system	10.6	Water	Soft Tissue	Yes
N/A	9.6	Appetite	Hard tissue	No
N/A	9.3	Appetite	Hard Tissue	No
Smart CO2	10.6	Water	Soft Tissue	Yes
CO2DENTA	10.6	Water	Soft Tissue	Yes

Table 1. Examples of CO2 lasers available in market for dental use

Due to the lack of tactile sensation, their use in hard-tissue applications is not favorable. With new technologies, dental laser manufacturers now claim to have shorter pulse durations (as short as 150 microsecond pulse duration) with beam diameters of as small as 100 microns. This allows for cooling of tissues between pulses and results in minimal thermal damage. These lasers are now marketed for soft tissue intraoral procedures as described earlier. The laser is usually equipped with various hand pieces and tips of differing diameter for tissue ablation as shown in Figure 1. The hand piece is usually the size of a dental drill and the spot size that is emitted from these hand pieces allows for greater accuracy resulting in minimal damage to the surrounding tissues. As these lasers operate the best in pulse or super pulse infrared mode, they are able to remove precise amount of tissues with each pulse emission.

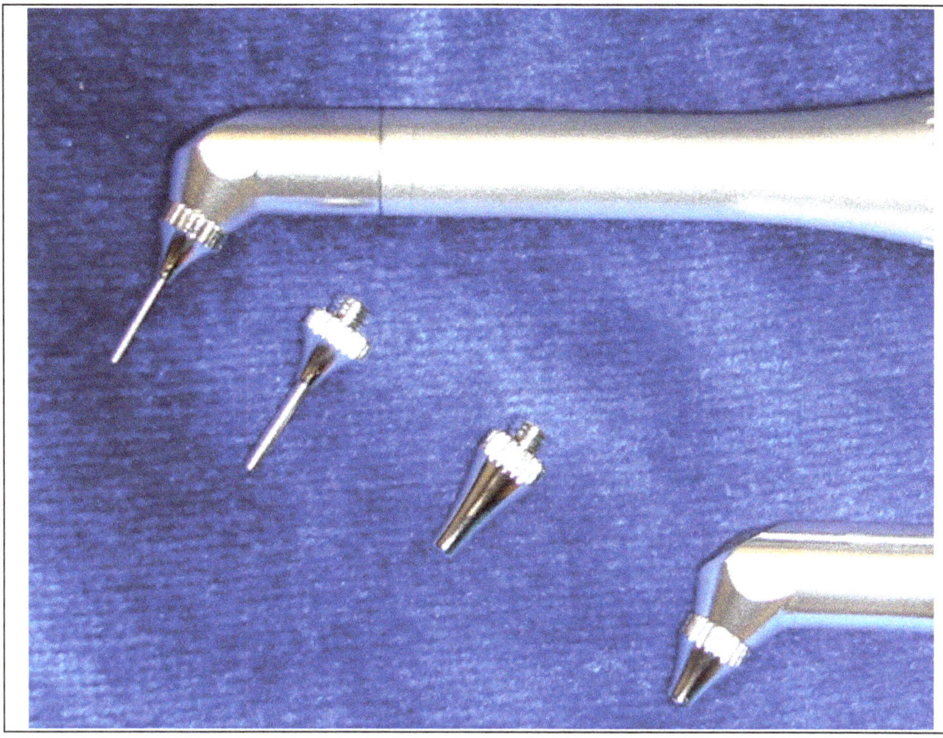

Fig. 1. CO2 laser hand piece with different tips marketed by GPT Dental

3. Evidence based clinical applications

Numerous studies have been done pre and post FDA approval to improve the technique and provide best practice guidelines for the clinicians. A report published by American Dental Association (ADA) in 2001 describing the challenges in the future of oral health care mentioned the role of laser applications. The report specifically mentioned that more clinical research and technical developments in CO2 laser delivery systems will promise to expand its clinical applications beyond soft tissue procedures (Seldin, 2001). Although CO2 laser 10.6micron wavelength is absorbed by water and even though 9.3micron is absorbed in hydroxyapetite, it is primarily a soft tissue laser (Convissar & Goldstein, 2003). Even before CO2 laser received FDA approval for soft tissue procedures in 1990's, many studies have looked at its hard tissue applications in 1980's. Table 2. Lists CO2 laser application in various dental procedures (Sulewski JG).

3.1 Soft tissue procedures

One of the earlier case series by Pick and colleagues reported soft tissue procedures using Sharplan 743 CO2 and Xanar Ambulase lasers. In a Clinical trial 250 patients were treated for conditions ranging from, gingival hyperplasia, benign and malignant lesions (along with conventional surgery), incisional & excisional biopsy, red-white lesions, and haemorrhagic and coagulation disorders. They concluded that CO2 laser provided bloodless field, less post-operative discomfort, tissue coagulation, and better accessibility in some areas of oral cavity compared to scalpel surgery (Pick & Pecaro, 1987b). The advantages compared to scalpel wounds also included site-specific wound sterilization; minimal swelling and scarring but slower healing; reduced necessity for suturing; decreased incidence of mechanical trauma; shorter operative time; favorable patient acceptance; decreased use of local anesthesia; and little or no postoperative pain (Pick & Powell, 1993; White et al., 2002; Wigdor et al., 1995). Literature also reported increased levels of hyaluronic acid in laser wounds compared to scalpel wounds, a chemical that plays a key role in wound repair (Pogrel, Pham, Guntenhoner, & Stern, 1993). With increased use of CO2 lasers clinically, adjacent tissue interaction and damage has been an issue. Studies have reported on chemical and thermal interaction of CO2 lasers with surrounding tissue. In vitro study using 9.3 micrometers Duolase CO2 laser (Medical Optics Inc.) investigated variations in incision depth and width, collateral damage, and bone charring using continuous mode (1-9W average power; 1-10W peak power; 0.5-500Hz; 1, 20, 200miliseconds), superpulse (1-7W average power; 20W peak power; 170-1170hz; 300microseconds) and optipulse (0.72-1.20W average power; 60-100W peak power; 10-40Hz; 300microseconds) mode with various parameter combinations. They concluded that superpulse and optipulse mode with lower average powers and higher peak powers created narrow and deep cuts. Also, almost no charring was noticed with optipulse mode. Optipulse mode reduced the collateral damage by the factor of 10 compared to continuous mode (Wilder-Smith et al., 1997).

Area of Dentistry	Procedure
Oral & maxillofacial surgery	Abscess incision and drainage, Hemostatic assistance, Fibroma removal, Oral papillectomy, Exposure of nonerupted or partially erupted teeth, Implant recovery, Lesion (tumor) removal, Vestibuloplasty, Frenectomy, Frenotomy, Operculectomy, coagulation
Oral Medicine	Aphthous ulcer treatment, Biopsies (incisional/excisional), Leukoplakia
Pre-prosthetic procedures	Crown lengthening (soft tissue only), Tissue retraction for impression
Periodontal procedures	Sulcular debridement, Gingival excision/incision, laser assisted new attachment procedure, Gingivectomy/gingivoplasty
Endodontics	Pulpotomy, Pulpotomy, as an adjunct to root canal therapy and retreatment cases, Removal of filling material such as gutta-percha or resin
Restorative uses	resin curing, teeth whitening agent activation, caries detection, pit and fissure sealants, enamel treatment to increase caries resistance, enamel etching for resin bonding procedures, caries removal, tissue ablation

Table 2. CO2 Laser soft tissue applications

3.2 Endodontics (apicoectomy, root canal debridement)

Hard tissue applications of continuous wave CO2 lasers have been limited due to thermal damage, charring effect and resultant rough tooth surface. Due to its high absorption overlap with phosphate in enamel apetite crystal, all radiation is absorbed in thin enamel (<10 um). This makes heat transfer as the main way of energy transport leading to thermal damage to pulp (Wigdor et al., 1995). Conversely studies have shown that high reflectivity (9%-50%) at 9.3- and 9.6um wavelengths may pose a safety concern and it requires accurate knowledge of radiation dose while doing treatment (Fried, Glena, Featherstone, & Seka, 1997). In 1990's TEA (Transversely excited atmospheric pressure) 10.6um pulsed (0.1-2usec) CO2 laser had the best reported success in ablating dental hard tissue. However, high plasma induction with TEA CO2 laser posed problems with decreased ablation efficiency and damage due to shock wave rendering it unacceptable for clinical use (Wigdor et al., 1995). Since the 9.6micrometer CO2 laser wavelength is highly absorbed in appetite crystals, it presents a future potential for its applications in cavity preparation, apicectomies and other hard tissue procedures. In vitro study using Scanning Electron Microscopic (SEM) images reported cleaner dentinal surface with fusion and recrystallized dentine following apicectomy and root treatment with pulsed TEA 9.6micrometer CO2 laser (Gouw-Soares et al., 2004). Contrasting results were reported in a more recent SEM analysis study. The study compared the marginal permeability and dentinal surface texture following apicectomy performed with burs and CO2 laser. Authors attributed rougher surface and less favourable marginal fit following CO2 laser treatment to the use of continuous wave mode, no cooling agent and less experience of the operator with CO2 laser (Lustosa-Pereira et al., 2011).

3.3 Periodontal procedures

Advantages of using CO2 laser for periodontal procedures have been accepted by American Academy of Periodontology in its position paper. Its ability to provide dry surgical field and haemostasis has been proven useful in periodontal surgical procedures. Additionally CO2 laser use has shown mixed results when used for periodontal pocket debridement in addition to mechanical debridement, pocket reduction, attachment gain, decreased microorganisms, and guided tissue regeneration cases (Convissar & Goldstein, 2003; Matthews, 2010; Wigdor et al., 1995). Porcine mandible study evaluating efficacy of newer micropulse 10.6um CO2 laser showed clinically acceptable results in coagulation, incision depth and width, time required to perform procedure, with minimal hard tissue damage on accidental exposure but surface melting with direct exposure to laser (Vaderhobli, White, Le, Ho, & Jordan, 2010). Other studies have also reported thermal side effects like dentin cracking, carbonization, and melting following CO2 laser use on root surfaces (Matthews, 2010).

3.4 Other restorative uses

Preventive uses of CO2 laser have been well researched. Literature suggests that 9.3 and 9.6um wavelength (pulse width <100usec) at a specified pulse rate has higher efficiency than 10.6um in heating dentin/enamel surface leading to desired crystallization and fusion of surface layer for sealing effect (McCormack, Fried, Featherstone, Glena, & Seka, 1995; Wigdor et al., 1995). A case report by Dederich, suggested to use 15W for 0.2sec duration to achieve dentinal sealing effect with CO2 laser without detrimental effect to pulpal tissue (Dederich, 1999). Furthermore, researchers showed that pulsed CO2 laser produced >1000 celsius temperature increase at the surface, enough to melt and recrystalize enamel and minimal changes deeper than 40um which is critical to avoid collateral damage (J. D. Featherstone & D. G. Nelson, 1987; Wigdor et al., 1995). For dental decay, CO2 lasers have mixed results ranging from thermal damage, dentin/pulp sterilization, and mineralization under treated surface (Melcer, 1986). Surface etching with CO2 laser showed 300% increased in dentin resin bond but no change in enamel bonding (Obata et al., 1999; Wigdor et al., 1995). Superpulse CO2 laser produced fastest debonding of orthodontic brackets (Obata et al., 1999). CO2 laser has also been used in otherwise hopeless prognosis cases of vertical root fracture with radiographical success at one-year follow-up. Teeth bleaching agent activation with CO2 laser causes higher temperature changes and due to lack of controlled clinical trials, ADA does not support its clinical use (Luk, Tam, & Hubert, 2004; N/A, 1998).

4. Future of dental applications of CO2 lasers

The future looks promising for CO2 laser use in the field of dentistry. We need more clinical research specifically randomized clinical trials to evaluate effectiveness of CO2 laser compared to traditional methods. The evidence will help develop standard clinical guidelines for practicing dentists.

5. Conclusion

Currently, CO2 lasers have been used widely in dentistry for soft tissue procedures. More research will help provide practice guidelines. Clinical research including randomized trials are needed to provide specifications for parameter settings, delivery mode, and other guidelines for hard tissue procedures. More information regarding shorter wavelength CO2 lasers in recent years makes future of CO2 laser promising in dentistry.

6. Acknowledgment

We would like to thank Lutheram Medical Center- Department of Dental Medicine for providing support for this project.

7. References

Boehm, R., Rich, J., Webster, J., & Janke, S. (1977). Thermal stress effects and surface cracking associated with laser use on human teeth. J Biomech Eng, 99, 189-194.

Convissar, R. A., & Goldstein, E. E. (2003). An overview of lasers in dentistry. [Case Reports, Review]. Gen Dent, 51(5), 436-440.

Dederich, D. N. (1999). CO2 laser fusion of a vertical root fracture. J Am Dent Assoc, 130(8), 1195-1199.

Featherstone, J. D., & Nelson, D. G. (1987). Laser effects on dental hard tissues. [Research Support, U.S. Gov't, P.H.S.]. Adv Dent Res, 1(1), 21-26.

Featherstone, J. D. B., & Nelson, D. G. A. (1987). Laser effects on dental hard tissues. Advances in Dental Research, 1(1), 21-26.

Fried, D., Glena, R. E., Featherstone, J. D., & Seka, W. (1997). Permanent and transient changes in the reflectance of CO2 laser-irradiated dental hard tissues at lambda = 9.3, 9.6, 10.3, and 10.6 microns and at fluences of 1-20 J/cm2. [In Vitro Research Support, U.S. Gov't, P.H.S.]. Lasers Surg Med, 20(1), 22-31.

Gouw-Soares, S., Stabholz, A., Lage-Marques, J. L., Zezell, D. M., Groth, E. B., & Eduardo, C. P. (2004). Comparative study of dentine permeability after apicectomy and surface treatment with 9.6 microm TEA CO2 and Er:YAG laser irradiation. [Research Support, Non-U.S. Gov't]. J Clin Laser Med Surg, 22(2), 129-139. doi: 10.1089/104454704774076190

Leighty, S. M., Pogrel, M. A., Goodis, H. E., & White, J. M. (1991). Thermal effects of the carbon dioxide laser on teeth. Lasers in the life sciences, 4(2), 93-102.

Luk, K., Tam, L., & Hubert, M. (2004). Effect of light energy on peroxide tooth bleaching. J Am Dent Assoc, 135(2), 194-201; quiz 228-199.

Lustosa-Pereira, A. C., Pozza, D. H., Cunha, A., Dedavid, B. A., Duarte-de Moraes, J. F., & Gerhardt-de Oliveira, M. (2011). Analysis of the morphology and composition of tooth apices apicectomized using three different ablation techniques. Med Oral Patol Oral Cir Bucal, 16(2), e225-230.

Malmström, H. S., McCormack, S. M., Fried, D., & Featherstone, J. D. B. (2001). Effect of CO2 laser on pulpal temperature and surface morphology: an in vitro study. Journal of Dentistry, 29(8), 521-529.

Matthews, D. C. (2010). Seeing the Light--the truth about soft tissue lasers and nonsurgical periodontal therapy. J Can Dent Assoc, 76, a30.

McCormack, S. M. (1995). Scanning electron microscope observations of CO2 laser effects on dental enamel. Journal of Dental Research, 74(10), 1702-1708.

McCormack, S. M., Fried, D., Featherstone, J. D., Glena, R. E., & Seka, W. (1995). Scanning electron microscope observations of CO2 laser effects on dental enamel. J Dent Res, 74(10), 1702-1708.

Melcer, J. (1986). Latest treatment in dentistry by means of the CO2 laser beam. Lasers Surg Med, 6(4), 396-398.

N/A. (1998). Laser-assisted bleaching: an update. ADA Council on Scientific Affairs. J Am Dent Assoc, 129(10), 1484-1487.

Obata, A., Tsumura, T., Niwa, K., Ashizawa, Y., Deguchi, T., & Ito, M. (1999). Super pulse CO2 laser for bracket bonding and debonding. Eur J Orthod, 21(2), 193-198.

Pick, R. M., & Pecaro, B. C. (1987a). Use of the CO2 laser in soft tissue dental surgery. Lasers in Surgery and Medicine, 7(2).

Pick, R. M., & Pecaro, B. C. (1987b). Use of the CO2 laser in soft tissue dental surgery. Lasers Surg Med, 7(2), 207-213.

Pick, R. M., & Powell, G. L. (1993). Laser in dentistry. Soft-tissue procedures. Dent Clin North Am, 37(2), 281-296.

Pogrel, M. A., Pham, H. D., Guntenhoner, M., & Stern, R. (1993). Profile of hyaluronidase activity distinguishes carbon dioxide laser from scalpel wound healing. Ann Surg, 217(2), 196-200.

Seldin, L. W. (2001). Future of Dentistry: Today's Vision: Tomorrow's Reality: American Dental Association, Health Policy Resources Center.

Stern, R. H., & Sognnaes, R. F. (1964). Laser beam effect on dental hard tissues. J Dent Res, 43(5).

Stern, R. H., Vahl, J., & Sognnaes, R. F. (1972). Lased enamel: ultrastructural observations of pulsed carbon dioxide laser effects. Journal of Dental Research, 51(2), 455-460.

Sulewski JG, (2010). Making the most of 16th Annual conference and exhibition: A practical orientation for attendees. Academy of Laser Dentistry, Miami, Florida.

Taylor, R., Shklar, G., & Roeber, F. (1965). THE EFFECTS OF LASER RADIATION ON TEETH, DENTAL PULP, AND ORAL MUCOSA OF EXPERIMENTAL ANIMALS. Oral Surg Oral Med Oral Pathol, 19, 786-795.

Vaderhobli, R. M., White, J. M., Le, C., Ho, S., & Jordan, R. (2010). In vitro study of the soft tissue effects of microsecond-pulsed CO(2) laser parameters during soft tissue incision and sulcular debridement. [In Vitro, Research Support, Non-U.S. Gov't]. Lasers Surg Med, 42(3), 257-263. doi: 10.1002/lsm.20888

White, J. M., Chaudhry, S. I., Kudler, J. J., Sekandari, N., Schoelch, M. L., & Silverman Jr, S. (1998). Nd: YAG and CO2 laser therapy of oral mucosal lesions. Journal of clinical laser medicine & surgery, 16(6), 299.

White, J. M., Gekelman, D., Shin, K. B., Park, J. S., Swenson, T. O., Rouse, B. P., . . . Oto, M. G. (2002). Laser interaction with dental soft tissues: What do we know from our years of applied scientific research?

Wigdor, H. A., Walsh, J. T., Jr., Featherstone, J. D., Visuri, S. R., Fried, D., & Waldvogel, J. L. (1995). Lasers in dentistry. [Case Reports, Research Support, Non-U.S. Gov't, Research Support, U.S. Gov't, Non-P.H.S., Research Support, U.S. Gov't, P.H.S., Review]. Lasers Surg Med, 16(2), 103-133.

Wilder-Smith, P., Dang, J., & Kurosaki, T. (1997). Investigating the range of surgical effects on soft tissue produced by a carbon dioxide laser. J Am Dent Assoc, 128(5), 583-588.

15

Clinical Application of CO_2 Laser

Hyeong-Seok Oh and Jin-Sung Kim
Wooridul Spine Hospital, Seoul
Korea

1. Introduction

The carbon dioxide (CO_2) laser was first introduced in 1964 by Patel and has been extensively used in the next two decades as an incision tool in increasingly wide areas, such as neurosurgery, dermatology and plastic surgery, otorhinolaryngology, ophthalmology, gynecology, and general surgery. In 1984, its reliability resulted in its approval by the U.S. Food and Drug Administration, and thus, medical use of lasers became more prevalent. Currently, the CO_2 laser is considered an indispensable piece of diagnostic and therapeutic equipment.

The CO_2 laser produces a beam of infrared light with the principal wavelength bands centering at 10,600 nanometers. Collisional energy transfer between the nitrogen and the carbon dioxide molecule causes vibrational excitation of the carbon dioxide, with sufficient efficiency to lead to the desired population inversion necessary for laser operation. It is easy to actively Q-switch a CO_2 laser by means of a rotating mirror or an electro-optic switch, giving rise to Q-switched peak powers up to gigawatts (GW) of peak power.

CO_2 lasers are attracting attention as cutting tools. They are able to seal lymphatic and blood vessels less than 0.5-mm wide and can reduce intraoperative bleeding and the occurrence of postoperative swelling. CO_2 lasers emit a longer wavelength than those transmitted by other types of lasers. Their penetration depth of 0.03 mm is very safe. Coagulation in small blood vessels, as well as sealing of lymphatic and small peripheral nerves, have been reported in experimental studies using CO_2 lasers; this sealing alleviates postoperative pain.

The CO_2 laser also offers more comfort to patients by reducing intraoperative bleeding and postoperative edema, facilitating the process of wound healing after surgery. The boundaries between the tissues receiving heat damage and the surrounding intact tissue are very well defined. A CO_2 laser can evaporate through the surrounding tissue without physical force, sealing the vessel and minimizing bleeding; thus, it is useful when a bloodless view is required during surgery. Moreover, wounds can be treated in a sterile manner because of high-temperature evaporation of tissue lesions.

Regarding its disadvantages, the equipment is expensive, operators require time to become familiar with it, and the sophisticated operation is technically difficult. Therefore, more repetitions are required to gain the necessary experience and practice. In addition, there is a risk of fire if the laser is used improperly. It can also damage the cornea; thus, eye protection is needed for the surgeon and the patient. Because the gas discharged from the vaporization of tissue contains an excess of CO_2 or virus particles, it can be harmful to the human body.

2. Clinical application in neurosurgery

The CO_2 laser is most widely used in the field of neurosurgery for removal and evaporation of tumors located in difficult surgical fields, such as the base of the skull, ventricles, brainstem, and spinal cord.

2.1 Brain tumor surgery

The CO_2 laser has been used in brain microsurgery after Steller et al. (Stellar, Polanyi et al. 1970) had first successfully used it in removing a recurrent glioma in 1969. The most ideal treatment of a brain tumor is minimizing damage to the normal brain tissue and removing only the tumor area. To overcome the surgical difficulty of avoiding damage to the brain tissue, a special instrument was developed. Theoretically, lasers have several advantages. First, although the surgical field is narrow, it makes surgery possible. Other small-sized surgical approaches are facilitated to minimize injury to normal brain tissue. Second, brain retraction is minimized, thus causing less damage to normal brain tissue. Third, laser beam minimizes injury to surrounding tissues and enables removal of a tumor with less thermal injury. Fourth, lasers have a coagulating property that lessens bleeding of the surgical field. Fifth, operation time is shortened (Tew and Tobler 1983; Krishnamurthy and Powers 1994).

The CO_2 laser is the main instrument used in brain surgery. It has the advantage of rapidly removing separated tumor cells and exact irradiation of target cells by a microsurgical technique where the CO_2 laser is installed with a microscope. However, as energy cannot pass through an optical fiber, it is inconvenient to use the equipment. It has limited function in bleeding control, as control of bleeding is not possible in a vessel with a diameter 0.5 mm, necessitating the use of the equipment in conjunction with other equipments for severe bleeding management (Heppner 1978; Ascher and Heppner 1984; Deruty, Pelissou-Guyotat et al. 1993).

The CO_2 laser is most widely used in the field of neurosurgery, and it is mainly used in the removal of tumors by evaporation where surgical approach of the tumor site is difficult. It is common opinion that the CO_2 laser is most effective with skull base, ventricular, brainstem, and spinal cord tumors (Powers, Cush et al. 1991; Origitano and Reichman 1993). In particular, it is most effective in removing a meningioma that is relatively hard or has less vascular distribution to be calcified. In addition, it is suitable for removing a low-grade glioma that is relatively rigid (Deruty, Pelissou-Guyotat et al. 1993).

The Nd:YAG laser has the advantage that energy can be passed to thinner fiberoptic cables and excellent clotting function is possible at a 3-mm vessel. Therefore, it has been reported as a valid technique of removing brain tumors having greater vascular distribution and cerebral vascular malformation (Beck 1980). The combolaser has been developed in recent years by Fasano et al. (Glasscock, Jackson et al. 1981) and has been applied in surgery. It is composed of CO_2 and Nd:YAG lasers, combining the advantages of both. It works by first emitting Nd:YAG energy to the tumor for clotting, followed by tumor removal by evaporation using the CO_2 laser (Beck 1980; Glasscock, Jackson et al. 1981).

2.2 Spine surgery

Since the first trial of Nd:YAG in a lumbar disk surgery in 1986 (Choy, Case et al. 1987), there have been many reports about the usefulness of different kinds of lasers in disk surgery (Nerubay, Caspi et al. 1997; Hellinger 1999; Houck 2006). Nerubay et al. reported that 50 patients with low back and radicular pains were successfully treated by percutaneous laser nucleolysis using a CO_2 laser (Nerubay, Caspi et al. 1997), and successful vaporization of the disk was accomplished in animal models (Stein, Sedlacek et al. 1990).

Considering the similarity between the disk and the meniscus (Whipple, Caspari et al. 1984), we cite studies on the effect of the CO_2 laser on the meniscus. According to these research results, there was a considerable proliferation of cells resembling chondrocytes after 2 weeks of the CO_2 laser treatment and there was definitely an increase in the production of ground substance and immature collagen fibers after 4 weeks; the collagen had become well reorganized into a logical orientation, resembling the normal architecture of fibrocartilage, after 10 weeks (Benjamin, Qin et al. 1995).

These animal and clinical studies strongly support the claim that CO_2 lasers can safely and feasibly be used for the removal of protruded disks and discal cysts. Moreover, the CO_2 laser, when attached to an operating microscope, allows for quick and easy removal of the discal cyst and, if needed, easy vaporization of disk material.

2.2.1 Disk herniation

Laser removes disk material by vaporization (Stein, Sedlacek et al. 1990) and consequently lowers intradiskal pressure (Gropper, Robertson et al. 1984). In spine surgery, the use of a laser has advantages over scalpel use in terms of precision; the ability to be used on delicate tissues; minimal tissue manipulation; and less bleeding, swelling, and trauma (Jeon, Lee et al. 2007). It is especially useful in the small spaces involved in herniated disks (Kim, Choi et al. 2009). Therefore, a laser is an effective tool for performing a minimally invasive spinal surgery with percutaneous and open spinal procedures (Ahn, Lee et al. 2005; Lee, Ahn et al. 2006; Lee, Ahn et al. 2006; Jeon, Lee et al. 2007; Lee, Ahn et al. 2008; Kim, Choi et al. 2009; Kim and Lee 2009).

In the Wooridul Hospital, CO_2 laser-equipped surgical microscopes have been used for open lumbar microdiscectomy since December 1991 (Fig. 1). These microscopes coaxially align the invisible CO_2 laser beam with a visible helium-neon laser beam and can focus exactly on and evaporate the target disk material by the commonly used 20- to 30-W single-pulse mode laser. Therefore, we aimed to determine whether a CO_2 laser-equipped surgical microscope is a useful tool for microdiscectomy.

Lee et al. (Lee and Lee 2011) reported that the CO_2 laser-assisted microdiscectomy could be an effective alternative to conventional microdiscectomy techniques. Because the CO_2 laser enabled effective removal of extraforaminal lumbar disk herniation(EFLDH) via a narrow extraforaminal operative corridor without excessive loss of the facet joint and/or the par interarticularis, a thorough decompression of the extraforaminal and/or the foraminal zone was achieved while preserving spinal stability (Fig. 2). Thirty-one patients exhibited a marked reduction in leg pain immediately after the surgery. No patient complained of persistent severe leg pain in the perioperative period. In the present study, reherniation occurred in 1 patient (3.6%) at the 1-year follow-up. The CO_2 laser is also believed to decrease reherniation after discectomy owing to laser-induced metaplasia. (Kim, Choi et al. 2009)

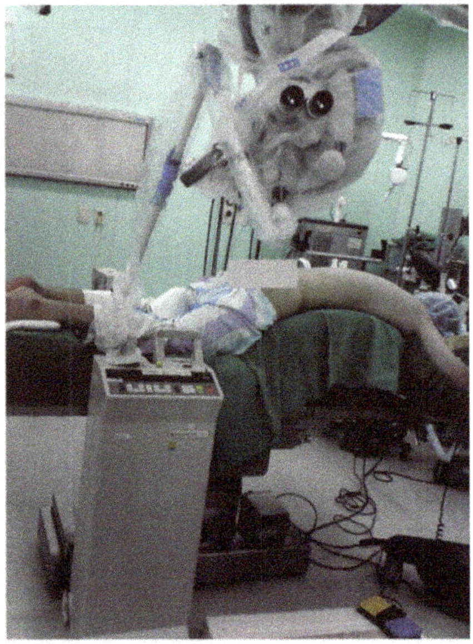

Fig. 1. Photograph of a CO2 laser-equipped surgical microscope.

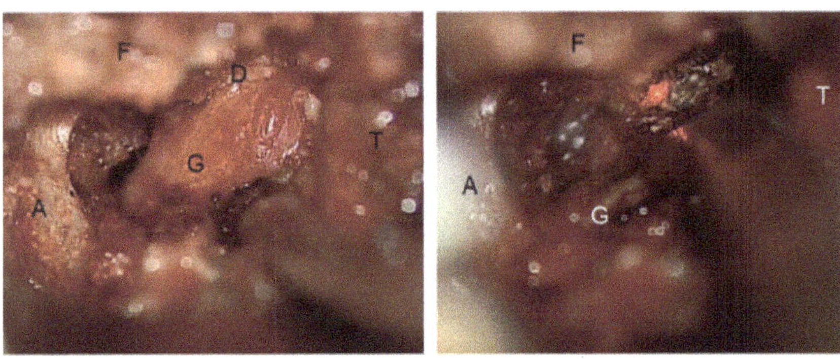

Fig. 2. Intraoperative photomicrographs depicting CO2 laser-assisted microdiscectomy for EFLDH at the L5/S1 level. Left. Photomicrograph taken after exposure of the L5 dorsal root ganglion (A:upper border of the sacral ala; D:herniated disc; F: the lateral L5-S1 facet joint, G:the L5 dorsal root ganglion; and T:the lower border of the L5 transverse process). Right: EFLDH being removed by CO2 laser with gentle retraction of L5 dorsal root ganglion. Note the red He-Ne beam in the surgical field.

2.2.2 Recurrent disk herniation

There are various surgical treatments for recurrent lumbar disk herniation (rLDH), including revision microdiscectomy, lumbar fusion with or without instrumentation (Choi, Lee et al. 2008), and recently, some minimally invasive methods, such as percutaneous endoscopic lumbar discectomy (PELD) (Ahn, Lee et al. 2004), have also been developed. They noted that favorable pain relief was achieved in most patients through this procedure.

Owing to the steep learning curve of PELD (Lee and Lee 2008), the modified microdiscectomy is still more popularity.

In the study by Kims et al (Kim 2010), 21 cases of rLDH, which caused the same symptoms and signs as those of virgin lumbar disk herniations, were excised successfully with modified microdiscectomy using a CO_2 laser.

The author used CO_2 laser during modified lumbar microdiscectomy and reported that using the technique, surgeons can focus the laser beam exactly on the target adhesion scar for adhesiolysis and vaporization and then quickly and easily dissect the adhesion scar tissue. In his results, no approach-related or CO_2 laser-related complications developed. In our opinion, the reason that no incidental durotomy occurred in our series is the precise and gentle dissection using the CO_2 laser (Fig. 3 A.B).

A. 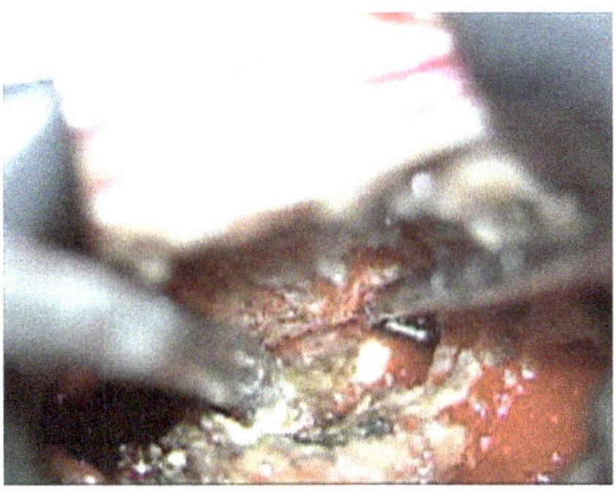 B. 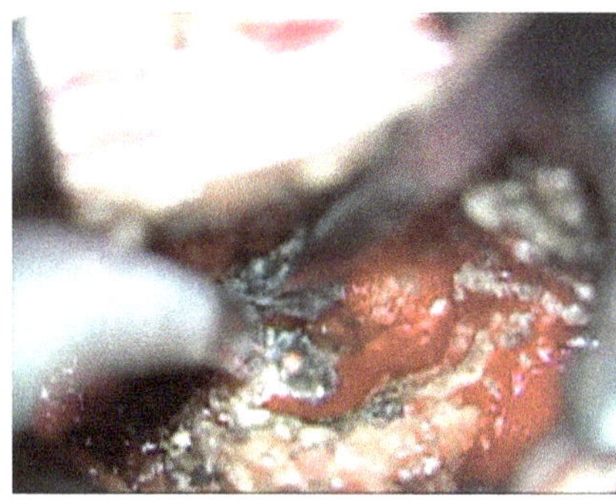

Fig. 3. A. Operative view showing granulation tissue and recurrent lumbar disc herniation located ventromedially to the L5 nerve root (black asterisk) B. The small tip of the CO2 laser could be seen on the protruded disc (black arrow).

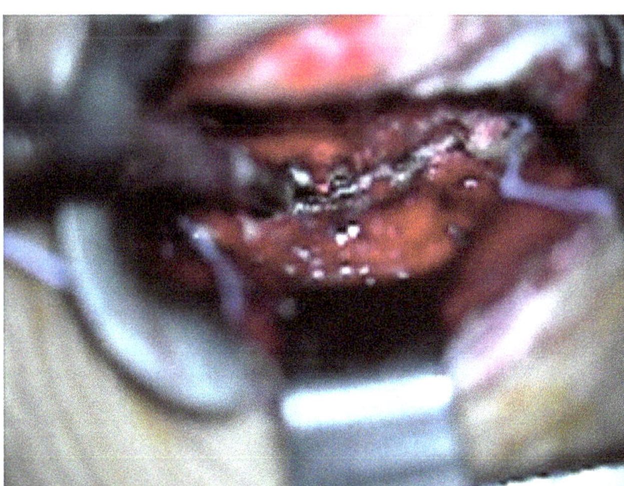

Fig. 4. Operative view presenting easily access narrow ventral part of nerve root using CO2 laser where blunt scalpel couldn't access, with a slight gentle retraction of nerve root

Because an epidural or perineural scar tissue may hinder the dissection using the modified microdiscectomy, increasing the risk of incidental durotomy or iatrogenic neural injury, the CO_2 laser can help surgeons make more precise and safe dissections of the scar tissue than when using a blunt scalpel. Calcifications around recurrent disk fragments are often seen, which may also hinder surgeons to dissect safely. However, with the aid of the CO_2 laser, surgeons can evaporate the calcified portion of the disk without excessive retraction of the nerve root via a narrow operative corridor (Lee, Ahn et al. 2008). Moreover, with a slight gentle retraction of the nerve root, surgeons can easily access the narrow ventral part of the nerve root using the CO_2 laser, where a blunt scalpel could not (Fig. 4).

2.2.3 Discal cyst

Many kinds of surgical methods have been introduced for the treatment of discal cysts. Most discal cysts reported have been treated by open surgical excision (Chiba, Toyama et al. 2001; Lee, Lee et al. 2006) or with some direct intervention, such as computed tomography-guided aspiration and steroid injection (Kang, Liu et al. 2008). Recently, it was reported that a discal cyst was treated with a minimally invasive technique using PELD (Min 2006).

Kim et al. (Kim and Lee 2009) reported that the CO_2 laser, when attached to an operating microscope, allows for quick and easy removal of a discal cyst and, if needed, easy vaporization of disk material. In his study, 14 cases of discal cyst that caused the same symptoms and signs as those of lumbar disk herniations were excised successfully by open surgery using a CO_2 laser.

After the intraoperative removal of the discal cyst, the authors found the communication hole between the cyst and the protruded disk. They then used the heat energy produced by CO_2 lasering and removed the pulled-out disk fragment, if any existed, after pushing into the disk space with a right-angled probe (Fig. 5 A.B).

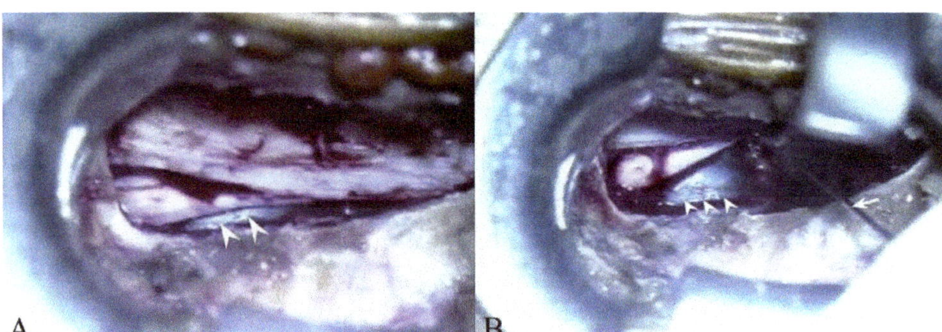

Fig. 5. (A) Distinct communication-like hole between the cyst and intervertebral disc (white arrow) and (B) small tip of laser on the protruded disc (white arrow)

2.2.4 Cervical ossification of ligamentum flavum (OPLL)

The choice of a surgical approach for multi-level cervical OPLL is still a controversial issue.

Direct anterior decompression by corpectomy followed by fusion should be the proper choice of surgical treatment of this multi-level OPLL than indirect decompression by posterior laminoplasy.

The rationale of preferring the anterior approach is based on evidence that the compressive elements are located anterior to the spinal cord in 75% of cases, and therapeutic benefit can be obtained by directly approaching these lesions(Cusick 1991). The degree of cervical myelopathy caused by OPLL is also reported to be influenced not only by static compression from the ossification mass, but also by abnormal intervertebral mobility at the responsible level(s)(Onari, Akiyama et al. 2001).

Despite these theoretical advantages, anterior corpectomy has been reported to be fraught with iatrogenic deterioration of the neurological state, and complications such as spinal fluid fistula or graft problem. Naturally, it will be more technically demanding if the OPLL is involved at multiple cervical levels, and treatment success will depend heavily on a less traumatic manipulation.

In Lee at al report, the authors concluded that direct anterior cervical corpectomy using the CO2 laser resulted in a better recovery of neurological deficit, and adequate decompression of the spinal canal and maintenance of cervical regional lordosis at the operated level for patients with multilevel cervical OPLL. Assuming the surgeon can employ safe anterior microsurgical tools combined CO2 laser and decompression method, proceeding with direct decompressive corpectomy rather than indirect, inadequate laminoplasty is recommended if the patient's preoperative status is appropriate.

They expected that a focused laser beam could vaporize the OPLL and even produce a positive effect. A 5-W pulse, single-pulse mode laser was sufficient to vaporize a thinned OPLL or an osteophyte, as it is known to penetrate only the outer table of the bone(Neblett 1992).(Fig.6)

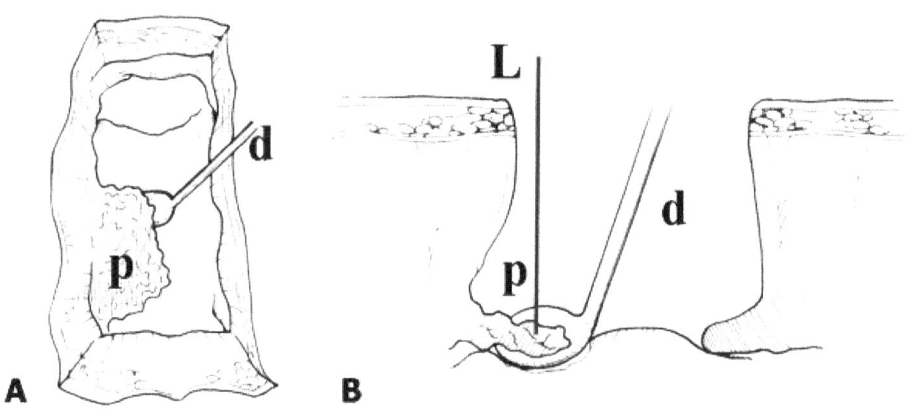

Fig. 6. An illustration showing the tope view (A) and side view(B) of the surgical technique used to remove the densely adhered OPLL using the CO2 laser (L) with the angled microdissector (d), which is held between the OPLL. (p) and the dura to avoid laser-induced damage to the dura.

2.2.5 Complication

Previously, a case of major vessel injury involving perforation of the iliac artery during CO2 laser-assisted lumbar microdiscectomy, caused by prolonged irradiation of the CO2 laser into the deep anterior disc space, has been reported.(Jeon, Lee et al. 2007) Avoiding point focusing of the CO 2 laser on the surface of the anterior annulus, as well as injecting a small amount of saline at the bottom of the intradiscal space during laser ablation, can prevent the occurrence of such a complication. (Jeon, Lee et al. 2007)

And during this procedure, keeping the surgical field moistened was important to minimize the risk of inadvertent injury, as water can absorb CO2 laser energy immediately (Choi, Lee et al. 2005). (Fig.7)

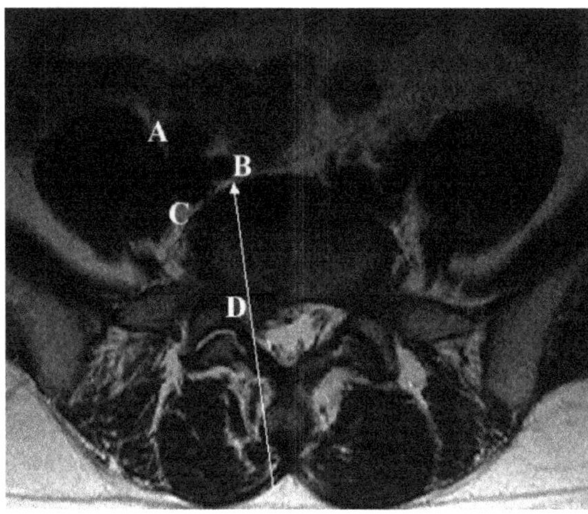

Fig. 7. Preoperative axial MRI at the L5-S1 level. White arrow indicates the direction of the carbon dioxide laser beam. A, External iliac artery. B, Internal iliac artery. C, Common iliac vein. D, Herniated disc fragment.

3. Clinical application in dermatology (Jung 2008)

3.1 Evolution of rejuvenation using lasers. (Alexiades-Armenakas, Dover et al. 2008; Jih and Kimyai-Asadi 2008)

It is described about the evolution of CO2 fractional lasers utilized in the treatment of aging on a typical appliance, equipmental characteristics, clinical utilization in the future development direction,

3.1.1 Historical background of fractional photothermolysis(FP)

1. The period of introduction of concept of SPTL

Since Dr Rox Anderson represented the concept of selective photothermolysis(SPTL) In 1983, specific treatment methods of selectively targeting chromophore like melanin and hemoglobin have been developed clinically.

2. The period of CO2 laser resurfacing

After destructive treatment by CO2 laser at skin tumor was universal, it was known that aging skin was reformed as causing shrinkage in the recovery process after resurfacing. Laser resurfacing has been used as representative of treatment of aging of the skin in spite of long downtime from early 1990.

3. Introduction of NAR

There was effort to reduce the inconvenience of resurfacing by conventional laser using for rejuvenation and minimized the downtime. In 1991 Dr Shimon Dckhouse developed intense pulsed light(IPL) emiting to single pulse from multiple optical energy and introduced advantage for various clinical effect. So it was introduced the concept of non-ablative rejuvenation(NAR).

3.1.2 The development and change of fractional photothermolysis(FP)

Since introduction of the infrared wavelength range of equipment in 2006, AFR concept of a number of devices are being launched as merging ablation and FP using CO2 and Er:YAG laser. Recently, laser equipment of IR, CO2, and Er:YAG are coexisting. The concept of FP having advantage of being safely usable of high powered energy is proliferated broadly to all territory of laser.

3.2 Treatment principle of CO2 fractional laser

CO2 laser is used to treatment of antiaging as removing by ablation of aging tissue and accelerating to regeneration of dermis by transmission of thermal stimulus.

Though resurfacing using existing CO2 laser has many discomfort as ablation of total skin, CO2 fractional laser is focus to treat only fine territory partially. So it is enable to treat safely epidermis and dermis though more high energy than existing treatment is transferred. Because the wound can be restored quickly and easily from surrounding normal skin though injured at epidermis and dermis by laser (Fig. 8.9).

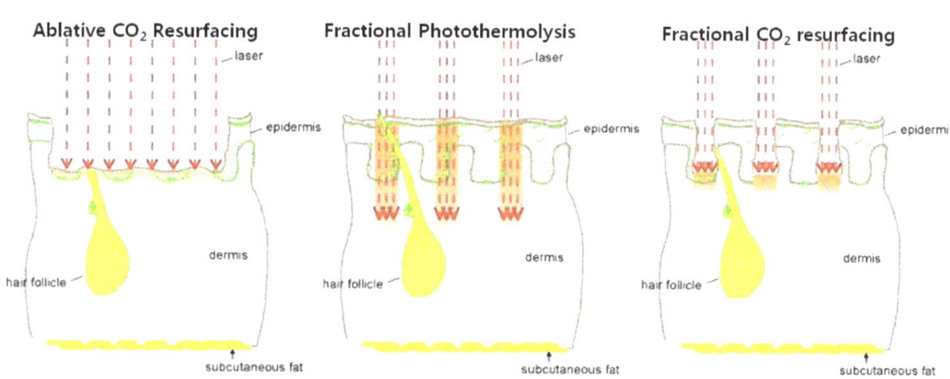

Fig. 8. The basic concept of Fractional Photothermolysis

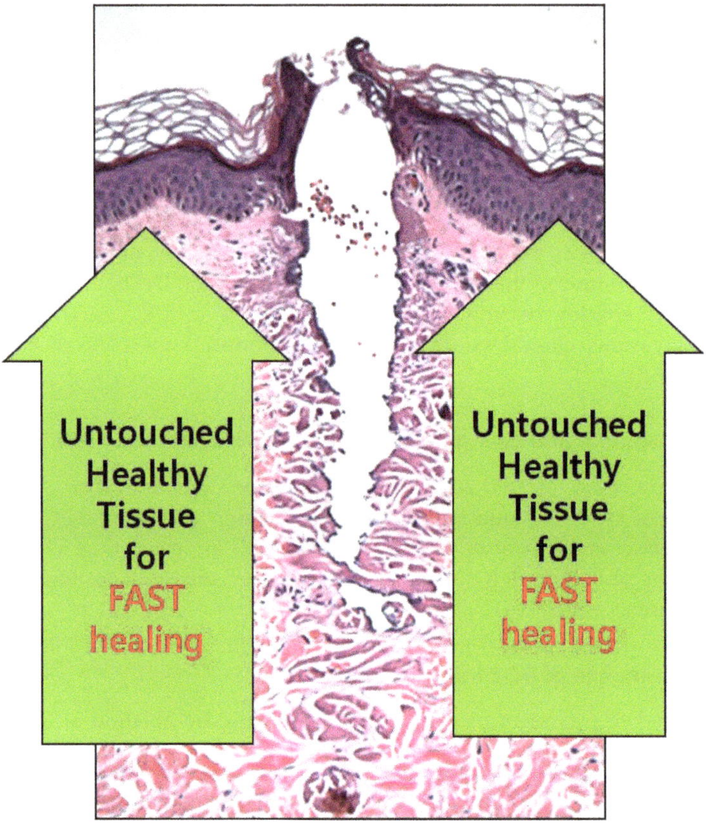

Fig. 9. The basic concept of Fractional CO2 resurfacing: advantages.

You shall awaken warning to attach a laser cover at an operating room entrance as you use a CO2 laser if you enforce an operation.

An enough exhaust device shall install because a lot of extensions occur, and you disturb an operation visual field, and you pollute air when you vaporize an organization.

3.2.1 Differences between the fractional infrared laser (1064~1600nm) and the CO2 fractional laser (10,600nm)

CO2 laser ablates tissue, such as the epidermis and dermis, resulting in tissue damage. This outcome differs completely from the Fraxal tissue reaction, which occurs when a cut is treated by the existing infrared laser (Fig. 10). However, no comparative study has been conducted on the recovery process of infrared rays (IR) and CO2 laser. We know that CO2 laser damages the dermo-epidermal junction, causing severe inflammation at the outset, which in turn results in edema and erythema reactions. We also understand that the lesions damaged by CO2 laser ablation is first filled with keratonocyte within 48 hours and replaced by dermis through the remodeling process, a process that can be continued even after three months(Hantash, Bedi et al. 2007) (Fig. 11).

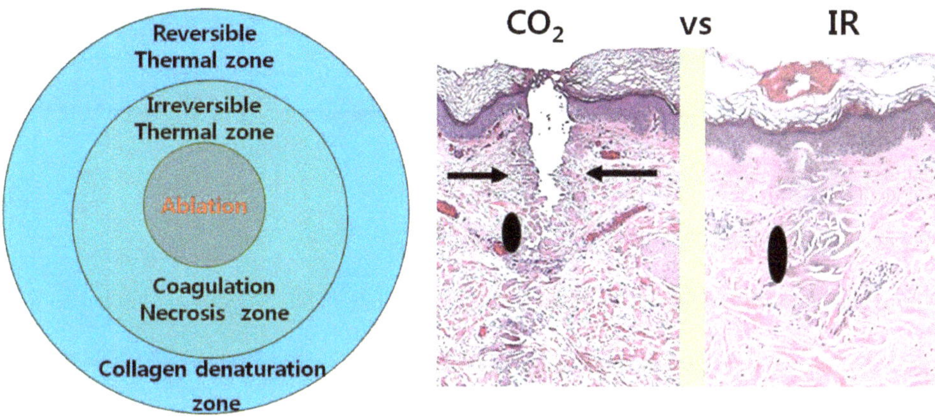

Fig. 10. The main difference between fractional infrared and fractional CO2 laser.

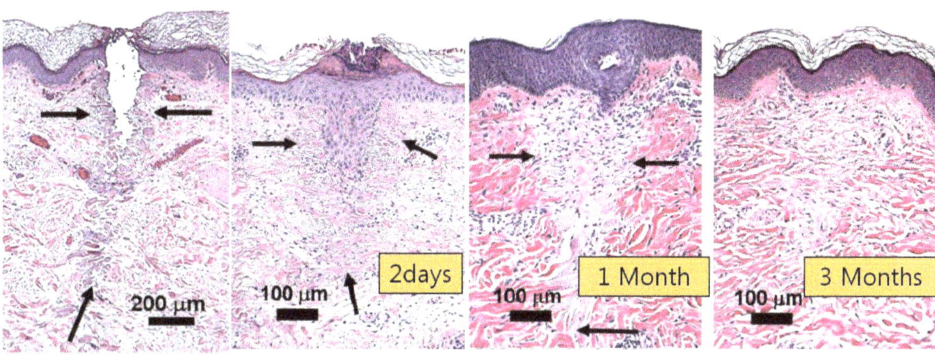

Fig. 11. The wound healing process of fractional CO2 laser treatment

3.2.1.1 Difference in tissue reaction according to wavelength

Infrared (Er:Glass, 1,400-1,600nm)
CO2: 10,640 nm
Er:YAG: 2,940 nm

1. Water absorption

The water absorption rate is in the order of Er: YAG, CO2, and IR (Fig. 12)

2. Lateral Heat Diffusion

CO2 and Er:YAG lasers have higher water absorption rates compared to IR (1064~1600nm) equipment under the same condition (Fig. 12), and most of the energy disappears during the ablation and vaporization process. Therefore, they have less lateral heat diffusion to surrounding tissues when the laser is irradiated and can minimize heat accumulation inside the dermis. In other words, IR equipment accumulates relatively more heat in dermis tissues.

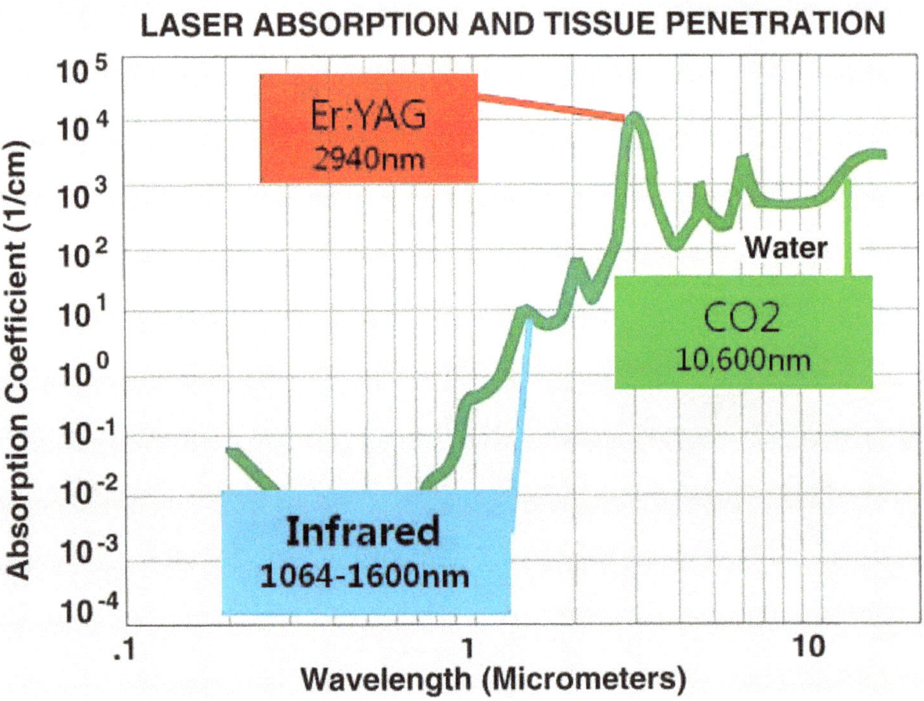

Fig. 12. Water absorption curve.

3. Shrinkage (Rosenberg, Brito et al. 1999; Fitzpatrick, Rostan et al. 2000; Jun, Harris et al. 2003; Zelickson, Kist et al. 2004; Hantash, Bedi et al. 2007)

Shrinkage is a tissue reaction related to the tightening effect that occurs due to tissue shortening after laser irradiation. So far, it is known to progress in the collagen denaturation zone (reversible thermal damage zone) (Fig. 10). Even though no comparative studies have been conducted on this process, IR equipment accumulates more heat inside dermis tissues under the same condition, which is why we think shrinkage will happen more with IR equipment. However, it is hard to estimate if more shrinkage is directly related to the rejuvenation effect, such as tissue tightening or lifting.

4. Penetration Depth

IR can penetrate 1.0~1.5 mm maximum. However, it is believed that the Er:YAG or CO2 lasers penetrate less than the IR laser since they absorb more water. According to recent studies, the CO2 laser can penetrate as deeply as the IR laser if the wattage is increased, the size of the beam is reduced, and the quality beam is irradiated vertically on skin. Due to the CO2 laser's characteristics, if the laser is repeatedly stacked on the same area, tissues can be constantly vaporized so that the laser can penetrate deeply enough. However, repetitive and deep penetration performed several times can cause unwanted excessive heat stimulation, leading to a higher chance of side effects. Therefore, it is right to consider the depth of single beam irradiation as a standard.

5. Risk of erythema and pigmentation

The potential of erythema and pigmentation is related to the damage level of the epidermis and dermis-epidermis joint area as well as the inflammatory reaction due to heat stimulation on the dermis. The risk decreases as fractional treatment convergence is decreased, with the proper convergence of fractional treatment being 20%. Since the risk increases in proportion to the level of heat damage, the CO2 fractional laser can theoretically reduce epidermis damage and dermis heat damage because it has less lateral heat diffusion compared to the IR laser, which in turn will reduce the risk of erythema and pigmentation if it can produce a quality laser beam and be irradiated.

6. Level of Pain

Theoretically, the CO2 fractional laser minimizes pain during procedures. Since most of the heat is lost during the tissue ablation process performed after laser irradiation, it has less lateral heat diffusion, which in turn reduces heat accumulation in tissues compared to IR equipment. Therefore, we expect less pain if we use the CO2 fractional laser. In addition, short pulse duration, smaller spot size, and shorter irradiation time on the skin can reduce pain. Therefore, it is theoretically possible to actualize the CO2 fractional laser with very little pain. However, this laser can damage the epidermis and dermis joint, which will cause more severe initial inflammation reaction and edema and burn feeling.

3.2.1.2 Can parameter be independently controlled?

Since the CO2 laser is a continuous wave laser, output is written in watts. Most CO2 lasers can control parameters such as watt and pulse duration independently. Therefore, the same amount of energy (J) can be irradiated while the depth is controlled by watt level and the coagulation range (heat damage range) can be controlled by pulse duration (Fig. 13). These are very important strong points of the CO2 laser, differing from IR equipment, which changes power and pulse duration according to the J (energy) level. That is, IR equipments cannot independently control factors since power and pulse duration is simultaneously increased as J (energy) is increased.

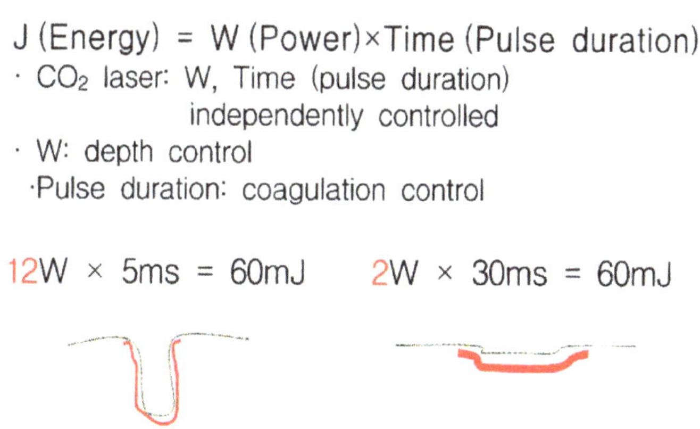

J (Energy) = Watt (Power)* Time (pulse duration)

Fig. 13. Parameter controllability

Due to these characteristics, the CO2 fractional laser can treat very deep layers and offers a better recovery process after treatment as it allows high power (deep penetration) and short pulse duration (minimal lateral heat diffusion to the surrounding area), creates a very narrow vertical ablation zone, and forms a limited lateral heat diffusion zone (Fig. 14).

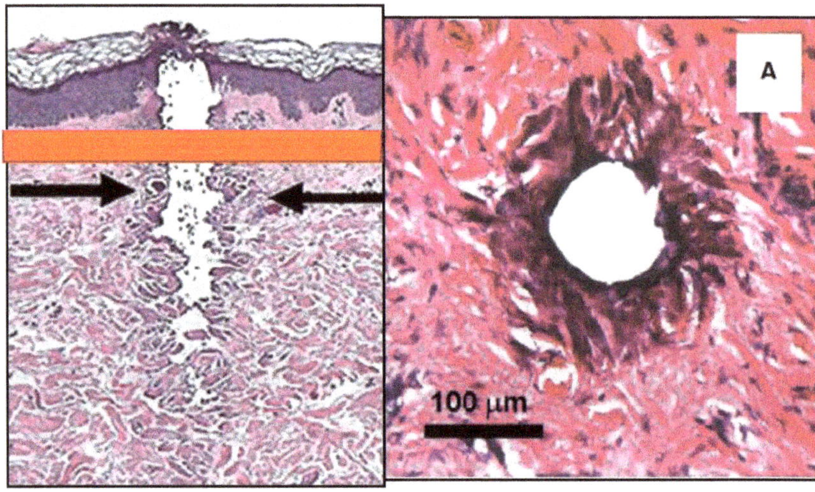

Fig. 14. Vertical and horizontal view of Fractional CO2 laser.

3.2.2 Basic considerations for the fractional laser

3.2.2.1 Spot size

The spot size of the laser refers to the diameter of irradiated beam. As the spot gets bigger, the re-epithelization process takes longer, which in turn cases a longer downtime. In fact, if the spot size is 140um, re-epithelization of the dermo-epidermal junction takes less than 36 hours. It takes two to four days for 300μm, three to five days for 500μm, and five to ten days for over 1.25mm. Therefore, it is important to make the spot small to facilitate safe treatment and fewer inconveniences. Since there is a limit to the minimum spot size that can be actualized physically, it is impossible to reduce the size below a certain point. If the spot is too small, a lot more lasers should be irradiated on a certain area to get converge that is required for proper treatment. In that case, we cannot exclude excessive heat accumulation, which is why we cannot say a smaller spot size is always advantageous.

3.2.2.2 Laser beam irradiation methods

There are various methods of irradiating several spots on a certain area. They can be classified into the moving method and the stamp method according to the type of equipment being utilized.

1. Moving method

The moving method is used for Fraxal repair that has a function that can realize regular spot converge by controlling beam irradiation speed regardless of moving speed and eCO2 equipment that has fixed beam irradiation speed by moving the handpiece constantly.

2. Stamp method: micro lens array and scanner type

After Fraxal acquired a patent on moving method irradiation, the equipment produced later on mostly adapted the stamp method. This method can be classified into two divisions. The most common method is micro lens array (the beam goes through the lens and is simultaneously irradiated after being divided into several fine beams, like LUX1540 and Affirm) that treats the parts by stamping microbeams. The other one is also a stamp method, but uses the scanner method rather than lens array and irradiates beams in order. The scanner method can be classified in two manners. The sequential type irradiates beams sequentially adjacently. Alternative type irradiates one line and skips the close line and irradiates the next line. Random type irradiates with no order. Theoretically, the sequential type includes a high possibility of heat accumulation due to adjacent irradiation. The random type can be a safer treatment since it minimizes heat accumulation. However, it is not clear if it has clinical significance.

3.2.2.3 Density

Density is the portion of treated area where the laser beam is irradiated. Density is considered to be low when there are many normal tissues left around the area. If density is too low, it is safe but can be ineffective. However, when density is high, there are less normal tissues, which is meaningless since fractional technology's strength is safe and fast recovery. Therefore, we need to determine the density within proper range, considering the purpose of the treatment as well as the characteristics of the equipment. If more than two passes of treatment is done right after the first pass, density increases in proportion. However, it can weaken its safety as it causes excessive heat accumulation and repetitive irradiation on the same area. Therefore, it is safer to obtain the intended density from the first pass rather than acquiring it from repetitive passes.

3.2.2.4 Characteristics of the beam

We can observe various equipment characteristics, such as spot size, pitch, pitch control possibility and range, controllable range of irradiation time, watt range, scanning method scanning range, and scanning shape. However, the most important evaluation factor should be: "How superior is each beam's characteristic?" In fact, the users are not fully aware of CO2 laser equipment's characteristics. For example, if this equipment includes a scanner with various modes with an inconsistent quality beam, inconsistent irradiation direction or penetration depth, it can be a low-priced CO2 laser with a scanner. It is too much to expect this type of equipment to irradiate the quality beam uniformly on skin with detailed control over lateral heat diffusion and to penetrate into the skin with the depth one wants and obtain satisfactory results safely.

• Type of resonator and beam (*Longitudinally excited, Transversely excited, Gas dynamic, Waveguide laser*)
• Durability of resonator
• Beam window
• Calibration
• Auto-detection
• Individual control of pulse duration

Table 1. Factors Affecting the Beam Quality of CO_2 Laser

Therefore, when comparing CO2 fractional laser performance, the most important factor that shows the biggest deviation is quality of beam. The following are the major factors that determine beam quality (Table 1).

3.3 Clinical application of CO2 fractional laser

Laser induced regeneration by vaporizing aging tissue at the epidermis and dermis. It can effectively lead to regrowth and remodeling in the process, so it can be applied broadly to a variety of issues (scar, pigmentation, texture so on) at epidermis and dermis and it was also effective to improve the aging skin because laser can treated from 0.2mm in depth to 1mm or more at dermis.

However, it is limited to case reflected in the previously mentioned characteristics of the CO2 when the CO2 laser beam to penetrate the organization is elaborately controlled.

Because it can not be told that there was less possibility of erythema and pigmentation if not uncontrolled beam.

By such advantage, CO2 fractional laser can be utilized to various indications of epidermis and dermis. (Table 2)

Since it was initially introduced in 2006, CO2 fractional laser has been made up a large development and evolution for a short period of time, and future developments are expected as follows. (Table 3)

1. Superficial problem	
• Pigmentary lesions	• Pore
• Texture & aging skin	• Laxity(hand, neck)
• Wrinkles	
2. Surface and deeper structure	
• Scar	• Striae distensa
3. Dermal problem	
• Pigmentary lesions: refractory PIH, cloasma, tattoo	

Table 2. Clinical Indications of CO_2 Fractional Laser Treatment

• Controllable, smaller spot	• Powerful
• Variable parameter	• Surface cooling for epidermal preservation
• Increasing dermal shrinkage	• Deeper penetration
• Pattern of beam	• Quality of beam

Table 3. Future Development of CO_2 Fractional Laser

4. Clinical application in plastic surgery

You decrease a pain after operations by a laser in case of constancy as the way that used a CO2 laser compares to the way that used a knife as you seal the peripheral nerve edge. And because CO2 laser can seal the lymphatics located at cutting plane and directly seal at peripheral vessel small sized less than 0.5mm, you can support patient to more comfort by accelerating wound healing process and by decreasing intraoperative bleeding and postoperative edema, contusion.

This CO2 laser has definite advantage that it is very clear between the tissue which has thermal injury and no damaged tissue located at surrounding area, and it enable to cut tissue by not putting to physical stress to surrounding tissue. It has profit at the procedure by need of bloodless surgical field because of minimizing bleeding. And it enable to treat antiseptically to wound as vaporize the tissue by high temperature.

In palatoplasty and pharyngeal flap operation, compared to conventional method, complication caused by intra- and postoperative bleeding was decreased and there was no difference between CO2 using and conventional method in aspect of wound healing. And CO2 laser has merits to decrease hospitalization (Song IC 1998).

In patients of shortening of frenulum, CO2 laser is available to frenotomy (Yoon CH 1998)

In blepharoplasty using CO2 laser, it was first reported by Baker in 1984(Baker, Muenzler et al. 1984). Using CO2 laser, blepharoplasty was processed safely and more enhanced to decrease bleeding, operation time, edema, time to heal. Mittelman et al (Mittelman and Apfelberg 1990) evaluated as safe method in spite of having risks of eyeball injury, breakaway from its course, fire at operation room, burn injury of skin.

Extramammary Paget's disease is eczema like disease accompanied mainly by itching of anus and genitalia. It was found very much among a middle age or patients of prime of manhood with the past history which a treatment of skin clinic was failed in during long periods. About this disease, invasive method was surgical excision and topical ointment using 5-fluorouracil, radiotherapy. Recently it is the trend that an interest of the treatment using CO2 laser and Nd:YAG laer is rising (Ewing 1991; Yoon ES 2000)

In wound healing, Low-powered CO2 laser helps to induce the synthesis of DNA by give effects to permeability of cell membrane, and to activate fibroblast and condrocyte, and to accelerate to absorp the hematom and to remove necrotic tissue, and to help to healing process of bone and cartilage (Tsai, Huang et al. 1997).

In atropic scar, pulsed CO2 laser having high power was utilized.

In acne scar, dermabrasion using CO2 laser minimizing the common complication by conventional method and is enable to control the depth of peeling by depth of acne scar. So is enable to avoid to complication of hypertrophic scar and keloid etc.. It is available to process peeling safely and deeply. So recovery toward a daily life is fast as managements after operations is convenient because of decreasing of postoperative edema and pain by disconnecting at nerve terminals (Yoon ES 1998).

5. Clinical application in otorhinolaryngology

5.1 Laryngeal microsurgery using CO2 laser

CO2 laser is infrared having 10,600 nm wavelength. It has vibrational energy and enable to control to transfer through the mirror. The diameter of focus is about 1mm by handpiece and about 200-250μm by micromanipulator from focus distance of 400mm. It was used by controlling the focus distance by using purpose, and it enable to cut like common scalpel. If defocused beam is emitted from more longer distance than focus distance, tissue coagulation was available as decreasing per unit of energy (Ossoff, Coleman et al. 1994).

Laryngeal microsurgery using CO2 laser has many advantages more than conventional surgical method. CO2 laser enables the surgeon to remove the exact desired pathology through controlling of power and emission time, which is necessary to preserve normal function. In situations where it is necessary to remove the pathology of a tubed organ, direct contact of the excised area and disturbance of the surgical field can be avoided. A bloodless operation is possible by the unique ability of the CO2 laser to cauterize even arteries and veins as small as 0.5mm in diameter. Safety margins can be secured more easily when removing a tumor. A local inflammatory response is extremely small after an operation, and there is a little hyperplasia of a granulation tissue and scar formation during healing processes. In case of a huge tumor, CO2 laser can help reduce the amount of tissue removed, which in turn can assist in maintenance of laryngeal function (Hall 1971; Norris and Mullarky 1982).

Possible anesthesia methods for laser microlaryngosurgery are intubation and non-intubated anesthesia etc. Rarely it is possible by Jet ventilation or tracheostomy. In cases of intubated anesthesia, protective tubes such as laser-Shield, Bivoan, and Malincrodt tubes can be used. The cuff of the intubation tube must be protected with a cottonoid soaked in a solution of normal saline during an operation. Non intubated anesthesia may be used when the visual field during an operation is obstructed by an anesthesia tube. This is achieved by removing the tube during surgery at 100% O2 saturation, and quickly reinserting the tube after 2-3 minutes when the O2 saturation begins to fall. Tumors located posterior to the larynx may be removed by inserting a laryngoscope behind the intubation tube (Fried 1984; Shapshay, Beamis et al. 1989).

When using a laser for laryngomicrosurgery, laser power should be set between 1 to 10 watts according the type of tissue to be removed or type of surgery. The most frequently used setting is a microscope magnified 16 to 25 times, laser power of 2 watts, and focus of 250 to 400μm. If the target tissue is too hard and blood vessels are too small, super pulse or ultrapulse methods should be used with a smaller focus (Ossoff, Coleman et al. 1994).

Carbonization of cancer pathology can be visualized more clearly during an operation with a continuous mode laser and can help discriminate between normal and cancer cells. A laryngeal forceps or suction tube should be used to pull the target tissue before using a laser to cut more cleanly while reducing carbonization or thermal injury.

CO2 laser can be used in a variety of laryngeal diseases. First, benign laryngeal diseases such as laryngeal nodule, Reinke's edema, laryngeal cyst, granuloma, papilloma, angioma, septum and so on. Secondly, pre-cancerous pathology like white keratosis. Thirdly, it is partly applied in a treatment of malignant laryngeal pathology.

CO2 laser can also be used to prevent asphyxia in various laryngeal diseases causing airway obstruction such as laryngeal and tracheal stenosis.

6. Conclusion

CO2 laser has been developed until nowadays. At present, CO2 laser is considered as essential instrument in medicine. In future, we think that CO2 laser may be more developed for achieving the goal and more generalized.

7. Acknowledgment

Thank to corresponding Dr. Jun-Sung Kim for helping to support academic basis. And thank to Dr. Sang-Ho Lee for permitting the chance to writing in Wooridul Hospital. Also thank to Dr. Chan-Woo Jung at Leaders dermatologic clinics in South Korea for supporting dermatologic description and figure. Special thank to Mrs. Park Min for supporting dedication of this manuscript.

8. References

Ahn, Y., S. H. Lee, et al. (2004). "Percutaneous endoscopic lumbar discectomy for recurrent disc herniation: surgical technique, outcome, and prognostic factors of 43 consecutive cases." *Spine (Phila Pa 1976)* 29(16): E326-332.

Ahn, Y., S. H. Lee, et al. (2005). "Percutaneous endoscopic cervical discectomy: clinical outcome and radiographic changes." *Photomed Laser Surg* 23(4): 362-368.

Alexiades-Armenakas, M. R., J. S. Dover, et al. (2008). "The spectrum of laser skin resurfacing: nonablative, fractional, and ablative laser resurfacing." *J Am Acad Dermatol* 58(5): 719-737; quiz 738-740.

Ascher, P. W. and F. Heppner (1984). "CO2-Laser in neurosurgery." *Neurosurg Rev* 7(2-3): 123-133.

Baker, S. S., W. S. Muenzler, et al. (1984). "Carbon dioxide laser blepharoplasty." *Ophthalmology* 91(3): 238-244.

Beck, O. J. (1980). "The use of the Nd-YAG and the CO2 laser in neurosurgery." *Neurosurg Rev* 3(4): 261-266.

Benjamin, M., S. Qin, et al. (1995). "Fibrocartilage associated with human tendons and their pulleys." *J Anat* 187 (Pt 3): 625-633.

Chiba, K., Y. Toyama, et al. (2001). "Intraspinal cyst communicating with the intervertebral disc in the lumbar spine: discal cyst." *Spine (Phila Pa 1976)* 26(19): 2112-2118.

Choi, K. B., D. Y. Lee, et al. (2008). "Contralateral reherniation after open lumbar microdiscectomy : a comparison with ipsilateral reherniation." *J Korean Neurosurg Soc* 44(5): 320-326.

Choi, S., S. H. Lee, et al. (2005). "Factors affecting prognosis of patients who underwent corpectomy and fusion for treatment of cervical ossification of the posterior longitudinal ligament: analysis of 47 patients." *J Spinal Disord Tech* 18(4): 309-314.

Choy, D. S., R. B. Case, et al. (1987). "Percutaneous laser nucleolysis of lumbar disks." *N Engl J Med* 317(12): 771-772.

Cusick, J. F. (1991). "Pathophysiology and treatment of cervical spondylotic myelopathy." *Clin Neurosurg* 37: 661-681.

Deruty, R., I. Pelissou-Guyotat, et al. (1993). "Routine use of the CO2 laser technique for resection of cerebral tumours." *Acta Neurochir (Wien)* 123(1-2): 43-45.

Ewing, T. L. (1991). "Paget's disease of the vulva treated by combined surgery and laser." *Gynecol Oncol* 43(2): 137-140.

Fitzpatrick, R. E., E. F. Rostan, et al. (2000). "Collagen tightening induced by carbon dioxide laser versus erbium: YAG laser." *Lasers Surg Med* 27(5): 395-403.

Fried, M. P. (1984). "Limitations of laser laryngoscopy." *Otolaryngol Clin North Am* 17(1): 199-207.

Glasscock, M. E., 3rd, C. G. Jackson, et al. (1981). "The argon laser in acoustic tumor surgery." *Laryngoscope* 91(9 Pt 1): 1405-1416.

Gropper, G. R., J. H. Robertson, et al. (1984). "Comparative histological and radiographic effects of CO2 laser versus standard surgical anterior cervical discectomy in the dog." *Neurosurgery* 14(1): 42-47.

Hall, R. R. (1971). "The healing of tissues incised by a carbon-dioxide laser." Br J Surg 58(3): 222-225.

Hantash, B. M., V. P. Bedi, et al. (2007). "In vivo histological evaluation of a novel ablative fractional resurfacing device." *Lasers Surg Med* 39(2): 96-107.

Hellinger, J. (1999). "Technical aspects of the percutaneous cervical and lumbar laser-disc-decompression and -nucleotomy." *Neurol Res* 21(1): 99-102.

Heppner, F. (1978). "[The laser scalpel in the nervous system]." *Wien Med Wochenschr* 128(7): 198-201.

Houck, P. M. (2006). "Comparison of operating room lasers: uses, hazards, guidelines." *Nurs Clin North Am* 41(2): 193-218, vi.

Jeon, S. H., S. H. Lee, et al. (2007). "Iliac artery perforation following lumbar discectomy with microsurgical carbon dioxide laser: a report of a rare case and discussion on the treatment." *Spine (Phila Pa 1976)* 32(3): E124-125.

Jih, M. H. and A. Kimyai-Asadi (2008). "Fractional photothermolysis: a review and update." *Semin Cutan Med Surg* 27(1): 63-71.

Jun, J. H., J. L. Harris, et al. (2003). "Effect of thermal damage and biaxial loading on the optical properties of a collagenous tissue." *J Biomech Eng* 125(4): 540-548.

Jung, C. W. (2008). "Understanding of CO2 laser in dermatology." *Journal of dermatology korean society for laser medicine and surgery* 12(1): 90-99.

Kang, H., W. C. Liu, et al. (2008). "Midterm results of percutaneous CT-guided aspiration of symptomatic lumbar discal cysts." *AJR Am J Roentgenol* 190(5): W310-314.

Kim, J. S. (2010). "Facility of the removal of recurrent disc herniation due to use of laser in primary operation." *Journal of the korean society for laser medicine and surgery* 14(1): 83-87.

Kim, J. S., G. Choi, et al. (2009). "Removal of a discal cyst using a percutaneous endoscopic interlaminar approach: a case report." *Photomed Laser Surg* 27(2): 365-369.

Kim, J. S. and S. H. Lee (2009). "Carbon dioxide (CO2) laser-assisted ablation of lumbar discal cyst." *Photomed Laser Surg* 27(6): 837-842.

Krishnamurthy, S. and S. K. Powers (1994). "Lasers in neurosurgery." *Lasers Surg Med* 15(2): 126-167.

Lee, D. Y., Y. Ahn, et al. (2006). "Percutaneous endoscopic lumbar discectomy for adolescent lumbar disc herniation: surgical outcomes in 46 consecutive patients." *Mt Sinai J Med* 73(6): 864-870.

Lee, D. Y. and S. H. Lee (2008). "Learning curve for percutaneous endoscopic lumbar discectomy." *Neurol Med Chir (Tokyo)* 48(9): 383-388; discussion 388-389.

Lee, D. Y. and S. H. Lee (2011). "Carbon dioxide (CO2) laser-assisted microdiscectomy for extraforaminal lumbar disc herniation at the L5-S1 level." *Photomed Laser Surg* 29(8): 531-535.

Lee, H. K., D. H. Lee, et al. (2006). "Discal cyst of the lumbar spine: MR imaging features." *Clin Imaging* 30(5): 326-330.

Lee, S. H., Y. Ahn, et al. (2006). "Immediate pain improvement is a useful predictor of long-term favorable outcome after percutaneous laser disc decompression for cervical disc herniation." *Photomed Laser Surg* 24(4): 508-513.

Lee, S. H., Y. Ahn, et al. (2008). "Laser-assisted anterior cervical corpectomy versus posterior laminoplasty for cervical myelopathic patients with multilevel ossification of the posterior longitudinal ligament." *Photomed Laser Surg* 26(2): 119-127.

Min, J., Chung, B.J., and Lee, S.H. (2006). "Endoscopically managed synovial cyst of the lumbar spine." *Korean Journal of Spine* 3: 242-245.

Mittelman, H. and D. B. Apfelberg (1990). "Carbon dioxide laser blepharoplasty--advantages and disadvantages." *Ann Plast Surg* 24(1): 1-6.

Neblett, C. R. (1992). *Laser and the cervical spine*. Philadelphia, Lippincott Williams & Wilkins.

Nerubay, J., I. Caspi, et al. (1997). "Percutaneous laser nucleolysis of the intervertebral lumbar disc. An experimental study." *Clin Orthop Relat Res*(337): 42-44.

Norris, C. W. and M. B. Mullarky (1982). "Experimental skin incision made with the carbon dioxide laser." *Laryngoscope* 92(4): 416-419.

Onari, K., N. Akiyama, et al. (2001). "Long-term follow-up results of anterior interbody fusion applied for cervical myelopathy due to ossification of the posterior longitudinal ligament." *Spine (Phila Pa 1976)* 26(5): 488-493.

Origitano, T. C. and O. H. Reichman (1993). "Photodynamic therapy for intracranial neoplasms: development of an image-based computer-assisted protocol for photodynamic therapy of intracranial neoplasms." *Neurosurgery* 32(4): 587-595; discussion 595-586.

Ossoff, R. H., J. A. Coleman, et al. (1994). "Clinical applications of lasers in otolaryngology--head and neck surgery." *Lasers Surg Med* 15(3): 217-248.

Powers, S. K., S. S. Cush, et al. (1991). "Stereotactic intratumoral photodynamic therapy for recurrent malignant brain tumors." *Neurosurgery* 29(5): 688-695; discussion 695-686.

Rosenberg, G. J., M. A. Brito, Jr., et al. (1999). "Long-term histologic effects of the CO2 laser." *Plast Reconstr Surg* 104(7): 2239-2244; discussion 2245-2236.

Shapshay, S. M., J. F. Beamis, Jr., et al. (1989). "Total cervical tracheal stenosis: treatment by laser, dilation, and stenting." *Ann Otol Rhinol Laryngol* 98(11): 890-895.

Song IC, P. W., Kil MS (1998). "Laser assisted palatoplasty and pharyngeal flap." J Korean *Soc Plast Reconstr Surg* 25: 252.

Stein, E., T. Sedlacek, et al. (1990). "Acute and chronic effects of bone ablation with a pulsed holmium laser." *Lasers Surg Med* 10(4): 384-388.

Stellar, S., T. G. Polanyi, et al. (1970). "Experimental studies with the carbon dioxide laser as a neurosurgical instrument." *Med Biol Eng* 8(6): 549-558.

Tew, J. M., Jr. and W. D. Tobler (1983). "The laser: history, biophysics, and neurosurgical applications." *Clin Neurosurg* 31: 506-549.

Tsai, C. L., L. L. Huang, et al. (1997). "Effect of CO2 laser on healing of cultured meniscus." *Lasers Surg Med* 20(2): 172-178.

Whipple, T. L., R. B. Caspari, et al. (1984). "Laser subtotal meniscectomy in rabbits." *Lasers Surg Med* 3(4): 297-304.

Yoon CH, R. Y., Park HS, Kim HJ (1998). "Comparative study between using CO2-laser and classic method in frenulotomy." *J Korean Soc Plast Reconstr Surg* 25: 1475.

Yoon ES, C. J., Han SK, Kim WK (2000). "Treatment of extramammary Paget's disease using the CO2 laser." *J Korean Soc Plast Reconstr Surg* 27: 169.

Yoon ES, K. S., Ahn DS, Park SH (1998). "CO2 laser resurfacing of acne scar." *J Korean Soc Aesth Plast Reconstr Surg* 4: 381.

Zelickson, B. D., D. Kist, et al. (2004). "Histological and ultrastructural evaluation of the effects of a radiofrequency-based nonablative dermal remodeling device: a pilot study." *Arch Dermatol* 140(2): 204-209.

16

Protons Acceleration by CO_2 Laser Pulses and Perspectives for Medical Applications

Pasquale Londrillo, Graziano Servizi, Andrea Sgattoni, Stefano Sinigardi, Marco Sumini and Giorgio Turchetti
Università di Bologna, INFN Sezione di Bologna
Italy

1. Introduction

The acceleration of electrons with the high electric fields generated in a plasma by a very intense laser pulse was proposed over forty years ago [Tajima & Dawson (1979)], but only the advent of the chirped pulse amplification CPA [Mourou et al. (2006)] allowed to increase the laser power and intensity up to the required values. The continuous progress, since a decade, of compact Ti:Sa lasers allowed 1 GeV good quality electron beams to be generated [Leemans et al. (2006)]. The optical acceleration of protons and ions has been also actively investigated. The highest energy of protons, 60 MeV, has been reached with short high energy pulses of Nd:Yag lasers, developed for inertial fusion [Snavely (2000)]. With compact ultrashort Ti:Sa laser pulses intensities of 10^{21} W/cm^2 are reached and proton beams with energy up to a few tens of MeV are obtained [Zeil et al. (2010)].

The targets are typically thin metal foils and the acceleration is achieved in the TNSA regime (Target Normal Sheath Acceleration). The laser, interacting with an overcritical plasma, cannot propagate through it and heats the electrons on the surface of the target. A large number of "hot" electrons is hence produced and they are are accelerated in the forward direction and can cross the target. When reaching the rear surface they create an intense electrostatic field which accelerates the protons present on the surface [Passoni & Lontano (2004)]. The energy spectrum is exponential and the angular spread is significant so that the beam is not suitable for free propagation. Energy selection and collimation may reduce the intensity below the threshold required for any application. However other acceleration mechanisms have been considered such as the radiation pressure dominated regime RPA (Radiation Pressure Acceleration), where two distinct mechanism act depending on the thickness h of the target. If $h \sim \lambda$ the hole boring regime with break-up of the electron density wave is active [Macchi et al. (2005)], whereas for ultrathin targets the acceleration mechanism is the same as for the relativistic mirror [Londrillo et al. (2010); Macchi et al. (2009)] and high efficiencies can be reached. This regime was recently experimentally observed using a few nanometers carbon targets, a circularly polarized laser pulse with $\lambda \sim 1\mu$m and very high contrast [Henig (2009)].

The efficiency of the RPA should be higher than TNSA and the proton bunches should have a small energy and angular spread; however the requirements of circular polarization and high contrast of the laser pulse render this regime not easily achievable. The subcritical or quasi-critical targets are also a possible alternative because the laser-plasma energy coupling can be very high with a much higher energy transfer from the laser pulse to the plasma: a different regime may be achieved where the laser drills a channel and a strong electron current is created on its trail. At the exit from the plasma it creates a magnetic azimuthal field and a longitudinal electric field. As a consequence, the protons are efficiently accelerated and nicely collimated [Bulanov (2010); Nakamura, Bulanov, Esirkepov & Kando (2010); Naumova & Bulanov (2002); Yogo (2008)]. Experimental results with near critical targets confirmed the theoretical and simulation results on the enhancement of maximum energy and reduction of angular spread with respect to the TNSA acceleration mechanism [Fukuda & Bulanov (2009)].

The possibility of reaching energies close to the threshold of 60 MeV for cancer therapy with compact Ti:Sa laser system has stimulated several projects dedicated to medical applications. Indeed the protons or ions deposit most of their energy at the end of their range and are biologically more effective with respect to electrons or X rays since they allow to spare nearby healthy tissues. However the cyclotrons and synchrotrons currently used require large and expensive infrastructures. The use of laser acceleration opens a perspective for more compact and cheaper devices suitable to be installed on a regional scale. Two possible strategies are being considered:

A) increase the power of the laser system in order to reach energies in the 100-200 MeV energy range [Bulanov (2008); Hofmann (2011); Murakami (2008)]

B) use the laser system as injector into a DTL linac to increase the energy starting from 10 MeV [Antici (2011)] or to inject a 30 MeV laser accelerated protons bunch, into a high field compact linac in order to raise the energy up to 100 or 200 MeV [Londrillo et al. (2011)].

The hybrid acceleration scheme does not require to develop new laser systems but only the improvement of targets and the design of a transport system capable of shaping the beam in such a way to render it suitable for injection. Simulations and experiments are presently being developed to explore the feasibility of transport of an optically accelerated protons bunch reaching the beam quality required for irradiation and for injection into a post-acceleration device [Melone (2011); Nishiuchi (2010); Schollmeier (2008)]. The reduction of the energy spread with a longitudinal phase space rotation provided by a synchronized RF was also proved [Noda (2008)].

Recently, new protons acceleration experiments have been performed taking advantage of short pulses of long wavelength $\lambda = 10\mu$m produced by CO_2 lasers. This approach provides a parallel research pathway which offers some advantages. Being the wavelength one order of magnitude larger than the optical values typical of Ti:Sa or Nd laser, the plasma critical density for $\lambda = 10\mu$m is about 10^{19} cm^{-3} which can be reached ionizing a supersonic gas-jet. The CO_2 lasers deliver a pulse with a native quasi circular polarization which is interesting for triggering the RPA regime at lower intensities.

Recent experiments showed that using a 1 TW pulse of 1 J interacting with a solid target, protons can be accelerated up to 1 MeV with exponential energy spectrum and wide angular spread typical of TNSA regime [Pogorelsky (2010b)]. On a gas jet at the critical density a quasi monochromatic beam of protons at 2 MeV was obtained suggesting that a RPA mechanism is dominating [Pogorelsky (2010a)]. With a train of short pulses and 100 J total energy a different acceleration mechanism was achieved and protons of 25 MeV, even though a low number, were accelerated with a very small energy spread [Haberberger, Tochitsky, Fiuza, Gong, Fonseca, Silva, Mori & Joshi (2011); Haberberger, Tochitsky, Gong & Joshi (2011)]. This result inserts the CO_2 lasers among the possible candidates for the production of proton beams for medical therapy, possibly combined with a post acceleration device. The CO_2 lasers have high efficiency and produce almost circularly polarized light which is suited for acceleration mechanisms like RPA. In particular the possibility of using gas jets at under-critical or slightly overcritical density opens very interesting perspectives because of the high repetition rate allowed, the absence of debris (opposed to the case of solid thin targets) jointly with the advantages of circular polarization.

We can now study the very same physical problem using a CO_2 laser instead of a Ti:Sa laser keeping the dynamics of the interaction unchanged. This can be done keeping unchanged the adimensional parameters which characterize the laser-plasma: the ratio of the plasma density over critical density (n/n_c) and the normalized vector potential (a). This correspond to consider a CO_2 laser pulse with the same peak power but ten times longer (same number of wave cycle) and a laser waist proportional to the wavelength. The plasma density is hundred times lower and the total volume interested by the acceleration is three order of magnitude larger than the case of a Ti:Sa laser. If we assume a definite fraction of the protons in the corresponding volume are accelerated and the density is kept at the critical value, then the number of accelerated protons is proportional to λ. If coupled with a high repetition rate, long wavelength pulses may offer a further advantage to reach the doses required by therapy.

In the present note we shall review the basic mechanisms for laser acceleration to present the related scaling laws and compare the results one expects from small (1 μ) and large (10 μ) wavelength pulses. Systematic 2D and 3D simulations were performed with the high order PIC code `ALaDyn` [Benedetti et al. (2008)] developed by the university of Bologna to provide quantitative results in addition to the qualitative results of scaling laws. We shall also discuss the transport of a protons beam through an optical system.

The paper consists of six sections: after this introduction, in section 2 we recall the basic features and parameters of the laser beam, in section 3 the TNSA regime is reviewed, in section 4 the RPA regime is presented, in section 5 the acceleration on under-critical target is discussed, in section 6 we discuss the transport of the optically accelerated proton bunch, in section 7 we analyze the perspectives for therapy.

2. Laser beam interaction with matter

A laser pulse is described by an electromagnetic wave packet, which is a solution of Maxwell's equations

$$\begin{aligned} \operatorname{rot}\mathbf{E} &= -\frac{1}{c}\frac{\partial}{\partial t}\mathbf{B} & \operatorname{div}\mathbf{B} &= 0 \\ \operatorname{rot}\mathbf{B} &= \frac{4\pi}{c}\mathbf{j} + \frac{1}{c}\frac{\partial}{\partial t}\mathbf{E} & \operatorname{div}\mathbf{E} &= 4\pi\rho \end{aligned} \tag{1}$$

in the vacuum where source charges and currents are absent $\rho = 0$, $\mathbf{j} = 0$. The sources arise when the pulse propagates in material medium creating a plasma. The scalar and vector potential defined by

$$\mathbf{B} = \mathrm{rot}\,\mathbf{A} \qquad \mathbf{E} = -\mathrm{grad}\,\Phi - \frac{1}{c}\frac{\partial}{\partial t}\mathbf{A} \tag{2}$$

identically satisfy he first two equations. Choosing a gauge such that

$$\mathrm{div}\,\mathbf{A} + \frac{1}{c}\Phi = 0 \tag{3}$$

the last two equations become the wave equations

$$\begin{aligned} \Delta\mathbf{A} - \frac{1}{c^2}\frac{\partial^2}{\partial t^2}\mathbf{A} &= -\frac{4\pi}{c}\mathbf{j} \\ \Delta\Phi - \frac{1}{c^2}\frac{\partial^2}{\partial t^2}\Phi &= -4\pi\rho \end{aligned} \tag{4}$$

In the vacuum $\rho = 0$ and we may choose $\Phi = 0$. The gauge equation in this case reads div $\mathbf{A} = 0$ and the electric field is given by $\mathbf{E} = -c^{-1}\partial\mathbf{A}/\partial t$. If we consider a two dimensional wave propagating in the x, z plane we may set $A_x = \partial A/\partial z$ and $A_z = -\partial A/\partial x$ where $A = A(x, z)$ so that div $\mathbf{A} = 0$ is identically satisfied. In this case the electric field is given by

$$E_x = -\frac{1}{c}\frac{\partial}{\partial t}\frac{\partial}{\partial z}A \qquad E_y = 0 \qquad E_z = \frac{1}{c}\frac{\partial}{\partial t}\frac{\partial}{\partial x}A \tag{5}$$

and the magnetic field

$$B_x = 0 \qquad B_y = \left(\frac{\partial^2}{\partial x^2} + \frac{\partial^2}{\partial z^2}\right)A \qquad B_z = 0 \tag{6}$$

Given an initial Gaussian wave field specified by

$$A(x,z,0) = A_0(x,z) \equiv \frac{e^{-x^2/2\sigma_x^2}}{\sqrt{2\pi\sigma_x^2}}\frac{e^{-z^2/2\sigma_z^2}}{\sqrt{2\pi\sigma_z^2}}\cos(k_0\,z) \tag{7}$$

The evolution at time t is obtained by computing its Fourier transform and by taking into account that A satisfies the wave equation

$$\Delta A - \frac{1}{c^2}\frac{\partial^2}{\partial t^2}A = 0 \tag{8}$$

The result for the propagating wave packet is given by

$$A(x,z,t) = \frac{1}{(2\pi)^2}\int_{-\infty}^{+\infty} dk_x \int_{-\infty}^{+\infty} dk_z\, e^{-\sigma_x^2 k_x^2/2}\, e^{-\sigma_z^2 (k_z - k_0)^2/2}\cos(xk_x + zk_z - \omega t) \tag{9}$$

When the wave interacts with a medium it ionizes it if the intensity is sufficiently high and the charged particles move according to the equations of motion

$$\frac{d\mathbf{r}}{dt} = \frac{\mathbf{p}}{m\gamma} \qquad \frac{d\mathbf{p}}{dt} = e\mathbf{E} + \frac{e}{mc\gamma}\mathbf{p}\times\mathbf{B} \tag{10}$$

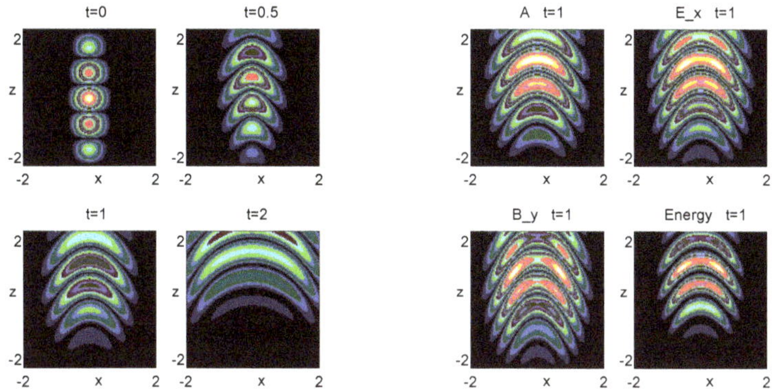

Fig. 1. Gaussian wave packet for the function A defined by eq. (7, 9), for $\sigma_x = 0.25$, $\sigma_z = 1$, $k_0 = 4$, $c = 1$, at different times (left figures). Electric field E_x, magnetic field B_y and energy at time $t = 1$ (right figures).

where e is the electric charge and γ is the relativistic factor

$$\gamma = \left(1 + \frac{\mathbf{p}^2}{m^2c^2}\right)^{1/2}$$

Given N particles we introduce the phase space density $f(\mathbf{r}, \mathbf{p}, t)$, which evolves according to the Liouville equation

$$\frac{\partial f}{\partial t} + \frac{\mathbf{p}}{m\gamma}\frac{\partial f}{\partial \mathbf{r}} + \left(e\mathbf{E} + \frac{e}{mc\gamma}\,\mathbf{p} \times \mathbf{B}\right)\frac{\partial f}{\partial \mathbf{p}} = 0 \tag{11}$$

where the fields are solution of the Maxwell's equations, the sources being defined by

$$\rho(\mathbf{r}, t) = e \int f(\mathbf{r}, \mathbf{p}, t)\, d\mathbf{p} \qquad \mathbf{j} = e \int \frac{\mathbf{p}}{m\gamma} f(\mathbf{r}, \mathbf{p}, t)\, d\mathbf{p} \tag{12}$$

The set (1), (11), (12) forms the Maxwell-Vlasov equations and provides the dynamical setting to investigate the laser plasma interaction. In actual computations the number N of numerical particles, used to sample the phase space density, is considerably lower than the number N_{ph} of physical particles and the masses and charges are N_{ph}/N times larger with respect to the masses and charges of the physical particles.The equations of motion are not affected since they depend only on the ratio e/m. The computation of charge densities and currents requires an interpolation procedure with smooth functions like splines,

2.1 Basic parameters

The basic parameter of the electromagnetic wave is a dimensionless quantity which gives the ratio between the electromagnetic energy and the electron (or proton) rest mass energy

$$a = \frac{eA}{mc^2} \qquad a_p = \frac{eA}{m_pc^2} \tag{13}$$

When $a \sim 1$ the electron becomes relativistic since the energy acquired from the wave is comparable with its rest energy. The intensity is related to the electromagnetic energy by

$$\frac{I}{c} = \frac{B^2 + E^2}{8\pi} = \pi \frac{A^2}{\lambda^2} \tag{14}$$

and consequently letting $r_c = e^2/(mc^2) = 3\,10^{-13}$ cm be the classical electron radius we have

$$a^2 = \frac{r_c}{\pi\, mc^3}\,\lambda^2 I = \frac{I}{mc^3 n_c} \tag{15}$$

where n_c is the critical density defined below (19). A frequently used formula follows

$$a = 0.85\ 10^{-9} I^{1/2}(\mathrm{W/cm^2})\,\lambda(\mu\mathrm{m}) \tag{16}$$

The most relevant plasma parameter is the electron density which determines the plasma oscillations frequency ω_p

$$\omega_p^2 = \frac{4\pi e^2 n}{m} = 4\pi r_c c^2 n \tag{17}$$

where $n = \rho/e$ is the electron density, and ρ is the charge density. The plasma is an active optical medium and its refraction index is

$$n_{\mathrm{refr}} = \left(1 - \frac{\omega_p^2}{\omega^2}\right)^{1/2} \tag{18}$$

For $\omega_p < \omega$ the medium is transparent. The density n_c at which the medium becomes opaque is called critical density and is given by $\omega_p = \omega$ namely

$$n_c = \frac{\omega^2}{4\pi c^2 r_c} \simeq \frac{\pi}{\lambda^2 r_c} = \frac{10^{21}}{\lambda^2(\mu\mathrm{m})}\ \mathrm{cm}^{-3} \tag{19}$$

When the plasma is overcritical the wave becomes evanescent and decays exponentially. The decay length, called skin depth, is given by

$$\ell_s = \frac{\lambda}{2\pi}\left(\frac{\omega_p^2}{\omega^2} - 1\right)^{-1/2} = \frac{\lambda}{2\pi}\left(\frac{n}{n_c} - 1\right)^{-1/2} \tag{20}$$

If we assume the waist to be a fixed multiple of the wavelength, $w = \kappa\lambda$ then a^2 is proportional to the pulse power and does not depend on the wavelength. On the contrary the critical density in proportional to λ^{-2}, and is $n_c = 10^{-19}$ cm^{-3} for a CO_2 laser pulse. This means that a gas jet, from which a plasma density in the range of $10^{18} \div 10^{20}$ cm^{-3} can be obtained, provides a medium with quasi critical electron density. Moreover since the pulse length is in the range of hundreds of microns, the millimetric thickness of a gas jet is adequate for protons acceleration. Since the pulse durations for a Ti:Sa and CO_2 laser are 30 fs and 1 ps respectively, in order to have the same power the ratio of the energies must be 1/30. For the same power and the same value of a, the proton energy should be the same but their number would be higher for a CO_2 pulse.

3. The TNSA regime

This regime is observed when the laser beam interacts with a metallic foil whose electron density is largely overcritical $n \gg n_c$, the thickness h of the foil is large with respect to the skin depth $h \gg \ell_s$ and the polarization is linear. When the laser pulse interacts with the overcritical plasma it is reflected by the target. A sizable fraction of its energy can be absorbed by the electrons of the plasma by linear or nonlinear mechanisms. These electrons can travel through the target and expand around both the front and the rear side. The thickness of the electron cloud is estimated equal to some Debye lengths λ_D where $\lambda_D^2 = k_B\Theta/(4\pi e^2 n)$ and Θ refers to the temperature of the electrons whose density is n. The protons which are present on both surfaces as part of the contaminants deposited on the target (hydrocarbons, water vapors) are accelerated by the electrostatic field built up by the expanding electron cloud preferentially along the normal to the surfaces. If the laser contrast is high enough the front side of the target may be preserved intact until the main part of the pulse interacts with the plasma and a substantially symmetric acceleration takes place in both forward direction (from the rear side) and backward direction (protons from the front side). On the other hand, if the front surface is destroyed by the prepulses or by the laser pedestal, only the forward acceleration is observed.

We can consider the motion of a single electron in a plane e.m. wave. Assuming that the propagation is along the z axis and the vector potential has only the A_y component, defining $a = eA_y/(mc^2)$ the generalized momenta are $P_x = 0$, $P_y = p_y - mca$, $P_z = p$. If we have a particle in an external field then $P_y = 0$ because it is so initially and it is conserved. When collective effects are present we consider a fluid approximation assuming that $\langle P_y \rangle = 0$ which implies $\langle p_y \rangle = mca$. The longitudinal motion is a coherent one given by the ponderomotive force, whereas the transverse one is random and we may assume the temperature Θ to be given by

$$k_B\Theta = T = mc^2[(1+a^2)^{1/2} - 1] \tag{21}$$

so that for $a \gg 1$ we have $k_b\Theta \simeq a\, mc^2$. As a consequence the Debye length λ_D is given by

$$\lambda_D^2 = \frac{k_B\Theta}{mc^2}\,\frac{1}{4\pi\, r_c\, n} = \frac{a}{4\pi\, r_c\, n} \tag{22}$$

where n_0 denotes the electron density. An estimate of the electrostatic field is obtained by supposing the electrons charge distribution obeys Boltzmann statistics so that $n = n_0\, e^{eV/(k_B\Theta)}$ and V satisfies the Poisson-Boltzmann equation

$$\frac{d^2V}{dz^2} = 4\pi\, n_0\, e\, \exp\left(eV/(k_B\Theta)\right) \tag{23}$$

which must be solved with the conditions $V(h) = V'(h) = 0$. The solution is given by

$$V = \frac{T}{e}\log\left(1 + \tan^2\left(\frac{h-z}{\lambda_D\sqrt{2}}\right)\right) \tag{24}$$

The longitudinal electric field is given by $E_z = -V'(z)$; the maximum protons energy $E_{\rm max}$ is simply the potential energy at the origin and can be expressed as

$$E_{\max}(\text{MeV}) = \frac{E_{\max}}{2mc^2} \simeq \frac{a}{2}\log\left(1+\tan^2\sqrt{2}\right) \simeq 2a \tag{25}$$

This result is compatible with experiments which show a linear dependence of $E_{\max}$ with $I^{1/2}$. However for very short and very collimated laser pulses the experimentally observed scaling law is $E_{\max} \propto I^{0.8}$. More refined theoretical models agree with this scaling law [Zani et al. (2011)].

The efficiency of TNSA acceleration can be enhanced if the efficiency of the energy transfer from laser to the target can be increased. If the laser interacts with a near critical density plasma the energy coupling of the laser with the target is considerably increased, comparing with the case of highly overcritical plasma, and a higher number of "hot" electrons can be obtained. The pre-pulse induced ionization creates a pre-plasma and improves the energy transfer from the laser to the electrons letting the laser interact with a plasma at lower density. The characteristics of the preplasma are not easily controlled in a metallic target being the control of the laser-pedestal and pre-pulses very difficult. However a different design of the target may be considered where a foam layer is deposited on the thin metal foil [Nakamura, Tampo, Kodama, Bulanov & Kando (2010)]. Such a target leads to a considerably greater laser energy absorption and to possibly an improved control of the laser target interaction. Systematic 2D and 3D PIC simulations have shown that the presence of a foam increases the maximum protons energy..

In figure 2 we show a comparison of the results obtained from a fully 3D simulation of the interaction of a laser beam with $a = 10$ with a thin metal foil with and without the coating of a slightly overcritical foam layer. The saturation of the maximum energy in the considered time interval is quite evident and the gain with the foam layer is almost a factor 3.

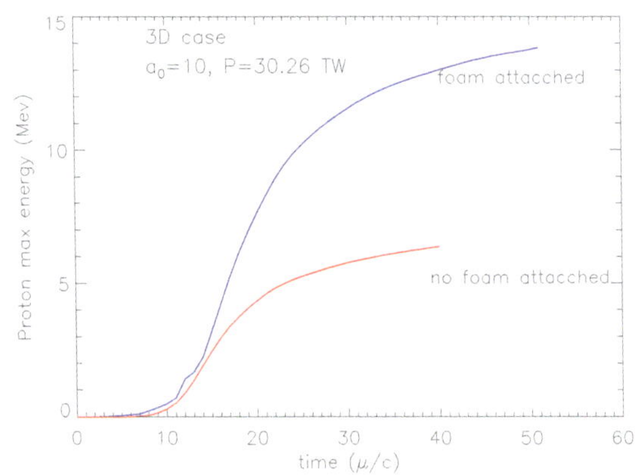

Fig. 2. Comparison of the maximum proton energy rise with time for a metal foil (red curve) with a foil on which a foam is superimposed (blue). The laser pulse has $\lambda = 0.8\,\mu$m, its duration is 25 fs, the waist is 3 μm and the power is $W = 30$ TW so that $a_0 = 10$. The foil is $0,5\,\mu$m thick with a $n = 40\,n_c$ whereas the foam layer is 2μ thick and $n = 2n_c$. The polarization is linear and the incidence is normal. The accelerated protons come form the contaminants layer which is modeled as an ultra-thin H layer of 50 nm at density $n = 9n_c$.

The main feature of TNSA is that the energy spectrum has an exponential decay with a clear cut-off. Letting $N(E)$ be the number of protons having energy in the range $[0, E]$ the spectrum is given by

$$\frac{dN}{dE} = \frac{N_0}{E_0} e^{-E/E_0} \quad 0 \le E < E_{\max} \qquad \frac{dN}{dE} = 0 \quad E > E_{\max} \tag{26}$$

where $E_{\max}$ is the maximum energy where the exponential distribution is cut off. If $E_0 \ll E_{\max}$, so that the error in normalization while replacing the integration upper bound $E_{\max}$ with ∞ is negligible, then E_0 is precisely the average energy and N_0 is the total protons number. For instance suppose we have $E_{\max} = 20$ MeV and and $E_0 = 2$ MeV with $N_0 = 10^{12}$. Such a number would be obtained with a 6 J pulse if 0.3 J are transferred to the protons. We should notice that the number of protons with a narrow energy in the range $[E, E + \Delta E]$ would be

$$N([E, E + \Delta E]) \simeq N_0 \frac{\Delta E}{E_0}\, e^{-E/E_0} \tag{27}$$

and consequently choosing $E = 10$ MeV, $\Delta E = 0.1$ MeV we would have $N = 3\,10^8$ protons. The angular spread of the protons is important, so that after collimation with an iris in order to allow focusing with a quadrupole or a solenoid the number would be further reduced, possibly below 10^7, which is rather low (but might be acceptable) in view of possible applications.

In figure 3 we show the energy spectrum for a metal foil and a foam layer whose parameters are the same as in figure 2. The maximum energy is $E_{\max} \simeq 14$ MeV and the average energy is 1.8 MeV.

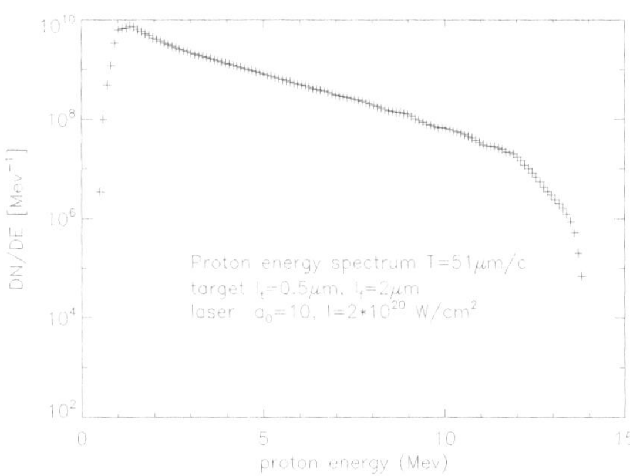

Fig. 3. Energy spectrum for metal foil with foam and the same parameters and the same laser pulse as figure 2.

In figure 4 we show the scaling of the maximum energy and average energy at different a. In this case, the average E is not computed on all the particles, but instead only on a subset that linearly fits the logarithmic plot, where the beginning and the end of the spectrum are excluded. It turns out that the average energy, in 3D simulations, is $\sim 1/7$ of the maximum energy, both for the bare target and a target with a foam.

We noticed that a good linear fit holds in both cases and that for 2D simulations the maximum value of energy $E_{\max}$ is about twice the energy obtained for 3D simulations.

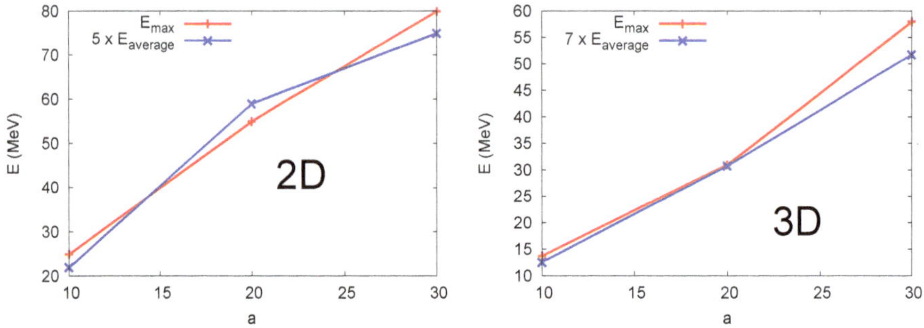

Fig. 4. Scalings for the maximum and average energies computed from linear fit on a logarithmic plot for 2D and 3D simulations for a target with a metal foil 0.5 μm thick and density $n = 80n_c$, a foam layer of 2 μm and density $n = 2n_c$, a layer of contaminants of 50 nm and density of $n = 9n_c$.

3.1 CO_2 results

Experiments in Brookhaven with 1 TW CO_2 laser pulses have shown that protons with maximum energy of 1 MeV can be obtained [Pogorelsky (2010b)]. The scaling laws obtained from both experiments and theoretical/numerical work does not offer significant perspectives to reach proton energies interesting for hadron therapy purposes unless powers in the PW range can be reached. As a consequence TNSA accelerated protons might be proposed for medical applications only coupled with a post acceleration device. Even in this case energies above 10 MeV should be reached. The use of targets with a coating of of 5-10 microns of a silicon foam with near critical density on the illuminated surface may increase the maximum energy by a significant factor (2-4) as simulations and experiments have shown [Nakamura, Tampo, Kodama, Bulanov & Kando (2010)]. As a consequence the injection energy may be reached with compact lasers. This improved type of targets may be used for CO_2 laser pulses as well. The most promising scenarios however are met when quasi critical are used, typically provided by a gas jet. In this case however we face a different acceleration regime which is dominated by the radiation pressure if the electron heating is modest as in the case of a circular polarization.

4. The RPA regime

The radiation pressure becomes the dominant mechanism in the acceleration of protons when $I > 10^{23}$ W/cm^2. However the effect of radiation pressure prevails over electrostatic acceleration even at lower energies when the electrons heating is decreased using a circularly polarized light. Given a vector potential **A** the relativistic Hamiltonian of a charged particle is

$$H = mc^2(\gamma - 1) \qquad \gamma = \left(1 + \frac{\mathbf{p}^2}{m^2c^2}\right)^{1/2} \qquad \mathbf{p} = \mathbf{P} - \frac{e}{c}\mathbf{A} \tag{28}$$

where $\mathbf{p}$ and $\mathbf{P}$ are the ordinary and generalized momenta. For a wave propagating on the z direction $\mathbf{A} = \mathbf{A}(z - ct)$ an averaging procedure with respect to the explicit time dependence of the Hamiltonian can be carried out. For a plane wave the time average vanishes $\langle \mathbf{A} \rangle = 0$. As a consequence under suitable conditions it can be easily shown that

$$\langle H \rangle \simeq mc^2 \left(1 + \frac{\mathbf{P}^2 + \frac{e^2}{c^2} \langle \mathbf{A}^2 \rangle}{m^2 c^2} \right)^{1/2} \tag{29}$$

This is the case in the non relativistic limit since the Hamiltonian becomes quadratic or if the vector potential has a single component $\mathbf{A} = A_y(z - ct)\, \mathbf{e}_y$. In this case supposing $p_y^2/(p_x^2 + p_z^2) \ll 1$ averaging after a first order expansion in this variable shows that (29) is still valid. As a consequence the equations of motion read

$$\frac{d\mathbf{r}}{dt} = \frac{\mathbf{P}}{\gamma m} \qquad \frac{d\mathbf{P}}{dt} = -\frac{e^2}{2mc^2 \gamma} \operatorname{grad} \langle \mathbf{A}^2 \rangle = -\frac{mc^2}{2\gamma} \operatorname{grad} \langle \mathbf{a}^2 \rangle \tag{30}$$

The averages of generalized and ordinary momentum are the same and the gradient of the squared electromagnetic potential is the ponderomotive force in this approximation.

This force dominates when the laser pulse is circularly polarized, because the electrons heating is strongly suppressed. The features of the acceleration mechanism depend on the target geometry.

4.1 Hole Boring

For targets of thickness comparable with the pulse wavelength, we have the *hole boring* regime. The electrons density wave brakes on a distance comparable with the skin depth and the fluid approximation is no longer applicable. A piecewise constant approximation leads to a linear electric fields and the protons maximum energy can be easily evaluated. The expression one obtains is

$$E_{\text{max}}(\text{MeV}) = \frac{E_{\text{max}}}{2mc^2} \simeq a^2 \, \frac{n_c}{n} \tag{31}$$

and for an ion of charge Z the energy is multiplied by Z. The angular spread of the beam is significantly lower with respect to the TNSA acceleration but the factor n_c/n strongly reduces the energy value. Even though the scaling with a is quadratic rather than linear, only at very high values of the intensity the energy overcomes the value reached with TNSA. The only solution is to avoid the use of solid targets and work with a near critical density. This condition is naturally met for a wavelength of 10 μm since the critical density $n_c = 10^{19}$ cm^{-3} is met on gas jets. An experiment recently performed confirms that such a regime can be met and protons with a narrow energy spectrum can be accelerated [Palmer et al. (2010); Pogorelsky (2010a)].

4.2 Relativistic mirror

For ultra-thin targets of thickness comparable with the skin depth the radiation pressure is able to push all the electrons of the foil which create a huge charge separation and all the ions are promptly accelerated. The result is that the target is practically accelerated as a whole as a rigid object behaving like a mirror whose equations of motion are

$$\frac{dx}{dt} = c\beta \qquad \frac{d\beta}{dt} = \frac{2I}{\mu c^2}\frac{1-\beta}{1+\beta}\,(1-\beta^2)^{3/2} \tag{32}$$

where μ is the surface density of the mirror and $I = I_0 f\left((t - x/c)/\tau_{\text{laser}}\right)$ is the laser pulse. The function $f(s)$ vanishes except for $|s| < 1$ where it is positive. The equations of motion 30 have a first integral of motion. Setting

$$t' = \frac{t}{\tau_{\text{laser}}}, \qquad x' = \frac{x}{c\tau_{\text{laser}}}, \qquad w = t' - x' \qquad \chi = \frac{2I_0\,\tau_{\text{laser}}}{\mu\, c^2} \tag{33}$$

the equations of motion become

$$\frac{d\beta}{dt'} = \chi\, f(w)\,\frac{1-\beta}{1+\beta}\,(1-\beta^2)^{3/2} \qquad \frac{dw}{dt'} = 1 - \beta \tag{34}$$

and introducing the integrating factor $C = (1+\beta)(1-\beta)^{-1}(1-\beta^2)^{-3/2}$ the differential form $dH = C\left[(1-\beta)d\beta - \chi\, f(w)\ \ (1-\beta^2)^{3/2}(1-\beta)/(1+\beta)\,dw\right]$ becomes exact and the first integral is

$$H = \chi\int_{-1}^{w} f(w')dw' - \left(\frac{1+\beta}{1-\beta}\right)^{1/2} \tag{35}$$

The initial condition corresponds to $w = -1$, $\beta = 0$ so that $H = -1$. At the end of the pulse we have $w = 1$ and $\beta = \beta_*$ which is the highest speed value. Denoting by $F = \int_{-1}^{1} f(w)dw$ the fluence we have $\chi F - [(1+\beta_*)/(1-\beta_*)]^{1/2} = -1$. We obtain β_* and γ_* as a function of $\alpha = \chi F$ which is given by

$$\alpha = \chi F = \frac{2I_0\,F\,\tau_{\text{laser}}\,S}{\mu\,c^2\,S} = \frac{2E_{\text{laser}}}{E_{\text{rest mirr}}} \tag{36}$$

Expressing $\gamma_* = (1-\beta_*^2)^{-1/2}$ as a function of α we obtain the expression for the kinetic energy of the ion which is given by

$$E_{\text{max}} = Am_pc^2(\gamma_* - 1) = Am_pc^2\frac{\alpha^2}{2+2\alpha} = \frac{E_{\text{laser}}}{N}\,\frac{\alpha}{1+\alpha} \tag{37}$$

where (36) has been used taking into account $E_{\text{rest mirr}} = NAm_pc^2$ where N is the number of ions in the mirror. As a consequence the efficiency of the acceleration process is given by

$$\eta = \frac{E_{\text{mirr}}}{E_{\text{laser}}} = \frac{\alpha}{1+\alpha} \tag{38}$$

From equation (36) it appears that the thinner is the mirror the higher is the efficiency and the protons energy because $\mu = AZ^{-1}\,m_p\,h\,n$. However a limit is imposed by the transparency limit Macchi et al. (2009). The target remains opaque and is accelerated as a mirror provided that

$$a \le \zeta = \pi\frac{n}{n_c}\,\frac{h}{\lambda} \tag{39}$$

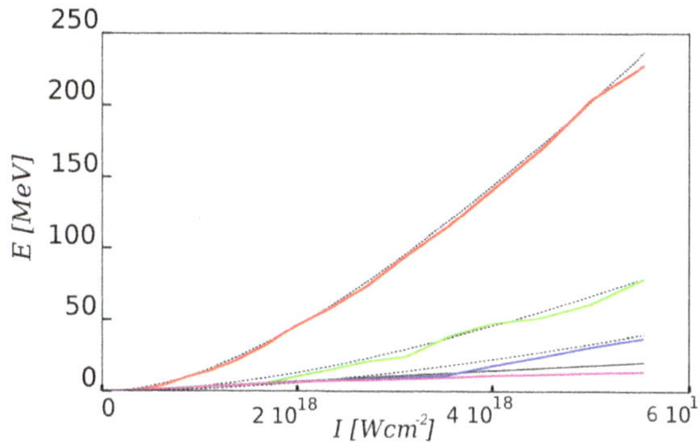

Fig. 5. Protons energy in MeV versus intensity for a pulse with $\lambda = 10\mu$m. various foil thickness are considered: $\ell = 5\mu$m (red), $\ell = 10\mu$m (green), $\ell = 15\mu$m (Blue), $\ell = 30\mu$m (purple). The 1D PIC results are compared with the analytical results for the RPA hole boring (solid black line) and the relativistic mirror (dotted line).

where n denotes the electrons density. The number N of ions in the target and the minimum number N_* below which transparency is induced and the corresponding minimum thickness h_* of the target are given by

$$N = \frac{n}{Z} h\, S = \frac{S}{Z}\,\frac{\zeta}{\lambda r_c} \qquad N_* = \frac{S}{Z}\,\frac{a}{\lambda r_c} \qquad h_* = \frac{N_*}{n\, S} = \frac{a}{Z\,\lambda\, r_c\, n_e} \tag{40}$$

In this regime, keeping the ratio h/λ fixed, the volume of the accelerated ions increases as λ^3, their number as λ if n/n_c is also kept fixed. As a consequence, at the transparency limit the parameter a and the laser power have a fixed value. The parameter α is constant if the energy is kept proportional to λ. In this case the total energy and the proton number increase with λ, whereas the energy of each proton does not vary. If the laser energy and the pulse duration both increase with λ the power remains constant. In Figure 5 we compare the analytic scaling provided by equations (31) and (37) with 1D PIC simulations for a circularly polarized 10μm pulse. The transition between these regimes is evident and the agreement between the analytical and numerical result is quite good. The RPA regime is well suited for CO_2 lasers because the native polarization is circular. Slightly overcritical targets are provided by gas jets and in the hole boring regime the choice $n \simeq n_c$ is necessary in order to achieve high energies, according to (31). In addition a quasi monochromatic spectrum is obtained at moderate intensities typical of CO_2 lasers. In this case the hole boring scenario is the most promising as recent experiments have shown [Palmer et al. (2010); Pogorelsky (2010a)].

The relativistic mirror model is very attractive on the basis on the analytical results obtained from the 1D model. However 2D and 3D simulations show that a deterioration occurs when the dimensionality is increased, due to the onset of Rayleigh-Taylor like instabilities. Considering also the difficulties met in the preparation of ultrathin targets in order to be close to the transparency limit and the requirements on the contrast, this regime is not likely to be interesting for application, even using long wavelength pulses, in a near future.

5. Near critical targets

The acceleration of protons on targets with a nearly critical electron density has been recently investigated on several experiments and PIC simulations. Positive features of this type of targets are a better efficiency in the energy transfer from the laser to the electrons and the absence of debris since the medium is transparent. The optimal length of the target is a few times the length of the laser pulse which drills a channel and accelerates longitudinally the electrons which create a magnetic field circulating around the beam axis. The magnetic field moves behind the laser pulse until it exits in the vacuum where it expands; the electrons, whose energy is dissipated, are displaced by the magnetic field and create a quasi static electric field. The ions coming from the filament around the axis are accelerated and collimated. More specifically the mechanism controlling the proton acceleration is provided by the formation of a slowly evolving magnetic dipole (a toroidal configuration in 3D geometry) behind the leading laser pulse. This structure is generated by the coherent return axial current due to the accelerated electron beam, which contains a large fraction of the laser pulse energy. The magnetic vortex, when exiting on a low density (or a vacuum) region, expands symmetrically thus creating a strong induction axial electric field. At higher electron density $n_c < n < 3n_c$ this mechanism is the most effective in the acceleration process. At lower density $n < n_c$, a significant contribution comes also from the electrostatic field due to charge separation at the channel rear side, much alike the TNSA regime.

The maximum energy of protons depends on the target thickness and density and a scaling law is obtained by equating the laser energy to the electrons energy, following the waveguide model, provided that the length of the plasma channel h is much larger than the length of the laser pulse $h \gg L_p$ to insure that the depletion of the laser energy is complete. Using equation (15), $I = a^2\, mc^3 n_c$, the laser energy in a channel reads

$$E_{\text{laser}} = \pi R^2\, \tau I = \pi R^2 L_{\text{laser}}\, a^2\, mc^2\, n_c \tag{41}$$

The electrons energy is given by

$$E_{\text{el}} = \pi R^2\, h\, n\, a\, mc^2 \tag{42}$$

and equating the energies we obtain

$$a \sim \frac{nh}{n_c L_{\text{laser}}} \tag{43}$$

Another scaling provides the optimal channel length which is given by

$$n \sim h^{3/2} \tag{44}$$

If the transition to the vacuum is not abrupt but the target with electron density n and length h continues with a decreasing density before reaching the vacuum, some improvements on the top energy and the collimation can be obtained. The main advantages with respect to the TNSA regime are that the energies reached are two or three times higher, the collimation is improved and the efficiency is higher. For wavelength in the micron range the problem is to find the right targets. This is solved naturally for pulses with wavelength in the 10 μm range, since gas jets can be used. Indeed promising results have been obtained in recent experiments. We have performed several 2D and 3D PIC simulations of quasi critical targets. The electrons and protons density at two different times exhibiting the formation of the channel are shown in figures 6 and 7 for a 3D simulation. The spectrum of the protons is still exponential and

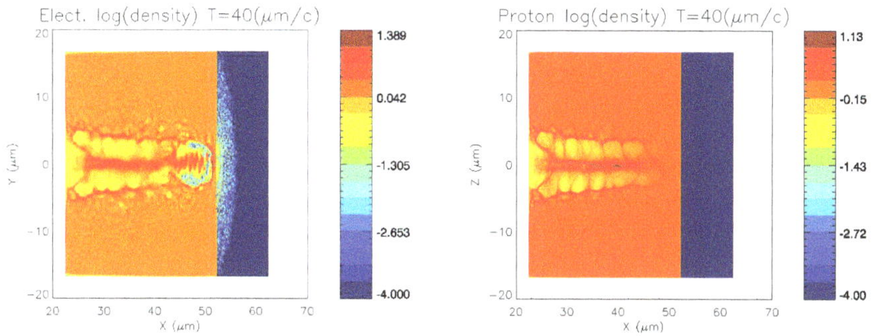

Fig. 6. Electrons (left frame) and protons (right frame) density in the case of a laser pulse of 200 TW, with wavelength $\lambda = 0.8\,\mu$m, pulse duration $\tau = 25$ fs and $a = 32$, incident on a target 30 μm thick of density $n = n_c$ at time $t = 40\mu/c$ from the the beginning of the laser target interaction. The simulation is 3D.

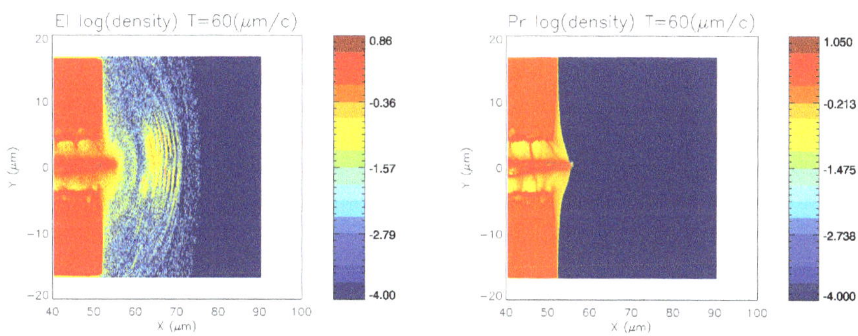

Fig. 7. The same as figure 6 at time $t = 60\mu/c$

the angular dispersion is better than for TNSA. In figure 8 we show the energy spectrum for a 200 TW laser pulse, the same as for figures 6 and 7. When a target is slightly overcritical it is transparent if the intensity is high enough. Indeed one has to take into account the self induced transparency since the critical density becomes

$$n_c = \frac{\pi}{\lambda^2 r_c}\gamma \simeq \frac{\pi}{\lambda^2 r_c}\,a \tag{45}$$

As a consequence by increasing the intensity the medium becomes transparent and the laser drills a channel. If the intensities are moderate this effect is small and a medium at a few times the critical density remains opaque and the hole boring acceleration prevails. Using a CO_2 laser with a gas jet one can have both regimes. By increasing the density at a fixed intensity the medium looses its transparency and the highest protons energy decreases reaches a minimum and then increases because the RPA hole boring regime sets in [Willingale (2009)].

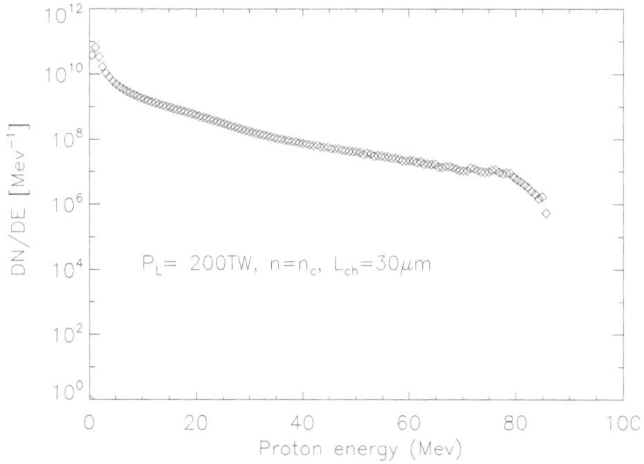

Fig. 8. Energy spectrum for a target at critical density for a laser pulse with $W = 200$ TW for the same parameters as figures 6

6. The transport

The use of protons or ions beams for medical therapy has to face severe constraints, which are not yet met by present laser produced ion bunches. Indeed the energy range is $60 \leq E \leq 250$ MeV, the average overall dose is 60 Gray (1 Gy corresponds to 1 mJ per gram) and the full dose delivery is reached in several treatments. Assuming that the dose for a single proton session is 10 Gy and that it is delivered in 2 minutes with 10 Hz pulses, the number of protons per shot to reach this dose on a 1 g tissue would be 10^6. For a TNSA beam with a maximum energy slightly above 60 MeV, such an intensity is not easily achievable because the energy spectrum is exponential and the average energy is much lower (1/7 in 3D simulations) than the maximum energy. The situation for beams obtained from the interaction with a quasi critical target is more favorable, since the collimation is better and the maximum energy is higher. The energy currently reached with compact Ti:Sa lasers are around 20 MeV. This value has been recently overcome with a CO_2 laser working with a gas target, where a monochromatic protons bunch was produced. As a consequence, presently the conditions to use for therapy the proton beams accelerated by compact high repetition lasers are not yet met. The increase of power from 100-200 TW to 1 PW is likely to allow to reach the threshold of 60 MeV of maximum energy for therapeutic use even though the intensity might be lower than required.

A possible alternative consists in maintaining the energy in the 10-30 MeV range and post-accelerating the beam. Suitable devices have already been developed to accelerate a proton beam coming from a cyclotron in this energy range. The injection energy varies from 10 MeV for a rather large DTL device [Antici (2011)] to 30 MeV for compact high field linacs like ACLIP [Amaldi et al. (2009)]. Preliminary experiments and some simulations have been already carried out. The injection of a beam in a RF cavity at 1 Hz for monochromatizing it has been experimentally proved even though at low energy (2 MeV) [Nishiuchi (2010)]. Several experiments on beam transport have been performed using quadrupoles or solenoids to focus it. A beam with $E \leq 14$ MeV has been transported through a line formed by two collimators and two permanent magnetic quadrupoles [Schollmeier (2008)]. An experiment of transport was performed with a proton bunch accelerated by the laser PHELIX with an energy spectrum up to 30 MeV using quadrupoles or a solenoid, which proved to keep the emittance to a lower value [Hofmann (2009; 2011)]. A proposal was made for a 100 MeV device capable of delivering 10^5 protons per shot on the tissues starting from a 300 TW laser beam and $a = 60$ so that a dose of 40 Gy could be delivered on a target tissue of 0.03 g in 2 minutes at 10 Hz [Sakaki et al. (2009)] using a gantry with quadrupoles and bending magnets. Radiobiology experiments were carried out on cancer cells irradiating them with proton bunches of 0.8-2.4 MeV, obtained from laser accelerations, and the break-up of DNA double strands was observed [Yogo (2009)]. A design study for post-acceleration of a 10 MeV beam into a DTL was carried out showing that with a moderate power laser $W \leq 100$ TW and the use of microlenses right after the interaction region an injectable beam with parameters suitable for therapy could be obtained [Antici (2011)]

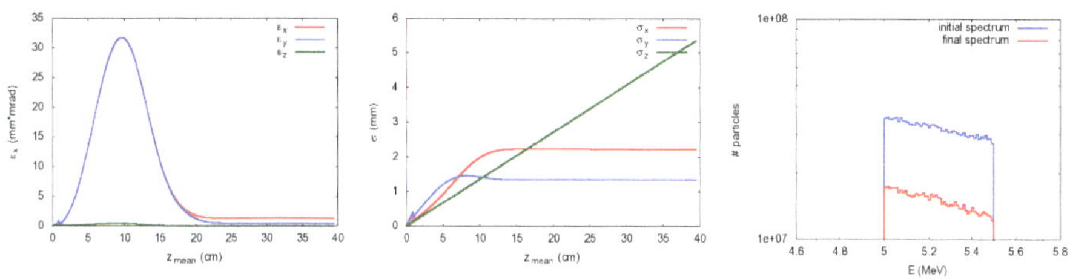

Fig. 9. Left frame: emittances ϵ_x (red), ϵ_y (blue) and ϵ_z (green) for a beam produced with a 30 TW laser pulse transported through an iris with $r = 0.5$ mm and a solenoid, starting right after the iris, of 11.7 cm, $B = 10$ Tesla and $\lambda = 2$ cm, where the B field is described by a function like $B(z) = 1/(1 + e^{-z/\lambda})$. An energy cut is set at $5 < E < 5.5$ MeV. Middle frame: envelopes σ_x (red), σ_y (blue) and σ_z (green). Right frame: initial (blue) and final (red) energy spectra. The laser has the same parameters as figure 2.

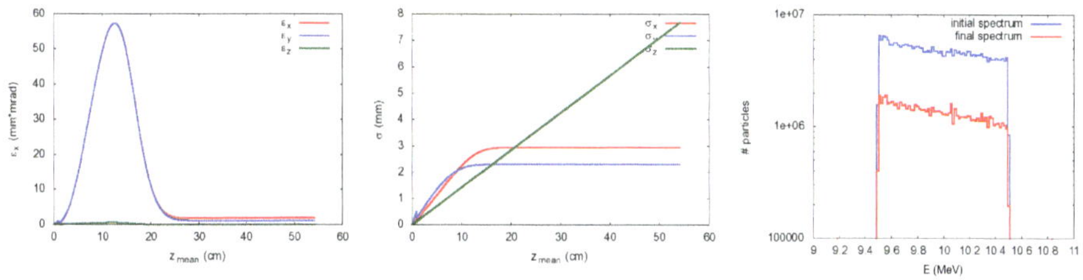

Fig. 10. Left frame: emittances ϵ_x (red), ϵ_y (blue) and ϵ_z (green) for a beam produced with a 30 TW laser pulse transported through an iris with $r = 0.5$ mm and a solenoid, starting right after the iris, of 15.6 cm, $B = 10$ Tesla and $\lambda = 2$ cm. An energy cut is set at $9.5 < E < 10.5$ MeV. Middle frame: envelopes σ_x (red), σ_y (blue) and σ_z (green). Right frame: initial (blue) and final (red) energy spectra

We have simulated the transport of a proton beam produced with a 30 TW laser pulse having a waist of 3 μm so that $I = 2\,10^{20}$ W/cm^2 and $a \simeq 10$. The simulated target was 0.5 μm thick, orthogonal to the z propagation axis, with $n = 40n_c$, with a 2 μm coating with $n = 2n_c$ and a 50 nm layer of contaminants with $n = 9n_c$ on the opposite side. The maximum energy was $\sim$ 14 MeV. We have placed a collimator formed by a screen with a hole of 0.5 mm radius at 1 cm from the interaction region followed by a solenoid.

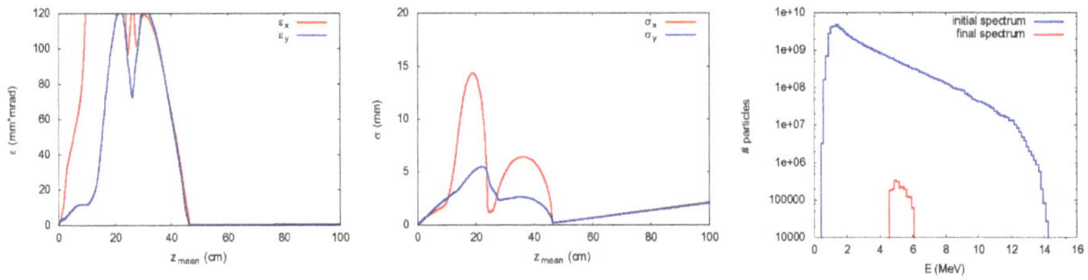

Fig. 11. Left frame: emittances ϵ_x (red) and ϵ_y (blue) for a beam produced with a 30 TW laser pulse transported through a chicane (selecting around 5 MeV), an iris with $r = 0.5$ mm and a solenoid, starting right after the iris, of 11.7 cm, $B = 10$ Tesla and $\lambda = 2$ cm. Middle frame: envelopes σ_x (red) and σ_y (blue) Right frame: initial (blue) and final (red) energy spectra

In figure 9 we show the emittance and the envelope for a small fraction of the bunch, with a cut in the energy spectrum between 5 and 5.5 MeV. We have considered also the transport of the fraction of the beam with an energy between 9.5 MeV and 10.5 MeV. The results are comparable and the emittance and envelopes are shown in Figure 10. An energy selection can be made with a physical device such as a chicane or an RF cavity to achieve the rotation in the longitudinal phase space. The chicane applied to the full bunch leads to an emittance growth along the axis where the dipoles bend the beam. By applying suitable collimators the emittance and envelopes can be reduced to reasonable values but only a small fraction of the beam reaches the end of the transport line. In Figure 11, we show the full bunch propagating along a chicane and then focusing in a solenoid with field of $B = 10$ Tesla.

Other more effective methods based on selection of particles at the desired energy by putting collimator at the corresponding focus point of a solenoid are under investigation.

No transport experiments have been performed with CO_2 laser accelerated proton beams, but the situation is certainly much more favorable if a quasi-monochromatic peak can be obtained with adequate intensity [Haberberger, Tochitsky, Fiuza, Gong, Fonseca, Silva, Mori & Joshi (2011); Haberberger, Tochitsky, Gong & Joshi (2011)]. In this case no energy selection is required and the transport on the beam becomes quite easy because tight focusing can be achieved preventing emittance increase and keeping the beam size to small values suitable for injection in a linac.

7. Conclusions: Protons and ions therapy

Cancer treatment, a priority for health care in advanced countries, is based on surgery, chemo-therapy and radiation therapy. Early detection of tumors increases the survival period but there are limits to massive and frequent screening. Unlike chemotherapy radiation therapy killing of malignant cells is quite homogeneous and can be effective even for massive tumors. The most common treatment is based on X rays, driven by electron accelerators. The reason is compactness of the accelerating device and moderate cost, affordable by medium size hospitals. However most of ionizing radiations, including X rays, exhibit a peak in dose deposition close to the entry point and a subsequent exponential decay. As a consequence the dose deposition in healthy tissues is important, even when this undesirable effect is minimized by irradiating from different directions and modulating the intensity (IMRT). The dose deposition mechanism for protons and ions is different and the dose curve as function of depth exhibits a sharp peak, known as Bragg peak. Healthy tissues are spared and this therapy is applicable to the most severe cases which are not suitable for surgery. The protons are usually accelerated by cyclotrons and a large gantry is needed to rotate the beam around the patient. Carbon ions are accelerated by synchrotrons, which have a larger size, require heavy transport lines and a very large gantry. Even though the number of centers for protons and ions therapy is increasing they are up to now limited to national facilities. A reduction of size and cost of a protons accelerator for therapy would allow the creation of regional centers extending this treatment to a larger fraction of patients. Devices based on innovative techniques such as the superconducting cyclotron or the dielectric wall accelerator have been proposed but conclusive results have not yet been achieved. Laser acceleration of protons has entered this competition even though several years are needed before the feasibility is actually demonstrated. The typical dose for therapy is 60 Gy which means 60 mJ on a mass of 1 g. Split over 6 session this amounts to 10 mJ and corresponds to 10^9 protons of 60 MeV which is the threshold for very superficial tumors. Supposing each bunch contains 10^6 protons at 10 Hz repetition rate this dose is delivered in a couple of minutes. The major problem is that the energy and intensity required can hardly be reached with compact existing Ti:Sa lasers. Suppose that a bunch of 10^{12} protons is accelerated by TNSA having an average energy of 10 MeV and maximum energy of 70 MeV, according to previous scaling, then the total proton energy is 1.6 J and supposing a 10% efficiency the laser energy would be 16 J. Since the spectrum is exponential the fraction of protons at 60 MeV with $\Delta E/E = 1\%$ would be $1.5\,10^8$. This is a very demanding requirement on the laser. If we choose instead an average energy of 5 MeV we would have the same number of protons at an energy of 30 MeV, for the same $\Delta E/E$.

After post-acceleration one could reach not only the 60 MeV threshold but also higher energies, suitable for deep tumors.

The real problem with TNSA or improved TNSA, achieved with a foam deposition at quasi critical density on the illuminated surface, is that the protons having the desired energy are a very small fraction of the total and carry out a very small fraction of the total energy. Increasing the repetition rate form 10 to 100 Hz would help but would not solve the problem. The way out is to produce a quasi monochromatic spectrum. This has been achieved with a CO_2 laser on a gas jet and this result is of extreme interest. The use of ultrathin nanometric targets allows to obtain quasi monochromatic beams via RPA, but in spite of the scientific interest this regime is still very far out from possible applications due to the the extremely high contrast required on the laser beam. The use of gas targets and a suitably shaped beam pulse seem to be the corner stones in the production of a quasi monochromatic beam. In this respect the CO_2 laser beams have an advantage with respect to pulses of shorter wavelength. For a quasi monochromatic beam the intensity is still rather low because only a small fraction of the laser energy is transferred to the beam. Once the energy and intensity requirements are satisfied other conditions have to be satisfied to render the proton beam suitable for therapy: the shot to shot stability must be kept within a narrow range and suitable dose control systems have to be developed [Linz & Alonso (2007)]. As a consequence it will take several years before laser accelerators can meet the requirements for clinical use. During this period the laser performance will be improved, new targets will be developed, transport and post acceleration systems will be tested and beam quality and stability will be pursued. Even though most of the research activity will be devoted to short wavelength lasers, the development of long wavelength system such as CO_2 lasers will hopefully continue because they may provide a very valuable alternative and solve some of the most critical problems such as monochromaticity and proton beam quality.

8. Acknowledgments

We would like to thank the Italian Ministry of Foreign Affairs (MAE) for a grant we received for the scientific cooperation with Japan to develop the research project PROMETHEUS, devoted to a research infrastructure on laser driven proton sources for biomedical applications. We thank the Fondazione del Monte di Bologna e Ravenna for a grant devoted to the feasibility study of a hybrid accelerator devoted to biomedical Research within the framework of PROMETHEUS. We acknowledge the Alma Mater Foundation for the governance of the PROMETHEUS project.

9. References

Amaldi, U., Braccini, S. & Puggioni, P. (2009). High frequency linacs for hadrontherapy, *Reviews of Accelerator Science and Technology* 2(111131).

Antici, P. (2011). A compact post-acceleration scheme for laser-generated protons, *Physics of Plasmas* 18.

Benedetti, C., Sgattoni, A., Turchetti, G. & Londrillo, P. (2008). Aladyn: A high accuracy pic code for the maxwell-vlasov equations, *IEEE Transactions on Plasma Science* 36(4).

Bulanov, S. (2008). Accelerating protons to therapeutic energies with ultraintense, ultraclean, and ultrashort laser pulses, *Medical Physics* 35.

Bulanov, S. S. (2010). Generation of gev protons from 1pw laser interaction with near critical density targets, *Physics of Plasmas* 17(043105).

Fukuda, Y. & Bulanov, S. (2009). Energy increase in multi-mev ion acceleration in the interaction of short laser pulse with a cluster gas target, *Phys. Rev. Letters* 103(165002).

Haberberger, D., Tochitsky, S., Fiuza, F., Gong, C., Fonseca, R. A., Silva, L. O., Mori, W. B. & Joshi, C. (2011). Collisionless shocks in laser-produced plasma generate monoenergetic high-energy proton beams, *Nature Physics* .

Haberberger, D., Tochitsky, S., Gong, C. & Joshi, C. (2011). Production of 25 mev protons in co_2 laser plasma interactions in a gas jet, *Proceedings of 2011 Particle Accelerator Conference, New York, NY, USA* .

Henig, A. (2009). Radiation-pressure acceleration of ion beams driven by circularly polarized laser pulses, *Phys. Rev. Lett.* 103(245003).

Hofmann, I. (2009). Laser accelerated ions and their potential for therapy, *Proceedings of HIAT09, Venice, Italy* .

Hofmann, I. (2011). Collection and focusing of laser accelerated ion beams for therapy applications, *Phys. Rev. Spec. Top. Acc. and Beams* 14(031304).

Leemans, W. P., Nagler, B., Gonsalves, A. J., Tóth, C., Nakamura, K., Geddes, C. G. R., Esarey, E., Schroeder, C. B. & Hooker, S. M. (2006). Gev electron beams from a centimetre-scale accelerator, *Nature Physics* 2: 696–699.

Linz, U. & Alonso, J. (2007). What will it take for laser driven proton accelerators to be applied to tumor therapy, *Phys. Rev. Special Topics Accel. and Beams* 10(094801).

Londrillo, P., Benedetti, C., Sgattoni, A., Turchetti, G. & Lucchio, L. D. (2010). Comparison of scaling laws with pic simulations for protons accelerated with long wavelength pulses, *Nucl. Instr. and Meth. in Phys. Res. A* 620: 51–55.

Londrillo, P., Sgattoni, A., Sumini, M. & Turchetti, G. (2011). Optical acceleration and perspectives for cancer therapy with proton beams, *Proceedings del workshop OSCM - Oncogenesi tra scienza e clinica medica - Frascati 10-11 giugno 2010 - ENEA report* .

Macchi, A., Cattani, F., Liseykina, T. V. & Cornolti, F. (2005). Laser acceleration of ion bunches at the front surface of overdense plasmas, *Phys. Rev. Lett.* 94(165003).

Macchi, A., Veghini, S. & Pegoraro, F. (2009). Light sail acceleration reexamined, *Phys. Rev. Lett.* 103(085003).

Melone, J. J. (2011). In situ characterisation of permanent magnetic quadrupoles for focussing proton beams.
URL: *http://arxiv.org/abs/1104.1932v1*

Mourou, G. A., Tajima, T. & Bulanov, S. V. (2006). Optics in the relativistic regime, *Rev. of Mod. Physics* 78: 309–371.

Murakami, M. (2008). Radiotherapy using a laser proton accelerator.
URL: *http://arxiv.org/abs/0804.3826*

Nakamura, T., Bulanov, S., Esirkepov, T. & Kando, M. (2010). High-energy ions from near-critical density plasmas via magnetic vortex acceleration, *Phys. Rev. Lett.* 105(135002).

Nakamura, T., Tampo, M., Kodama, R., Bulanov, S. V. & Kando, M. (2010). Interaction of high contrast laser pulse with foam-attached target, *Physics of Plasmas* 17(113107).

Naumova, N. M. & Bulanov, S. V. (2002). Polarization and anisotropy in three dimensional relativistic self focusing, *Physical Review E* 65(045402).

Nishiuchi, M. (2010). Measured and simulated transport of 1.9 mev laser-accelerated proton bunches through an integrated test beam line at 1 hz, *Phys. Rev. Spec. Top. Acc. and Beams* 13(071304).

Noda, A. (2008). Quality improvement of laser produced protons by phase rotation and its possible exyension to high energies, *Proceedings of LINAC08, Victoria, BC, Canada* .

Palmer, C. A. J., Dover, N. P., Pogorelsky, I., Babzien, M., Dudnikova, G. I., Ispiriyan, M., Polyanskiy, M. N., Schreiber, J., Shkolnikov, P., Yakimenko, V. & Najmudin, Z. (2010). Monoenergetic proton beams accelerated by a radiation pressure driven shock.
URL: *http://arxiv.org/abs/1006.3163*

Passoni, M. & Lontano, M. (2004). One dimensional model of the electrostatic ion acceleration in altradense laser-solid acceleration, *Laser and particle beams* 22.

Pogorelsky, I. V. (2010a). Laser energy conversion to solitons and monoenergetic protons in near critical hydrogen plasma, *Proceedings of IPAC10, Kyoto, Japan* .

Pogorelsky, I. V. (2010b). Ultrafast co_2 laser technology. application to ions acceleration, *NIM A* 620: 67–70.

Sakaki, H., Hori, T., Nishiuchi, M., Bolton, P., Daido, H., Kawanishi, S., Sutherland, K., Souda, H., Noda, A., Iseki, Y. & Yoshiyuki, T. (2009). Designing integrated laser-driven ion accelerator systems for hadron therapy at pmrc (photo medical research center), *Proceedings of the 23rd Particle Accelerator Conference, Vancouver, Canada* .

Schollmeier, M. (2008). Controlled transport and focusing of laser-accelerated protons with miniature magnetic devices, *PRL* 101(055004).

Snavely, R. A. (2000). Intense high-energy proton beams from petawatt-laser irradiation of solids, *PRL* 85: 2945–2948.

Tajima, T. & Dawson, J. M. (1979). Laser electron accelerator, *Phys. Rev. Lett.* 43(4): 267–270.

Willingale, L. (2009). Characterization of high intensity laser propagation in the relativistic transport regime through measurements of energetic proton beams, *PRL* 102(125002).

Yogo, A. (2008). Laser ion acceleration via control of the near-critical density target, *Phys. Rev. E* 77(016401).

Yogo, A. (2009). Application of laser-accelerated protons to the demonstration of dna double-strand breaks in human cancer cells, *Appl. Phys. Lett.* 94(181502).

Zani, A., Sgattoni, A. & Passoni, M. (2011). Parametric investigations of target normal sheath acceleration experiments, *Nuclear Instruments and Methods in Physics Research A* 653: 94–97.

Zeil, K., Kraft, S. D., Bock, S., Bussmann, M., Cowan, T. E., Kluge, T., Metzkes, J., Richter, T., Sauerbrey, R. & Schramm, U. (2010). The scaling of proton energies in ultrashort pulse laser plasma acceleration, *New Journal of Physics* 12(045015).

17

Non-Thermal, Non-Ablative CO_2 Laser Therapy (NACLT): A New Approach to Relieve Pain in Some Painful Oral Diseases

Nasrin Zand
Iranian Center for Medical Lasers (ICML), Academic Center for Education, Culture and Research (ACECR), Tehran Iran

1. Introduction

CO_2 laser has been used as a very useful device in surgery for ablation, coagulation and cutting the tissues for the last four decades. It is interesting to know that this high power laser can also be used as a therapeutic laser for immediate pain reduction in some oral lesions without any visible side effects such as ulceration, erosion formation and even erythema.

Recently few case reports and clinical trials have been published about using CO_2 laser in non-ablative manner to reduce pain in oral lesions. In these studies, the oral painful lesions were irradiated through a layer of transparent, non-anesthetic gel with high water content to reduce the beam absorption by the soft tissue. The patients reported immediate and significant pain relief after laser irradiation. The procedure was painless and anesthesia was not required. This technique was called non-thermal, Non-Ablative CO_2 Laser Therapy (NACLT). The results of powermetry and thermometry demonstrated the low power nature of NACLT. However there are some differences between analgesic effects of NACLT and the other classical low power lasers which will be discussed in the next sections.

To provide a comprehensive understanding of NACLT, this chapter is organized in several sections. First, due to low level therapeutic nature of NACLT, conventional low power therapeutic lasers, their biological effects and their pain relieving properties are reviewed. Then, a discussion about the interesting analgesic effects of CO_2 lasers is presented. In the next section, NACLT as a new low level laser therapy procedure and its pain relieving applications in painful oral lesions is discussed. Finally, the presumed mechanisms of analgesic effects of NACLT are covered.

2. Low level laser therapy (laser phototherapy)

2.1 History

Low-level laser (or light) therapy (LLLT) has been investigated and used clinically for over 40 years. However, it is only in relatively recent times that LLLT has become scientifically and clinically accepted by even a fraction of the medical community (Hamblin 2010).

The history of the use of laser phototherapy in medicine goes back to the late 1960s, only eight years after the invention of the first laser (Ruby laser) by Theodore Maiman. In 1967, Endre Mester in Semmelweis University, Budapest, Hungary decided to test if laser radiation might cause cancer in mice. He shaved the dorsal hair of the mice, divided them into two groups and irradiated the shaved areas with a low powered ruby laser (694-nm) in one group. They did not get cancer and to his surprise the hair on the treated group grew back more quickly than the untreated group. This was the first demonstration of "laser biostimulation" (Hamblin, Waynant et al. 2006).

In early 1960's, the first low level laser, Helium-Neon was developed by Professor Ali Javan. It emits visible, red light with a wavelength of 632.8nm. This low power laser has been used extensively in experimental and therapeutic studies. Today, the semiconductor lasers, including InGaAlP lasers (633-700nm), GaAlAs lasers (780-890nm, invisible, near infrared area), GaAs laser (904nm, invisible, near infrared area) are widely used by researchers and clinicians.

LLLT originally thought to be a peculiar property of laser light (soft or cold lasers); the subject has now broadened, using non-coherent light (light-emitting diodes, LEDs). Today, medical treatment with coherent-light sources (lasers) or noncoherent light (LEDs) has passed through its childhood and adolescence (Hamblin, Waynant et al. 2006). Currently, low-level laser (or light) therapy (LLLT) is practiced as part of physical therapy in many parts of the world. Although LLLT was used mainly for wound healing and pain relief, the medical applications of LLLT have broadened to include diseases such as stroke, myocardial infarction, and degenerative or traumatic brain disorders (Hashmi, Huang et al. 2010).

Although many experimental and clinical studies have reported the positive effects of phototherapy to promote wound healing , pain relief and anti-inflammatory effects, some negative reports also have been published, further confounding the issue (Demidova-Rice, Salomatina et al. 2007), for instance regarding the application of laser phototherapy on wound healing (Posten, Wrone et al. 2005). This controversy seems to be due to two main reasons; first of all, the basic biochemical mechanisms underlying these biological effects are not completely understood. Secondly, the complexity of rationally choosing amongst a large number of laser irradiation parameters (such as wavelength, fluence, power density, pulse structure and treatment timing), inappropriate anatomical treatment location and concurrent patient medication (such as steroidal and non-steroidal anti-inflammatories which can inhibit healing) has led to conflicting results and publication of a number of unfavourable, as well as many favourable, studies. In particular a biphasic dose response has been frequently observed where low levels of light have a much better effect than higher levels (Gigo-Benato, Geuna et al. 2005; Aimbire, Albertini et al. 2006; Hamblin, Waynant et al. 2006; Goncalves, Souza et al. 2007; Huang, Chen et al. 2009; Hamblin 2010).

It should be noticed that LLLT has a diversified terminology. It is also called "cold laser", "soft laser", "biostimulation" , "photobiomodulation", "low intensity laser therapy" , "low energy laser therapy", "laser phototherapy (LPT)", "laser therapy", and "non-ablative irradiation". Some investigators state that using frequent terms, such as "low power laser therapy" is misleading, since high power lasers, too, can be used for laser phototherapy (Tuner and Hode 2010) (as we will discuss in the next sections, CO_2 laser is applied as a low level (therapeutic laser) in NACLT, too). Some of the researchers favour the term "laser phototherapy (LPT)" which is an emerging terminology (Tuner and Hode 2010).

2.2 A brief review on biological effects of low level therapeutic lasers

Low level laser (or light) therapy (LLLT) is the application of light (usually a low power laser or LED in the range of 1mW - 500mW) to a pathology to promote wound healing and tissue repair, reduce inflammation and relieve pain. The light is typically of narrow spectral width in the red or near infrared spectrum (600nm - 1000nm); at power densities (between 1mw-5W/cm2) (Huang, Chen et al. 2009), not associated with macroscopic thermal effects, in contrast to thermally mediated surgical applications (Chow, David et al. 2007). In using high power surgical lasers, the collimation of laser light leads to the emission of a narrow, intense beam of light and is used for precise tissue destruction (photothermal effect). However, in LLLT, light radiation intensities are so low that the resulting biological effects are ascribable to physical or chemical changes associated with the interaction of cells and tissues with the laser radiation, and not simply to a result of heating (Snyder, Byrnes et al. 2002; Gigo-Benato, Geuna et al. 2005).

The main areas of medicine where laser phototherapy has a known and major role are as follows: promoting wound healing, tissue repair and prevention of tissue death, relief of inflammation in chronic diseases and injuries with its associated pain and edema, relief of neurogenic pain and some neurological problems (Hamblin, Waynant et al. 2006).

The first law of photobiology states that for low power visible light to have any effect on a living biological system, the photons must be absorbed by electronic absorption bands belonging to some molecular photoacceptors, or chromophores (Sutherland 2002; Huang, Chen et al. 2009). Red and near infrared light is absorbed by photoreceptors contained in the protein components of the respiratory chain located in mitochondria, in particular cytochrome c oxidase and flavoproteins like NADH-dehydrogenase. This can lead to a short time activation of respiratory chain and oxidation of NADH pool leading to changes in the redox state of both mitochondria and cytoplasm, leading to increased ATP production, and biological responses at the cellular level through cascades of biochemical reactions (Karu 1989; Karu, Pyatibrat et al. 2004; Karu and Kolyakov 2005). These effects in turn lead to increased cell proliferation and migration, modulation in levels of cytokines, growth factors, inflammatory mediators, and increased tissue oxygenation. The results of these biochemical and cellular changes in animals and patients lead to valuable biological effects such as promoting wound healing and tissue repair, relief of inflammation, pain reduction, and amelioration of damage after heart attacks, stroke, nerve injury and even retinal toxicity (Hamblin, Waynant et al. 2006).

2.3 Pain relieving effects of low level therapeutic lasers

Low-level laser therapy (LLLT) is increasingly recognized as an appropriate option for pain relief. In fact, it is for this indication that biostimulative lasers have been approved for marketing by the U.S. Food and Drug Administration through the premarket notification/510(k) (Gigo-Benato, Geuna et al. 2005). Many studies have demonstrated the efficacy of phototherapy in various pain syndromes (Tuner and Hode 2010). Responding to the increasing levels of evidence, the World Health Organization's Committee of the Decay of the Bone and Joint has also recently incorporated LLLT into guidelines for treatment of neck pain (Haldeman, Carroll et al. 2008; Chow, Armati et al. 2011).

> *"Efficacy of LLLT in painful clinical conditions has been established by several recent systematic reviews and meta-analyses [level 1 evidence, according to the Australian Government, NHMRC (1999)] (Chow, David et al. 2007). This level of evidence relates to chronic neck pain(Chow and Barnsley 2005), tendinitis (Bjordal, Couppe et al. 2001), chronic joint disorders (Bjordal, Couppe et al. 2003), and chronic pain (Enwemeka, Parker et al. 2004) . Randomized controlled trials (RCTs) provide level II evidence for the efficacy of laser therapy in chronic low back pain (Umegaki, Tanaka et al. 1989; Soriano and Rios 1998; Basford, Sheffield et al. 1999) . In other reviews of laser therapy for painful conditions such as rheumatoid arthritis (Brosseau, Robinson et al. 2005) and musculoskeletal pain (Gam, Thorsen et al. 1993; De Bie, De. Vet et al. 1998), the evidence is equivocal. Such variability in outcomes may be due to the multiplicity of parameters used, including wavelengths, energy, and power densities, with differing frequencies of application(Chow and Barnsley 2005)."* ***(Chow, David et al. 2007)***

2.4 Mechanisms of analgesic effects of low level laser therapy

The basic biological mechanisms behind the analgesic effects of conventional LLLT are not completely understood. Some of the explanations for these pain relieving effects are as follows:

- Reversible blockage of action potential generation of nociceptive signals in primary afferent neurons and specific reversible inhibition and functional impairment of Aδ and C fibers, which transmit nociceptive stimuli (Wakabayashi, Hamba et al. 1993; Kasai, Kono et al. 1996; Orchardson, Peacock et al. 1997; Chow, Armati et al. 2011).
- Increase in β-endorphin synthesis and release (Labajos 1988; Montesinos 1988; Hagiwara, Iwasaka et al. 2007).
- Inhibiting cyclooxygenase, interrupting conversion of arachidonic acid in to prostaglandins, especially prostaglandin E2 (PGE2) (Shimizu, Yamaguchi et al. 1995; Mizutani, Musya et al. 2004).
- Suppression of Substance P, a neuropeptide associated with nociception (Ohno 1997).
- Suppression of bradykinin activity, a pro-inflammatory neuropeptide that irritates nociceptors and is a key element in clinical pain and the associated inflammation (Maeda 1989; Jimbo, Noda et al. 1998)
- Increased production of serotonin, affecting negatively neurotransmission (Tuner and Hode 2010)
- Increased synaptic activity of acetylcholine esterase (Simunovic 2000)
- Involvement of nitric oxide in analgesic effects of therapeutic lasers (Mrowiec 1997)
- Single oxygen production, which in small amounts, is very important in biochemical processes and may be important in biostimulation
- Decreased inflammation and subsequently decreased inflammatory sensitization of small-diameter afferent nerve endings
- Improvement of local microcirculation, increased tissue oxygenation, shift of metabolism from anaerobic to aerobic pathways, decreased production of acidic metabolites which stimulate the pain receptors
- Increased lymphatic flow and consequently reducing edema, which leads to decreased sensitization of pain receptors

- Systemic effect, some researchers propose that laser phototherapy has both a local effect in the area treated by laser light, and a systemic effect through the release of metabolites (Tuner and Hode 2010).

3. Pain relieving effects of carbon dioxide lasers

Laser phototherapy is a relatively new application of carbon dioxide lasers, in spite of the fact that papers on the subject were published as early as the mid-eighties (Tuner and Hode 2010). CO_2 laser biostimulative and pain relieving effects can be assessed in the following two main groups:

- The lower post operative pain following CO_2 laser surgery compared to conventional surgery, which is attributed to the simultaneous low level laser therapy (photobiomodulation) effects of high power CO_2 laser irradiation (Tuner and Hode 2010).
- Application of CO_2 laser, as a phototherapeutic laser. "For using CO_2 laser as a low power therapeutic laser, one needs to make the beam wide enough not to burn. Another option is to scan rapidly over the lesion with a narrow beam. Therefore the power density or average power is kept low enough to avoid burning and their incident energy and power density are set within the low intensity biostimulative laser therapy range" (Tuner and Hode 2010). NACLT is a new technique, in which CO_2 laser is used in non-thermal, non-ablative manner as a biomodulative, low intensity laser for immediate and significant pain relief. (Zand, Ataie-Fashtami et al. 2009; Zand, Mansouri et al. 2009; Zand, Najafi et al. 2010).

This section is organized as follows. At first some clinical studies which demonstrate the pain relieving effects of high power surgical CO_2 lasers are reviewed, and then low power biostimulative CO_2 laser studies are briefly reviewed. In the next section, NACLT as a new low power, biomodulative laser therapy protocol for pain reduction is introduced.

3.1 Pain relieving effects of carbon dioxide laser as a high power surgical laser

Some investigators who used the classical thermal effects of CO_2 laser (vaporization, cutting and coagulation) reported less post-operative pain following CO_2 laser surgery, and a reduced requirement for post-operative analgesics (Duncavage and Ossoff 1986; Colvard and Kuo 1991; Demidov, Rykov et al. 1992; Chia, Darzi et al. 1995; Kaplan, Kott et al. 1996; Andre 2003).

Kaplan, one of the pioneers of CO_2 laser surgery, attributed the excellent healing and lower post operative pain experienced with CO_2 laser surgery compared to conventional surgery to the simultaneous laser therapy effects of CO_2 laser irradiation. Kaplan stated that laser surgery and laser therapy should be regarded *as two sides of the same coin* (Kaplan, Kott et al. 1996; Tuner and Hode 2010).

Duncavage reported that the advantages of the CO_2 laser surgery included homeostasis, precise visualization, and less edema and pain than the conventional techniques (Duncavage and Ossoff 1986).

Colvard and Kuo used high-power, ablative CO_2 laser at a power output of 4 W under local anesthesia for painful minor aphthous ulcers of 14 patients. In all, 88.8% of the patients in the study were completely pain free following anesthetic resolution, and none of the patients required post-operative medication for pain relief. (Colvard and Kuo 1991).

Demidov and his colleagues used high-energy CO_2 laser as a laser scalpel in 120 cases of breast surgery including 70 operations for cancer. They reported reduced pain sensitivity in the region of the wounds in the postoperative period (Demidov, Rykov et al. 1992).

In another study Chia reported that high-power CO_2 laser haemorrhoidectomy was associated with a reduced requirement for post-operative analgesia (Chia, Darzi et al. 1995).

Andre used classical high power CO_2 laser at 10–15 W in continuous mode under local anesthesia for treating ingrowing nails of 302 patients. He reported that ingrowing nails were easily operated with the CO_2 laser; bleeding was minimal, infection was rare, the wounds healed without exhudative drainage and cosmetic results were good. In addition the immediate post-operative pain was less severe than after classical nail surgery with scalpel (Andre 2003).

In another study, Tada et al. compared the clinical effects and postoperative course of the scanning CO_2 laser surgery and conventional surgical method to evaluate the clinical effectiveness of the former for the treatment of ingrowing nail deformities. They demonstrated that statistically, the operating time and the duration of postoperative pain were reduced significantly by the scanning CO_2 laser. Furthermore, patients treated with CO_2 laser were able to return to daily life significantly sooner (Tada, Hatoko et al. 2004).

Kaviani et al. investigated whether the CO_2 laser surgery was superior to conventional surgical techniques for minor breast surgery in a randomized clinical trial. They demonstrated that application of the CO_2 laser in breast mass biopsy had some advantages, including a lower requirement for local anesthetic and a lower rate of intraoperative bleeding; however it did not reduce the postoperative pain grade severity (Kaviani, Fateh et al. 2008).

Demetriades used ablative CO_2 laser in painful oral aphthous ulcer of a patient with Behçet's Syndrome. His experience showed transient pain relief following ablative CO_2 laser irradiation (Demetriades, Hansford et al. 2009).

3.2 Pain relieving effects of carbon dioxide laser as a low level (therapeutic) laser

It is interesting to know that in addition to classical low level therapeutic lasers, surgical lasers could also be used as therapeutic instruments, for example; defocused CO_2 laser 10,600 nm, defocused ruby laser 694 nm and defocused Nd:YAG laser 1064 nm can be used for photobiomodulation. "When high power laser are used for biomodulation, one only needs to make the beam wide enough not to burn. The patient will then feel only a mild heat. An alternative is to scan rapidly over the lesion with a narrow beam. Therefore the power density or average power is kept low enough to avoid burning and their incident energy and power density are set within the low intensity laser therapy range" (Tuner and Hode 2010).

The famous investigation of Mester with a low powered ruby laser (694-nm) on the shaved areas of the mice and quickly growing back of hairs can be an example of using a surgical laser as a therapeutic, biostimulative one (please see 2.1. History).

At Uppsala Academic Hospital, a CO_2-laser was tested successfully for biostimulative treatment of epicondylitis. This method was called EDL (Emitted Defocused Laser-light). It should be noted, however, that CO_2-lasers were not used as surgical lasers in this study;

their incident energy and power density were set within the laser therapy range by spreading out the beam over such a large surface that the laser did not cause burning (Tuner and Hode 2010).

Nicola used CO_2 low power laser treating chronic pharyngitis. 85 patients with non-specific chronic pharyngitis were elected to be treated: Group I, 40 patients with predominance of hyperaemic aspect; and group II, 45 patients, predominance of hypertrophied aspect. Both groups were treated for 8 to 10 sessions. They concluded that this method was very suitable for the treatment of chronic pharyngitis (Nicola and Nicola 1994).

In another study, 846 patients with different types of fibromyositic rheumatisms were submitted to defocalized laser therapy from 1980 to 1995. They employed Diodes and CO_2 lasers. Control groups were used to compare results with those of traditional methods. Results were evaluated on the basis of subjective (such as local pain) and objective criteria. On the whole, results were positive in comparison with other methods both as regards recovery time and persistence of results. Approximately 2/3 of the patients benefited from the treatment indicated that there were greater advantages in use of laser over other presently available methods. Longo and his collogues recommended that standardalization of treatment protocols deserves further studies (Longo, Simunovic et al. 1997).

The CO_2-laser can also be used as an acupuncture tool. Simulation of acupuncture points has been carried out both with biostimulating power densities (e.g.100mW/cm^2) and burning/coagulation/ evaporation power densities. Some clinics state that CO_2 lasers give better results on acupuncture points than HeNe lasers. "As the CO_2 laser's beam cannot penetrate more than around 0.5 mm into tissue, the effects must be due to the influence of the laser energy on the cells encountered, so that signal substances are released and then circulate in the organism. This indirectly confirms the hypothesis that conventional laser therapy has both a local effect in the area treated by laser light, and a systemic effect through the release of metabolites. It is well known that these kinds of secondary effect also occur at the traditional wavelengths of 633, 830, and 904 nm" (Tuner and Hode 2010).

4. NACLT (Non-Ablative CO_2 Laser Therapy)

Recently, there have been few reports about using CO_2 laser in non-ablative manner to reduce pain in painful mucosal lesions (Elad, Or et al. 2003; Sharon-Buller and Sela 2004; Zand, Ataie-Fashtami et al. 2009; Zand, Mansouri et al. 2009; Zand, Najafi et al. 2010).

In a report, Elad et al suggested that CO_2 laser treatment could be of benefit to control pain in severe oral chronic graft-versus-host-disease (Elad, Or et al. 2003). In this study, the oral lesions of four patients were irradiated by CO_2 laser during 17 sessions. The CO_2 laser was applied, over mucosal lesions, using 1W power for 2-3s/1mm^2. The treated mucosa was kept wet during the process. The treatment was pain free, and anesthesia was not required. The average VAS scores before, during, and immediately after CO_2 laser treatment were 8.09, 3.47, and 4.88 respectively. There was no visual damage to the oral mucosa and no aggravation of the oral lesions (Elad, Or et al. 2003).

In another case report, aphthous ulcers of two patients were irradiated with CO_2 laser at 1.0-1.5 W power, with a defocused hand piece for 5 seconds. Before laser irradiation, a thin film of Elmex Gel (a transparent gel with high water content) was placed on the lesions to reduce the beam absorption by the soft tissue. Anesthesia was not required since the treatment was not painful. The patients reported immediate pain relief after laser irradiation (Sharon-Buller and Sela 2004).

In these two reports, water (Elad, Or et al. 2003) and transparent gel with high water content (Sharon-Buller and Sela 2004) were used to reduce the beam absorption by the soft tissue.

These interesting results encouraged us to conduct a randomized controlled clinical trial to confirm the pain-relieving effect of CO_2 laser in minor aphthous ulcers as a prototype of painful oral lesions (Zand, Ataie-Fashtami et al. 2009). The results of this clinical trial demonstrated that a single session of low-intensity, non-thermal, CO_2 laser irradiation could reduce pain in minor aphthous ulcers immediately and significantly, with no visible side effects (Zand, Ataie-Fashtami et al. 2009), the technique was called NACLT (Non-Ablative CO_2 Laser Therapy) afterwards.

In order to use the CO_2 laser as a phototherapeutic laser for NACLT, the CO_2 laser beam is irradiated through a thick layer of transparent gel with high water content to reduce the beam absorption by the tissue. In addition, the CO_2 laser is operated with a de-focused hand piece 5–6 mm distant from the mucosal surface, scanning rapidly over the lesion with circular motion (Zand, Ataie-Fashtami et al. 2009; Zand, Mansouri et al. 2009; Zand, Najafi et al. 2010). The results of the powermetry have shown that the final laser power output, after passing through the gel, is reduced to 2-5 mW, which is in the range of low power lasers. Thermometry has also shown no significant temperature rise on the surface of the ulcers and under the gel layer during the laser irradiation, supporting the low power nature of the applied CO_2 laser in NACLT(Zand, Ataie-Fashtami et al. 2009). It appears that due to high water content of the gel, it absorbs CO_2 laser irradiation considerably, resulting in significant drops in the power output, by a factor of 200-500. In fact by irradiation of CO_2 laser through a transparent gel with high water content, CO_2 laser can be used as a non-destructive, non-thermal laser to reduce pain in some oral lesions immediately and significantly. This technique was called non-ablative CO_2 laser therapy (NACLT), in order to avoid any confusion with classical high power thermal CO_2 laser effects. This technique could also be called non-thermal CO_2 laser therapy (NTCLT) to avoid misinterpretation with fractional non-ablative lasers used for skin rejuvenation (Zand, Ataie-Fashtami et al. 2009).

NACLT is a pain free procedure and neither systemic nor topical anesthesia is required. The patients don't feel warmth in their lesions during the procedure, in contrast to conventional defocused CO_2 laser therapy in which the patients feel mild warmth. Up to now, in the series of studies about the analgesic effects of NACLT, we have observed no visual effects of thermal damage to the oral mucosa such as tissue ablation, ulceration, erythema or aggravation of the lesions following the careful application of the technique (Zand, Ataie-Fashtami et al. 2009; Zand, Mansouri et al. 2009; Zand, Najafi et al. 2010). So that the before-after NACLT pictures of the lesions cannot be differentiated from each other easily (Figure 1). Since there is no tissue ablation and plume formation during NACLT, in contrast to the classical ablative CO_2 laser surgery, it seems rational to conclude that this procedure has no potential for carrying viral particles to the surgeon and other operating room staff (Zand, Ataie-Fashtami et al. 2009).

There are some differences between analgesic effects of NACLT and the other classical low power lasers. The analgesic effect in LLLT is usually gradual, cumulative, and multi-session (Pinheiro, Cavalcanti et al. 1998; Gur, Karakoc et al. 2002; Gur, Sarac et al. 2004; Nes and Posso 2005; Chow, Heller et al. 2006; Djavid, Mehrdad et al. 2007; Bjordal, Bensadoun et al. 2011; Iwatsuki, Yoshimine et al. 2011; Ribeiro, de Aguiar et al. 2011). In contrast, the pain relieving effect of NACLT is immediate, dramatic and more sustained than conventional phototherapeutic lasers, so that immediately after NACLT, the patients of the studies have been able to eat and drink easily (Zand, Ataie-Fashtami et al. 2009; Zand, Mansouri et al. 2009; Zand, Najafi et al. 2010).

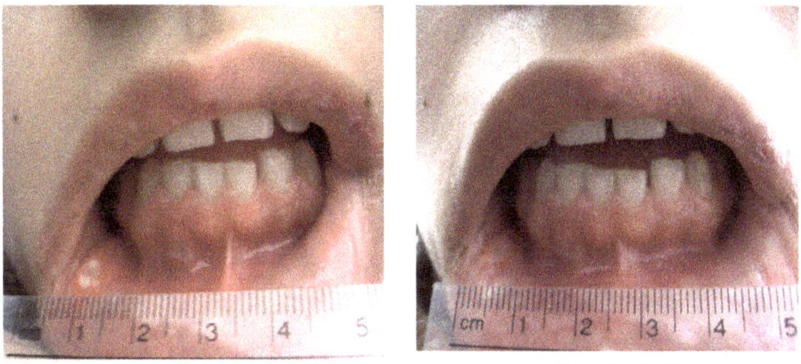

Before NACLT   After NACLT

Fig. 1. Minor aphthous ulcers before and immediately after NACLT

4.1 NACLT protocol

At first, all standard precautions of laser utilization should be considered. Before laser irradiation, the patient and surgical staff should be given appropriate protective eye shields and eye glasses matched to the laser wavelength (10,600 nm) to protect inadvertent laser impact.[1] Before laser irradiation, a layer of a transparent, non-anesthetic gel with high water content is placed on the lesion. In our studies, we use a transparent gel (Abzar Darman Co., Iran) with 87.5% water content, with a thickness of 3-4 mm on the lesion, is used. The CO_2 laser is operated at 1W power, with a de-focused hand piece, 5-6 mm distant from the mucosal surface, in continuous mode, scanning rapidly over the lesion with a circular motion. The irradiation time depends on the size of the lesion. For example, in our studies the irradiation time for minor aphthous ulcers is about 5-10 seconds. When using NACLT for larger lesions, such as; pemphigus vulgaris, every one centimeter square of the lesion has been irradiated for 5 seconds in each pass, and repeated immediately if the contact pain of the lesion persists. Between the passes, the prior gel should be wiped gently and replaced by a new layer of gel, otherwise the water content of the gel will decrease which may lead to increasing the beam absorption by the lesion and subsequent tissue injury and even burning.

---

[1] In NACLT studies, we used eye glasses matched to the CO_2 laser wavelength (10,600 nm), but we presume that it might be safer to use eye glasses matched to both the 10,600 nm and the guiding beam to protect the eyes from the reflected beam from the surface of the gel, the presumption that should be evaluated in further studies.

4.2 NACLT applications in clinical studies

4.2.1 Recurrent oral aphthous stomatitis

4.2.1.1 Definition

Recurrent aphthous stomatitis (RAS) is a common oral disorder of uncertain etiopathogenesis (Scully, Gorsky et al. 2003), characterized by painful, round or ovoid ulcers with circumscribed margins, yellowish fibrinoid base, surrounded with erythematous haloes. The lesions typically first presenting in childhood or adolescence, recur at varying intervals throughout life (Jurge, Kuffer et al. 2006). The frequency and severity of the ulcerations usually decreases with age (Arikan, Birol et al. 2006). RAS occurs worldwide although it appears most common in the developed world (Jurge, Kuffer et al. 2006).

Recurrent aphthous stomatitis (RAS) is classified into three clinical forms, namely minor (miRAS), major (maRAS) and herpetiformis. Minor aphthous ulcers, which comprise over 80-90% of cases (Shashy and Ridley 2000), are less than 1 cm in diameter, last up to 7-14 days, and they heal without scar formation. Major aphthous ulcers are over 1 cm in diameter, and their healing may take 20 to 30 days at a time and often heal with scarring. Herpetiform ulcers (HUs) are multiple, clustered, 1-3 mm lesions that may coalesce into larger ulcers. They typically heal within 15 days (Prolo P 2006).

Although the exact underlying pathophysiology of RAS is not completely known, some evidences propose that aphthous ulcers are related to a focal immune dysfunction in which T lymphocytes have a significant role (Shashy and Ridley 2000; Jurge, Kuffer et al. 2006). Many etiologic, predisposing, and precipitating factors, such as genetic factors, immunologic problems, trauma, hypersensitivity to foods and drugs, hormonal changes, hematological deficiencies, cessation of smoking, and psychological stresses have been propsed (Shashy and Ridley 2000; Arikan, Birol et al. 2006).

Since there is no consensus regarding the cause of recurrent oral aphthous ulcers, it is difficult to have completely effective treatment or prevention (Shashy and Ridley 2000). There are currently few agents that have been found in randomized controlled clinical trials to cure aphthous ulcers (Jurge, Kuffer et al. 2006). As a result, the management of RAS is directed largely toward symptomatic relief. The main problem with aphthous ulcers is their pain which may be so severe. Many different therapeutic agents, including topical corticosteroids, mouth rinses, antibiotics, local anesthetic gels or pastilles, and treatment modalities, such as silver nitrate cautery and cryotherapy, have been tried for pain control in miRAS patients (Alidaee, Taheri et al. 2005; Arikan, Birol et al. 2006).

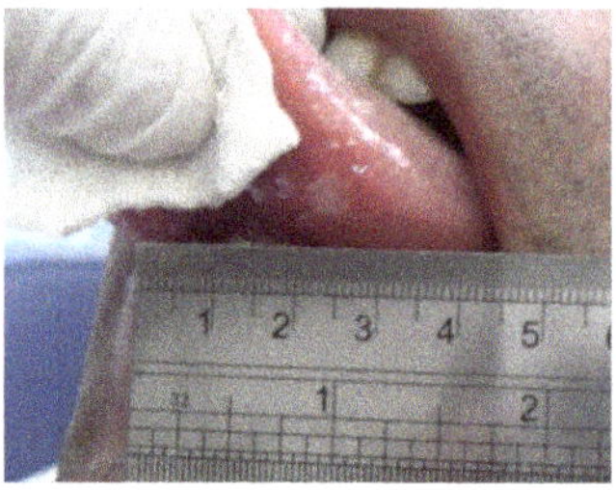

Fig. 2. Minor aphthous ulcer

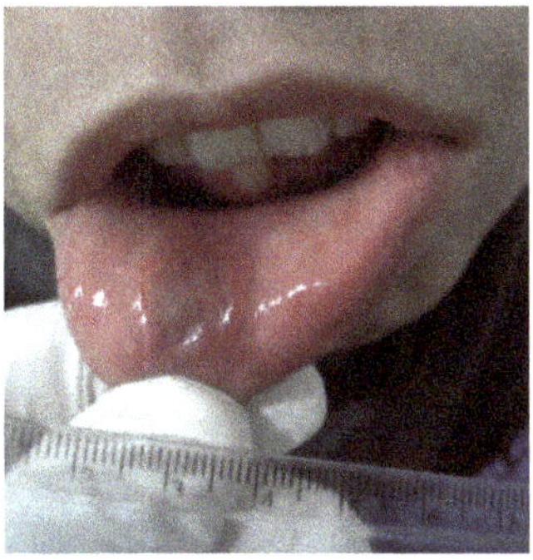

Fig. 3. Major aphthous ulcer

4.2.1.2 NACLT and minor oral aphthous stomatitis

A randomized controlled clinical trial was designed to evaluate the pain relieving effects of a single-session of NACLT in minor recurrent aphthous stomatitis as a prototype of painful oral ulcers. Fifteen patients, each with two discrete aphthous ulcers, were included. One of the ulcers was randomly allocated to be treated with NACLT and the other one served as a placebo. In each patient, the laser lesion was treated with NACLT, while the placebo lesion was irradiated with the same laser, but with an inactive probe. The patients scored and recorded the pain severity of their lesions on a 10-grade visual analogue scale (VAS) up to 4 days post operatively. In the laser group, the pain severity scores of the lesions were dramatically declined immediately after irradiation ($p<0.001$), whereas there were no changes in the mean scores in the placebo lesions at the same time. The reduction in pain scores was significantly greater in the laser group than in the placebo group in all of the follow up periods ($p<0.001$). The procedure itself was not painful, so anesthesia was not required. The patients reported no warmth in their lesions during laser treatment. There was no visual effect of thermal damage to the oral mucosa such as ablation, coagulation or erythema. The results showed that a single-session of NACLT reduced pain in minor aphthous ulcers immediately and significantly, without any visible side effects (Zand, Ataie-Fashtami et al. 2009).

4.2.1.3 NACLT and major oral aphthous stomatitis

A pilot randomized controlled clinical trial was designed to evaluate the analgesic effects of a single-session of NACLT in major recurrent aphthous ulcers. Five patients, each with two discrete major aphthous ulcers were included. One of the ulcers was randomly allocated to be treated with NACLT and the other one served as a placebo. The lesions in laser group were irradiated with CO_2 laser (λ = 10,600 nm; Lancet-2, Russia) through a thick layer of transparent, non-anesthetic gel (Abzar Darman Co., Iran) with 87.5% water content, with a thickness of 3-4 mm. The CO_2 laser was operated at 1W power, with a de-focused hand piece, 5-6 mm distant from the mucosal surface, in continuous mode, scanning rapidly over the lesion with circular motion. The patients' idiopathic (non-contact) and contact pain severity scores were recorded before and immediately after NACLT. These scores were also recorded up to 4 days post- operatively. The results of the study demonstrated that in the laser group, both the non-contact and contact pain severity scores of the lesions were dramatically declined immediately after irradiation ($p<0.001$), whereas there were no changes in the mean scores in the placebo lesions at the same time. The reduction in pain scores was significantly greater in the laser group than in the placebo group in all of the follow up periods ($p<0.001$). There were not any visible side effects following NACLT. None of the patients reported warmth feeling in their lesions during laser treatment. The results of the study suggested that a single-session of NACLT could reduce pain in major aphthous ulcers immediately and significantly, without any visible side effects (Zand, Ataie-Fashtami et al. 2009). This study is in progress.

4.2.1.4 Literature review

Colvard and Kuo evaluated the potential efficacy of the high-power, surgical CO_2 laser for pain relief in 28 painful minor aphthous ulcers of 14 patients. Their anesthetic protocol included pre-operative pain medication (oral administration of ketoprofen) and local anesthesia by infiltration of 1:200,000 2% isocaine with 1:200000 neocobefrin to overcome the painful nature of the procedure. During the procedure, CO_2 laser was used as a classical, ablative manner with power output 4 W and as much necrotic tissue as possible was removed. Over all 88.8% of the patients were completely pain free following anesthetic resolution, and none of the patients required post-operative medication for pain relief. The authors concluded that CO_2 laser should be included as an alternative modality for the treatment of miRAS, due to its ability to reduce or eliminate pain (Colvard and Kuo 1991). In this study, CO_2 laser was used in classical, high power ablative manner. However the post operative analgesic effects of the procedure demonstrated the simultaneous biomodulative effects of CO_2 laser irradiation. The same concept Kaplan stated that laser surgery and low level laser therapy should be regarded as two sides of the same coin.

Fekrazad et al. evaluated the effects of Nd: YAG laser (power: 3 W, energy: 100 mj, pulse repetition rate: 30Hz, irradiation time: 60 sec) in 138 patients with aphtous ulcers. The patients were randomly assigned into three groups, as follows: (1) treatment with a focalized beam; (2) treatment with a non-focalised beam and (3) placebo treatment. In group (1) the laser beam was administered from a distance of 6 mm from the centre of the ulcer without using a clear and defined point of irradiation. In group (2) a well defined point beam of the laser was irradiated from a distance of 2 mm from the center of the ulcer, in a helical fashion. In group (3) the HeNe Laser was used as placebo with inactive probe. In group (1) and (2) a significant reduction of pain was observed compared to group (3). The duration of pain and the duration of recovery period were shortest in group (2) (Fekrazad, Jafari et al. 2006).

De Souza TO et al. assessed the effect of low-level laser therapy on pain control and the repair of recurrent aphthous stomatitis. Twenty patients with recurrent aphthous ulcers were divided into two groups. The patients in Group I ($n = 5$) treated with topical triamcinolone acetonide and the patients in Group II ($n = 15$) received laser treatment with an InGaA1P diode laser (670 nm, 50 mW, 3 J/cm^2 per point) in daily sessions on consecutive days. All patients were assessed daily, and the following clinical parameters were determined during each session: pain intensity before and after treatment and clinical measurement of lesion size. The results showed that 75% of the patients reported a reduction in pain in the same session after laser treatment, and total regression of the lesion occurred after 4 days. Total regression in the corticosteroid group was from 5 to 7 days. They concluded that LLLT with these laser parameters demonstrated analgesic and healing effects with regard to recurrent aphthous stomatitis (De Souza, Martins et al. 2010).

Title	Author	Study	Type of irradiation	Need to Anesthesia	Number of patients
Managing aphthous ulcers: Laser Treatment Applied	Colvard M, Kuo P	Before-after clinical trial 1991	Ablative CO_2 laser surgery	+	18
CO_2-laser treatment of ulcerative lesions	Sharon-Buller A, et al.	Case report 2004	CO_2 laser irradiation of the lesions through a thin film of transparent gel with high water content	_	2
Relieving pain in minor aphthous stomatitis by a single session of non-thermal carbon dioxide laser irradiation	Zand N, Ataie-fashtami L. et al.	Randomized controlled clinical trial (RCT) 2008	CO_2 laser Irradiation of the lesions through a thick (3-4mm) layer of transparent gel with high water content; (NACLT)	_	15
Analgesic effects of single session of Non-Ablative CO_2 Laser Therapy (NACLT) in major aphthous ulcers: (a preliminary study)	Zand N, Ataie-fashtami L. et al.	Randomized controlled clinical trial (RCT) 2009	CO_2 laser Irradiation of the lesions through a thick (3-4mm) layer of transparent gel with high water content; (NACLT)	-	5

Table 1. Irradiation of aphthous ulcers with CO_2 laser

4.2.2 Behçet's disease

4.2.2.1 Definition

Behçet's disease (BD) which is classified among vasculitides is a complex, multisystem inflammatory disease characterized by oral and genital aphthae, cutaneous lesions, arthritis, ocular, gastrointestinal, and neurologic manifestations (Meador, Ehrlich et al. 2002; Suzuki Kurokawa and Suzuki 2004; Lin and Liang 2006).

The most common clinical feature is the presence of recurrent and usually painful mucocutaneous ulcers (Lin and Liang 2006). Oral aphthosis is the most frequent and constant manifestation of Behçet's disease (Davatchi, Shahram et al. 2005) and usually the initial presenting symptom in most, if not all, patients (Lin and Liang 2006) . The distinct difference between the clinical features of aphthous ulcers of RAS and Behçet's Syndrome remains unclear. The aphthous ulcers of Behçet's disease are typically painful punched-out ulcers with a white yellowish fibrinoid base, surrounded with erythematous halo. They range in size from a few millimeters to 2 cm. These ulcers typically heal spontaneously within 1 to 3 weeks,usually without scarring (Ghate and Jorizzo 1999; Lin and Liang 2006).

Genital ulceration occurs in approximately 75% of the patients with Behçet's disease (Lin and Liang 2006). The genital aphthous ulcers are morphologically similar to the oral ulcers, except that lesions are usually larger, more painful, heal more slowly, recur less frequently and can have scarring tendency (Davatchi, Shahram et al. 2005; Lin and Liang 2006). In females they are often larger than 10 mm, and deeper than oral lesions. They are localized on the vulva, vagina, and rarely cervix. The giant aphthous lesion of the vulva is frequent, causing dysfunction and leaving sometimes indelible cicatrix. In males, genital aphthosis is often seen on the scrotum, but may be seen also on the shaft of penis or on the meatus. Sometimes they become giant lesions (Davatchi, Shahram et al. 2005) . Genital ulcerations of Behçet's disease may be very painful, exert a negative impact on the patient's quality of life, and these lesions are often refractory to multiple treatments(Kasugai, Watanabe et al. 2010).

Treatment of Behçet's disease is based on the clinical symptoms and severity of systemic involvement, including topical therapies as well as colchicine, dapsone, thalidomide, and immunosuppressants, interferon-alpha/beta, anti-tumor necrosis factor antibody, the latter specially in treatment for the cases with poor prognosis including eye, intestine, vessel and central nervous system involvement (Suzuki Kurokawa and Suzuki 2004).

The mucosal lesions, especially genital lesions can often become refractory to multiple treatments and present challenges to physicians. Topical or intralesional corticosteroids, oral pentoxifylline, sucralfate, dapsone, colchicine, and systemic low-dose corticosteroids, used either alone or in combination, are safe and having varying evidence for effect in mild to moderate mucocutaneous disease. Azathioprine or methotrexate can be used if the lesions are refractory to the previously mentioned therapies. Tumor necrosis factor (TNF) inhibitors such as infliximab or etanercept should be considered as the next step treatments . Tacrolimus, cyclosporine, and interferon-alpha-2a should be used generally only if TNF inhibitors have failed as a result of their toxicities (Lin and Liang 2006).

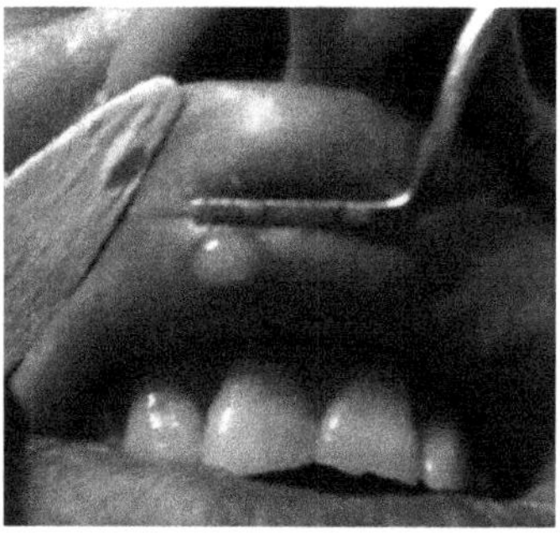

Fig. 4. Behçet's Disease

4.2.2.2 NACLT and oral aphthous ulcers of Behçet's disease

A pilot before-after clinical trial was designed to evaluate the analgesic effects of a single-session of NACLT in painful aphthous ulcers of Behçet's disease. Up until the time of this publication,, three patients with known Behcet's disease have been eligible and consented to participate in the study according the inclusion/exclusion criteria.

Four painful oral aphthous ulcers of the three patients were treated by NACLT. The pain severity of the lesions were dramatically declined immediately after irradiation ($p<0.001$). This analgesic effect was consistently sustained during the five days follow-up periods. Up until the time of this publication, the results of this pilot study suggest that a single session of NACLT could relieve pain in oral aphthous ulcers of Behçet's disease immediately and significantly without visible side effects of thermal damage or aggravation of the lesions. Similar to the other NACLT studies, the procedure itself was pain free and no anesthesia was required. The study is in progress.

4.2.2.3 NACLT and genital aphthous ulcers of Behçet's disease

In another case report that is being published, the extremely painful genital aphthous ulcers of a 23-year-old female with Behcet's disease were irradiated by NACLT. Before laser irradiation the pain of the lesions was so severe which impeded daily functions, such as sitting, walking, and even sleeping and did not respond to conventional analgesics. The non-contact and contact visual analogue scale (VAS) pain scores of the left genital ulcer were 8 and 10 and the scores of the right sided ulcer were 6 and 10 respectively. Immediately after NACLT of the ulcers and its surrounding erythematous rim, the contact pain of the lesions relieved completely (so that she could even walk downstairs without difficulties). Similar to the other prior NACLT investigations, there were no visual side effects of thermal damage to the lesions, such as tissue ablation or aggravation of the lesions following NACLT. The procedure was painless and neither systemic nor local anesthesia was required. This analgesic effect of NACLT was sustained during the healing period and she experienced no problem in daily functions and did not require topical or systemic analgesics. She just had mild burning sensation during urination for the first three days which relieved after healing of the lesions. It should be noted that concomitantly, treatment with prednisolone 30mg/day and colchicine 2mg/ day was initiated in the hospital. Additionally, the depth of the genital ulcers decreased remarkably two days after NACLT. The ulcers healed completely within 11 days which seemed to be much shorter than what was expected. Interestingly in spite of the large size of the ulcers, they left a very small (6mm) scar. The results of this case report suggest that NACLT could be potentially considered as an alternative method for pain relief in painful genital aphthous ulcers of Behçet's disease without any complications.

It should be noticed that Behçet's syndrome is a serious multisystem disease, which in some cases it may lead to systemic complications such as; severe ocular problems (even blindness), intestinal, central nervous system,... involvement. Therefore the patients must be warned that NACLT should not substitute the systemic therapy of the disease in spite of its significant pain relieving effect.

Certainly, controlled clinical trials with larger sample sizes are necessary to further evaluate the efficacy and safety of NACLT in reducing the pain of oral and genital aphthous ulcers of

Behcet's disease. In addition such studies can demonstrate whether NACLT could accelerate wound healing in these lesions and specially prevent scar formation in genital aphthous ulcers of Behcet's disease or not.

4.2.2.4 Literature review

Demetriades used ablative CO_2 laser in four painful oral aphthous ulcers of a patient with Behçet's Syndrome. Before laser irradiation, the lesions were infiltrated with a minimal amount of lidocaine 2% with 1:100,000 epinephrine. A CO_2-laser set at 2W superpulse mode with a 0.4 mm ceramic tip was used, in a defocused way to lightly char the surface of the ulcers. The patient tolerated the procedure well. On subsequent follow-up, one week after the procedure, the patient reported considerable relief of symptoms on most of the treated ulcers. The oropharyngeal ulcer displayed only moderate response, but the patient reported an overall improvement of his quality of life (Demetriades, Hansford et al. 2009).

Title	Author	Study	Type of irradiation	Need to Anesthesia
General manifestations of Behçet's syndrome and the success of CO_2-laser as treatment for oral lesions: A review of the literature and case presentation	Demetriades M et al	Case report 2009	Ablative defocused CO_2-laser irradiation	+
Relieving pain in painful genital ulcers of Behcet's disease by a single session of non thermal, Non-Ablative CO_2 Laser Therapy(NACLT): A Case Report	Zand N, Fateh M. Et al.	Case report, under publish	NACLT	–
Immediate pain relief of oral aphthous ulcers of Behcet's disease by non-thermal, Non-Ablative CO_2 Laser Therapy (NACLT)	Zand N, Fateh M. Et al.	Pilot before-after clinical trial/ under publish	NACLT	-

Table 2. Irradiation of aphthous ulcers of Behcet's disease with CO_2 laser

4.2.3 Pemphigus vulgaris

4.2.3.1 Definition

Pemphigus Vulgaris (PV) is a rare, potentially life-threatening, autoimmune blistering disease of the skin and mucous membranes. Although the disease can affect anyone, it is most prevalent in people of Mediterranean or Jewish ancestry(Bystryn and Rudolph 2005). The prevalence of the disease is 30/100,000 and annual incidence has been reported between 1 and 5 in 100,000 according to different studies in Iran (Chams-Davatchi, Valikhani et al. 2005; Asilian, Yoosefi et al. 2006). The lesions are characterized by intra-epidermal vesicles with acantholysis and an intact basal layer. In the majority of patients, painful mucous membrane erosions are the presenting sign of pemphigus vulgaris and may be the only sign for weeks to months before any bullous skin lesion develops. The mucous membranes most often affected are those of the oral cavity, in which intact blisters are rare, probably because they are fragile and break easily, leaving scattered and often extensive erosions. The lesions are usually multiple, superficial, and irregular in shape, and arise from mucosa of healthy appearance. Although any surface can be involved, the most common sites are the buccal and labial mucosa, the palate, and the tongue (Bystryn and Rudolph 2005).

Oral lesions in pemphigus vulgaris may be so painful during the active period of the disease that may interfere with their eating, drinking and even speaking (Black, Mignogna et al. 2005; Bystryn and Rudolph 2005).

High doses of systemic corticosteroids plus immunosuppressive agents have dramatically declined the mortality rate of the disease. Understandably, owing to the life threatening nature of PV, the main focus of the peer reviewed literature has been on suppression and remission of PV (Rashid and Candido 2008). However, remission is not instantaneous and takes time to achieve. This delay in remission allows ample opportunity for complications to develop, secondary to the pain associated with PV. This can be highlighted by cases of repeated dehydration and malnutrition seen in PV patients (Rashid and Candido 2008). Therefore it seems necessary to obtain new modalities for pain control of these oral lesions during conventional systemic therapy.

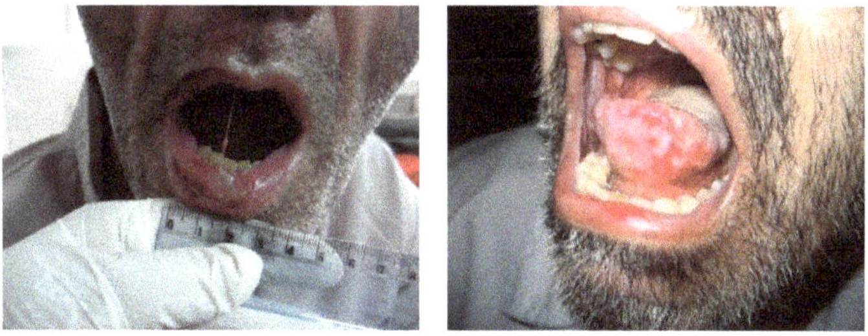

Fig. 5. Pemphigus vulgaris

4.2.3.2 NACLT and oral lesions of pemphigus vulgaris

A pilot before-after clinical trial was designed to evaluate the analgesic effects of application of a single session of NACLT in oral lesions of PV. Thirty eight painful oral lesions of ten patients with PV were irradiated with CO_2 laser by NACLT protocol. The patients scored and recorded the pain severity of their lesions on a visual analogue scale (VAS) up to 7 days post operatively. Immediately after NACLT, the severity of idiopathic (non-contact) and contact pain were dramatically declined ($p<0.001$), so that the patients could eat and drink without any difficulties. This analgesic effect was sustained during follow-up periods. There was no visual effect of thermal damage to the oral mucosa or aggravation of the lesions. The results of this pilot study suggested that a single session of NACLT could reduce pain in oral lesions of pemphigus vulgaris immediately and significantly, without visible side effects (Zand, Mansouri et al. 2009). We recommend that in further studies, the pain severity of the lesions should be followed up for longer periods of time.

It should be noted that due to the life threatening nature of PV without appropriate systemic treatment, the patients must be warned that NACLT should not alter their conventional treatment at all, in spite of its significant analgesic effect, as we instructed our patients to comply with their prescribed medical regimen.

4.2.3.3 Literature review

In a case report, the oral lesions of two patients with recalcitrant oral pemphigus vulgaris (who were under systemic treatment) were irradiated with CO_2 laser at 1-1.5 W in a defocused mode for 5-10 seconds. The patients reported no pain after treatment. Recall examination after 1 month, 3 months and 5 month revealed complete healing of the lesions with no recurrence (Bhardwaj, Joshi et al. 2010).

The pictures of the paper demonstrate the themal, ablative nature within the procedure. Its pain relieving effects can be explained by its simultaneous biomodulative effects of CO_2 laser irradiation.

Title	Author	Study	Type of irradiation	Need to Anesthesia	Number of patients
Relieving pain in painful oral lesions of pemphigus vulgaris by a single session, Non-ablative 10600 nm CO_2 Laser irradiation (pilot study)	Zand, N., Mansouri, P. et al.	Before-after clinical trial 2009	NACLT	–	Ten patients/38 lesions
Management of recalcitrant oral pemphigus vulgaris with CO_2 laser- Report of two cases	Bhardwaj, A. et al.	Case report 2010	CO_2 laser Irradiation of the lesions in a defocused mode (thermal)	?	Two patients/? lesions

Table 3. Irradiation of oral lesions of pemphigus vulgaris with CO_2 laser

4.2.4 Post chemotherapy oral mucositis

4.2.4.1 Definition

Oral mucositis is a common, debilitating, and potentially serious complication of chemoradiotherapy. Studies have shown that mucositis will develop in about 40% of chemotherapy patients, 80% of bone marrow transplant patients, and 100% of patients treated with radiotherapy to the head and neck (Berger & Kilroy 1997; Sonis et al. 1999).

It presents as erythema, edema, ulceration, bleeding along with pain. The pain of the lesions is aggravated by the patient's swallowing and normal oral functioning. Consequently, oral intake difficulties lead to loss of weight. The progression of oral lesions and their impact on the patient's general condition may require nasogastric tube feeding or temporary discontinuation of the treatment or modification of the therapeutic plan (Arora, Pai et al. 2008).

Pathologic evaluation of mucositis reveals mucosal thinning leading to a shallow ulcer thought to be caused by inflammation and depletion of the epithelial basal layer with subsequent denudation and bacterial infection. The wound healing response to this injury is characterized by inflammatory cell infiltration, interstitial exudate, fibrin and cell debris producing a "pseudomembrane" (Sonis 2004).

Various preventive measures causing alteration of mucosa (cryotherapy, allopurinol, pilocarpine, leucovorin), modification of mucosal proliferation (beta-carotene, glutamine, cytokines), and antimicrobial or anti-inflammatory action (chlorhexidine, corticosteroids) have been tried. General oral care, diet, topical mucosal coating agents (sucralfate, magnesium hydroxide), topical anesthetics, and systemic analgesics (opiods and non opioids) (Arora, Pai et al. 2008), recombinant human keratinocyte growth factor (palifermin) and Amifostine have also been suggested (Kuhn, Porto et al. 2009). However, currently no definitive preventive or therapeutic intervention exists that is completely successful at preventing oral mucositis and treatment for this complication has thus been symptomatic (Arora, Pai et al. 2008).

Fig. 6. Post Docetaxole chemotherapy oral mucositis

4.2.4.2 NACLT and mild to moderate post chemotherapy oral mucositis

A pilot before-after clinical trial was designed to evaluate the effects of single session of non-thermal, Non-ablative CO_2 Laser therapy (NACLT) to reduce pain in mild to moderate oral mucositis following breast cancer chemotherapy with Docetaxole. Six patients were included and their oral lesions were irradiated by NACLT. The patients reported the idiopathic (non-contact) and contact pain of their lesions on VAS (visual analogue scale) before and immediately after laser and up to 7 days post operatively. The results of the study showed that dramatically after NACLT, the severity of pain declined immediately and it was sustained during follow-up periods ($P<0.001$). Similar to the other NACLT clinical trials, the procedure itself was painless and anesthesia was not required. There was no visible side effect such as ulceration, erosion and even erythema following NACLT. The results of the study suggested that single session of NACLT could reduce pain in lesions of mild/moderate post Docetaxole oral mucositis immediately and dramatically without visible side effects (Zand, Najafi et al. 2010).

4.2.4.3 A brief literature review

Some studies have shown that laser phototherapy (LPT) can be useful in prevention or treatment of oral mucositis. The principle behind using laser phototherapy (LPT) is that it accelerates wound healing and has anti-inflammatory effects (Arora, Pai et al. 2008). In addition, pain relieving properties of low power lasers seem to be of value in management of painful lesions of oral mucositis (Arora, Pai et al. 2008).

Some researchers state that prophylactic laser application seems more successful than curative laser application, although the reason is not entirely clear (Arora, Pai et al. 2008). Most studies of LLLT in cancer patients have focused on oral mucositis prevention (Bensadoun, Franquin et al. 1994; Bensadoun, Franquin et al. 1999; Rubenstein, Peterson et al. 2004; Bensadoun, Le Page et al. 2006; Kuhn, Porto et al. 2009; Clarkson, Worthington et al. 2010; Bjordal, Bensadoun et al. 2011) . Since the subjects of this section refers to analgesic effects of low power lasers in established chemotherapy-induced oral mucositis (COM), we briefly review some articles in which LLLT has been used for relieving pain in patients with chemotherapy-induced oral mucositis and not the prophylactic laser protocols.

Cauwels and Martens evaluated the capacity of analgesic effect and wound healing of low level laser therapy in 16 children suffering from chemotherapy-induced oral mucositis. All children were treated using a GaAlAs diode laser (wavelength: 830 nm, potency: 150 mW). The energy released was adapted according to the severity of the oral lesions. The same protocol was repeated every 48 hours until healing of each lesion occurred. The results of the study demonstrated that immediately after irradiation of the oral mucositis, pain relief was noticed. Depending on the severity of oral mucositis, on average, 2.5 treatments per lesion in a period of 1 week were sufficient to heal a mucositis lesion. They concluded that LLLT could reduce the severity and duration of mucositis and to relieve pain significantly (Cauwels and Martens 2011).

Nes and Posso investigated the pain relieving effect of LLLT among 13 patients who have developed moderate chemotherapy-induced oral mucositis. The laser used was GaAlAs (830 nm, power: 250 mW, energy density: 35 J cm^{-2}). The patients were treated during a 5-day period, and the pain was measured before and after each laser application. There was a significant ($P= 0.007$) 67% decrease in the daily average experience of pain felt before and after each treatment, confirming that LLLT can relieve pain among patients who have developed chemotherapy-induced oral mucositis (Nes and Posso 2005).

4.2.5 Other NACLT studies

We have evaluated the pain relieving effects of NACLT in some other painful mucosal lesions such as painful oral lesions of Stevens Johnson Syndrome, etc. as case reports, the results of which are being published. In addition we have used NACLT in few ulcerated and non-ulcerated skin diseases with variable degrees of success. In some lesions, such as post herpetic neuralgia, NACLT acts like other conventional therapeutic lasers, in which the pain relieving effect is completely short standing and needs the several sessions of NACLT to be performed frequently.

In addition, we have evaluated the effects of NACLT in promoting wound healing in few studies with variable degrees of success. Although in few studies, NACLT has shown some valuable results in this field we are not yet ready to express our view in this regard. Certainly, controlled clinical trials with larger sample sizes are necessary to further evaluate the promoting wound healing effects of NACLT.

It should be mentioned that since the biological effects of NACLT and their mechanisms are not fully known, it seems ethically questionable to use NACLT in diseases with malignant potential, such as oral lichen planus, consequently we have not assessed the effects of NACLT in such illnesses.

4.3 Presumed mechanisms of pain relieving effects of NACLT

In order to develop an understanding of the mechanisms of analgesic effect of NACLT, powermetry and thermometry were performed in prior studies, the results of which demonstrated the low power nature of the applied CO_2 laser (Zand, Ataie-Fashtami et al. 2009). (Please see also 4. NACLT)

Since the analgesic effect of NACLT is immediate, we assume that at least in part, physiological neural changes such as blockage of action potential generation and conduction of nociceptive signals in primary afferent neurons might take part in this analgesic effect (Zand, Ataie-Fashtami et al. 2009). Destruction of nerve endings is less probable to be induced by NACLT, because, even in the studies in which CO_2 laser has been used as a surgical scalpel, there have been no statistically significant differences in the number of intact peripheral nerve structures in laser-treated sites in comparison with sites treated with electrocautery and scalpel (Rocha, Pinheiro et al. 2001). It is not known, whether the other mechanisms such as increase in β-endorphin synthesis and release, changes in bradykinin, prostaglandins, substance P, serotonin, acetylcholine, nitric oxide, singlet oxygen production, and the other biochemical events- which have been proposed to play a part in pain relieving effect of conventional low power lasers - are responsible for analgesic effect of NACLT or not. (Please see also 2.4.Mechanisms of analgesic effects of low power laser therapy) Further basic studies are necessary to elucidate the mechanisms of this analgesic effect.

On the other hand, there are some differences between analgesic effects of NACLT and the other classical low power lasers. The analgesic effect in LLLT is usually gradual, cumulative, and multi-session (Pinheiro, Cavalcanti et al. 1998; Gur, Karakoc et al. 2002; Gur, Sarac et al. 2004; Nes and Posso 2005; Chow, Heller et al. 2006; Djavid, Mehrdad et al. 2007; Bjordal, Bensadoun et al. 2011; Iwatsuki, Yoshimine et al. 2011; Ribeiro, de Aguiar et al. 2011). In contrast, the pain relieving effect of NACLT is immediate, dramatic and more sustained than conventional low level therapeutic lasers (Zand, Ataie-Fashtami et al. 2009; Zand, Mansouri et al. 2009; Zand, Najafi et al. 2010). Therefore one could presume that the mechanisms of analgesic effect of NACLT might have some differences from that for conventional low power lasers, an assumption that should be assessed in further basic studies.

In some ulcerated oral lesions such as aphthous ulcers, the pain of the lesions derives from inflammatory sensitization of small-diameter afferent nerve endings that form a plexus at the junction of the epithelial and subepithelial layers. Branches of this plexus extend upward, into the epithelial layer; producing a superficial, focal, inflammatory lesion that is directly associated with exposed sensory nerve endings. Therefore, in such ulcers, CO_2 laser irradiation can reach the exposed nerve endings easily and as we assume, for example, block the action potential generation and conduction of nociceptive signals in primary afferent neurons. On the other hand, in other under-publish studies, we have used NACLT for reducing pain in few non-ulcerated lesions, such as pre-aphthous lesions with moderate-good results. As the CO_2 laser's beam has a very limited depth of penetration in to the tissue, explaining this analgesic effect of NACLT seems more complex.

Tuner and Hode state that the therapeutic effects of defocused CO_2 laser must be due to the influence of the laser energy on the cells encountered, so that signal substances are released and then circulate in the organism. This indirectly confirms the hypothesis that conventional laser therapy has both a local effect in the area treated by laser light, and a systemic effect through the release of metabolites (Tuner and Hode 2010). We don't know whether such mechanisms may at least in part, take part in the analgesic effect of NACLT or not.

Further fundamental studies are necessary to elucidate the mechanisms of analgesic effect of NACLT.

5. Conclusion

CO_2 laser has been used as a very useful high power, thermal laser in surgery for cutting, ablation and coagulation of the tissues for many years. In contrast, in non- thermal, Non-Ablative CO_2 Laser Therapy (NACLT), the CO_2 laser is used as a low level (phototherapeutic) laser to reduce pain in some oral mucosal lesions without any visual effects of thermal damage to the oral mucosa such as ablation, ulceration or aggravation of the lesions. The results of powermetry and thermometry have demonstrated the low power nature of the applied CO_2 laser in NACLT.

As discussed above, in order to use the CO_2 laser as a phototherapeutic laser for NACLT, the CO_2 laser beam is irradiated through a thick layer of transparent, non-anesthetic gel with high water content. In addition, the CO_2 laser is operated with a de-focused hand piece, scanning rapidly over the lesion with circular motion. With these considerations, CO_2 laser can be used as a non-destructive, non-thermal, phototherapeutic laser (NACLT) to reduce pain in some oral mucosal lesions immediately and dramatically, so that after NACLT, the patients of the studies have been able to eat and drink easily at once. So far, in the series of NACLT studies, we have not observed any visible side effects following careful performance of the technique.

Certainly, controlled clinical trials with larger sample sizes will be able to prove the analgesic effects of NACLT more definitely. We recommend that in further studies, the pain severity of the lesions would be followed up for longer periods of time.

In addition, it should be emphasized that in serious diseases such as pemphigus vulgaris, Behcet's disease, etc., the patients must be warned that NACLT should not alter their conventional systemic treatment in spite of its significant analgesic effect.

6. References

Aimbire, F., R. Albertini, et al. (2006). "Low-level laser therapy induces dose-dependent reduction of TNFalpha levels in acute inflammation." *Photomed Laser Surg* 24(1): 33-37.

Alidaee, M. R., A. Taheri, et al. (2005). "Silver nitrate cautery in aphthous stomatitis: a randomized controlled trial." *Br J Dermatol* 153(3): 521-525.

Andre, P. (2003). "Ingrowing nails and carbon dioxide laser surgery." *J Eur Acad Dermatol Venereol* 17(3): 288-290.

Arikan, O. K., A. Birol, et al. (2006). "A prospective randomized controlled trial to determine if cryotherapy can reduce the pain of patients with minor form of recurrent aphthous stomatitis." *Oral Surg Oral Med Oral Pathol Oral Radiol Endod* 101(1): e1-5.

Arora, H., K. M. Pai, et al. (2008). "Efficacy of He-Ne Laser in the prevention and treatment of radiotherapy-induced oral mucositis in oral cancer patients." Oral Surg Oral Med *Oral Pathol Oral Radiol Endod* 105(2): 180-186, 186 e181.

Asilian, A., A. Yoosefi, et al. (2006). "Pemphigus vulgaris in Iran: epidemiology and clinical profile." *Skinmed* 5(2): 69-71.

Basford, J. R., C. G. Sheffield, et al. (1999). "Laser therapy: a randomized, controlled trial of the effects of low-intensity Nd:YAG laser irradiation on musculoskeletal back pain." *Arch Phys Med Rehabil* 80(6): 647-652.

Bensadoun, R., J. Franquin, et al. (1999). "Low-energy He-Ne laser in the prevention of radiation-induced mucositis." *Support Care Cancer* 7: 244–252.

Bensadoun, R., J. Franquin, et al. (1994). "Low-energy He/Ne laser in the prevention of radiation-induced mucositis. A multicenter phase III randomized study in patients with head and neck cancer." *Support Care Cancer* 7: 244-252.

Bensadoun, R., F. Le Page, et al. (2006). "Radiation-induced mucositis of the aerodigestive tract: prevention and treatment. MASCC/ISOO mucositis group's recommendations." *Bull Cancer* 93: 201-211.

Bhardwaj, M., M. Joshi, et al. (2010). "Management of recalcitrant oral pemphigus vulgaris with CO2 laser-Report of two cases." *Journal of Indian Society of Periodontology* 14(2): 132-135.

Bjordal, J., C. Couppe, et al. (2001). "Low-level laser therapy for tendinopathy: evidence of a dose-response pattern." *Phys Ther Rev* 6: 91-99.

Bjordal, J. M., R. J. Bensadoun, et al. (2011). "A systematic review with meta-analysis of the effect of low-level laser therapy (LLLT) in cancer therapy-induced oral mucositis." *Support Care Cancer* 19(8): 1069-1077.

Bjordal, J. M., C. Couppe, et al. (2003). "A systematic review of low level laser therapy with location-specific doses for pain from chronic joint disorders." Aust J Physiother 49(2): 107-116.

Black, M., M. D. Mignogna, et al. (2005). "Number II. Pemphigus vulgaris." *Oral Dis* 11(3): 119-130.

Brosseau, L., V. Robinson, et al. (2005). "Low level laser therapy (Classes I, II and III) for treating rheumatoid arthritis." *Cochrane Database Syst Rev*(4): CD002049.

Bystryn, J. C. and J. L. Rudolph (2005). "Pemphigus." *Lancet* 366(9479): 61-73.

Cauwels, R. G. and L. C. Martens (2011). "Low level laser therapy in oral mucositis: a pilot study." *Eur Arch Paediatr Dent* 12(2): 118-123.

Chams-Davatchi, C., M. Valikhani, et al. (2005). "Pemphigus: analysis of 1209 cases." *Int J Dermatol* 44(6): 470-476.

Chia, Y. W., A. Darzi, et al. (1995). "CO2 laser haemorrhoidectomy--does it alter anorectal function or decrease pain compared to conventional haemorrhoidectomy?" *Int J Colorectal Dis* 10(1): 22-24.

Chow, R., P. Armati, et al. (2011). "Inhibitory effects of laser irradiation on peripheral mammalian nerves and relevance to analgesic effects: a systematic review." *Photomed Laser Surg* 29(6): 365-381.

Chow, R. T. and L. Barnsley (2005). "Systematic review of the literature of low-level laser therapy (LLLT) in the management of neck pain." *Lasers Surg Med* 37(1): 46-52.

Chow, R. T., M. A. David, et al. (2007). "830 nm laser irradiation induces varicosity formation, reduces mitochondrial membrane potential and blocks fast axonal flow in small and medium diameter rat dorsal root ganglion neurons: implications for the analgesic effects of 830 nm laser." *J Peripher Nerv Syst* 12(1): 28-39.

Chow, R. T., G. Z. Heller, et al. (2006). "The effect of 300 mW, 830 nm laser on chronic neck pain: a double-blind, randomized, placebo-controlled study." *Pain* 124(1-2): 201-210.

Clarkson, J. E., H. V. Worthington, et al. (2010). "Interventions for treating oral mucositis for patients with cancer receiving treatment." *Cochrane Database Syst Rev*(8): CD001973.

Colvard, M. and P. Kuo (1991). "Managing aphthous ulcers: laser treatment applied." *J Am Dent Assoc* 122(6): 51-53.

Davatchi, F., C. Shahram, et al. (2005). "Behcet disease." *Acta Medica Iranica* 43(4): 233-242.

De Bie, R., H. De. Vet, et al. (1998). "Efficacy of 904 nm laser therapy in the management of musculoskeletal disorders: a systematic review." *Phys Ther Rev* 3: 59-72.

De Souza, T., M. Martins, et al. (2010). "Clinical evaluation of low-level laser treatment for recurring aphthous stomatitis." *Photomed Laser Surg* 28 Suppl(2): 85-88.

Demetriades, M., Hansford, et al. (2009). "General Manifestations of Behçet's Syndrome and the Success of CO2-Laser as Treatment for Oral Lesions: A Review of the Literature and Case Presentation." *Journal of the Massachusetts Dental Society* 58 (3): 24-27.

Demidov, V. P., V. I. Rykov, et al. (1992). "[The use of the carbon dioxide laser in the surgical treatment of breast cancer]." *Vopr Onkol* 38(1): 42-50.

Demidova-Rice, T. N., E. V. Salomatina, et al. (2007). "Low-level light stimulates excisional wound healing in mice." *Lasers Surg Med* 39: 706–715.

Djavid, G. E., R. Mehrdad, et al. (2007). "In chronic low back pain, low level laser therapy combined with exercise is more beneficial than exercise alone in the long term: a randomised trial." *Aust J Physiother* 53(3): 155-160.

Duncavage, J. A. and R. H. Ossoff (1986). "Use of the CO2 laser for malignant disease of the oral cavity." *Lasers Surg Med* 6(5): 442-444.

Elad, S., R. Or, et al. (2003). "CO2 laser in oral graft-versus-host disease: a pilot study." *Bone Marrow Transplant* 32(10): 1031-1034.

Enwemeka, C. S., J. C. Parker, et al. (2004). "The efficacy of low-power lasers in tissue repair and pain control: a meta-analysis study." *Photomed Laser Surg* 22(4): 323-329.

Fekrazad, R., S. Jafari, et al. (2006). *Evaluation of the effects of pulsed Nd:YAG laser on RAU.* The 6th international congress of the World Association of Laser Therapy, Lissamol, Cyprus.

Gam, A. N., H. Thorsen, et al. (1993). "The effect of low-level laser therapy on musculoskeletal pain: a meta-analysis." *Pain* 52(1): 63-66.

Ghate, J. V. and J. L. Jorizzo (1999). "Behcet's disease and complex aphthosis." *J Am Acad Dermatol* 40(1): 1-18; quiz 19-20.

Gigo-Benato, D., S. Geuna, et al. (2005). "Phototherapy for enhancing peripheral nerve repair: a review of the literature." *Muscle Nerve* 31(6): 694-701.

Goncalves, W. L., F. M. Souza, et al. (2007). "Influence of He-Ne laser therapy on the dynamics of wound healing in mice treated with anti-inflammatory drugs." *Braz J Med Biol Res* 40(6): 877-884.

Gur, A., M. Karakoc, et al. (2002). "Efficacy of low power laser therapy in fibromyalgia: a single-blind, placebo-controlled trial." *Lasers Med Sci* 17(1): 57-61.

Gur, A., A. J. Sarac, et al. (2004). "Efficacy of 904 nm gallium arsenide low level laser therapy in the management of chronic myofascial pain in the neck: a double-blind and randomize-controlled trial." *Lasers Surg Med* 35(3): 229-235.

Hagiwara, S., H. Iwasaka, et al. (2007). "GaAlAs (830 nm) low-level laser enhances peripheral endogenous opioid analgesia in rats." *Lasers Surg Med* 39(10): 797-802.

Haldeman, S., L. Carroll, et al. (2008). "The Bone and Joint Decade 2000-2010 Task Force on Neck Pain and Its Associated Disorders: executive summary." *Spine (Phila Pa 1976)* 33(4 Suppl): S5-7.

Hamblin, M. R. (2010). "Introduction to experimental and clinical studies using low-level laser (light) therapy (LLLT)." *Lasers Surg Med* 42(6): 447-449.

Hamblin, M. R., R. W. Waynant, et al. (2006). "Mechanisms for Low-Light Therapy." *Proc. of SPIE* 6140 614001-614010.

Hashmi, J. T., Y. Y. Huang, et al. (2010). "Role of low-level laser therapy in neurorehabilitation." *PM R* 2(12 Suppl 2): S292-305.

Huang, Y. Y., A. C. Chen, et al. (2009). "Biphasic dose response in low level light therapy." *Dose Response* 7(4): 358-383.

Iwatsuki, K., T. Yoshimine, et al. (2011). "Percutaneous diode laser irradiation for lumbar discogenic pain: a clinical study." *Photomed Laser Surg* 29(7): 459-463.

Jimbo, K., K. Noda, et al. (1998). "Suppressive effects of low-power laser irradiation on bradykinin evoked action potentials in cultured murine dorsal root ganglion cells." *Neurosci Lett* 240(2): 93-96.

Jurge, S., R. Kuffer, et al. (2006). "Number VI - Recurrent aphthous stomatitis." *Oral Diseases* 12(1): 1-21.

Kaplan, I., I. Kott, et al. (1996). "The CO2 laser in the treatment of lesions of the eyelids and periorbital region." *J Clin Laser Med Surg* 14(4): 185-187.

Karu, T. (1989). "Photobiology of low-power laser effects." *Health Phys* 56(5): 691-704.

Karu, T. I. and S. F. Kolyakov (2005). "Exact action spectra for cellular responses relevant to phototherapy" *Photomed Laser Surg* 23(4): 355-361.

Karu, T. I., L. V. Pyatibrat, et al. (2004). "Photobiological modulation of cell attachment via cytochrome c oxidase."_*Photochem Photobiol Sci* 2004;3(2):211–216. 3(2): 211-216.

Kasai, S., T. Kono, et al. (1996). "Effect of low-power laser irradiation on impulse conduction in anesthetized rabbits." *J Clin Laser Med Surg* 14(3): 107-109.

Kasugai, C., D. Watanabe, et al. (2010). "Infliximab treatment of severe genital ulcers associated with Behcet disease." *J Am Acad Dermatol* 62(1): 162-164.

Kaviani, A., M. Fateh, et al. (2008). "Comparison of carbon dioxide laser and scalpel for breast lumpectomy: a randomized controlled trial." *Photomed Laser Surg* 26(3): 257-262.

Kuhn, A., F. A. Porto, et al. (2009). "Low-level infrared laser therapy in chemotherapy-induced oral mucositis: a randomized placebo-controlled trial in children." *J Pediatr Hematol Oncol* 31(1): 33-37.

Labajos, M. (1988). "β-endorphin levels modification after GaAs and HeNe laser irradiation on the rabbit. Comparative study " *Investigacion clinica laser* 1-2(6-8).

Lin, P. and G. Liang (2006). "Behcet disease: recommendation for clinical management of mucocutaneous lesions." *J Clin Rheumatol* 12(6): 282-286.

Longo, L., Z. Simunovic, et al. (1997). "Laser therapy for fibromyositic rheumatisms." *J Clin Laser Med Surg* 15(5): 217-220.

Maeda, T. (1989). "Morphological demonstration of low reactive laser therapeutic pain attenuation effect of the gallium aluminium arsenide diode laser" *Laser Ther* 1: 23-30.

Meador, R., G. Ehrlich, et al. (2002). "Behcet's disease: immunopathologic and therapeutic aspects." *Curr Rheumatol Rep* 4(1): 47-54.

Mizutani, K., Y. Musya, et al. (2004). "A clinical study on serum prostaglandin E2 with low-level laser therapy." *Photomed Laser Surg* 22(6): 537-539.

Montesinos, A. (1988). "Experimental effects of low power laser in encephalin and endorphin synthesis." *Journ Eur Med Laser Ass* 1(3): 2-7.

Mrowiec, J. (1997). *Analgesic effect of low-power infrared laser irradiation in rats*. SPIE.

Nes, A. G. and M. B. Posso (2005). "Patients with moderate chemotherapy-induced mucositis: pain therapy using low intensity lasers." *Int Nurs Rev* 52(1): 68-72.

Nicola, E. M. and H. Nicola (1994). *Low power CO2 laser in the treatment of chronic pharyngitis: a five- year experience* Proc. SPIE.

Ohno, T. (1997). "[Pain suppressive effect of low power laser irradiation. A quantitative analysis of substance P in the rat spinal dorsal root ganglion]." *Nippon Ika Daigaku Zasshi* 64(5): 395-400.

Orchardson, R., J. M. Peacock, et al. (1997). "Effect of pulsed Nd:YAG laser radiation on action potential conduction in isolated mammalian spinal nerves." *Lasers Surg Med* 21(2): 142-148.

Pinheiro, A. L., E. T. Cavalcanti, et al. (1998). "Low-level laser therapy is an important tool to treat disorders of the maxillofacial region." *J Clin Laser Med Surg* 16(4): 223-226.

Posten, W., D. A. Wrone, et al. (2005). "Low-level laser therapy for wound healing: mechanism and efficacy." *Dermatol Surg* 31(3): 334-340.

Prolo P, F. Z., Domingo D, Outhouse T, Thornhill M (2006). "Interventions for recurrent aphthous stomatitis (mouth ulcers)." *The Cochrane liberary* 3(10).

Rashid, R. M. and K. D. Candido (2008). "Pemphigus pain: a review on management." *Clin J Pain* 24(8): 734-735.

Ribeiro, A. S., M. C. de Aguiar, et al. (2011). "660 AsGaAl laser to alleviate pain caused by cryosurgical treatment of oral leukoplakia: a preliminary study." *Photomed Laser Surg* 29(5): 345-350.

Rocha, E. A., A. L. Pinheiro, et al. (2001). "Quantitative evaluation of intact peripheral nerve structures after utilization of CO2 laser, electrocautery, and scalpel." *J Clin Laser Med Surg* 19(3): 121-126.

Rubenstein, E. B., D. E. Peterson, et al. (2004). "Clinical practice guidelines for the prevention and treatment of cancer therapy-induced oral and gastrointestinal mucositis." *Cancer* 100(9 Suppl): 2026-2046.

Scully, C., M. Gorsky, et al. (2003). "The diagnosis and management of recurrent aphthous stomatitis: a consensus approach." *J Am Dent Assoc* 134(2): 200-207.

Sharon-Buller, A. and M. Sela (2004). "CO2-laser treatment of ulcerative lesions." *Oral Surg Oral Med Oral Pathol Oral Radiol Endod* 97(3): 332-334.

Shashy, R. G. and M. B. Ridley (2000). "Aphthous ulcers: a difficult clinical entity." *Am J Otolaryngol* 21(6): 389-393.

Shimizu, N., M. Yamaguchi, et al. (1995). "Inhibition of prostaglandin E2 and interleukin 1-beta production by low-power laser irradiation in stretched human periodontal ligament cells." *J Dent Res* 74(7): 1382-1388.

Simunovic, Z. (2000). pain and practical aspects of its management. *Lasers in medicine and dentistry.* Z. Simunovic. zagreb, AKD: 269-301.

Snyder, S., K. Byrnes, et al. (2002). "Quantification of calcitonin gene-related peptide mRNA and neuronal cell death in facial motor nuclei following axotomy and 633 nm low power laser treatment.." *Lasers Surg Med* 31: 216 –222.

Sonis, S. T. (2004). "The pathobiology of mucositis." *Nat Rev Cancer* 4(4): 277-284.

Soriano, F. and R. Rios (1998). "Gallium arsenide laser treatment of chronic low back pain: a prospective, randomized and double blind study." *Laser Ther* 10: 175-180.

Sutherland, J. C. (2002). "Biological effects of polychromatic light." *Photochem Photobiol* 76(2): 164-170.

Suzuki Kurokawa, M. and N. Suzuki (2004). "Behcet's disease." *Clin Exp Med* 4(1): 10-20.

Tada, H., M. Hatoko, et al. (2004). "Clinical comparison of the scanning CO2 laser and conventional surgery in the treatment of ingrown nail deformities." *J Dermatolog Treat* 15(6): 387-390.

Tuner, J. and L. Hode (2010). Biostimulation, Laser therapy with high output lasers. *The new laser therapy handbook* Prima books AB: 67-147.

Tuner, J. and L. Hode (2010). The mechanisms, effects on pain. *The new laser therapy handbook,* Prima books AB: 557-559.

Tuner, J. and L. Hode (2010). Medical indications. *The new laser therapy handbook,* Prima Books: 149-372.

Tuner, J. and L. Hode (2010). Some basic laser physics,Therapeutic lasers. *The new laser therapy handbook,* Prima books AB: 1-47.

Umegaki, S., Y. Tanaka, et al. (1989). "Effectiveness of low-power laser therapy on low-back pain - double-blind comparative study to evaluate the analgesic effect of low power laser therapy on low back pain." *Clin Rep* 23: 2838–2846.

Wakabayashi, H., M. Hamba, et al. (1993). "Effect of irradiation by semiconductor laser on responses evoked in trigeminal caudal neurons by tooth pulp stimulation." *Lasers Surg Med* 13(6): 605-610.

Zand, N., L. Ataie-Fashtami, et al. (2009). "Relieving pain in minor aphthous stomatitis by a single session of non-thermal carbon dioxide laser irradiation." *Lasers Med Sci* 24(4): 515-520.

Zand, N., L. Ataie-Fashtami, et al. (2009). "Analgesic effects of single session of Non-Ablative CO2 Laser Therapy (NACLT) in major aphthous ulcers: (a preliminary study) " *Lasers in medicine* 6(4): 6-12.

Zand, N., P. Mansouri, et al. (2009). Relieving pain in painful oral lesions of pemphigus vulgaris by a single session, Non-ablative 10600 nm CO_2 Laser irradiation (pilot study). *The 29th Annual conference of the American Society for Lasers in surgery and medicine*. Harbor. 41: 67-68.

Zand, N., S. Najafi, et al. (2010). "NACLT (Non-ablative CO2 laser therapy): a new approach to relieve pain in mild to moderate oral mucositis following breast cancer chemotherapy (a pilot study) " *EJC supplements* 8(3): 166.

Permissions

The contributors of this book come from diverse backgrounds, making this book a truly international effort. This book will bring forth new frontiers with its revolutionizing research information and detailed analysis of the nascent developments around the world.

We would like to thank Dan C. Dumitras, for lending his expertise to make the book truly unique. He has played a crucial role in the development of this book. Without his invaluable contribution this book wouldn't have been possible. He has made vital efforts to compile up to date information on the varied aspects of this subject to make this book a valuable addition to the collection of many professionals and students.

This book was conceptualized with the vision of imparting up-to-date information and advanced data in this field. To ensure the same, a matchless editorial board was set up. Every individual on the board went through rigorous rounds of assessment to prove their worth. After which they invested a large part of their time researching and compiling the most relevant data for our readers. Conferences and sessions were held from time to time between the editorial board and the contributing authors to present the data in the most comprehensible form. The editorial team has worked tirelessly to provide valuable and valid information to help people across the globe.

Every chapter published in this book has been scrutinized by our experts. Their significance has been extensively debated. The topics covered herein carry significant findings which will fuel the growth of the discipline. They may even be implemented as practical applications or may be referred to as a beginning point for another development. Chapters in this book were first published by InTech; hereby published with permission under the Creative Commons Attribution License or equivalent.

The editorial board has been involved in producing this book since its inception. They have spent rigorous hours researching and exploring the diverse topics which have resulted in the successful publishing of this book. They have passed on their knowledge of decades through this book. To expedite this challenging task, the publisher supported the team at every step. A small team of assistant editors was also appointed to further simplify the editing procedure and attain best results for the readers.

Our editorial team has been hand-picked from every corner of the world. Their multi-ethnicity adds dynamic inputs to the discussions which result in innovative outcomes. These outcomes are then further discussed with the researchers and contributors who give their valuable feedback and opinion regarding the same. The feedback is then collaborated with the researches and they are edited in a comprehensive manner to aid the understanding of the subject.

Apart from the editorial board, the designing team has also invested a significant amount of their time in understanding the subject and creating the most relevant covers. They scrutinized every image to scout for the most suitable representation of the subject and create an appropriate cover for the book.

The publishing team has been involved in this book since its early stages. They were actively engaged in every process, be it collecting the data, connecting with the contributors or procuring relevant information. The team has been an ardent support to the editorial, designing and production team. Their endless efforts to recruit the best for this project, has resulted in the accomplishment of this book. They are a veteran in the field of academics and their pool of knowledge is as vast as their experience in printing. Their expertise and guidance has proved useful at every step. Their uncompromising quality standards have made this book an exceptional effort. Their encouragement from time to time has been an inspiration for everyone.

The publisher and the editorial board hope that this book will prove to be a valuable piece of knowledge for researchers, students, practitioners and scholars across the globe.

List of Contributors

Dan C. Dumitras, Ana Maria Bratu and Cristina Popa
Department of Lasers, National Institute for Laser, Plasma, and Radiation Physics, Bucharest, Romania

Vladimir Petukhov and Vadim Gorobets
Institute of Physics of National Academy of Sciences of Belarus, Belarus

Rakesh Kumar Soni
Raja Ramanna Centre for Advanced Technology, Indore (M.P.), India

Mikhail N. Polyanskiy and Marcus Babzien
Brookhaven National Laboratory, USA

Akira Endo
Research Institute for Science and Engineering, Waseda University, Tokyo, Japan

A. D. Pryamikov, A. F. Kosolapov, V. G. Plotnichenko and E. M. Dianov
Fiber Optics Research Center of Russian Academy of Sciences, Russian Federation

Keiichiro Urabe and Kunihide Tachibana
Kyoto University, Japan

Joanna Radziejewska
Institute of Fundamental Technological Research, Polish Academy of Sciences, Warsaw, Poland

G. Vasconcelos and D. C. Chagas
Institute of Advanced Studies - IEAv, EFO-L, S. J. dos Campos, SP, Brazil

A. N. Dias
University of Vale do Paraíba - UNIVAP, S. J. dos Campos, SP, Brazil

Afia Kouadri-David
PSM Team, European University of Brittany, France, INSA of Rennes, LGCGM, France

Rui F. M. Lobo
Group for Nanoscale Science and Nanotechnology (GNCN), Physics Department, Faculdade de Ciencias e Tecnologia - New University of Lisbon, Caparica, Portugal
Institute for Science and Technology of Materials and Surfaces (ICEMS), Portugal

Mohammadreza Riahi
Shahid Beheshti University/Laser and Plasma Research Institute, Iran

Pinalben Viraparia
University Of California San Francisco, USA
Lutheran Medical Center-Advanced Education in General Dentistry, USA

Ram M. Vaderhobli
Lutheran Medical Center/UCSF, Dept. Preventive and Restorative Dental Sciences, UCSF, USA

Joel M. White
University Of California San Francisco, USA

Hyeong-Seok Oh and Jin-Sung Kim
Wooridul Spine Hospital, Seoul, Korea

Pasquale Londrillo, Graziano Servizi, Andrea Sgattoni, Stefano Sinigardi, Marco Sumini and Giorgio Turchetti
Università di Bologna, INFN Sezione di Bologna, Italy

Nasrin Zand
Iranian Center for Medical Lasers (ICML), Academic Center for Education, Culture and Research (ACECR), Tehran, Iran

www.ingramcontent.com/pod-product-compliance
Ingram Content Group UK Ltd.
Pitfield, Milton Keynes, MK11 3LW, UK
UKHW022206051026
15624UKWH00001B/13

9 781632 400413